Electron Microscopy

The Jones and Bartlett Series in Biology

Electron Microscopy
Principles and Techniques for Biologists

John J. Bozzola, PhD, MS
Center for Electron Microscopy
Southern Illinois University
Carbondale, Illinois

Lonnie D. Russell, PhD, MS
Laboratory of Structural Biology
Southern Illinois University, School of Medicine
Carbondale, Illinois

JONES AND BARTLETT PUBLISHERS
BOSTON

Editorial, Sales, and Customer Service Offices
Jones and Bartlett Publishers
20 Park Plaza
Boston, MA 02116

Library of Congress Cataloging-in-Publication Data

Bozzola, John J.
    Electron microscopy : principles and techniques for biologists / John J. Bozzola,
Lonnie D. Russell.
      p.   cm.
    Includes bibliographical references and index.
    ISBN 0-86720-126-6
    1. Electron microscopy. I. Russell, Lonnie Dee. II. Title.
QH212.E4B69   1991
578′.45—dc20   91-19743
                                               CIP

Cover art shows a scanning electron micrograph of *Candida albicans*, a pathogenic yeast isolated from human source.

Printed in the United States of America
95 94 93 92 91   10 9 8 7 6 5 4 3 2 1

To Our Parents

John, Sr.   *Lonnie, Sr.*
Angeline   *Jean*

Contents

CHAPTER 6
The Transmission Electron Microscope

CHAPTER 7
The Scanning Electron Microscope

Preface

This textbook was written out of a desire to teach electron microscopy to biologists in as simple a way as possible. It is an introductory textbook designed for those who are entering the field. In our own introductory electron microscopy courses, we have used several of the many excellent reference volumes available for electron microscopy. For beginners, however, the costs of several texts were prohibitive and the information was overwhelming. Naturally, reference books tend to emphasize detail and theoretical aspects in place of the practical concerns that a beginner may have. Often, the various tissue preparation methodologies are given equal treatment, although there might be just a few methods which are used most of the time. It was our goal to provide a starting point for beginners and to direct the student toward any appropriate reference sources.

In writing this text we have emphasized mainstream methods in the hope that the forest will be seen and the trees ignored, at least for the time being. Our philosophy has been to introduce the topic and at the end of the chapter provide references that deal more extensively with the topic. We have attempted to achieve a balance of theory and practical applications of electron microscopy. In each instance, our criteria for inclusion of theory in the text was the down-the-road application of this theory which would either lead to an understanding of the major principles involved in the discipline or be a necessary prerequisite to perform biological electron microscopy. Thus, we may have deleted someone's favorite technique for the sake of an overall understanding of the topic.

We realized at the onset of this venture that our interests in electron microscopy are complementary. One of us is more interested in the theory, the techniques, and ways of using the electron microscope to achieve the desired result. The other is interested primarily in the applications and the biological implications of the use of the electron microscope. From the start it was clear to both of us which chapters were our forté and how the material should be divided. In reading each others chapters, we were able to offer a perspective from an orientation different than the writer's. This helped us to balance microscope theory with biological application.

We have not only described the workings of the electron microscope and the major techniques for biological specimen preparation but we have illustrated, in many instances, how these techniques have contributed to the biological sciences. Chapter 19, "A Survey of Biological Ultrastructure," serves as an introduction to a wide variety of animal and plant components for the novice, yet emphasizes mammalian tissues where the bulk of biological electron microscopy is currently applied. Chapter 18, "Interpretation of Micrographs" serves as a tool for the novice to understand how various factors can influence what is seen in the micrograph. Both chapters are profusely illustrated. In making the decision to write these chapters, we reasoned that knowledge of the microscope, microtome, and methods to prepare tissues was not sufficient to introduce a person to electron microscopy. Some experience in interpretation and examination of electron micrographs is also important. A novice can hardly begin to interpret a micrograph, whereas an investigator with considerable experience can glean a great deal from a micrograph. Those individuals experienced in viewing micrographs and interpreting them take for granted the time spent in becoming familiar with electron micrographs.

It is the reader who will eventually determine if our philosophy in writing this text was a sound one. Any first edition text is examined critically by all that come in contact with it. As students, the authors were of the opinion that everything that went into print must be faultless. As the years went by, this proved not to be a correct assumption. It is in this vein that we, the authors, solicit your comments and suggestions on how to improve the text or to let us know what you found enjoyable about the text. We welcome exceptional micrographs for possible inclusion in subsequent editions of the text. If this book is instrumental in your career decisions or facilitates your research progress in electron microscopy, we would love to hear about it. That was our original intention.

Acknowledgments

We would like to express our thanks to the people who helped in so many ways in the production of this book. First of all, we thank our colleagues who shared with us their electron micrographs. Cindy Claybough, Karen Fiorini, and Karen Schmitt drew and helped improve upon most of the original line drawings and artwork. Dee Gates, Sushmita Ghosh, Cheri Kelly, Scott Pelok, Steve Schmitt, and Randall Tindall helped in the preparation of the illustrative electron micrographs and environmental photographs, and often served as models. John Richardson provided local advice on publication and graphics considerations.

We would also like to acknowledge the many reviewers who critiqued the manuscript in its various stages of development. We benefited greatly from their long experience in teaching electron microscopy to both undergraduate and graduate students.

We are very grateful for the comments and suggestions of: Professor Richard F. E. Crang, University of Illinois at Urbana-Champaign; Professor William J. Dougherty, Medical University of Southern Carolina; Professor Laszlo Hanzely, Northern Illinois University; Professor Julian P. Heath, Baylor College of Medicine; Professor Harry T. Horner, Iowa State University; Dr. Morton D. Maser, Woods Hole Educational Associates; Dr. Judy A. Murphy, R. J. Lee Group; Professor Jerome J. Paulin, University of Georgia; Professor Lee D. Peachey, University of Pennsylvania; Professor David Prescott, University of Colorado; and Steve Schmitt, Southern Illinois University.

The editorial guidance offered by Joseph Burns, Paula Carroll and Judy Salvucci of Jones and Bartlett, and Joyce Jackson of Impressions was very much appreciated.

Finally, we thank our colleagues and supervisors in the Office of Research Development and Administration in the Graduate School, and the Physiology Department in the School of Medicine and College of Science for the freedom and understanding so essential to complete this project.

Electron Microscopy

The Past, Present, and Future of Electron Microscopy

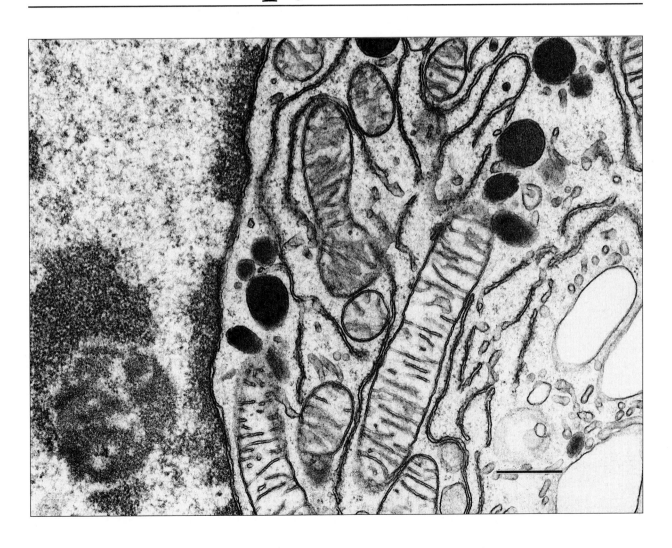

Rarely are new vistas opened to the human eye. The undersea world and perhaps air flight, outer space, or the moon have provided us with fascinating visual encyclopedias. The invention of the compound *light microscope* by the Janssens, manufacturers of eyeglasses, in 1590 opened the door to the microscopic world. Their microscope magnified objects up to 20 to 30 times their original size. In the next century, Antonie van Leeuwenhoek developed a simple (one lens) microscope, which represented a tremendous improvement in lens fabrication and permitted magnifications up to 300 times. By the beginning of the 20th century, objects could be magnified up to 1,000 times their original size, and particles that were only 0.2 μm apart could be distinguished from one another. Biological materials revealed a substructure as complex, variable, and dynamic as then could be imagined. In the 1930s another vista, as exciting as any of the rest, began to open: the submicroscopic world as viewed by the *electron microscope* (Figures 1-1, 1-2, and 1-3). The electron microscope took advantage of the much shorter wavelength of the electron. With the electron microscope, another thousand-fold increase in magnification was made possible, accompanied by a parallel increase in res-

olution capability allowing biologists to both define and expand the world of light microscopy (Figure 1-4). Viruses, DNA, and many smaller organelles were visualized for the first time.

The electron microscope has profoundly influenced our understanding of tissue organization and especially the cell. It has given us the capability to visualize molecules (Figure 1-5) and even the atom.

Electron microscopy is defined as a specialized field of science that employs the electron microscope as a tool. Numerous techniques for tissue preparation and tissue analysis are included under the broad heading of electron microscopy. Although major contributions have been made utilizing the electron microscope, it is only one of many tools used to solve biological problems. *Electron microscopists* are those who use electron microscopy.

Historical Perspective

It took 300 years to perfect the basic light microscope, but less than 40 years to refine the electron microscope. The world opened by electron microscopy has been an extremely exciting visual experience for biologists; however, it now is clear that

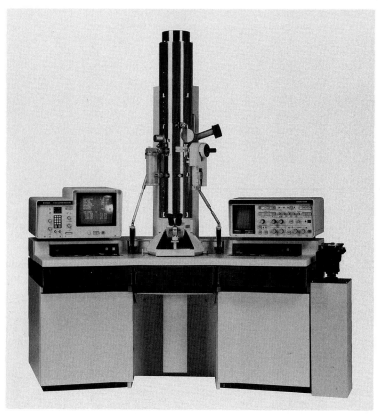

Figure 1-1 Modern electron microscope. (Courtesy of Hitachi Scientific Instruments.)

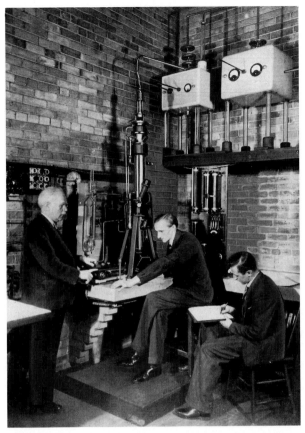

Figure 1-2   This early electron microscope, known as the "Toronto Microscope," was operating in 1939 and was the first in North America. (From a March 1940 issue of the *New York Times*, courtesy Ladd Research Industries.)

Figure 1-3   This electron microscope, built by W. Ladd, was used during World War II to examine rubber from German tires. This instrument helped the Allies build a tire that would compete with the much superior German tire. The instrument is now in the museum at the Armed Forces Institute of Pathology, Washington, D.C. (Courtesy of Ladd Research Industries.)

the *descriptive age* of biological electron microscopy has passed. Beginning in the 1940s and continuing well into the 1970s, descriptive biological electron microscopy thrived. Investigators were preoccupied with the discovery of the cell components. Discoveries were often so frequent and revolutionary that investigators continually questioned whether their findings were artifact or real. Most findings proved to be real representations of cell structure. Some did not. But the period of questioning was essential in order to establish this new science as a credible one.

Ernst Ruska and Max Knoll are credited with developing the first electron microscope. Ruska's efforts were recognized in 1986 with the Nobel Prize in Physics. Shortly after the invention of the electron microscope, many individuals were occupied primarily with developing the instrument. Fascination with developing the instrument itself gradually gave way to using the instrument as a tool to answer biological questions, as biologists became increasingly interested in the applications of electron

microscopes. A now classic paper by Porter and colleagues (1945) clearly demonstrated early on that the electron microscope could be used to study cells in detail.

The kinds of questions addressed by electron microscopists have matured. Initially, descriptive microscopy prevailed, but with time gave way to *experimental* approaches. The field of *cell biology*, heavily dependent on electron microscopy for its origins and early growth, has now matured into a multidisciplinary field that utilizes electron microscopy as one of several tools.

Most of the biologists who pioneered biological electron microscopy are still living. Some of these pioneers include Albert Claude, Don Fawcett, Earnest Fullam, Charles Leblond, John Luft, George Palade, Daniel Pease, Keith Porter, and Fritiof Sjostrand. Albert Claude and George Palade were awarded the 1974 Nobel Prize in Medicine for their accomplishments in cell biology employing electron microscopy. For accomplishments of a similar nature, Keith Porter received the Presidential Medal.

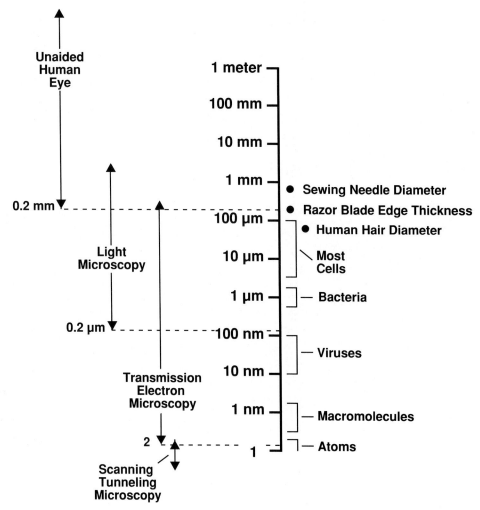

Figure 1-4   Log scale (middle) of the range of resolving power of various magnifying tools (left) and the structures they are capable of resolving (right).

It is still possible to hear firsthand about the excitement in the early days at hotbeds of electron microscopy such as the Rockefeller Institute.

Biological electron microscopists are, by and large, *anatomists* and *cell biologists*. However, individuals in many disciplines have come to use electron microscopy as a tool.

The techniques of basic biological electron microscopy were, at first, in the hands of a few investigators. Methods for preparing and sectioning tissue varied greatly among laboratories. Now, although minor variations exist, most techniques are standardized, freely available to everyone, and may be found in reference books. In spite of the numerous specialized texts on methodology, nothing has replaced or is soon to replace practical, hands-on training in the laboratory with the guidance of knowledgeable, skilled personnel.

Electron microscope technicians have specialized training to carry out the day-to-day operations of the laboratory (Figure 1-6). In most instances, they have experience in biological aspects of electron microscopy, and many may have limited knowledge of the microscope's workings, which enables them to repair this equipment and its electrical components. Although technical personnel can handle more routine maintenance, most labs now have service contracts for outside specialists to perform periodic maintenance or make emergency service calls.

Electron microscopy in many universities and industrial settings has often become *centralized*. Centralization allows for many users of the same equipment and for a higher level of training of personnel as well as increases the breadth of the services that can be offered. The individual investigator then has time to concentrate on the biological problems rather than the technical problems of tissue preparation and equipment maintenance.

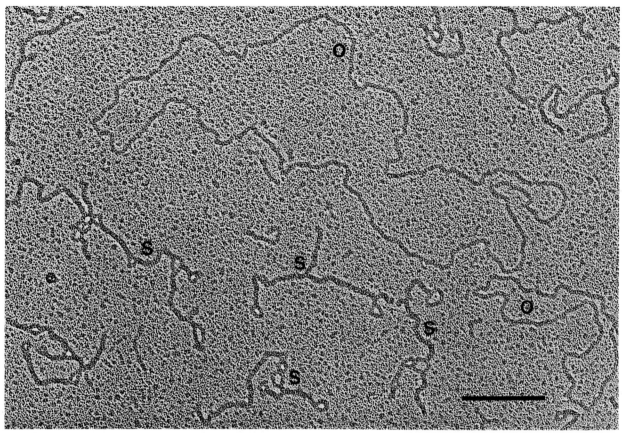

Figure 1-5 Micrograph of isolated DNA molecules that have been stained with uranyl acetate and shadowed with a thin coat of platinum. The micrograph shows open circular (O) and super-coiled (S) forms of DNA from plasmids. (Courtesy of L. Coggins.) Bar = 0.25 μm.

Electron microscopy has crossed disciplines to the degree that no single discipline can claim ownership of this tool. Anatomy, biochemistry, botany, cell biology, forensic medicine, microbiology, pathology (especially renal and tumor pathology), physiology, and toxicology are biological and biomedical fields that rely heavily on the electron microscope. To some degree, virtually all biologically related journals publish electron micrographs. Some journals are devoted almost exclusively to papers on fine structure, while others emphasize technical advances in the field. The *Electron Microscopy Society of America (EMSA)* meets yearly and serves as a format for the presentation of scientific papers and the introduction of commercial products.

Development of the Electron Microscope

Two basic types of instruments are called electron microscopes (Figure 1-7). Both were invented at about the same time, but they have fundamentally different uses. The *transmission electron microscope* (TEM) projects electrons through a very thin slice of tissue (specimen) to produce a two-dimensional image on a phosphorescent screen. The brightness of a particular area of the image is proportional to the number of electrons that are transmitted through the specimen. The *scanning electron microscope* (SEM) produces an image that gives the impression of three dimensions. This microscope uses a 2 to 3 nm spot of electrons that scans the surface of the specimen to generate secondary electrons from the specimen that are then detected by a sensor. The image is produced over time as the entire specimen is scanned. A third, less used type of electron microscope, the *scanning transmission electron microscope,* (STEM) has features of both the transmission and scanning electron microscopes. The STEM uses a scanning beam of electrons to penetrate thin specimens.

Naturally, the development of the microscope preceded the development of techniques to be used with the microscope. The first electron microscop-

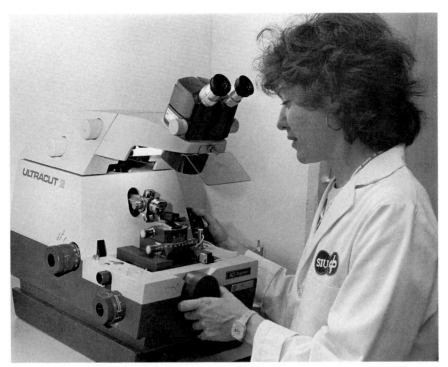

Figure 1-6   An electron microscope technician at work in a modern laboratory.

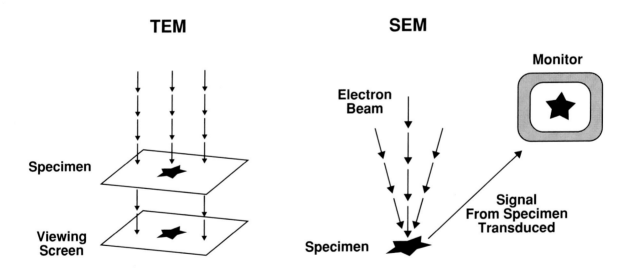

Two Basic Types of Microscopes

Figure 1-7   The basic differences between (TEM) transmission and (SEM) scanning electron microscopes.

ists, being engineers and physicists, were generally inexperienced in biological applications. In fact, the first microscope had no place for a specimen. Next, biological applications were sought, and the instrument was modified accordingly. However, even with capable instruments, major advances in biological applications did not happen immediately, but awaited the development of tissue preparation techniques, which were inadequate until the early 1950s.

A chronological listing of important historical events related to the development of the electron microscope follows:

1873 Abbe and Helmholtz independently showed that resolution depends on the wavelength of the energy source. This finding provided the theoretical promise of developing an electron microscope although, at this time, electrons had not been discovered.

1924 De Broglie (1929 Nobel Laureate in Physics) demonstrated that electrons have properties of waves.

1926 Busch demonstrated that the path of electrons could be deflected by magnetic lenses in the same way that light could be deflected by an optical lens.

1932 Knoll and Ruska developed the first electron microscope in Germany. No biological application was envisioned.

1935 The resolution of the electron microscope surpassed that of the light microscope.

1937 Metropolitan Vickers was the first commercial enterprise to develop a prototype of an electron microscope.

1938 von Ardenne constructed the first scanning electron microscope.

1938–39 The Siemens Corporation introduced the first commercial transmission electron microscope.

1940–41 The RCA Corporation sold the first commercial transmission electron microscope in the United States. The subsequent generations of RCA microscopes proved invaluable to advances in microscopy in North America.

1941–63 Continued improvements were made in both resolution and convenient use of the transmission electron microscope, such that 0.2 to 0.3 nm resolutions were achieved. Improvements were made in the scanning electron microscope to achieve a resolution of about 10 nm.

1954 The Siemens Elmscope I electron microscope was introduced, an extremely popular and useful instrument that provided excellent transmission results in the hands of biologists.

1958 The Stereoscan, a scanning electron microscope readily used by biologists, was introduced by Cambridge Instruments.

1980s The scanning tunneling electron microscope was developed by H. Rohrer and G. Binnig (1986 Nobel Prize Laureates).

1990 Commercial transmission electron microscopes currently marketed: Hitachi (Japan), ISI (Korea), JEOL (Japan), Philips (Holland), Zeiss (Germany).

Commercial scanning electron microscopes currently marketed: AmRay (United States), Cambridge (United Kingdom), Gatan (United Kingdom), Hitachi (Japan), ISI (Korea), JEOL (Japan), Philips (Holland), Tracor Northern (United States), Wintron (USSR), Zeiss (Germany).

Development of Preparative Techniques

About fifteen years after the development of the electron microscope by Knoll and Ruska, the first serious efforts were made to apply this technology to biological problems. Significant historical developments in tissue preparation techniques follow:

1934 Marton published the first electron micrograph of biological tissue. The image was inferior to conventional light microscope images (Figure 1-8).

1947–48 Claude introduced osmium fixation, a crude ultramicrotome and naphthaline embedding.

1949 Methacrylate was introduced as an embedding medium.

1950 Latta and Hartmann developed glass knives.

1952 Palade employed a buffering system to fix tissue in osmium tetroxide. Rapid progress in biological observation had its beginnings in the early 1950s.

1953 Porter and Blum introduced the first widely used microtome (Sorval MT-1). Fernandez-Moran first used diamond knives to make ultrathin sections.

1956 Epoxy resin was introduced as an embedding medium. Luft introduced potassium permanganate fixation. Palade and Siekevitz used the electron microscope to analyze cell fractions.

1957–63 Freeze fracture was developed.

1958 Watson introduced staining with heavy metals (lead and uranium).

1961 Epon (Shell Oil Co.) was introduced as an epoxy embedding medium. Epon has recently been taken off the market and replaced by a comparable embedment.

1963 Sabatini and coworkers introduced glutaraldehyde as a primary fixative.

Figure 1-8 The first electron micrographs of a biological specimen were made by Dr. L. Marton in 1934. Plant tissue (*Drosera intermedia* of the sundew family) has been impregnated with osmium and is seen lying on a grid. The figure on the right is a magnified image of the one on the left. (From Marton's original article in the Belgium Royal Academy of Science, 1934. Used with permission of the publisher.)

Contributions to Biology and the Future of Electron Microscopy

Electron microscopy has contributed immensely to the field of biology. Just as we take for granted the basic descriptions that were important to the field of biology, it is easy to forget that virtually all organelles and cell inclusions were either discovered or resolved in finer detail using the electron microscope. Such descriptions have laid the foundation for experimental manipulations directed at unraveling cell function and understanding how cellular structure varies in normal, experimental, and diseased states. Electron microscopy is used not only to visualize biological materials, but also to analyze the chemical makeup and physical properties of biological materials (*analytical electron microscopy*).

The electron microscope has revolutionized our concept of cell structure and of how cells interact. Today's electron micrographs (Figure 1-9) are superior to the early images obtained. The techniques that apply to this instrument have multiplied and are undergoing refinement. Thousands of microscopes are used for biological applications in a variety of disciplines throughout the world.

As important as the development of the electron microscope itself are the various specialized techniques that are associated with electron microscopy and that have been developed to prepare and evaluate tissue. These include *autoradiography, freeze fracture, immunocytochemistry, electron diffraction, cryoelectron microscopy, elemental analysis, quantitative microscopy, high voltage microscopy,* all of which are covered in more detail in subsequent chapters. These techniques, along with experimental approaches, have largely replaced descriptive studies. Specialized techniques will undoubtedly prove worthwhile for a number of years to come.

The future of electron microscopy is bright. Computer-assisted imaging, image enhancement, and image storage are beyond the developmental stages and are now being viewed as essential accessories to the microscope. Automated quantitation, also in final developmental stages, will allow a more objective assessment of biological samples. It is a future goal to have the routine capability of imaging *living* systems at high resolution! Electron microscopy is expected to continue to meet the submicroscopic imaging needs of science and medicine.

Journals Devoted Primarily to Electron Microscopy

Electron Microscopy Reviews (Pergamon Press)
Journal of Electron Microscopy (Japanese)
Journal of Electron Microscopy Technique
Journal of Microscopy (English)
Biology of the Cell (formally *Journal de Microscopie* or *Biologie Cellulaire;* French)
Journal of Ultrastructural Pathology
Scanning Electron Microscopy
Ultramicroscopy

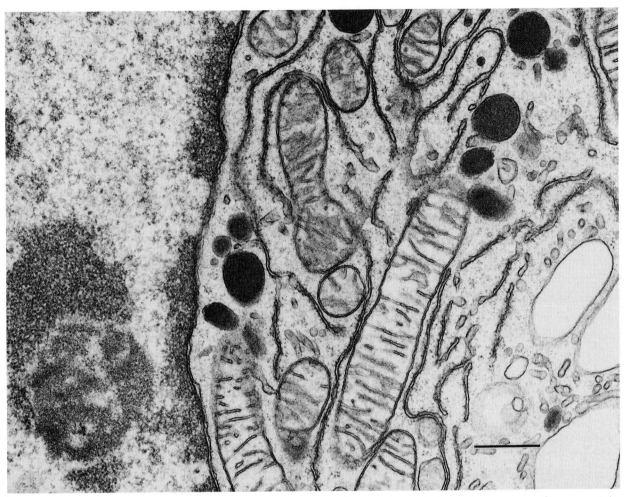

Figure 1-9   A recently taken electron micrograph showing the clarity of detail obtained by state-of-the-art microscopes and preparative techniques. A macrophage from mouse tissue. Bar = 0.5 μm.

Selected Journals Publishing Electron Micrographs

Developmental Dynanmics (formerly *American Journal of Anatomy*)
Anatomical Record
Cell and Tissue Research

Journal of Anatomy
Journal of Cell Biology
Journal of Histochemistry and Cytochemistry
Tissue and Cell
Journal of Ultrastructure and Molecular Structure Research (formerly *Journal of Ultrastructural Research*)

References

Classic References

Busch, H. 1926. Berechnung. der bahn von kathodenstrahlen in axialsymmetrischen electromagnitischen felde. *Ann Physik* 81:974–93.

De Broglie, L. 1924. Researches sur la theorie des quanta. Thesis, Paris: Masson and Cie. Also 1925 *Ann. de Physique* 3:22–128.

Knoll, M., and E. Ruska. 1932. Beitrag zur geometrischen elektronoptik. *Ann Physik* 12:607–61.

Latta, H., and J. F. Hartmann. 1950. Use of a glass edge in thin sectioning for electron microscopy. *Proc Soc Biol Med* 74:436–9.

Marton, L. 1934. La microscopie electronique des objects biologiques. *Bull Classe Sci Acad Roy Belg,* Series 5, 20:439–46.

Palade, G. E. 1952. Study of fixation for electron microscopy. *J Exptl Med* 95:285–98.

Palade, G. E., and P. Siekevitz. 1956. Liver microsomes: An integrated morphological and biochemical study. *J Biophys Biochem Cytol* 2:171–98.

Porter, K. R., and J. Blum. 1953. A study of microtomy for electron microscopy. *Anat Rec* 117:685–710.

Porter, K. R., et al. 1945. A study of tissue culture cells by electron microscopy. *J Exptl Med* 81:233–46.

Sabatini, D. D., et al. 1963. Cytochemistry and electron microscopy. The preservation of cellular ultrastructure and enzyme activity by aldehyde fixation. *J Cell Biol* 17:19–57.

Watson, M. L. 1958. Staining of tissue sections for electron microscopy with heavy metals. II Application of solutions containing lead and barium. *J Biophys Biochem Cytol* 4:727–9.

Selected Historical References

Hawkes, P. W. 1985. The beginnings of electron microscopy. In *Advances in electronics and electron physics,* Suppl 16, P. W. Hawkes, ed. Orlando: Academic Press.

Marton, L. 1968. *Early history of the electron microscope.* San Francisco: San Francisco Press Inc., pp 1–56.

Pfefferkorn, G. E. 1984. The early days of electron microscopy. *Scanning Electron Microscopy,* SEM Inc. 1:1–8.

Preven, D. R., and J. D. Gruhn. 1985. The development of electron microscopy. *Arch of Pathol Lab Med* 109:683–91.

Specimen Preparation for Transmission Electron Microscopy

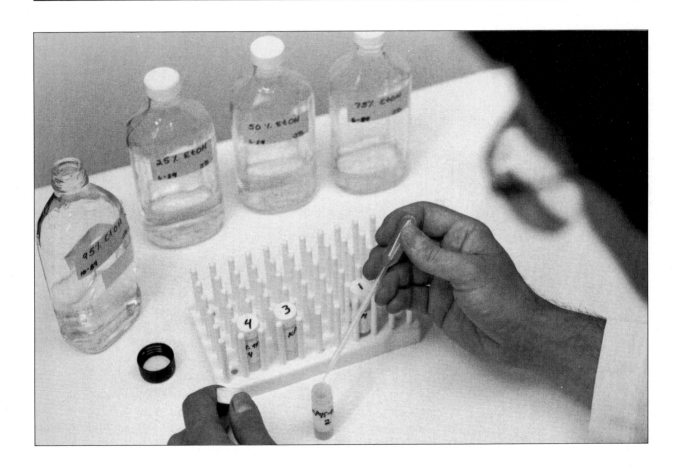

In preparing specimens for transmission electron microscopy, virtually every step of the procedure affects the quality of the final electron micrographs. Therefore, it is important to process tissue according to prescribed methods and to understand what is happening to the sample in processing. Specimen processing should begin with careful planning and proceed with meticulous attention to detail. Since tissue processing represents considerable effort, it is better to execute it properly the first time rather than to repeat the entire procedure.

Perhaps the least forgiving of all the steps is the tissue processing that occurs prior to sectioning. A poorly prepared specimen is useless to the investigator, whereas misadventures during the sectioning or staining process are somewhat more easily remedied. More sections can be cut and stained without a great amount of wasted time. Once tissue is processed in an acceptable manner and properly embedded in plastic, the intended result should be obtainable.

Tissue preparation for transmission electron microscopy can be divided into eight major steps: *primary fixation, washing, secondary fixation, dehydration, infiltration with transitional solvents, infiltration with resin, embedding, and curing.* The process begins with living hydrated tissue and ends with tissue that is virtually water-free and preserved in a static state within a plastic resin matrix. The plastic resin mixture permeates the tissue, replacing all water within the cell and making the cell firm enough for sections to be cut.

There are many ways to prepare tissue for electron microscopy. This chapter will emphasize *mainstream* protocols. Less used variations on the general protocol are numerous and have been catalogued in reference books (Glauert, 1975; Hayat, 1981). As an overview of what follows, Table 2-1 shows a general tissue preparation protocol commonly used in laboratories where tissues are prepared for electron microscopy.

Fixation

Ideally, one purpose of fixation is to *preserve the structure of living tissue with no alteration from the living state.* This means that an ideal fixative must also halt potentially destructive autolytic processes at the time of fixation. Additionally, *fixation should protect tissues against disruption during embedding*

and sectioning and subsequent exposure to the electron beam. Fixatives always fall short of these ideals; there are always some artifacts induced by the fixation process (see Chapter 18). Often one selects a particular fixation protocol for its ability to preserve one ultrastructural feature over another. For example, membrane structure may be desired. In some instances, artifacts or poor fixation of particular structures are tolerated at the expense of other beneficial features.

Historically, osmium tetroxide, which had limited use in light microscopy, was the first fixative used for electron microscopy (Claude, 1948). Tissues were either immersed in the fixative or exposed to osmium tetroxide vapors. Osmium tetroxide was later used in a physiological buffering system (Palade, 1952). Subsequently, potassium permanganate was introduced as a fixative (Luft, 1956). It was not until 1963 that the standard glutaraldehyde-osmium tetroxide protocol, used in most laboratories today, was developed (Sabatini et al., 1963). Primary fixation with glutaraldehyde was followed by secondary exposure (postfixation) to osmium tetroxide.

Most fixation protocols developed subsequently are modifications of the two-step basic procedure described above. A primary fixative was developed that combined glutaraldehyde and a low concentration of formaldehyde (up to 4%), which allowed more rapid initial fixation of the tissue because formaldehyde penetrates tissue more readily than glutaraldehyde does (Karnovsky, 1965). A secondary fixative using an osmium tetroxide solution that is reduced with ferrocyanide was introduced to enhance preservation of membranes and glycogen (Karnovsky, 1971). In the following years, numerous fixative variations based on the glutaraldehyde-osmium tetroxide protocol have been developed for particular cells or tissues.

One major historical development in fixation methodology should be noted: the use of an intact living animal's vascular system to introduce the fixative into the tissues. The introduction of vascular or organ *perfusion* fixation (Palay, 1962) allowed rapid exposure of fixatives to tissue elements and more closely achieved the goals set for a fixative.

The Mechanism of Chemical Fixation for Electron Microscopy

Glutaraldehyde
The structure of glutaraldehyde (MW 100.12) is shown below.

$$CHO - CH_2 - CH_2 - CH_2 - CHO$$

Table 2-1   General Tissue Preparation Scheme for Electron Microscopy

Activity	Chemical	Time Involved*
Primary Fixation	tissue is fixed with 2–4% glutaraldehyde in buffer	1–2 hr
Washing	buffer (3 changes, 1 of which may be overnight)	1–12 hr
Secondary Fixation	osmium tetroxide (1–2%: usually buffered)	1–2 hr
Dehydration	30% ethanol	5 min
	50% ethanol	5–15 min
	70% ethanol	5–15 min
	95% ethanol (2 changes)	5–15 min
	absolute ethanol (2 changes)	20 min ea
Transitional Solvent	propylene oxide (3 changes)	10 min ea
Infiltration of Resin	propylene oxide: resin mixtures gradually increasing concentration of resin	overnight–3 d
Embedding	pure resin mixture	2–4 hr
Curing (at 60–70° C)		1–3 d

* The specified times do not include the time involved in preparation of chemicals.

Glutaraldehyde is a five-carbon compound containing terminal aldehyde groups. Glutaraldehyde's attribute as a fixative is in its ability to cross-link protein by virtue of the terminal aldehyde groups that make the molecule bifunctional. Specifically, the aldehyde groups react with a-amino groups of lysine in adjacent proteins, thereby cross-linking them. The general reaction that takes place between protein and glutaraldehyde is as follows:

protein ×2 + glutaraldehyde
yields
protein – glutaraldehyde – protein + water

or

(COOH – protein – NH₂) ×2
+
CHO – CH₂ – CH₂ – CH₂ – CHO
yields
COOH – protein – N = CH – CH₂ – CH₂ – ⌐
└ CH₂ – CH = N – protein – COOH
+
2H₂O

In the above reaction, it is not necessarily assumed that both ends of the glutaraldehyde molecule react with the protein simultaneously. Some glutaraldehyde terminal groups may remain unreacted.

Other possible types of glutaraldehyde and protein reactions have been proposed. In many instances, the simplified reaction shown above may turn out to be more complex and involve amino acids other than lysine. Glutaraldehyde will react to some degree with lipids, carbohydrates, and nucleic acids; thus its specificity within the cell may not be limited to protein. Regardless of the nature of the reactions taking place, the overall cross-linking of components within the cell is usually widespread.

Because protein is a universal constituent of cells, present in the cytosol and membrane constituents and within cytoskeletal elements, one could envision that glutaraldehyde could cross-link soluble proteins to each other and link these to fixed cytoskeletal and membranous constituents throughout the cell. Thus, the constituents of the cell would be unified into a single interlocking structure or meshwork held together by a multitude of glutaraldehyde molecules. If some constituents were not cross-linked, they would nevertheless be trapped in the meshwork if they are too large to escape. In the process of cross-linking, glutaraldehyde also will denature protein. Low concentrations of glutaraldehyde have less denaturing effects than higher concentrations. The extent of protein denaturation has a profound effect on many localization protocols for most proteins (see Chapters 9 and 10).

The rate of penetration of glutaraldehyde into tissues is very slow. The deeper the penetration of glutaraldehyde into a tissue, the slower it will advance from that point on. Much depends on the nature of the tissue being used; glutaraldehyde penetrates more slowly into compact tissue with multiple membrane layers to penetrate than it does into tissue that has large fluid spaces. Glutaraldehyde can be expected to penetrate at a rate of less than 1 mm per hour if compact tissue is immersed into the glutaraldehyde solution. Therefore, to obtain rapid immersion fixation, the immersed tissue should be under 1mm in thickness in at least one dimension.

Pale-colored tissue fixed with glutaraldehyde will turn a pale yellow with time. The change is extremely gradual and difficult to notice unless one compares the color of unfixed tissue with the fixed tissue. A gradual and progressive change in firmness of the tissue always accompanies glutaraldehyde fixation. This is especially well demonstrated when tissues are perfused with glutaraldehyde (see below).

CAUTION: Handle glutaraldehyde carefully. Its ability to fix tissues is not restricted to the tissues being prepared for electron microscopy. Exposure to skin will cause the skin to turn yellow and harden. Usually, the exposed (yellowed) skin will be sloughed off in a few days. Avoid prolonged breathing of glutaraldehyde and contact with the eyes. Repeated exposure to glutaraldehyde may cause contact dermatitis. Glutaraldehyde should be used under a well-ventilated hood and disposed of in a proper manner.

Osmium Tetroxide

Osmium tetroxide, often erroneously called "osmium" in the electron microscopy vernacular, is a compound with a molecular weight of 254.2. The osmium tetroxide molecule is symmetrical and contains four double-bonded oxygen molecules.

$$O = Os = O$$

with the double-bonded oxygens above and below osmium as well.

Osmium tetroxide works as a secondary fixative of tissue structure by reacting primarily with lipid moieties. It is widely believed that the unsaturated bonds of fatty acids are oxidized by osmium tetroxide, with the osmium tetroxide being reduced to black metallic osmium. This reduced heavy metal adds density and contrast to the biological tissue.

Osmium tetroxide has a number of stable oxidation states and is soluble in polar and nonpolar solvents. Numerous other types of reactions of osmium tetroxide have been described that explain the stabilization of many cell components by osmium. Furthermore, the molecular weight of osmium tetroxide (254.2) is sufficiently high to be effective in scattering electrons. Thus, osmium tetroxide is also an important stain. With time, tissues exposed to osmium stain intensely black as viewed with the naked eye. Osmium tetroxide also acts as a *mordant* (a substance capable of combining with stain or dye at a later time). For electron microscopy, osmium tetroxide is a mordant for enhancing lead staining (see Chapter 5).

The penetration rate of osmium tetroxide is often slower than that of glutaraldehyde. Osmium tetroxide generally will not penetrate compact tissues of more than 0.5 mm in one hour. Very little additional penetration of tissue occurs after one hour. Specimen cubes that are much larger often show a core of unblackened tissue that is virtually useless for electron microscopy (Figure 2-1). The concentrations of both glutaraldehyde and osmium tetroxide are diluted within the specimen as these fixatives penetrate tissues. Usually, the quality of fixation at the center of a large tissue block cannot be expected to be as good as the peripheral area of the tissue. Furthermore, peripheral tissues may be over fixed. Osmium tetroxide is a nonpolar compound. However, it penetrates both charged and uncharged surfaces and is soluble in both polar and nonpolar tissue components.

The length of time tissue is fixed in osmium tetroxide has an effect on the appearance of the tissue. Too short exposure to osmium tetroxide results in under fixation. Prolonged exposure results in extraction of tissue components, especially later during the dehydration process. Exposures of tissues to osmium tetroxide should be generally limited to under 1.5 hours; however, notable exceptions exist (impervious seeds, botanicals, and some bacteria, for example, need longer exposure times).

CAUTION: Be extremely careful in handling osmium tetroxide, whether it is in crystalline or

|⊢—2mm—⊣|

Figure 2-1 Drawing of a tissue section incompletely penetrated by osmium tetroxide.

solution form. The fixative properties of osmium can be seen not only on tissue samples prepared for electron microscopy, but also on the living tissues of the individuals performing the fixation. Osmium tetroxide vapors are dangerous to the conjunctivum and the cornea of the eye and to the respiratory and alimentary membranes. Osmium tetroxide should always be used under a properly ventilated hood. Osmium tetroxide is moderately volatile. Osmium tetroxide vapors smell similar to chlorine bleach, a smell not readily forgotten in the event of accidental exposure to the substance. Avoid direct contact by wearing impervious plastic or rubber gloves.

Used or excess osmium tetroxide should be properly sealed and disposed of according to current safety standards. Osmium tetroxide should not be stored in plastic containers. Spills can be cleaned up with absorbent material (wood shavings or cat litter will do). Osmium tetroxide can be reduced by corn oil to a less harmful substance and then given to a properly licensed agency for final disposal.

Selection of a Fixative and a Buffer

Fixative

In selecting a fixative, it is important to determine what has worked in the past for the particular tissue under consideration with the particular conditions employed. Literature is generally available for the particular tissue to be prepared, and it should be consulted prior to making a final decision on which fixative to use. Usually, recent literature is the most helpful because the methods of fixation have improved continually over the years, and the standards by which fixation quality is judged have become higher with time. Current literature sources often will refer the reader to other sources for the detailed methodology.

It is important to determine what tissue features are to be emphasized in the study. If ribosomes are to be examined, then a fixative should be used that emphasizes ribosomes in the tissues, such as glutaraldehyde followed by osmium tetroxide. In such an instance, permanganates or osmium-ferrocyanide mixtures should be avoided because ribosomes are extracted with these protocols.

Buffers

If a buffering system is not used with the primary fixative, the pH of the tissue is drastically lowered during the fixation procedure. As a consequence,

numerous artifacts may be produced. Buffering systems that maintain physiologic pH (e.g., 7.2 to 7.4 for most mammalian tissues) result in fewer artifacts. Again, the literature should be consulted since the physiological pH for particular biological systems may vary from 5.5 to 7.5 or even higher.

There are several choices of buffering conditions available. Appropriate buffering systems for a particular biological tissue have usually been worked out, and the information is readily available in the literature. Common buffering systems in use are phosphate, s-collidine, tris-maleate, and cacodylate. Organic buffers such as PIPES and HEPES are often used in tissue culture work and may sometimes be used in the primary fixative. Phosphate buffer is often the buffer of choice since it is not itself toxic and shares certain properties with cells. A simplified guide to preparing some buffers is given in Table 2-2. When using Table 2-2, prepare solutions A and B and mix according to the formula given. In the formula, x refers to the amount of solution in the pH adjustment table (Table 2-3) that will yield a desired pH.

Just as important as the selection of the buffer is the *osmolarity* of the buffering system. The osmolarity of the buffer, as it contributes to the osmolarity of the overall fixative, is important because osmolarity changes induce shrinkage or swelling of tissue due to osmotic effects. Artifactual changes in cell size should be avoided or minimized. Literature sources are the best indication of the appropriate osmolarity for a tissue under consideration. The older literature shows that investigators paid meticulous attention to developing a buffering system with a physiologic osmolarity. Physiologic osmolarity (320 milliosmoles for most mammalian extracellular fluids) was at one time considered an ideal goal when preparing a fixative.

The total osmolarity of the fixative includes both the osmolarity of the buffer and any added substance(s) and that of the fixative (a 3% solution of glutaraldehyde alone is 300 milliosmoles). As it turns out, slightly hyperosmolar solutions often give the best empirical fixation results. Some tissues, however, appear to show little difference in preservation in buffering systems when molarities ranging from 0.05 to 0.2 are used, while others are very sensitive to buffer osmolarity. The osmolarity of the buffer and added substances (such as salts and sugars) are very important factors in assessing overall fixative osmolarity.

The fixatives (glutaraldehyde, formaldehyde, and osmium tetroxide) are thought to alter the

Table 2-2   Preparation of Common Buffers Utilized in Electron Microscopy

| Buffer | pH Range | Stock Solutions | | Formula |
		A	B	
Plumel's cacodylate	5.0–7.4	0.2 M sodium cacodylate (42.8 g Na [CH$_3$]$_2$AsO$_2$·3 H$_2$O) per liter of distilled H$_2$O	0.2 N HCl	25 ml A + x ml B made. Bring volume up to 100 ml with distilled H$_2$O.
Sorenson's phosphate	5.0–8.2	0.67 M monosodium phosphate (9.08 g NaH$_2$PO$_4$) per liter of distilled H$_2$O	0.67 M disodium phosphate (11.88 g Na$_2$HPO$_4$·4 H$_2$O)	x ml A plus (100-x) ml B
Gomori's tris-maleate	5.2–8.6	0.2 M tris acid maleate (24.2 g tris-[hydroxymethyl] amino-methane + 23.2 g maleic acid 19.6 g maleic anhydride per liter)	0.2 N NaOH	25 ml A + x ml B made up to 100 ml
s-collidine	6.0–8.0	2.67 ml pure s-collidine in 50.0 ml of distilled H$_2$O	~9.0 ml of 1.0 M HCl	A + B made up to 100 ml

Table 2-3   pH Adjustment

pH	Plumel's cacodylate	Sorenson's phosphate	Gomori's tris-maleate
5.0	23.5	98.8	
5.2	22.5	98.0	3.5
5.4	21.5	96.7	5.4
5.6	19.6	94.8	7.8
5.8	17.4	91.9	10.3
6.0	14.8	87.7	13.0
6.2	11.9	81.5	15.8
6.4	9.2	73.2	18.5
6.6	6.7	62.7	21.3
6.8	4.7	50.8	22.5
7.0	3.3	39.2	24.0
7.2	2.1	28.5	25.5
7.4	1.4	19.6	27.0
7.6		13.2	29.0
7.8		8.6	31.8
8.0		5.5	34.5
8.2		3.3	37.5
8.4			40.5
8.6			43.3

ticular buffering system over another is often based on the results obtained after examination of the tissue.

The osmolarity of the fixative can be adjusted by increasing the osmolarity of the buffer or by adding substance(s) to the buffer such as sucrose, glucose, or sodium chloride. Calcium or magnesium chloride is often added to the buffer to preserve certain features within the tissue (such as membranes and DNA).

CAUTION: Some common buffers used in electron microscopy are extremely toxic and carcinogenic. For example, cacodylate buffer contains arsenic. S-collidine buffer, besides having a disagreeable odor, is also extremely toxic because of its pyrimidine components. Veronol-acetate buffer has barbiturates. Exercise extreme care when preparing a buffer. Avoid breathing the buffer salt powders and avoid contact of the buffer constituents with the skin. Use gloves and a fume hood and a safe method of disposal. Do not allow buffers to dry in glassware. They become airborne and are easily inspired.

permeability of the cell membrane to some degree. The osmotic effect of the fixative on the cell appears not to have a large impact on the cell. It may be that the rapid penetration of the buffer and its contents is what affects cells initially. *The selection of one par-*

Obtaining and Preparing Buffered Glutaraldehyde Fixative

Assuming a standard glutaraldehyde prefixation and osmium tetroxide postfixation combination is to be used, the preparation of a solution of 3% buffered glutaraldehyde and another of 1% osmium tetroxoide will be described.

Electron microscopy suppliers sell glutaraldehyde in two grades. The one most commonly used is *electron microscopy grade* solutions of glutaraldehyde, which vary in glutaraldehyde content from 8% to 70% and are usually packed in sealed ampoules (Figure 2-2) or screw cap bottles. Usually, 100 ml of electron microscope grade solutions are the maximum quantities available in a single container. The cost for a 10 ml ampoule of 50% electron microscopy grade glutaraldehyde in 1991 was about $7 and a 100 ml bottle was about $30. The use of electron microscope grade of glutaraldehyde is reasonably cost efficient if tissues are to be immersed in glutaraldehyde. If perfusion fixation of large animals is to be undertaken, *biological grade glutaraldehyde* is available in large volumes at a considerably lower price (1 gal or 3.8 l of a 50% solution costs about $50). Biological grade glutaraldehyde is less pure, but is considered adequate for most perfusion fixation purposes.

Any glutaraldehyde that is not sealed in an ampoule will polymerize with time and produce a whitish haze when added to the buffer. If this occurs, it should be discarded. Glutaraldehyde not in ampoules should be stored in a refrigerator and used within a month of receipt from the supplier. Storage of glutaraldehyde at room temperature will lead to its rapid polymerization. Glutaraldehyde may be frozen, if desired, to extend its shelf life.

Buffer, made according to Table 2-2, is checked for the appropriate pH. The buffer should be made just before use. Using 50% glutaraldehyde is recommended because the final concentration of glutaraldehyde is readily calculated. The amount added to make 100 ml of fixative is twice the glutaraldehyde percentage needed (e.g., 6 ml of a 50% glutaraldehyde per 94 ml buffer will produce a 3% glutaraldehyde solution). After the solution is prepared, it is brought to the appropriate temperature where it is ready for use.

Obtaining and Preparing Osmium Fixative

Osmium can be purchased in ampoules, which are sealed under dry nitrogen, in either crystalline form or in aqueous solutions of about 4% (Figure 2-2). Properly sealed osmium tetroxide is stable indefinitely in either form. The cost of 1 gm of osmium tetroxide (in 1991) was approximately $37, and 20 ml of a 4% solution was approximately $45. Osmium tetroxide prices are known to vary widely.

From the crystalline form, one can prepare a 4% solution of osmium tetroxide in a container sealed with Parafilm and capped with a Teflon-lined screw cap. Such a solution is stable for many months when refrigerated. Osmium tetroxide dissolves slowly in distilled water at room temperature. It can be dissolved rapidly in distilled water warmed on a steam bath. Use of a magnetic stirring bar also will facilitate its entry into solution. Osmium tetroxide vapors have been known to escape from even fairly tightly sealed flasks or bottles and to cause darkening of the inside of a refrigerator. It is recommended that osmium tetroxide be stored in screwed down containers with Teflon cap liners. Even then, the sealed end should be wrapped in waxed film such as Parafilm and the entire container wrapped in foil and placed in a Ziplock bag. Another way to store osmium tetroxide is to place it in a sealed glass container that is then placed in a second sealed glass container.

Osmium tetroxide is prepared either in distilled water or in buffer. The use of buffer is often questioned because glutaraldehyde and osmium tetroxide destroy completely any gradient to ions, thus indicating that one no longer needs to maintain tonicity or pH of the postfixation solution.

Immersion and Perfusion Fixation

One can either immerse tissue into fixative or use the blood-vascular system of organ perfusion to fix tissue. If perfusion is used, the fixative is distributed to the tissues through the blood vessels. From there, fixatives will penetrate into the tissue spaces and to

Figure 2-2 Sealed ampoules of glutaraldehyde, 4% osmium tetroxide and, crystalline osmium tetroxide (left to right).

the various tissues. In some instances, there is no choice as to the method of fixation. There may be no accessible vascular system in certain tissues through which to perfuse them. The biological specimen may be so small as to render it impossible to perfuse. In such instances, the natural tissue size may permit fixation of the entire organism by immersion into the fixative solution.

Immersion

It is important to bear in mind the slow penetration rate of glutaraldehyde. If one elects the immersion technique, then the size of compact tissues must be *less* than 1 mm thick for glutaraldehyde to penetrate prior to the time that autolytic changes will take place. Usually, a sharp, clean razor blade is used to dice tissue into small cubes or slices. The tissue is cut on dental wax (Figure 2-3) or waxed paper with sawing motions of the razor blade so as not to compress the tissue too severely. Fixative is flooded over the tissue during the dicing procedure. At no time is the tissue allowed to dry! After the tissue is cut, the pieces are immersed in glutaraldehyde in a small vial (Figure 2-4). The volume of fixative should exceed the tissue volume by at least 5-fold and preferably 10-fold.

The minimum fixation period is usually an hour, but the fixation process has been extended for many hours, days, or even weeks. There is a controversy over how long material can stay in the primary fixative. Some believe that it is important to fix in glutaraldehyde for a few hours and to proceed imme-

diately to secondary fixation with osmium tetroxide. Their argument is that osmium tetroxide is a fixative as well, and time should not elapse needlessly in which tissues may undergo artifactual changes. Others believe that the specimens may stay in glutaraldehyde for prolonged periods without artifactual change. If there is any doubt, the tissue should be processed from the glutaraldehyde step into wash buffer without delay and then into osmium tetroxide as soon as the glutaraldehyde fixation is deemed optimal.

Fixation of pellets produced by centrifugation often presents a penetration problem when glutaraldehyde is used as a primary fixative. If the pellet is not loosened from the centrifuge tube or has been spun down to the bottom of a pointed centrifuge tube, glutaraldehyde will have poor access to the tissue. Essentially, access of glutaraldehyde to pelleted tissue is from one side of the pellet since the pellet is tightly adherent to the centrifuge tube. Pellets must be in the order of 0.5 mm or less in thickness for adequate penetration of glutaraldehyde. If they can be loosened from the centrifuge tube prior to glutaraldehyde fixation without resuspension of the tissue, their exposure to glutaraldehyde and subsequent chemicals is enhanced. Pelleted material may be fixed in suspension and pelleted at a later time. However, fixed suspended material often fails to form a solid pellet. Trial and error may be necessary to fix pelletable biologicals.

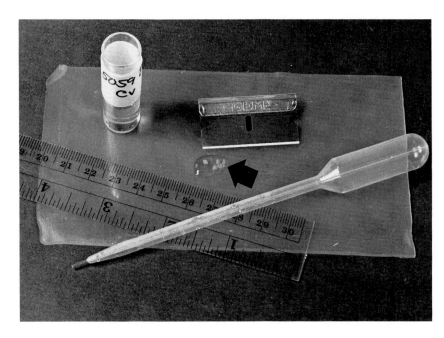

Figure 2-3 Setup appropriate for dicing tissue to be prepared for electron microscopy. Tissue (arrow) is placed on a sheet of dental wax and diced with a sharp razor blade. A ruler has been placed nearby to show the tissue block size, which should not exceed 1 mm after dicing.

Figure 2-4 Vial suitable for tissue processing for electron microscopy. Also shown is the vial cap and a dime for size comparison.

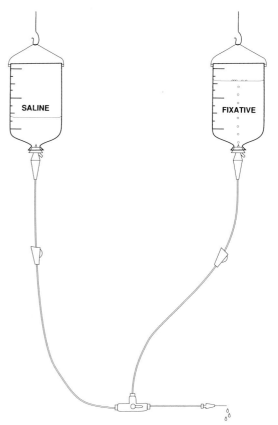

Figure 2-5 Drawing of gravity-fed perfusion apparatus. One of the two bottles contains saline to clear the vascular system and the other contains fixative. A valve allows only one of the solutions to flow at a time. A needle at the end of the tubing is used to tap into the animal's vascular system. (From Histological and Histopathological Evaluation of the Testis used with permission of Cache River Press.)

Perfusion Fixation

In most tissues, cells are only a few μm from their vascular supply. When possible, perfusion of fixative using the vascular system as a delivery route is usually deemed the optimal route for tissue fixation. Most perfusions can be accomplished without handling the tissue and without needing to cut fresh tissue, a manipulation that results in excess pressure on the fragile tissue and distortion of its components.

One must be knowledgeable of the circulatory system of the species under consideration. The most common perfusion scheme employs whole-body perfusion. In a mammal, an apparatus such as shown in Figure 2-5 is used to introduce solutions into the left ventricle of the heart or into a major blood vessel of an appropriately anesthetized animal. The perfusion bottles are placed about 4 feet above the animal being perfused. The general plan for vascular perfusion is to introduce fixative into the arterial system and to allow its egress from a cut made in the venous system. An outlet for blood and perfusion solutions (perfusate) is made in an area such as the

right atrium or ventricle. For example, a needle is inserted into the left ventricle and solution is allowed to flow from the left ventricle to the aorta where it is distributed to most of the capillary beds of the body. It returns to the heart via the venous system and exits the heart through a cut made in the right atrium or ventricle.

The solution initially introduced into the heart is usually a physiological solution such as saline (mammals) or Ringers (nonmammalian vertebrates), sometimes with an anticoagulant. It is perfused at room temperature to avoid changes in blood vessels that are related to temperature. Blood is cleared from the vascular system in this manner. The organ will usually blanch when it is cleared of blood. Exposure of glutaraldehyde to blood without prior clearing of the system will cause coagulation of blood and result in the failure of the perfusion. Most perfusion systems have a valving system (Figure 2-5) that allows switch over, at an appropriate time, from saline to glutaraldehyde. The exposure of the anesthetized animal to glutaraldehyde may cause involuntary twitching of the animal's muscles. The properly anesthetized animal is not conscious of pain during this procedure.

Perfusion is usually conducted for about 30 to 45 minutes. During this period, the animal's organs harden and take on a yellowish tint. After perfusion, the desired organ is removed and diced into small (under 1 mm) pieces with a sharp razor blade. While being cut, the tissue should be immersed in fixative on a piece of dental wax. It is common to place the small pieces of tissue in a vial containing glutaraldehyde fixative for an additional hour or two to insure that fixation is adequate (postperfusion immersion).

Perfusion methodology may not necessarily involve whole body fixation. It may be partial body fixation or simply organ perfusion. Often one can gain access to the vessels leading to an organ and introduce a perfusion needle into the vessel (preferably an artery). Egress of fixative is by cutting the final drainage vein from the organ. The literature is the best source of information on what has been successful historically in terms of perfusion for a particular tissue. In many cases, blood will not clear from the organ without addition of vasodilators or anticoagulants such as procaine hydrochloride or heparin, respectively.

Fixation Conditions

Many specific details become important in the fixation of various tissues. For almost every tissue, there are special needs that must be taken into account in order to obtain optimal fixation. These cannot be dealt with effectively within the space limitations of this chapter. For specific information about a variety of protocols that meet specific fixation needs, see Hayat's reference volume (1981), which is devoted entirely to the theoretical bases and practical methods of fixation for electron microscopy.

Primary fixation and secondary fixation of tissues in osmium tetroxide is often performed at 4° C. With respect to primary fixation, tissue autolysis and structural changes are minimized at this temperature. Thus, immersed tissue may be fixed prior to artifactual changes, although it is well-recognized that osmium tetroxide penetration at 4° C is slower than at room temperature. Secondary fixation in the

cold (4° C) is advocated for osmium tetroxide since it is thought that fewer constituents of the cell are leached from the tissue under these conditions. Empirically, the results may be better or unchanged for certain tissues that are fixed at room temperature as compared with fixation in the cold.

During primary fixation, the tissue should be gently and periodically agitated (for example every 10 minutes). After agitation, tissue surfaces that are up against the container vial will be dislodged and exposed to the fixative.

Less osmium tetroxide than glutaraldehyde is needed to fix tissues. The fluid level of osmium tetroxide in the vial should be two to three times the height of the tissue.

Some investigators will utilize the same buffering system during both the primary and secondary fixation protocols. The advantage in doing so is that the same ionic concentration and buffering tonicity are maintained until dehydration of the specimen. There may be little or no difference in the results obtained by using either a buffering system or distilled water as a vehicle for osmium tetroxide. Caution must be exercised when using some organic buffers for postfixation since they will react with osmium tetroxide.

Popular Fixation Protocols Other Than Glutaraldehyde-Osmium Tetroxide

Glutaraldehyde followed by osmium tetroxide is considered the standard fixation protocol, capable of stabilizing the maximum number of different cell components. Most specialized fixation techniques are only slight modifications of the glutaraldehyde-osmium protocol. A few are described.

Karnovsky's Fixative
This popular fixative uses a relatively low percentage of formaldehyde in the primary fixative. The use of formaldehyde with glutaraldehyde achieves a more rapid overall penetration of fixative (Karnovsky, 1971) than with glutaraldehyde alone. It is theorized that the formaldehyde temporarily stabilizes structures that are later more permanently stabilized by glutaraldehyde. Formaldehyde penetrates about five times faster than glutaraldehyde. Because of the addition of formaldehyde (and calcium chloride) to the glutaraldehyde fixative, the osmolarity of the total fixative is about 2110 milliosmoles, far in excess of what would be considered physiological osmolarity in mammalian systems. Since the effect of fixative osmolarity on tissue structure is far less than

that of the buffer and its added constituents, there seems to be negligible effect of the added formaldehyde with its high osmolarity on tissue structure.

Karnovsky's fixative is often used when penetration of fixatives is considered to be poor or when immersed tissue is of a larger size than usual. The formaldehyde-glutaraldehyde method has been used extensively for nervous tissue. The quality of tissues fixed with Karnovsky's method is similar to a good fixation with glutaraldehyde alone.

Formaldehyde (CH$_2$O) is a gas, but it can be obtained in liquid form in 37% to 40% solutions called *formalin*. When mixing fixatives, formalin is usually not employed because it contains numerous impurities such as methanol. Instead, powdered, polymerized formaldehyde or *paraformaldehyde* is used to prepare formalin. Since paraformaldehyde goes into solution with great difficulty at room temperature, the distilled water in which it will be dissolved is first heated to 60° to 80° C under a ventilated fume hood. A small amount of sodium hydroxide or preferably potassium hydroxide is added to the heated solution and stirred until the solution clears.

The following protocol should be followed when preparing 100 ml Karnovsky's fixative:

1. Dissolve 4 gm paraformaldehyde powder in 50 ml distilled water by heating to 60° to 70° C under a hood. Add two to four drops of 1N KOH, stirring until the solution clears.
2. After cooling add 10 ml of 50% glutaraldehyde.
3. Bring to 100 ml with 0.2 M sodium cacodylate or phosphate buffer (at the proper pH).
4. If cacodylate buffer is used, then 50 mg calcium chloride (CaCl$_2$) is added slowly to the solution.
5. Tissues are perfusion fixed routinely or immersion fixed for 2 to 5 hours. Standard washing and postfixation with osmium tetroxide and dehydration are employed.

Karnovsky's original fixative, containing 5% glutaraldehyde and 4% formaldehyde, is often modified to yield fixatives with lower concentrations of both formaldehyde and glutaraldehyde (e. g., 2% paraformaldehyde and 2.5% glutaraldehyde).

Osmium-reduced Ferrocyanide

This method is a slight modification of the standard glutaraldehyde-osmium tetroxide protocol (Karnovsky, 1971; Russell, 1978). It is especially suited for demonstration of membranes and glycogen. The drawbacks are that ribosomes are poorly preserved and the cytoplasmic matrix appears less dense than when observed with the traditional glutaraldehyde tissue preparation regimen. However, the added prominence of membranous elements within the cytoplasm gives an overall pleasing appearance to the tissue. Many of the micrographs used in this text are from tissues prepared using this method (see Chapter 19, Figures 19-1, 19-5, and 19-52).

Fixation with glutaraldehyde is performed according to standard methods published in the literature. The tissues are washed thoroughly in buffer overnight (three or four changes) and fixed in a freshly prepared mixture consisting of 1% osmium tetroxide and 1.25% potassium ferrocyanide (Russell, 1978). It is important to use ferrocyanide and *not* ferricyanide. To prepare the final solution, equal volumes of 2% osmium tetroxide and 2.5% potassium ferrocyanide are combined The final solution appears brown. The tissue may not blacken quite as nicely as seen with osmium tetroxide, but may turn a dark brown. It will later blacken during the standard alcohol dehydration steps.

Potassium Permanganate

Historically, the development of the permanganate fixatives (Luft, 1956) preceded the standard glutaraldehyde-osmium tetroxide fixation protocol (Sabatini et al., 1963). The popularity of permanganate fixatives has declined substantially in recent years, although they are used to fix certain plant specimens and yeast cells. Permanganate-fixed tissues, especially plant tissues, are noted for their preservation of membranous elements. The remainder of the tissue has a pleasing appearance, but upon close examination appears washed out with few nonmembranous elements visible or prominent. Permanganate is known to penetrate tissues more rapidly than glutaraldehyde. The mechanism of fixation is not known, although several theories are postulated (Hayat, 1981) to explain its reactivity with tissue components. An example of permanganate fixation is shown in Chapter 19, Figure 19-128.

Many protocols using permanganate are available; however, the most common is to use a buffered solution of 3% potassium permanganate (KMnO$_4$) for 2 hours in the cold (4° C). The specimens are then rinsed in several changes of buffer overnight and dehydrated in a standard protocol. Postfixation in osmium tetroxide is not used with potassium permanganate fixed tissues.

Fixative Additives

There are several substances that may be added to fixatives to enhance fixation of particular structures. For example, tannic acid added to the glutaraldehyde fixative will enhance membranes (Kalina and Pease, 1977), microtubules (Burton et al., 1975), and microfilaments (Goldman et al., 1979; Maupin and Pollard, 1983). Apparently tannic acid acts as a mordant to enhance later staining with heavy metals. The techniques to enhance particular features of the cell are numerous and are detailed in reference books (Glauert, 1975; Hayat, 1981).

Although not an additive per se, tissues may be fixed with glutaraldehyde while being exposed to short bursts of microwave energy. The potential advantage of such a technique is the rapid preservation of tissue processes that might not have otherwise been "captured" if the fixation were by glutaraldehyde alone (Login and Dvorak, 1985; Leong et al.,1985; van de Kant et al., 1990).

Fixation of Plant Tissues

In general, fixation of plant tissues is more difficult than animal tissues. Although a glutaraldehyde-osmium tetroxide combination is generally the most effective fixative for plant tissues, there are several minor modifications in this protocol that have their basis in differences in animal and plant specimens. There are also several reasons why plant tissues do not fix as well as animal tissues. Often plants are covered by waxy substances or cuticles that are hydrophobic and impede fixative penetration. The low protein content of plants renders glutaraldehyde less effective at cross-linking protein. The cell wall of plants presents a partial boundary to penetration of glutaraldehyde. The vacuole of plants rapidly dilutes the fixative concentration, necessitating the use of fixative of higher concentrations. A pH of 7.0 usually is used during fixation. Generally, when fixing plant tissues, it is not as important to begin fixation immediately after collecting the specimen (e.g., a leaf) since autolytic changes take place more slowly.

Preparation of plant cells may vary somewhat from the traditional methods employed for mammalian cells. For instance, while similar types of buffers may be employed (phosphates being the most common), the pH tends to be somewhat lower, at pH 7.0 to 7.2. Fixation times are usually longer (from 2 to 4 hours, on average), and dehydration and resin infiltration protocols are also lengthened. In some impenetrable specimens such as seeds, it is necessary to remove or abrade the seed coat in order to permit the reagents to enter. Some investigators may even resort to fixation times in glutaraldehyde and osmium tetroxide from 12 to 18 hours or more. This may seem excessive and even questionable; however, in

dormant specimens and exceptionally dense tissues the protracted fixation regimes may be necessary. Low viscosity embedding resins such as Spurr's epoxy resin or LR White acrylic resins are recommended.

HINT: A good resin formulation consists of taking one part of complete Spurr's resin and mixing it with one part of complete Epon 812, or substitute. The mixture yields a resin with good penetration and excellent contrast and stainability properties.

A commonly used protocol for plants follows.

Primary Fixation	immersion of small pieces of tissues in 2–4% phosphate buffered glutaraldehyde or formaldehyde/gluta-raldehyde mixtures	2–4 hr
Washing	buffer rinses (3 ea for 30 min) and possibly 1 overnight rinse	2–18 hr
Secondary Fixation	1–4% buffered osmium tetroxide	2-4 hr
Dehydration Series	25% ethanol	20 min
	50% ethanol	20 min
	75% ethanol	20 min to overnight
	95% ethanol	30 min
	100% (2–3 changes)	30 min ea
Transitional Solvent	propylene oxide (3 changes)	20 min ea
Resin Infiltration	2 part propylene oxide : 1 part Spurr's	1–2 hr
	1 part propylene oxide : 1 part Spurr's	1–4 hr
	1 part propylene oxide : 1 part Spurr's	1–4 hr
	complete Spurr's resin	1–4 hr to overnight
Embedding Curing	complete Spurr's resin 60° C	1–4 hr 48 hr

Washing

After primary fixation with glutaraldehyde, the tissue is usually washed in the same buffer vehicle used in the glutaraldehyde fixation step. Washing is extremely important because it eliminates any free unreacted glutaraldehyde that remains within the tissue. Aldehydes remaining from the primary fixation will be oxidized by osmium tetroxide. Some protocols call for one or two 10-minute washes of the tissue in buffer. Unreacted glutaraldehyde will diffuse as slowly outward from the tissue as inward so that a minimum of a few hours of washing with at least three changes of buffer is recommended. (Residual glutaraldehyde or partially polymerized aldehyde may generate a "peppery" background upon combination with osmium tetroxide.) Overnight washes in buffer (at least three changes) will eliminate most of the unreacted glutaraldehyde. Since small quantities of buffer are not costly, it is advisable to fill the vial to the top with buffer to extract as much glutaraldehyde from the tissue as possible. Washing is usually performed at 4° C overnight.

Dehydration

Dehydration is the process of replacing the water in cells with a fluid that acts as a solvent between the aqueous environment of the cell and the hydrophobic embedding media. Water is a highly polar molecule that is, by far, the major component of virtually all cells. Common dehydrating agents are ethanol or acetone. Ethanol is the most widely used dehydration agent. The general philosophy of the dehydration step is to replace water within the tissue gradually by using a graded series of dehydration agents. Usually 30 or 50% ethanol or acetone is the first solvent tissue is exposed to after secondary fixation followed by 70%, 85%, 95%, and absolute ethanol or acetone. As one reaches higher concentrations of the dehydrating agent, the time that tissue is exposed to the dehydration agent and the number of changes of the dehydration agent are increased in order to eliminate the small amount of water remaining in the tissue. Ethanol is preferred to acetone as a dehydrating agent because anhydrous acetone absorbs water from the atmosphere and is a more powerful extractor of lipids within the cell. Molecular sieves can be used to desiccate alcohol for ready use in dehydration protocols.

CAUTION: The fine grit from the ceramic molecular sieves may attach to the specimens and become embedded with them, damaging a glass or diamond knife at the time of sectioning.

The hygroscopic nature of dehydrants presents a problem because of their absorption of water from the air. Absolute ethanol or acetone left open for a short time will absorb water to the degree that it will not be capable of eliminating all of the water from the tissue. It is extremely important to keep dehydrants sealed tightly and to not open them too often. Bottles of dehydrants (100%) that have been opened and resealed and have been around the laboratory for months must be suspected of having absorbed water. They can be used to make the lower percentage (e.g., 50%) dehydrants used in the initial steps of dehydration. To avoid water absorption during the later phases of dehydration, it is important to use sealed vials. When replacing one dehydrant with another, it is necessary to do so as rapidly as possible.

Acetone, ethanol, and other dehydration agents tend to extract lipids from the tissue. Lipid droplets may appear non-homogeneous or may have lost their electron density due to dehydration steps. Osmium tetroxide used during tissue fixation will help stabilize unsaturated lipids during the dehydration process.

Use of Transitional Solvents

Replacement of the dehydration solution by another intermediary solvent that is highly miscible with the plastic embedding medium is usually necessary to interface with the embedding media. The standard solvent used is *propylene oxide*, a highly volatile and potentially carcinogenic liquid. It will also further dehydrate the tissue. Like absolute alcohol or ethanol, it is highly hygroscopic, and the same precautions mentioned previously for these substances should be taken with propylene oxide. Usually more than one change of propylene oxide is necessary to replace the alcohol. Although most embedding media are directly miscible with dehydrating agents, most protocols employ a transitional solvent between the dehydrant and the resin to speed up the infiltration process.

Infiltration of Resin

Infiltration is the process by which dehydrants or transition fluids are gradually replaced by liquid plastic monomers, which are very viscous like pancake syrups. Epoxy mixtures are introduced gradually into the tissue block after dehydration. The solvent, propylene oxide, is mixed with the epoxy and placed into vials with the tissue. Gradually, the epoxy-solvent ratio is increased until pure epoxy is used. The pure resin specimens are transferred into molds containing the liquid plastic and are finally placed into an oven where the epoxy components polymerize to form a solid. A typical infiltration schedule is outlined below.

1 part propylene oxide : 1 part epoxy	2 hours
1 part propylene oxide : 2 parts epoxy	2 hours
1 part propylene oxide : 3 parts epoxy	2 hours
1 part propylene oxide : 4 parts epoxy	2 hours
pure epoxy	1–3 hours
Place tissue and pure epoxy in embedding mold and place in oven.	

During infiltration by the epoxy, the vials of tissue are gently agitated on a turntable that positions the tissue vials at about 30° from vertical and slowly rotates the tissues (Figure 2-6). Tissue vials should be closed so that moisture from the air does not enter them.

Some tissues may be processed up to pure liquid resin and stored for weeks in the freezer. Under these conditions, the resin will polymerize very slowly. It is often wise to place some of the samples in epoxy and store them in the freezer. If there are problems with embedding, such as holes in the tissue due to poor infiltration, then the tissue may be pro-

Figure 2-6 Rotary mixer used to infiltrate tissues. Vials containing tissue are situated in holes of the mixer.

cessed backward (usually to propylene oxide), reinfiltrated, and then embedded again.

Embedding

The liquid embedding media must polymerize to form a solid matrix that thoroughly permeates the tissue. In common practice, this is accomplished by using epoxy monomers that are liquid when the components are mixed initially, but that polymerize with time and under certain curing conditions. Thus the tissue, which at one time was primarily hydrated, is solid and stable after embedding.

Epon Embedding

In the majority of instances, an epoxy embedding media is used. In the early 1960s, an epoxy embedding media, marketed by Shell Oil Co. under the brand name of Epon (or Epon 812), was introduced as an embedding medium (Luft, 1961). For many years, it was extremely popular and, therefore, used extensively. Epon, as such, is no longer available but has been replaced by similar types of epoxy resins. Although Epon is no longer in use, much of the embedding for electron microscopy used this embedment. Thus for historical purposes it is worthy of note.

The components of different epoxies are generally the same. Epon and its successors are made by thoroughly mixing the *resin*, the *hardeners*, and the *accelerator*. Using Epon embedding media as an example, these are as follows:

Resin	Epon 812 or substitute
Hardeners	DDSA (dodecenyl succinic anhydride) NMA (nadic methyl anhydride)
Accelerator	DMP-30 (2,4,6-tridimethylamino methyl phenol)

The hardness of the Epon mixture is controlled by the relative amounts of DDSA and NMA, the former producing softer blocks and the latter harder blocks.

Measuring Embedding Media

Because the ingredients of resin mixtures are usually highly viscous and difficult to clean from volumetric cylinders, it is not easy to mix them based on volume. The most convenient and accurate way to measure embedding media is to weigh the components on a top-loading balance. First, the container is tared (or zeroed), and the first ingredient is added to reach the weight called for in the instructions. The second ingredient is added to reach the cumulative weight total of the first and second ingredients. Most embedding media can be polymerized into materials of various hardness by varying the proportions of the ingredients. See the manufacturer's recommendations to obtain the necessary hardness.

CAUTION: All of the components of embedding media have the potential to cause skin irritation and rash. Use gloves when handling them and work in a well-ventilated room. Fumes must be avoided since epoxys are hyperallergenic and are suspected of being carcinogens.

Mixing Embedding Media

It is recommended that the resin and the hardener be thoroughly mixed in a disposable plastic beaker using a resin mixer, a glass rod, or a tongue depressor. The accelerator is then added and mixed thoroughly with the above. Mixing for at least 5 minutes will assure that polymerization will be adequate. Mixing that is too vigorous will incorporate numerous air bubbles into the plastic that will interfere with embedding. It is possible to remove the bubbles by placing the resin mix under vacuum. Resin is not easily cleaned from the container or from the stirring rod, so the use of disposable materials is suggested. Contaminated utensils and partially empty containers are placed at 60° C for 48 hours to polymerize the resin prior to disposal.

Other Embedments and Their Use

There are many types of embedding media. For the most part, they are referred to by their brand names, although in some instances there may be little difference between one brand and another. Some embedding media are chemically unique and have special uses, whereas others are replacements for Epon and show minor difference from the Epon formula. The three most common epoxy embedments in use today are *Araldite*, the epoxy replacements for Epon (see above), and *Spurr's*. Spurr's, formulated by Dr. A. R. Spurr, is an epoxy embedding medium of low viscosity. *Lowicryl* is a low viscosity acrylic embedding medium that can be used at low or high temperatures for immunocytochem-

istry. Denaturation of proteins is minimized at low temperatures, and the embedding medium will accept a small percentage of water remaining in the tissue. The use of propylene oxide thus may be avoided. *LR White* is a low viscosity embedding medium that can also be used for cytochemistry (see below). *Methacrylate* (GMA or HPMA) is a water-miscible embedding medium that is also often used for cytochemical studies, although it has often been used routinely for a variety of purposes. Water-soluble embedding media such as glycol methacrylates do not necessitate complete dehydration prior to embedding. They, like LR White described above, cause minimal denaturation of protein during tissue processing.

Suppliers provide specific instructions for mixing embedding media and curing of the embedment with the purchase of the embedding kits. As examples, recommended methods are provided here for four of the more common embedding media: EMbed 812, Araldite 502, Spurr's, and LR White.

Epoxy Resin 812
The standard replacement for Epon is a resin with the number 812, which is made by various manufacturers. Several brand names (Poly/Bed, EMbed, etc.) have been used to further characterize the mixture.

To make 812 embedding media:		
Resin	812 Resin	51.13 g (50 g)
Hardeners	DDSA	27.02 g (25 g)
	NMA	21.85 g (25 g)
Accelerator	DMP-30	1.5–2.0 ml

The ratio of components (DDSA and NMA) given above determine the overall hardness of the polymerized resin. The more NMA, the harder the final plastic; the more DDSA, the softer the final plastic. The numbers in parenthesis above indicate the amounts of resin and hardeners necessary to obtain a mixture that is slightly harder than that provided. It is important to match the hardness of tissues to the hardness of resin in order to facilitate sectioning. In addition, for extremely thin sections, as may be cut with a diamond knife, the resin mixture should be formulated to give a rather firm polymerized block.

For infiltration and embedding schedule for epoxy mixture 812, see schedule in Table 2-1.

Araldite
Araldite embedding media is an epoxy mixture characterized by uniform polymerizing and little shrinkage in the process. Sections show good stability un-

der the electron beam. In Europe, Araldite CY-212 is commonly used. In the United States, Araldite 502 is the most popular resin. Araldite can be combined with one of several other epoxy mixtures to derive the benefits of both.

To make Araldite 502 according to Luft (1961):

Resin	Araldite 502	54 ml
Hardener	DDSA	46 ml
Accelerator	DMP-30	1.5–2.0%

The resin and hardener can be mixed first and stored at 4° C in a tightly sealed container in the refrigerator for up to six months. After warming the closed container to prevent moisture condensation, the accelerator is added and the ingredients are mixed thoroughly just before use.

Infiltration and embedding schedule for Araldite 502:

propylene oxide (2 changes)	15 min ea
1 propylene oxide : 1 resin mixture	1 hr
pure resin mixture—add to 1:1 mixture used above	3–6 hr
pure resin mixture for polymerization	overnight 35° C
	next day 45° C
	next day 60° C

Satisfactory results can be obtained by overnight polymerization of resin mixture at 60° C.

Spurr's

The lowered viscosity of Spurr's embedding medium facilitates its penetration into a variety of tissues that are otherwise difficult to penetrate. Moreover, its routine use for tissues of all hardnesses is frequently advocated. The resin, vinylcyclohexane dioxide (VCD), is a di-epoxide that yields highly cross-linked polymers. The relative amount of flexibilizer (diglycidyl ether of polypropyleneglycol or DER 736) in the preparation regulates hardness, although a separate hardener (nonenyl succinic anhydride or NSA) is often used. The accelerator (dimethylaminoethanol or DMAE) governs the speed of the reaction.

To make Spurr's:

	Component	Firm Mixture	Soft Mixture	Rapid Cure
Resin	VCD	10.0 g	10.0 g	10.0 g
Flexibilizer	DER 736	6.0 g	7.0 g	6.0 g
Hardener	NSA	26.0 g	26.0 g	26.0 g
Accelerator	DMAE	0.4 g	0.4 g	1.0 g
Cure time (at 70° C)		8 hr	8 hr	3 hr

The medium is prepared by weighing components on a top-loading balance in a disposable plastic beaker. The accelerator should be added last after thoroughly mixing all of the other components. Then a final, thorough mixing should be undertaken. Embedding media is freshly prepared for each use or stored at approximately −20° C. The medium is compatible with ethanol or acetone dehydrants and with propylene oxide. The resin is very sensitive to traces of moisture (yields very brittle blocks), so anhydrous conditions must be scrupulously maintained.

CAUTION: Spurr's resin components, especially VCD, are quite toxic and are carcinogenic.

LR White

LR White is often used for enzyme and immunohistochemistry and cytochemistry. Its low viscosity allows rapid embedding, and its hydrophilic nature allows the penetration of water molecules that contain substrates or antibodies that may be brought into apposition with the sections. The fixation protocol is governed by the need to preserve enzymes and antigens for subsequent localization (see Chapters 9 and 10). Ethanol dehydration is preferred. Infiltration times in resin as short as 3 hours are possible, although overnight infiltration is suggested. Gelatin capsules are filled with the resin and sealed to keep oxygen from interfering with the polymerization process. The embedded tissue blocks are cured at 60° C with little to no temperature fluctuation for one day. A rapid tissue processing protocol for LR White is given later in this chapter.

Low Pressure Removal of Solvents and Air Bubbles

This optional step employs the use of a sealed container connected to a vacuum pump to remove infiltration solvents and air bubbles. Small amounts of infiltration material remaining within tissues are volatilized by this procedure. A large number of bubbles appearing at the surface indicate significant amounts of infiltration agent remaining in the tissue. In addition, small air bubbles within the embedment and clinging to the tissue will enlarge and surface under vacuum conditions.

Curing of the Embedment

The final step in embedding is curing the tissue blocks. Curing involves polymerization of the epoxy mixture. A chain of polymerized resin is formed and cross-linked to other chains of resin (except acrylics, which are not cross-linked) to form an integrated meshwork pattern of epoxy resins totally permeating the tissue. For most epoxy resins, increased temperature accelerates the rate of polymerization. Ultraviolet light can be used in combination with catalysts to polymerize certain resins such as Lowacryl. From 12 hours to 3 days are necessary to harden most embedments.

Embedding Containers

The resin is placed in polyethylene capsules, and the tissue cube is placed into the resin to gradually sink to the bottom. The capsules are a size that will fit most microtome chucks. The most common capsule goes by the brand name of BEEM (Better Equipment for Electron Microscopy; Figure 2-7). These polyethylene containers come in various configurations. Most have pointed tips to form a pyramid-shaped protrusion allowing the specimen to protrude from the "block," as the hardened epoxy mass is termed. The pointed tip facilitates trimming the block after the epoxy has polymerized (see Chapter 4). Special holders are manufactured to hold BEEM capsules (Figure 2-8). The holders can support about 25 capsules each and are convenient for moving capsules into an oven for polymerization.

Silicone rubber embedding molds, although incompatible with certain resins, are available in various shapes (Figure 2-8) to allow more accurate orientation of a specimen than in a BEEM capsule container. Tissues may be oriented to some extent in BEEM capsules by a process known as flat embedding (Figure 2-9). A razor blade is used to cut off the pointed end of a BEEM capsule. The cap to the BEEM capsule, which is rarely used anyway, is detached and the specimen is placed in the cap and oriented as desired. A small amount of epoxy mixture is added. The remaining cylinder is fit snugly into the cap and filled with epoxy mixture. One disadvantage of this procedure is that after hardening, large amounts of epoxy must be removed from the sides of the specimen to select the specimen for sectioning. A hand-held rotary grinding device is useful in removing large amounts of excess polymerized epoxy (Figure 2-10).

CAUTION: Dust from a rotary grinder can be dangerous. Use a fume hood when grinding specimens.

Embedding Labels

There must be some way to identify the specimens at a later time. In a large experiment, for example, one may embed 10 specimens of each type and have 40 or more treatment groups. These capsules may be easily confused with one another unless they are individually marked. If a BEEM capsule is used, then a narrow strip of paper may be inserted inside the top of the capsule (Figure 2-11). It should have a pencil or typewritten identification code. *CAUTION: Some carbon typewriter ribbons or inks from pens may dissolve upon contact with the epoxy mixture.* The code should designate the experiment (usually a number), the treatment (preferably a letter), and the block number (usually a number). It is time consuming to make labels by hand; however labels can be printed relatively easily on a computer and cut out by hand (Figure 2-11). Computers can be used to make and print over 400 labels in minutes.

Removal of Tissue Supports

BEEM capsules can be removed by using a razor blade to cut them free from the polymerized block, but this is a dangerous procedure that often results in cut fingers. BEEM capsule removers are an effective tool to accomplish the same purpose, and this equipment has a long life span (Figure 2-12). They are a worthwhile investment for anyone who uses BEEM capsules regularly. Exercise caution when using BEEM capsule removers with soft blocks, since they may distort the plastic and the tissue.

Tissues can be removed readily from silicone rubber mold holders, since the rubber is flexible and easily peeled away from the embedment.

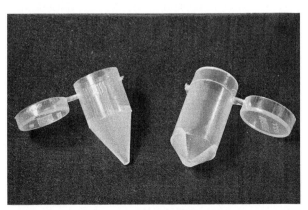

Figure 2-7 Two types of BEEM capsules.

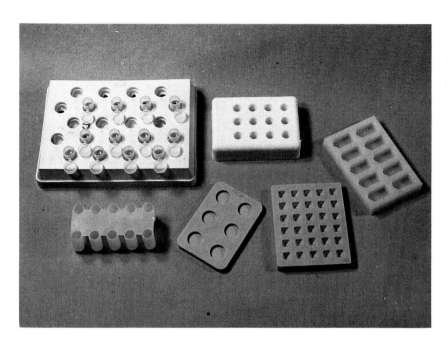

Figure 2-8 BEEM capsule holders containing BEEM capsules are shown at the top and bottom left. Several embedding molds of various types are shown.

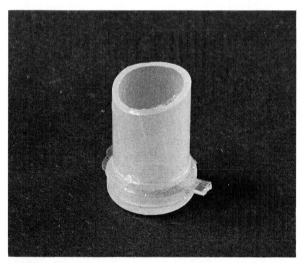

Figure 2-9 Flat embedding using BEEM capsules. The pointed tip of the BEEM capsule is cut off from the capsule, and the BEEM capsule cap is removed and placed over the cut end. The cap forms a flat surface for flat embedding of tissues.

Figure 2-10 Rotary grinding device for removing excess epoxy around a tissue block.

Special Tissue Preparative Needs

Cell Fractions

It was recognized early in its development that electron microscopy could be used to analyze *cell fractions* (Palade and Siekevitz 1956; Novikoff, 1956; de Duve and Beaufay, 1981) or to examine minute organisms that could be pelleted in a test tube. Elaborate fractionation schemes allowed the purification of plasma membranes and intracellular membranous components, and electron microscopy was often used to check the purity of such fractions or to perform cytochemical localization in pelleted material. Pelleted material is treated slightly differently than other tissue during tissue preparation. The pellet should be formed in a rounded, but not pointed, centrifuge tube (Figure 2-13). The pellet may be sufficiently tight to allow use of a flexible scraper to remove the unfixed pellet from the side of the tube, and to be able to chop the pellet into small fragments while, at the same time, avoiding resuspension of the material. If so, pellets can be processed routinely by gently adding and withdrawing

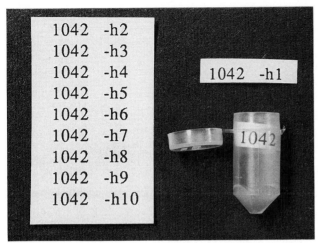

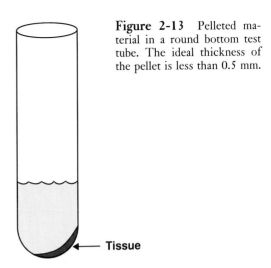

Figure 2-13 Pelleted material in a round bottom test tube. The ideal thickness of the pellet is less than 0.5 mm.

Figure 2-11 Computer-generated embedding labels and embedding label placed on a BEEM capsule.

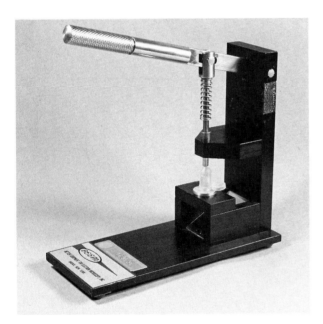

Figure 2-12 BEEM capsule remover.

agents during the tissue preparative scheme. An example of pelleted plasma membranes is shown in Figure 2-14.

Another method of dealing with cell pellets or cell suspensions is to fix the cell suspension in glutaraldehyde, centrifuge the cells, and then suspend the centrifuged cells in warm (45° C) 2% agar or agarose in buffer. Upon cooling to room temperature, the agarose may be cut into cubes with a razor blade and the cubes processed as usual (wash, osmicate, dehydrate, etc.).

It is likely that the fractions will be too fragile to remove from the tube. The pellet should be fixed in the centrifuge tube. The pellet thickness must be under 0.5 mm for glutaraldehyde and osmium tetroxide to penetrate rapidly. The pellet thickness is controlled by the amount of material placed in the centrifuge tube initially. Fluids are added gently by running them down the side of the tube. Most tissues that appear pale initially will yellow slightly under the influence of glutaraldehyde. When the surface of the tissue impacted against the tube has yellowed, one can be assured that glutaraldehyde has permeated the tissue.

Osmium tetroxide, an even slower penetrant than glutaraldehyde, is less likely to penetrate the pellet than glutaraldehyde. Often glutaraldehyde fixation imparts sufficient stability to the pelleted material to allow one to dislodge the pellet from the side of a test tube and process it intact. If the pellet is thin enough (under 0.5 mm), one can visualize the osmium tetroxide penetration by observing the darkening of the material on the side where the material impacts the container. The pellet may be able to be dislodged with a flexible scraper at this time. If not, it should be processed in its container through its dehydration, infiltration, transitional solvents, embedding, and curing steps. The container can usually be removed by a special method. Some microcentrifuge tubes and other tubes will fit snugly into microtome chucks.

Tissue Culture Cells

It is often desirable to process tissue culture cells in their original container if the container is not affected by any of the materials during tissue processing. Polystyrene, the most common type of plastic used for culture dishes and bottles, is dissolved

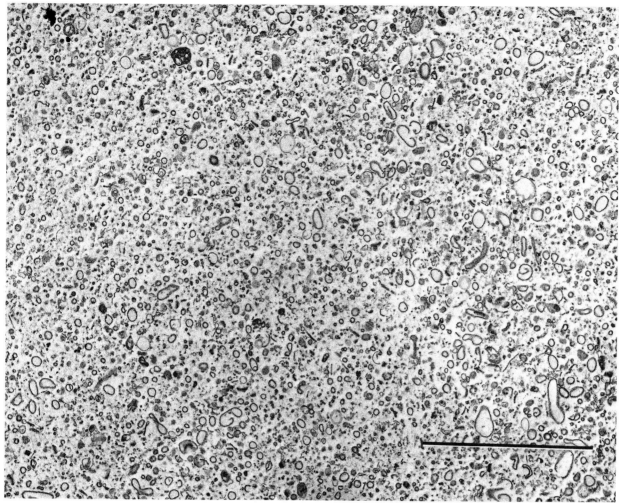

Figure 2-14 A section through a pellet of plasma membranes. The membranes were fragmented from the surface of cells, and in the process they formed the small vesicles seen in this figure. Vesicles were pelleted and processed for electron microscopy in the pelleted form. Bar = 0.75 μm.

by such tissue processing fluids as acetone or propylene oxide. Polypropylene containers are resistant to these chemicals. Fluids must be added gently to the containers during the processing steps in order to avoid disturbing the fragile monolayers. During the dehydration steps, the plastic covers or caps should be placed on the containers to prevent absorption of water by the alcohols.

It is also possible to grow cells on light microscope slides or coverslips that have been previously sterilized by autoclaving or by submersion in 70% ethanol for 20 minutes. After growing the cells on the microscope slide or coverglass, the substrate containing the cells is gently rinsed in physiologically balanced buffer and fixed by immersion in an aldehyde fixative for 15 to 30 minutes. Following several rinses in isotonic buffer, the cell/substrate

assembly is placed into an osmium fixative for 30 minutes, rinsed in distilled water several times, and dehydrated in an ethanol series of progressively increasing concentrations up to absolute ethanol. After several quick dips into propylene oxide (taking care not to allow the cells to dry out), the cells are covered with several drops of a 1:1 mixture of propylene oxide and embedding medium for 5 minutes. Care must be taken to cover the cells completely with this mixture and not to allow the cells to become dry. This procedure must be conducted in a fume hood with caution since the mixture is extremely flammable and toxic. Next, one fills a standard BEEM capsule with the plastic embedding medium and, after draining the monolayer of the 1:1 mixture, the capsule is inverted on top of the area containing the cells. The slide or coverslip containing the inverted BEEM capsule is then placed into a 60° C oven and polymerized for 24 hours. The BEEM capsule at 60° C (containing the partially poymerized resin) may be snapped off of the glass substrate and the specimen block placed back into the oven for another 24 hours. Usually, the capsule snaps away from the glass substrate cleanly, and no glass shards are removed with the embedded cells. If the capsule does not release smoothly, one may gently pry the capsule away by slipping a single-edged razor blade between the slide and the capsule. If problems are still encountered, some researchers claim that placing the slide onto a small block of dry ice and then prying with the razor blade will facilitate the release. One must be aware, however, that sometimes the two may not be separated cleanly by any method, and glass chips (that may damage a diamond knife) will be incorporated into the plastic.

With certain types of specimens (cell or tissue cultures) is possible to process the same tissues for viewing by light microscopy, as well as SEM and TEM, to obtain a correlative picture as described by Bozzola and Shechmeister (1973). Alternative methods for conducting correlative microscopy are given in the reference book by M. A. Hayat (1987).

Suitable Containers for Tissue Processing

A variety of clean containers can be used for tissue processing. Usually, glass vials about 10 to 30 mm in diameter with tight snap-on caps are the most convenient (Figure 2-4). It is important to economize on chemicals such as osmium tetroxide, so vials that are too large should be avoided. Labels must be securely fastened to the vials, and permanent markers should be used on the labels. It is wise to label the vial and *then* apply a tape over the label to protect it from solvents.

The vials and/or their lids should be compatible with the chemicals used to process tissues, otherwise they may dissolve in the process. For instance, propylene oxide will dissolve many types of culture dishes or plastic test tubes. Polystyrene tubes or culture dishes are dissolved by propylene oxide and acetone, whereas polypropylene or glass containers are safe to use. If there is any doubt, solvents should be placed in the container to test their suitability. The same is true for the vial caps or liners for vial caps.

Rapid Tissue Processing Protocols

Typically, processing tissue for ultrastructural examination takes a week or more. In many instances, it is necessary to process tissues rapidly for ultrastructural examination. For example, pathologists commonly process biopsy specimens rapidly to obtain a speedy diagnosis. Several protocols have been developed that shorten the times necessary for the various steps described previously.

A rapid method for processing *epoxy-embedded tissues* has been provided by Bencosme and Tsutsumi (1970) and Johannessen (1973). The entire process takes about 5 hours. The steps are listed as follows:

1. Fix tissue in fixative of choice for 30 minutes. The dimensions of the tissue block should not exceed 0.5 mm.
2. Rinse in 3 changes of 0.1 M sodium cacodylate buffer for 5 minutes each.
3. Postfix in 1% osmium in collidine-HCl buffer for 20 minutes.
4. Stain *en bloc* with uranyl acetate for 20 minutes.
5. Dehydration   50%   3 minutes
   in ethanol: 70%   3 minutes
                  95%   3 minutes
              100%   5 minutes (3 changes each)
6. Add 2 changes of propylene oxide for 5 minutes each.
7. Infiltration:
   1:1 mixture of epoxy and propylene oxide for 15 minutes
   3:1 mixture of epoxy and propylene oxide for 20 minutes
   pure epoxy for 10 minutes
8. Embed in fresh epoxy and place in oven at 75° C for 45 minutes. Subsequently, transfer capsules to a 95° C oven for 45 minutes. Allow capsules to cool at room temperature before sectioning.

NOTE: Steps 1–7 should be carried out with the tissue vials in a rotator to allow even and rapid penetration of chemicals.

A rapid method for processing LR White embedded tissues has been developed by David Leaffer. The method allows sectioning in as little as 5 hours.

1. Fix tissue in 4% glutaraldehyde in 0.1M cacodylate buffer for 1 to 2 hours. Tissue blocks under 1 mm across need less fixation time.
2. Wash in cacodylate buffer for about 2 minutes.
3. Postfix in 1% osmium tetroxide for 1 hour.
4. Stain *en bloc* with uranyl acetate for 1 hour.
5. Dehydrate: 30% ethanol   3 minutes
               50% ethanol   3 minutes
               80% ethanol   3 minutes
          100% ethanol   5 minutes
           (2 changes each)

6. Infiltration and embedding:
   1:1 mixture of LR White and ethanol   1 hour
   LR White   20 minutes (2 changes each)
7. Embed tissues in LR White and place in a 75° to 85°
   C oven for 30 minutes.

Automatic Tissue Processors

Automatic tissue processors are available for electron microscopy. Such devices are expensive and are justified only when the volume of tissue processing is great. Computerized programs allow fixation, dehydration, use of transitional solvents, infiltration, and resin infiltration to be carried out in a fully automated fashion (Figure 2-15).

Tissue Volume Changes During Specimen Preparation

Each chemical utilized in tissue processing has an effect on tissue volume. This problem cannot be avoided, but one should try to minimize volume changes as much as possible. For example, buffers are used with a range of osmolarities that maintain tissue volume (see above). When performing careful

morphometric studies (see Chapter 13), one should be aware of tissue volume changes.

The cell interior is maintained at a higher osmotic pressure than its exterior. The use of slightly hypertonic fixatives, as compared to the external environment of the cell, is advisable to prevent swelling of the cell. As a general rule, cells shrink slightly after glutaraldehyde fixation, but expand beyond their unfixed size in osmium tetroxide fixation. Subsequently, dehydration and embedding cause tissues to shrink to about 90% of their original size.

Judging Adequate Specimen Preparation

It is easier to describe artifacts in specimen preparation than it is to describe normal features of specimens. Chapter 18 deals with many of the common artifacts of specimen preparation from fixation onwards in the tissue preparation protocol. Experience is the best guide to judging adequate specimen preparation. Perhaps the simplest way to describe the well-prepared specimen is to say that it is aesthetically pleasing to the eye. The old adage "What looks good is good" often applies, although not always, since some tissue artifacts are also very pleasing to the eye. Good looking tissues are the combined product of many steps in the preparative

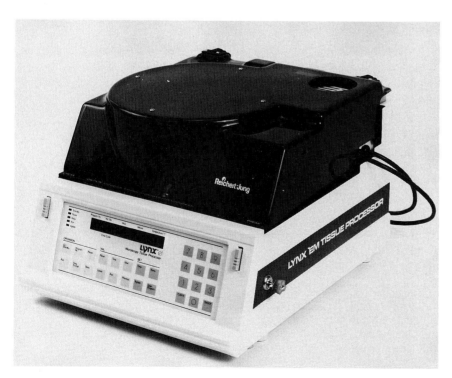

Figure 2-15 Automatic tissue processor. (Courtesy of Leica.)

scheme. Once one is used to viewing tissues that are adequately prepared, it is easier to judge them in the future.

Tissues that are well fixed show intact membranes. There is an evenness in the density of the cytoplasm that indicates no areas of clumping of cytoplasmic matrix. The cytoplasm should have a distinctly increased density as compared with the extracellular space or nontissue spaces. The smooth and rough endoplasmic reticulum are tubular and saccular, respectively. Completely round vesicles of endoplasmic reticulum are usually distortions of the natural form of the endoplasmic reticulum. The ribosomes are expected to be attached to the rough endoplasmic reticulum in tissues known to have plentiful bound endoplasmic reticulum. Mitochondria have a finely granular matrix and are not swollen, nor is their matrix rarefied (clear of any tissue constituent or cytoplasmic matrix). The membranes of the nuclear envelope are evenly spaced except at pore regions. Membranes and cellular constituents should appear prominent, especially if the fixation protocol was selected to highlight specific components.

Tissue should be easily sectioned. If holes are present in tissue blocks, then either air bubbles were embedded with the tissue, or dehydration or infiltration were inadequate. Water remaining in the specimen during dehydration will not allow most embedding media to infiltrate the specimen adequately. By examining the surface of the sectioned block with a dissecting scope under reflected light, it can be determined if holes remain in the embedment or if they are caused by the sectioning knife.

Tissue blocks may not harden sufficiently during the curing process, posing a sectioning problem. This usually is the result of improper mixing of the ingredients of the embedding mixture. Either ingredients were added in the wrong proportions or sufficient time was not spent in mixing them. In addition, defective or insufficient catalyst may prevent proper polymerization. Water may be present in certain anhydrous plastic monomers and will interfere with proper polymerization.

References

Bencosme, S. A., and V. Tsutsumi. 1970. A fast method for processing biological material for electron microscopy. *Lab Invest* 23:447–51.

Bozzola, J. J., and I. L. Shechmeister. 1973 *In situ* multiple sampling of attached bacteria for scanning and transmission electron microscopy. *Stain Technol* 48:317–25.

Burton, P. R., R. E. Hinkley, and G. B. Pierson. 1975. Tannic acid-stained microtubules with 12,13, and 15 protofilaments. *J Cell Biol* 65:227–33.

Claude, A. 1948. Studies on cells: morphology, chemical constitution of and distribution of biochemical functions. *Harvey Lecture* 43:1921–64.

de Duve, C., and H. Beaufay. 1981. A short history of tissue fractionation. *J Cell Biol* 91:293s–99s.

Glauert, A. M. 1975. *Fixation, dehydration, and embedding of biological specimens.* Amsterdam, The Netherlands: Elsevier, 207pp.

Goldman, R. E., B. Choojnacki, and M-J. Yerna. 1979. Ultrastructure of microfilament bundles in baby hamster kidney (BHK-21) cells: The use of tannic acid. *J Cell Biol* 80:759–66.

Hayat, M. A. 1981. *Fixation for electron microscopy.* New York: Academic Press.

Hayat, M. A. 1987. *Correlative microscopy in biology: Instrumentation and methods.* New York: Academic Press.

Johannessen, J. V. 1973. Rapid processing of kidney biopsies for electron microscopy. *Kidney International* 3:46–52.

Kalina, M., and D. Pease. 1977. The preservation of ultrastructure in saturated phosphatidyl cholines by tannic acid in model systems and type II pneumocytes. *J Cell Biol* 74:726–41.

Karnovsky, M. J. 1965. A formaldehyde-glutaraldehyde fixative of high osmolarity for use in electron microscopy. *J Cell Biol* 27:137A.

Karnovsky, M. J. 1971. Use of ferrocyanide-reduced osmium tetroxide in electron microscopy. In *Proc 14th Annu Meet Am Soc Cell Biol*, p 146. Abstract 284.

Leong, A. S. Y., M. E. Daymon, and J. Milios. 1985. Microwave irradiation as a form of fixation for light and electron microscopy. *J Path* 146:313–21.

Login, G. R., and A. M. Dvorak. 1985. Microwave fixation for electron microscopy. *Am J Path* 120: 230–45.

Luft, J. H. 1956. Permanganate: a new fixative for electron microscopy. *J Biophys Biochem Cytol* 2:799–802.

Luft, J. H. 1961. Improvements in epoxy embedding materials. *J Biophys Biochem Cytol* 9:409–14.

Maupin, P., and T. D. Pollard. 1983. Improved preservation and staining of HeLa cell actin filaments, clathrin-coated membranes, and other cytoplasmic structures by tannic acid-glutaraldehyde-saponin fixation. *J Cell Biol* 96:51–62.

Novikoff, A. B. 1956. Preservation of the fine structure of isolated liver cell particulates with polyvinylpyrollidone-sucrose. *J Biophys Biochem Cytol* 11:65–6.

Palade, G. E. 1952. A study of fixation for electron microscopy. *J Exptl Med* 95:285–98.

Palade, G. E., and P. Siekevitz. 1956. Pancreatic microsomes: an integrated morphological and biochemical study. *J Biophys Biochem Cytol* 2:171–200.

Russell, L. D., and S. Burguet. 1978. Ultrastructure of Leydig cells as revealed by secondary tissue treatment with a ferrocyanide:osmium mixture. *Tissue and Cell* 9:99–112.

Sabatini, D. D., K. Bensch, and R. J. Barrnett. 1963. Cytochemistry and electron microscopy. The preservation of cellular ultrastructure and enzymatic activity by aldehyde fixation. *J Cell Biol* 17:19–58.

van de Kant, H. J. G., D. G. de Rooij, and M. E. Boon. 1990. Microwave stabilization versus chemical fixation. A morphometric study in glycomethacrylate- and paraffin-embedded tissue. *Histochem. J.* 22:335–40.

Surface Cleaning

Buffers and Fixatives

Rinsing and Dehydration

Specimen Drying Techniques
Critical Point Drying
Freeze-Drying
Alternative Drying Procedures

Specimen Fracturing Procedures
Cryofracturing
Dry Fracturing

Replication Procedures
Negative Surface Replication Using
    Cellulose Acetate Film
Negative/Positive Replication Using
    Silicone, Resin
Corrosion Casting of Animal
    Vasculatures

Specimen Mounting

Specimen Coating for Conductivity
Sputter Coating Procedure
Noncoating Techniques

Specimen Storage

References

C H A P T E R 3

Specimen Preparation for Scanning Electron Microscopy

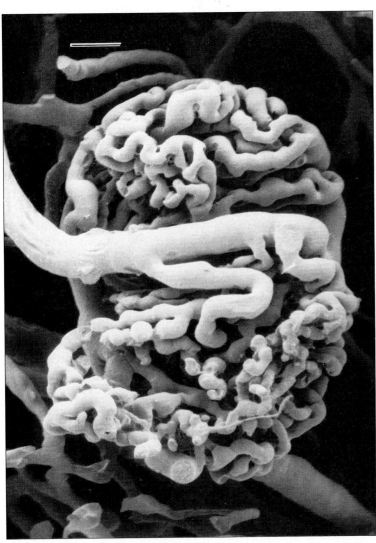

Courtesy of B. J. Moore and K. R. Holmes.

Many of the steps involved in preparing biological specimens for scanning electron microscopy (SEM) are similar to the initial steps used in preparing samples for transmission electron microscopy (TEM). As presented in Figure 3-1, a generalized sequence for preparing specimens for SEM involves rinsing surfaces to remove debris, stabilizing in an aldehyde fixative followed by osmium tetroxide, rinsing in distilled water, dehydrating, mounting the sample on a metal specimen stub, and coating the specimen with a thin, electrically conductive layer. There are, however, several significant differences in preparing specimens for SEM, as well as important choices to make prior to processing. These differences are discussed in this chapter.

Surface Cleaning

Since the surface of the specimen usually is the region of interest in SEM, one must first clean the surface of materials that might otherwise obscure these features. Materials such as mucus, secretions, red blood cells, bacteria, broken cell debris, silt, dust, and detritus must be removed *prior* to fixation or this material may be chemically fixed to the specimen surface and be impossible to remove later. The extent to which one cleans the specimen surface depends on the nature of the specimen, the chemical makeup of the surface, and the environment from which the specimen is removed. Under ideal conditions, the specimen is quickly rinsed in an appropriately buffered solution of the appropriate pH, temperature, and osmotic strength so as to mimic the natural milieu. One can then proceed to fix and further process the specimen. On the other hand, it may be necessary to clean the surface of obfuscating coatings as described in the following box.

How to Remove Surface Coatings

Even seemingly stubborn material can sometimes be removed from specimen surfaces by repeated gentle rinsing or flushing of the surface with a mild buffer. This may be accomplished by using a pipette, plastic squeeze bottle, or a dental irrigation device to direct gentle pulses of a rinse solution over the surface. When studying the structure of the endothelial cells lining the blood vessels in vascularized tissue, one should first perfuse the animal with physiologically balanced solutions to remove red blood cells and serum proteins and then perfuse the animal with the fixative. Otherwise, the cellular and proteinaceous components of the blood serum will be fixed onto the surface of the endothelium.

On difficult-to-remove coatings, one may incorporate various enzymes (mucinases, for example), mild detergents, or surfactants in the rinse solution prior to fixation. With hardy samples that are dry, it may be possible to use a camel's hair brush and mild bursts of microscopically clean, compressed air to dislodge debris. In all of these instances, one must treat all surfaces as gently as possible to avoid introducing artifactual changes. Always consult the published literature prior to using such agents in the rinsing solution.

Buffers and Fixatives

The same criteria for the selection of buffers and fixatives as described in Chapter 2 apply for SEM. Although some researchers feel that buffers used in SEM should be slightly hypotonic in contrast to the buffers used in TEM, this option is by no means clearly substantiated. When in doubt, rely on established SEM protocols until experience proves otherwise. If SEM protocols are not available, the same fixatives used in TEM studies will, in most cases, be suitable for SEM. These protocols use the standard, noncoagulating aldehyde-followed-by-osmium-tetroxide fixation regimes that have been used satisfactorily in TEM for many years. Under certain circumstances, it may be possible to use some types of coagulating fixatives (Craf's, Bouin's, FAA) that are used in light microscopy.

As detailed in the previous chapter on specimen preparation for TEM, one must consider pH, buffer type, tonicity, concentration of fixative, temperature, and time in fixative. Specimen size may be a lesser concern in SEM fixation, since one is generally interested in viewing the outermost structures, which contact the fixative first and are fixed immediately. Nonetheless, one should strive to maintain a size of no more than 1 mm in one of the dimensions. This will ensure not only that underlying tissues are adequately fixed to support the surface features, but that the dehydration steps will be more efficient. A listing of some commonly used fixatives is given in Table 3-1. For more information, consult the original references cited in the table. These articles have been assembled into a single volume compiled by Murphy and Roomans in 1984.

As can be seen in Table 3-1, various combinations of fixatives and buffers are possible. In addition, it is possible to add various chemicals (tannic

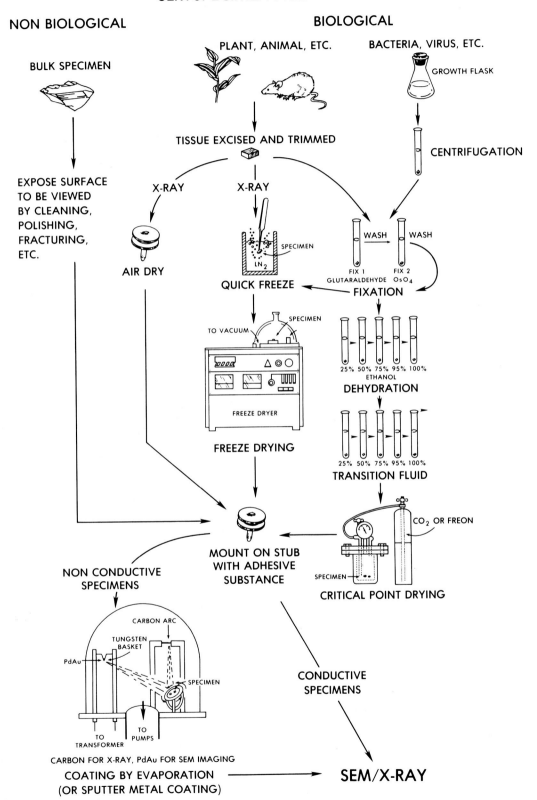

SEM SPECIMEN PREPARATION

Figure 3-1 Schematic showing sequence of events for processing biological specimens for SEM. (Courtesy of Judy Murphy.)

Table 3-1   Fixatives Commonly Used in SEM

Specimen	Fixative	Buffer System	Reference
Procaryotes	glutaraldehyde	cacodylate, phosphate	Watson et al. 1984
	osmium tetroxide	veronal-acetate	
	FAA (10% formalin, 85% ethanol, 5% glacial acetic acid)		
Fungi	glutaraldehyde/OsO$_4$ followed by OsO$_4$	cacodylate, phosphate	Watson et al. 1984
	OsO$_4$ vapors	none	
	glutaraldehyde followed by aqueous uranyl acetate	cacodylate	
Aquatic Organisms (protozoa, sponges, metazoa)	glutaraldehyde/ formaldehyde	cacodylate, collidine	Maugel et al. 1980
	Parducz (6 parts of 2% aqueous OsO$_4$ plus 1 part saturated aqueous HgCl$_2$— freshly prepared)	none	
	glutaraldehyde followed by OsO$_4$	phosphate, cacodylate sea or pond water	
Higher Plants	glutaraldehyde followed by OsO$_4$	phosphate buffer	Falk, 1980
	FAA alone or followed by OsO$_4$	phosphate or s-collidine	
	formaldehyde followed by freeze drying	none	
	osmium vapors	none	
Zoologicals	glutaraldehyde followed by OsO$_4$	cacodylate or phosphate	Nowell and Pawley, 1980
	OsO$_4$	cacodylate or phosphate	
	glutaraldehyde/ formaldehyde	various	
	glutaraldehyde or glutaraldehyde/formaldehyde followed by OsO$_4$	cacodylate or phosphate	
	FAA	none	

acid, lysine, ferrocyanide) or salts (MgCl$_2$, CaCl$_2$) to the fixative to effect a better preservation. More detail may be obtained by consulting the references by Murphy and Roomans (1984), Postek et al. (1980), and Revel et al. (1983) as well as the current scientific literature.

Working with Individual Cells or Macromolecules

During fixation, one normally works with multicellular organisms or tissues that may be handled as dissected cubes (described in the previous chapter). Small specimens (tissue culture or bacterial cells, viruses, individual molecules such as actin, etc.) pose the same problems they do in preparing them for TEM. A convenient way to deal with cells as small as 0.4 μm is to trap the cells on microfilters. In this procedure, a reusable cartridge is loaded with a filter of appropriately sized pores (the smallest is 0.2 μm), and the suspension of cells is gently passed through a syringe to settle on the filter (Figure 3-2). Alternatively, suction may be applied to draw liquid through the filter. In both cases, excessive force must not be applied or the cells will be damaged. After a brief rinse in the proper solution, the fixative is passed over the cells and left in contact with the monolayer for 15 to 60 minutes. After rinsing, the cells may be dehydrated by passing alcohols through the filter. The filter then is removed and dried prior to mounting, coating, and viewing in the SEM. To enhance conductivity, some microporous filters are constructed of silver.

Another method for dealing with cells or macromolecules is to use a substrate (such as a slide or coverglass) that has been coated with either poly-L-lysine (Mazia, 1975) or Alcian blue. The microscope slides are prepared for coating by first cleaning them using alcohol or a commercial acid cleaner. The cleaned slides are then coated by flooding them for 10 minutes with an aqueous solution containing 1 mg/ml of the reagent. After rinsing in distilled water, the slides may be used immediately or stored after drying. These reagents are effective adhesives for cells since they impart a positive charge to the negatively charged glass surfaces. Cells have a net negative charge and are therefore attracted to the now positive surface of the substrate. A fluid (culture medium or buffer) containing the specimen may subsequently be placed on the slide and the specimen allowed to settle by gravity onto the substrate. After rinsing and fixation, the adherent specimen is then processed (dehydrated, critical point dried, coated with metal, etc.) while attached to the substrate.

Rinsing and Dehydration

After fixation, tissue is rinsed in distilled water several times to remove buffer salts. The specimen is then either freeze-dried or dehydrated before being critical point dried (see next section). In rare instances, it may be possible to air-dry samples after fixation and rinsing, but *most biological tissues will collapse, flatten, or shrink under these conditions.* For instance, some specimens (such as embryonic tissues) will shrink to half of their normal volume when air-dried. A major problem inherent in air-drying involves the passage of the receding air/water interface through the specimen. During air-drying, the *surface tension* forces associated with this interface are approximately 2,000 psi. Most biological samples will be flattened by such forces. To prevent this collapse, one must somehow avoid the passage of the interface through the tissues. This may be ac-

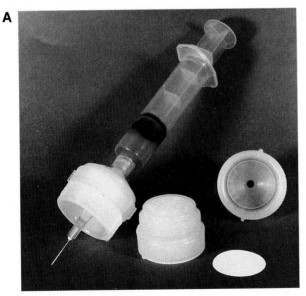

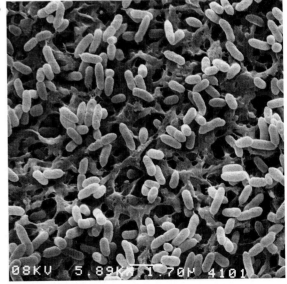

Figure 3-2 (A) Individual cells or tiny specimens suspended in a buffer system may be deposited onto microporous filters by passage through a filtering device, as shown. Fixatives are then syringed over the cells, followed by ethanolic dehydration. The filter holder is then opened, and the filter is removed, dried, and mounted onto a specimen stub. After coating for conductivity the filter surface is then examined as shown in Figure B. (B) Bacterial cells trapped on a microporous membrane filter and subsequently processed for SEM. Marker bar = 1.7 μm. (Courtesy of R. de la Parra, Millipore Corp.)

complished using freeze-drying, critical point drying, or other vacuum drying procedures as described in the following sections.

Specimen Drying Techniques

Critical Point Drying

Most soft biological specimens are dried using this procedure since it is fast and gives reliable, artifact-free results in the majority of cases. In a commonly followed protocol, specimens are dehydrated in the usual ethanol series and transferred into a dehydrant-filled, cooled vessel termed a "bomb" (e.g., a vessel for compressed gases). After sealing the vessel, the dehydrant (usually ethanol) is displaced with a pressurized *transitional fluid* such as liquid carbon dioxide or Freon. Following several changes of transitional fluid to ensure complete displacement of the dehydrant, the bomb is completely sealed and the temperature of the transitional liquid is raised slowly by the application of heat. Pressure inside the vessel begins to rise in response to the heat. Eventually, the transitional fluid will reach a *critical point*, or a particular temperature/pressure combination specific for the transitional fluid, at which point a transition occurs and the density of the liquid phase equals that of the vapor phase. Different transitional fluids have different combinations of temperature and pressure to achieve their respective critical points (Table 3-2). Water is unsuitable for use as a transitional fluid since it has a critical temperature of 374° C and a critical pressure of 3,184 psi (i.e.,

Table 3-2   Dehydrants and Transitional Fluids Used in Critical Point Drying

Dehydrant	Transitional Fluid	Critical Temp °C	Critical Pressure PSI
Ethanol, Amyl Acetate	Liquid CO_2	31.1	1,073
Acetone	Freon 116	19.7	432
Ethanol	Freon 23	25.9	701
Ethanol/Freon	Freon 13	28.9	561

The investigator can choose from several different transitional fluids based on the critical point desired. Liquid carbon dioxide is more commonly used (and probably less destructive to the ozone layer of our planet) than are the Freons (trademark of DuPont de Nemours and Company).

biological specimens would be destroyed under these conditions).

This technique is successful because, at the critical point, the specimen is totally immersed in a dense vapor phase devoid of the damaging liquid/air interface one wishes to avoid. After achieving the critical point, the heat is maintained at the critical temperature (to prevent condensation of the vapor back to a liquid), and the vapor is slowly released from the chamber until the vessel is at atmospheric pressure. The dried specimen may then be mounted on a specimen stub, coated with a metal for conductivity (see Chapter 5), and viewed in the SEM. Figure 3-3 is a diagram of a typical critical point drying apparatus; Figure 3-4 is a photograph of a commercially available unit.

When placing small chunks of tissue in the critical point dryer, one may wish to use special holders to minimize damage and to maintain the identities of the specimens. Such holders are readily fashioned from polyethylene BEEM embedding capsules (see Chapter 2) by perforating the plastic with a hot dissecting needle (Figure 3-5A). For smaller specimens that may be lost through such perforations, one may fashion another type of holder by first trimming off the sealed end of the capsule to form an open cylinder. BEEM capsule covers are then bored out to remove most of the plastic so that only a circular ring remains. This ring is used to hold a tightly woven nylon mesh in place by slipping it over the mesh on both ends of the cylinder (Figure 3-5B). Slips of paper containing specimen information written in pencil (NOT INK!) are placed inside the chambers for identification purposes. Specimens may be transferred into the chambers prior to or after dehydration. Care must be taken not to allow the specimen to air-dry, however. Commercial specimen holders are also available from original equipment manufacturers and from many EM supply houses.

For other types of specimens that may be attached to glass microscope slides, coverglasses, or on micropore filters, one may fashion special holders out of plastic syringe barrels, plastic tubing, or even metal washers. These special holders will prevent the specimens from shifting in the critical point dryer and possibly damaging the specimen. Figure 3-6 shows some of these devices.

The critical point drying procedure is not without problems. Some types of tissues may undergo significant shrinkage, ranging from 10% to 15% for nerve tissues to 60% for embryonic tissues (Boyde

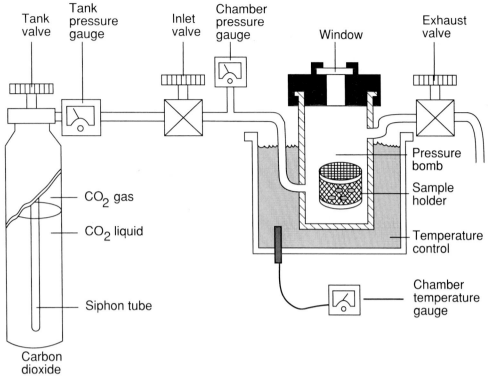

Figure 3-3 Diagram of critical point drying apparatus. The temperature of the pressure bomb may be regulated by a water bath or an electric heating element.

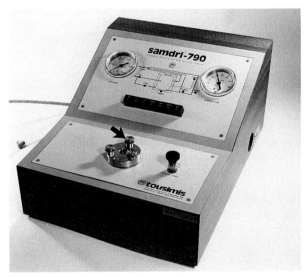

Figure 3-4 Photograph of a semiautomatic critical point drying apparatus. Specimens are placed into the recessed specimen chamber and the lid is sealed using the three thumb screws (arrow). Processing of the specimen is carried out by pressing a series of buttons (above the specimen chamber). The large dial on the left indicates the pressure inside the specimen chamber while the dial on the right indicates the temperature.

and Maconnachie, 1981). In addition, the transitional fluids are solvents that may extract steroids, carotenoids, porphyrins, and actin. Additional information about critical point drying can be found in the articles by Boyde (1978) and Cohen (1979).

Safety Precautions with Critical Point Drying

Dangerous pressures build up inside the bomb, and a few instances of the vessel cover rupturing have been reported. Modern instruments have protective discs that burst prior to the development of such pressures inside the chamber. Windows covering the chamber are of specially tested quartz or glass and should be checked for nicks or cracks before each use. It is recommended that the critical point dryer be placed inside of a fume hood or behind a shatter resistant window (plexiglass, for instance). One is cautioned not to look directly down into the pressurized chamber. Instead, a metal mirror (not glass) should be used to examine the chamber contents. The mirror should be mounted on a movable support that will fall away from the user in the event of a window failure. A well-ventilated room is necessary since the gases used in this process will displace oxygen and may asphyxiate or sicken unwary users.

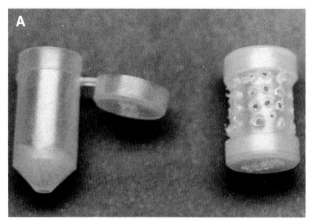

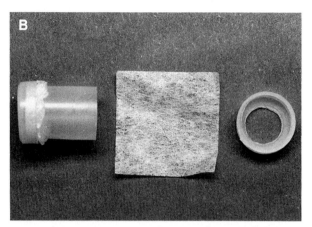

Figure 3-5 Devices for holding small specimens. (A) Unmodified (left) and modified (right) BEEM capsule used for embedding TEM specimens. Holes have been made in the side, and both ends have been capped off after insertion of specimen. (B) BEEM capsule with ends modified to accommodate either a screen mesh or micropore filter.

Figure 3-6 Commercially produced stainless steel holders for securing specimens during the critical point drying process.

Freeze-Drying

A second method for avoiding the water/air interface during specimen drying is to freeze the specimen rapidly while still in the aqueous phase. The frozen specimen is transferred into a special chamber designed to maintain the temperature lower than $-80°$ C, and the apparatus is rapidly evacuated into the 10^{-1} Pa range or higher. While frozen, the solid ice gradually undergoes sublimation to the gaseous phase and is either absorbed by desiccants placed into the chamber or removed by the vacuum system. Depending on the size of the specimen, temperatures used, and the vacuum level achieved, this process may take from several hours to several days to complete.

Although it is possible to freeze unfixed specimens and then to freeze-dry them, the freeze-drying process usually begins after the specimen has been fixed and rinsed extensively to remove traces of fixatives. Sometimes the specimen is infused with a cryoprotectant such as sucrose, glucose, glycerol, ethanol, DMSO, or dextran to reduce ice crystal formation and damage. Unfortunately, some of these materials will not be sublimed away and will remain behind obscuring surface features. Ethanol is probably the most useful cryoprotectant since it is removed by evacuation. Experimentation is in order to determine whether or not a cryoprotectant is needed when attempting freeze-drying a particular specimen for the first time.

Freeze-Drying Procedure

In a typical procedure, possibly after infusion with cryoprotectant, fresh or fixed, ethanol dehydrated tissue is plunged into a liquid nitrogen chilled fluid such as isopentane, liquid Freon, supercooled liquid nitrogen, or liquid propane. (*DANGER: Potentially explosive!*) These fluids, termed *quenchants*, are used instead of plunging the specimen directly into liquid nitrogen, which tends to boil vigorously when specimens are placed in it. Without a quenchant, the gaseous phase generated by the vigorous boiling of liquid nitrogen insulates the specimen and slows the rate of freezing so that damaging ice crystallization occurs.

NOTE: Extremely rapid freezing rates are important in order to avoid ice crystal formation. When freezing rates are faster than 140 K per second, water is converted into solid amorphous (non-crystalline) ice.

Quenchants wet the specimen surfaces rather than boiling vigorously, so that the heat will be rapidly removed from the specimen to effect a rapid freezing. Unfortunately, the freezing rate is rapid only to a depth of about 15 to 10 μm, so that underlying tissues will undergo extensive ice damage. Because the surface features are probably the ones being investigated, this may not be of consequence.

After the specimens have been frozen in the quenchant, they may be stored in liquid nitrogen indefinitely, or they may be transferred to the cooled stage of the freeze-dryer. The stage may be chilled with liquid nitrogen or cooled using an electronic system. Figure 3-7 shows a typical freeze-drying apparatus used in SEM specimen preparation.

After the specimen has been transferred onto the chilled stage rapidly (to avoid condensation of moisture), the apparatus is covered, evacuated to about 10^{-2} Pa, and maintained under vacuum/cold conditions as long as necessary for drying. After drying, the specimen stage is gradually warmed to room temperature, dry air is admitted to the chamber, and the specimen may be mounted onto the specimen stub for coating and viewing in the SEM. If the dried specimen must be stored for any period of time prior to coating and viewing, it must be stored in a desiccator.

NOTE: It is possible to fashion a simple freeze-dryer by cooling a large block of copper or brass in liquid nitrogen, placing the quenchant-frozen specimen into a recessed area

on the block, covering the recess with another recessed block, and placing the chilled device inside a vacuum evaporator. The block will remain cold inside of the vacuum for several hours, probably long enough for small specimens to freeze-dry. Covering the specimen is essential in this device in order to prevent condensation of water or vacuum pump oil onto the specimen surface.

One advantage of freeze-drying is that much less shrinkage occurs with most specimens compared to critical point drying. Disadvantages include the problems associated with rapid freezing and transfer into the apparatus, the time involved in drying the tissues (sometimes days versus hours in the critical point drying procedure), and problems with non-evaporable deposits left behind on surfaces. Boyde (1978) has discussed the pros and cons of freeze-drying in a separate paper referenced at the end of this chapter.

Alternative Drying Procedures

Under certain conditions, it may be possible to air-dry biological specimens. Usually air-drying of specimens that were in water, acetone, or alcohols results in excessive shrinkage, cracking, and collapse of fine structures such as cilia or filapodia. However, some solvents such as ethanol and hexamethyldisilazane may be used to air-dry certain types of tissues. For instance, the bacterial preparation shown in Figure 3-2 was prepared by fixing the cells in glutaraldehyde followed by osmium tetroxide. After a brief rinse in distilled water, the membrane filter was dehydrated in an ethanol series (35, 45, 55, 65, 70, 85, 95 and 100% for 5 minutes each) followed by immersion in hexamethyldisilizane for 5 minutes. The filters were then allowed to air-dry at room temperature for 30 minutes prior to mounting on an SEM stub, coating with Pd/Au, and examination in the SEM. A simlar procedure has been shown to work with sensitive cells as well; however, the following fluorocarbon procedure appears to be gentler with nearly all types of cells.

A new fluorocarbon compound, trade named Peldri II (Ted Pella Company), appears to offer good results with a wide variety of biological specimens (Kennedy et al., 1989).

Fluorocarbon Drying Procedure

In a standard protocol, the specimens are fixed in buffered glutaraldehyde followed by osmium tetroxide, rinsed in distilled water, and dehydrated in an acetone series. The

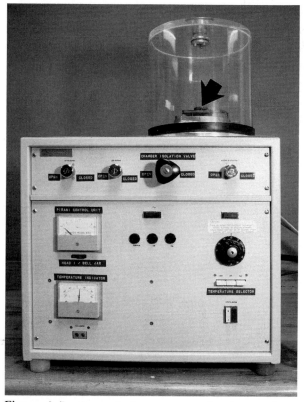

Figure 3-7 Freeze-drying apparatus showing specimen chamber where frozen hydrated specimens may be placed (arrow). The specimen is maintained at −80° C while under vacuum until all solidified water has been removed by sublimation.

fluorocarbon, which is solid at room temperature, is warmed to 40° C and mixed with an equal volume of acetone (*CAUTION: Flammable*) containing the specimens. After capping the vial and mixing gently, the sample is placed at room temperature to begin the exchange of acetone for fluorocarbon in the specimen. After 1 hour, the acetone/fluorocarbon mixture is removed and replaced with pure liquid fluorocarbon. The vial is capped, placed at room temperature, and allowed to solidify. The uncapped vial containing the solidified fluorocarbon and specimen is then placed into a vacuum desiccator or bell jar and evacuated using a rotary pump (Figure 3-8A) until all traces of the fluorocarbon are gone (one to several days). The dried specimens may then be treated as conventionally dried specimens. The advantage of this technique is obvious, since no expensive critical point or freeze-drying apparatus is needed. An example of the use of Peldri II for the drying of sensitive human monocyte-derived macrophages is shown in Figure 3-8B.

Peldri may also be used to freeze-dry specimens that have been frozen and fractured while submerged in liquid nitrogen. Such cryofractured specimens may be thawed, reacted with immune labels such as colloidal gold, and then dried in Peldri. The labeled surfaces may then be examined in a high resolution SEM or a platinum/carbon replica (Chapter 5) may be made of the surfaces.

Specimen Fracturing Procedures

If one wishes to expose the interior of biological specimens, it is possible to fracture the specimen using either a cryofracturing or a dry fracturing procedure as described in the following sections.

Cryofracturing

One method of exposing the interior of cells for viewing by SEM is to freeze the tissue in liquid nitrogen and fracture the frozen specimen while it is at liquid nitrogen temperatures. After the sample has been fractured, it is dehydrated, mounted onto a specimen stub, coated with a metal for conductivity, and examined in the SEM.

If one is working with unfixed tissues, it is desirable to cryoprotect them (10% glycerine or 10% to 25% dimethyl sulfoxide in buffer) prior to freezing in order to avoid ice crystal damage. Small pieces of tissue are essential in order to permit as rapid cooling rates as possible. Otherwise, even cryoprotected specimens may be damaged by ice crystal formation. (See Chapters 2 and 14 for information on rapid freezing and cryoprotection methodologies.) One reason for freezing unfixed rather than fixed specimens is to permit the washing out of soluble proteins, salts, and other solutes when the fractured cells are thawed. These intracellular materials might otherwise prevent viewing deeply into the cell interiors, or they may obscure such ultrastructural details as the fine filaments that make up the cytoskeleton. A second reason for working with unfixed specimens may involve the use of cytochemical or immunocytochemical probes on the fractured and thawed cells (i.e., some enzymes or antigens may be denatured by fixation).

Since the use of chemically unfixed materials in the cryofracturing technique may lead to the introduction of a number of artifacts (ice crystals, migration of unstabilized cellular components, etc.),

Figure 3-8(A) Sublimation apparatus used to dry specimens that have been dehydrated and treated with Peldri II fluorocarbon. A = container of $CaSO_4$ dessicant to maintain dryness of air, B = valve, C = vacuum desiccator, D = vacuum gauge, E = condensing coil to prevent reagents from going into vacuum pump, F = gas purifier column of activated charcoal and desiccant for further purification of gas being evacuated by vacuum pump, G = vacuum pump. (Courtesy of J. L. Pauly and J. Electron Microscopy Technique.)

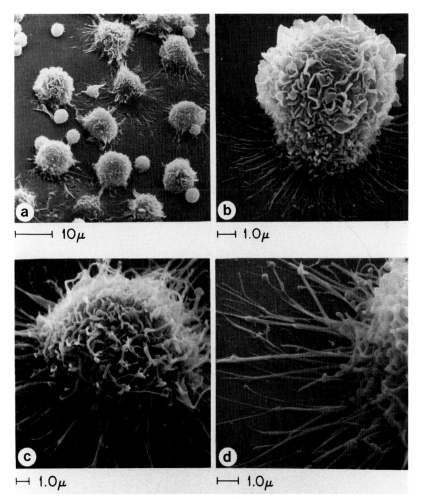

Figure 3-8(B) Morphology of human monocyte-derived macrophages that were cultivated for 11 days before harvesting and processing for SEM. (a) Low magnification of large adherent macrophages and small spherical lymphocytes. (b) Macrophages displaying typical ruffled membrane architecture. Note the filopodia that anchor the phagocytic cell to the substrate. (c) Pleomorphic nature of the surface membrane of macrophages is illustrated by this cell that displays numerous microvilli and a membrane architecture. (d) High magnification of filopodia illustrating the preservation of these delicate membrane structures obtained using Peldri II fluorocarbon procedure. (Courtesy of J. L. Pauly and J. Electron Microscopy Technique.)

one should first attempt the procedure using fixed specimens. If these prove unsuitable, then unfixed materials may be used. A good review of the cryofracturing procedures may be found in the articles by Tanaka (1981) and Haggis (1982). Tanaka's procedure is summarized in the box that follows. An example of the results obtained using this procedure is shown in Figure 3-9.

Cryofracturing Procedure of Tanaka (1989)

1. Excise small pieces of tissue (1 × 1 × 5 mm or less) and immediately place the tissue into 1% osmium tetroxide in a buffer of the appropriate type, pH, and molarity for the tissue being studied. Generally this is accomplished at 4° C for 60 to 90 minutes.
2. Rinse the tissue in isotonic buffer solution (three changes, 15 minutes each), and cryoprotect the fixed tissues by passage through 25 and 50% dimethyl sulfoxide (DMSO) solutions for 30 minutes each. The DMSO is diluted in buffer solution.

    CAUTION: DMSO is immediately absorbed through the skin and carries with it any toxic substances (i.e., arsenic salts of cacodylate buffers) directly into the blood stream. Wear rubber gloves and avoid splashing the DMSO solutions onto unprotected areas.
3. Rapidly freeze the tissue on a metal plate that was chilled with liquid nitrogen, and crack the frozen tissue block using a liquid nitrogen chilled scalpel and

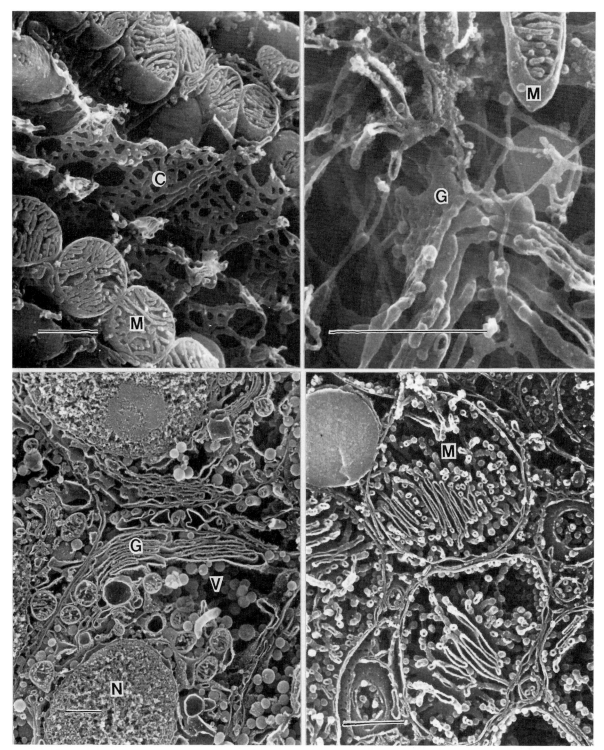

Figure 3-9 Examples of cryofractured tissues prepared by the technique of Tanaka (1981). Note the fine structural details present in the mitochondria and the net-like nature of the membranes (C) surrounding the mitochondria (M). Golgi (G) bodies are present in some cells and nuclei (N) and vesicles (V) are readily visible. Marker bars = 1 μm. (Courtesy of K. Tanaka.)

a hammer. This should be accomplished as rapidly as possible to prevent the formation of frost on the specimen surface. (Some investigators are able to fracture the specimen while it is submerged in liquid nitrogen. This is a demanding procedure, however, since the specimen is difficult to see in the rapidly boiling liquid nitrogen.)

4. Immediately place the cracked, frozen tissues into 50% DMSO to thaw.

5. Rinse several times in buffer solution to remove the DMSO, and place the tissues into 1% osmium tetroxide for 1 to 2 hours followed by 0.1% osmium tetroxide solution in the appropriate buffer. Keep in this solution for 24 to 72 hours at 20° C. (This step is used to remove cytoplasmic materials that might otherwise obscure ultrastructural details.)

6. Transfer the tissues into a 2% tannic acid solution for 2 hours followed by a 1% osmium tetroxide solution for 1 hour, followed by a buffered solution of 2% tannic acid for 12 hours, and then 1% osmium tetroxide for 1 to 2 hours. (This procedure builds up an electrically conductive coating of osmium.)

7. Rinse the tissue in distilled water (three changes, 15 minutes each), and dehydrate the specimens in an ethanol series.

8. Critical point dry the dehydrated tissues as described earlier in this chapter.

9. Carefully mount the specimens onto a specimen stub as outlined later in this chapter.

10. Coat the specimens with a heavy metal such as platinum (see Methods for Metal and Carbon Evaporation in Chapter 5) with a thickness of approximately 20 nm.

11. Observe in a high resolution SEM.

Dry Fracturing

Specimens that are not delicate may be fractured after fixation, dehydration, and critical point drying. Several methods may be used to fracture the dry specimen. If the tissue block was attached to a stub, the specimen may be fractured with a sharp scalpel or razor blade. In some tissues, double-stick tape may be pressed gently onto the surface of the specimen and pulled to strip away the uppermost layers. A fine needle may be used to dissect away tissue and expose the area to be studied.

Obviously this method may result in crushing or deforming delicate specimens, so it may be more desirable to embed the specimen in a supporting matrix such as paraffin prior to fracturing. In a procedure described by Shennawy et al. (1983), tissues are fixed, dehydrated, and embedded in paraffin using standard protocols employed by light microscope histologists. The paraffin block containing the specimen is trimmed to the level of the specimen, cooled, and the paraffin block is broken by snapping it in half. The two halves are deparaffinized using xylene, and the tissue blocks are placed into absolute ethanol, critical point dried, and examined in the SEM after mounting and coating with a heavy metal. Good results have been obtained using a variety of mammalian tissues (Figure 3-10).

Replication Procedures

If direct observation of a biological surface is not possible, a replica of the surface may be examined in the SEM. In instances where sacrifice of the live specimen is not feasible, or if one wishes to sequentially study changes on an identical specimen surface, or if the specimen undergoes unacceptable alteration when using more conventional procedures, then replication may be in order. Other reasons for using replicas are if a unique or rare specimen is too large to fit into the SEM specimen

Figure 3-10 Specimens may be embedded in paraffin and the hardened paraffin fractured following the procedure of Shennawy et al. (1983) to expose internal features. Mouse liver showing red blood cells inside of lacunar spaces. Magnification bar = 15.0 μm. The excessive brightness shown in some areas of the micrograph is due to the phenomenon termed "charging", resulting from the build up of high voltage static charge that repels excessive amounts of electrons into the secondary electron detector. (Courtesy of M. Doran and S. Pelok.)

chamber, if the specimen is too precious to subdivide, if it consists of an interior surface (airways, vasculature, etc.), or if the specimen is damaged by the vacuum or the electron beam. The three basic types of replicas are negative surface replicas, positive surface replicas, and corrosion casts.

Negative Surface Replication Using Cellulose Acetate Film

This simple but limited technique involves impressing an acetone-softened sheet of cellulose acetate film onto a surface and allowing the acetate to harden after the acetone evaporates. Since acetone is dropped onto the surface being replicated, this procedure can be conducted only on hard, nonliving surfaces such as bone, chitinous exoskeletons of insects, dried plant materials, etc. The specimen should have relatively low topography (since the procedure does not replicate high areas particularly well), and it should obviously not be damaged by the acetone. The technique has low resolution capabilities and is useful in magnifications up to several thousand times only (Figure 3-11A).

Cellulose Acetate Negative Replication Procedure

1. Cut a piece of cellulose acetate sheeting into a size suitable for the specimen (1″ square, for example).
2. Apply several drops of acetone to the dry surface of the specimen and immediately press the cellulose acetate film onto the wet surface. Hold it in position under the thumb for about 1 minute. Be careful not to slip the tape along the surface during the drying procedure or the replica will be damaged. The specimen surface should not be excessively moistened with acetone or the film will become wet on the side against the thumb.
3. Gently peel away the acetate film, and keep this first replica for study purposes. Generally, the first peel is used to clean particulate debris from the specimen surface and may be discarded. On the other hand, some materials (pollen, bacteria, fibers, etc.) that were extracted from the specimen surface during replication may be of interest to the investigator. This type of replica is called an extraction replica.
4. A second replica is made of the now cleaned surface, and the acetate film is allowed to dry for 10 minutes in a clean environment. Keep track of which side was in contact with the specimen surface.
5. Mount the replica on a specimen stub as described in the next section of this chapter. Care must be taken if glues or cements are used, since they will soften and distort the replica. It is best to use a double-stick tape

or transfer tab rather than a glue that is dissolved in an organic solvent.
6. Coat the mounted replica with a conducting metal by sputter coating or thermal evaporation (see Chapter 5).

Negative/Positive Replication Using Silicone, Resin

The use of silicone-based compounds for surface replication is more desirable than cellulose acetate because living specimens with high topography may be examined with better resolutions (Figure 3-11B). The technique is more complicated, however, and requires several types of ingredients—some of which may be toxic (i.e., epoxy resins). With this procedure, one first makes a negative replica using the silicone impression material and then fills this negative impression with a plastic monomer (Spurr's epoxy resin, for example) to make a positive replica. The replication materials (silicones, epoxy resins, etc.) can be obtained from most EM supply houses.

Silicone, Resin Positive/Negative Replication Procedure

1. Place the specimen to be replicated in a stable position. If animals or plants are used, it is important that movement of the specimen be minimal during replication.
2. Mix the silicone (polyvinylsiloxane or Xantropren Blue may be used) according to manufacturer's directions.

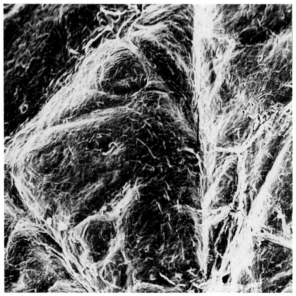

Figure 3-11(A) Cellulose acetate negative replication procedure of human skin surface. (Courtesy of O. Crankshaw and SEM, Inc.)

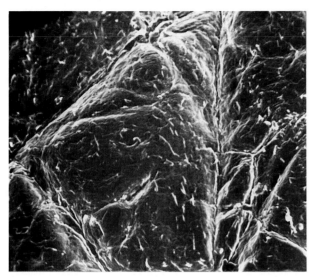

Figure 3-11(B) Silicone negative impression material of human skin surface. (Courtesy of O. Crankshaw and SEM, Inc.)

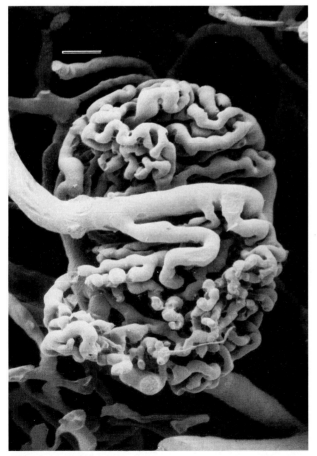

Figure 3-11(C) Corrosion cast of pig kidney glomerulus revealing formation of efferent arteriole within the glomerulus. Marker bar = 25 μm. (Courtesy of B. J. Moore and K. R. Holmes.)

3. Clean the surface to be replicated by swabbing it with lintless cloths soaked in buffer solutions and followed by 70% ethanol, if possible, to remove loose debris and oils.

4. Apply the silicone impression material to the surface. This should be done as quickly as possible since the silicone may harden rapidly. The material should be spread thinly and evenly over the surface using compressed air. A thin layer is preferred initially since it will conform better than an overly thick one.

5. Wait the recommended amount of time for the material to harden.

6. Apply a thicker layer of silicone material to support the initial thin layer and allow it to harden as well.

7. Gently peel away the silicone replica from the specimen surface.

8. Examine the replica under a dissecting microscope for the presence of air bubbles or other irregularities and repeat the procedure if necessary.

9. If one wishes to examine the negative replica, it should be mounted on a specimen stub and coated for electrical conductivity.

   If one wishes to examine a positive replica, it is first necessary to prepare a suitable low viscosity plastic resin such as Spurr's epoxy resin, Quetol, or one of the acrylics (see Chapter 2 for epoxy resin formulations).

10. Pour the plastic resin into the silicone negative impression carefully to avoid trapping air bubbles. Should bubbles be noticed, they may be dislodged using a small wooden probe, or the silicone and resin assemblage may be placed into a chamber under gentle vacuum to remove them.

11. Polymerize the plastic resin by placing the silicone and resin assemblage into an oven at 60° C overnight.

12. After cooling, gently peel the silicone away from the hardened positive plastic impression.

13. Mount the positive plastic replica on a specimen stub, coat with metal for conductivity, and examine in the SEM.

Corrosion Casting of Animal Vasculatures

One can make an extensive series of impressions of the endothelial lining of blood vessels of a recently killed animal by first perfusing the animal with a physiological fluid to flush away blood (see Chapter 2, perfusion fixation), followed by the injection of Batson's No. 17 anatomical casting compound (Batson, 1955). Upon polymerization of the casting compound, the tissue is dissolved in strong caustic solutions so that only the impression of the vessels remains (Figure 3-11C). This technique has been used for many years by medical schools to reveal the entire vasculature of animals and cadavers.

Corrosion Cast Replication Procedure

1. Kill the animal by injection and perfuse it with a phys-iologically balanced salt solution as described in Chapter 2.
2. After the blood vessels are totally clear and tissues have blanched, perfuse the vasculature with freshly prepared Batson's solution using a flow rate of 2 ml/min until the casting material totally fills the vessels.
3. Allow the animal to remain undisturbed until polymerization of the casting solution (usually 1 to 2 hours at room temperature). Follow manufacturer's directions for the exact timing and temperatures to use for the polymerization.
4. Place the tissues in a corrosion solution composed of 20% NaOH or KOH to dissolve the organic material. This may take 12 to 24 hours, depending upon the mass of tissue to be dissolved. The process may be accelerated by elevating the temperature of the solution to 50° C.
5. Rinse the vascular cast in distilled water. Extensive rinsing is necessary, and ultrasonication of the casts in distilled water is usually necessary to remove stubborn debris.
6. Dehydrate the cast by placing it into 95% ethanol and allowing it to air-dry.
7. Cut the cast into manageable sections, mount onto a specimen stub, coat with metal for electrical conductivity, and examine in the SEM.

A number of artifacts may be introduced during the replication process, so that familiarity with normally processed tissues is essential before one embarks on a study employing this method. It is obvious that examination of the actual specimen rather than a replica is more desirable; however, when this is not possible, replication does offer an alternative. For more information, see the references by Crankshaw (1984), Hodde and Nowell (1980), Pameijer (1978), Pfefferkorn and Boyd (1974), and Scott (1982).

Specimen Mounting

After specimens have been fixed, dehydrated, and dried using an appropriate protocol, they may be attached to a metallic *stub* (usually made of aluminum) and then coated with a metal prior to insertion into the SEM. Most biological specimens must be coated with conducting metals or carbon (Chapter 5) to prevent the buildup of high voltage static charges that will degrade the quality of the SEM image.

Although specimens are sometimes glued directly to the stub using an adhesive, some specimens may first be attached to a number of different substrates (glass microscope slides, coverglasses, cleaved mica sheets, microporous filters, polished metals such as aluminum, stainless steel, beryllium, etc.) and then attached to the stub.

An important first step in deciding which stub and glue to use is to determine the type of signal to be collected. In the standard viewing mode, secondary electrons are collected so that aluminum stubs and metallic coatings are permissible. On the other hand, if X-ray analysis or backscattered signals are to be used, then carbon stubs with nonmetallic, carbon-based glues and evaporated carbon coatings of the specimen are in order. Whenever possible, the surface of the stub or substrate should be as smooth and free of structure as possible to prevent confusing backgrounds. The stubs must be cleaned in organic solvents to remove oils used during the manufacturing process and handled with gloves or forceps to prevent the transfer of body oils onto the specimen and specimen holders.

As shown in Figure 3-12, stubs may be purchased or modified to fulfill different mounting requirements. One must always ascertain whether or not such stubs will fit into the SEM chamber and be manipulable without bumping into or damaging sensitive components (polepieces, wires, detectors) in the specimen chamber.

Criteria for Selecting Specimen Adhesives

Not many commercially available glues are adequate for use in SEM, since they do not fulfill some of the following important criteria. Suitable glues must:

- not damage specimens
- not contaminate the microscope
- provide good tackiness and still be easy to apply
- be resistant to the electron beam, heat, and vacuum
- provide a smooth, nonconfusing background.

For larger specimens, as well as for attaching substrates onto stubs, a number of glues and sticky tapes have been evaluated. Of the countless glues that have been tested, polyvinyl chloride (trade named "Microstik"), alpha-cyanoacrylate (Super Glue, Eastman 910), and cellulose nitrate (Duco china cement) combined with an equal volume of silver conductive paint or carbon paint appear to be the most useful. Care must be taken not to use too much glue, since it will extend the pumpdown time of the vacuum system and may wick into the specimen and damage it.

Adhesive tapes, especially those with adhesive on both sides, are sometimes useful for holding specimens and substrates. Some good double-stick tapes include

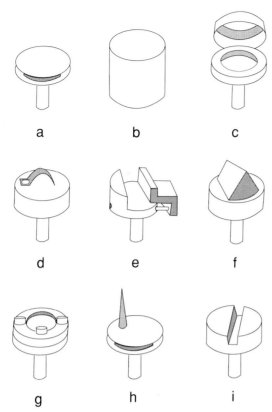

Figure 3-12 Specimen stubs used for mounting biological specimens for viewing in the SEM. The standard types are shown in a and b, while others are modifications to hold or pin down various types of specimens. (Courtesy of Judy Murphy and SEM, Inc. Redrawn with permission.)

Scotch Double-Coated Tapes #665 and 666 and Scotch Adhesive Tape #463. In general, tapes are less desirable than glues since they tend to outgas and break down under high beam currents.

A convenient way to quickly attach some specimens is to use transfer tabs available from some photo supply stores and EM supply houses (Figure 3-13). The tabs bear a dab of adhesive that may be transferred onto the stub by light pressure. Specimens may then be carefully placed onto the adhesive and pressed lightly, if possible, to ensure a good bond. If delicate specimens are being attached to the stub, then light bursts of compressed, clean air may be used to press the specimen into the adhesive.

Most biological specimens that have undergone drying are extremely brittle and prone to damage unless handled extremely carefully. Among the tools that may be used to pick up and mount specimens onto substrates or stubs are fine-pointed jeweler's forceps, wooden applicators or toothpicks, dissecting needles, eyelash probes (a single eyelash mounted on the end of a toothpick), and micropipettes. A handy device for picking up specimens is

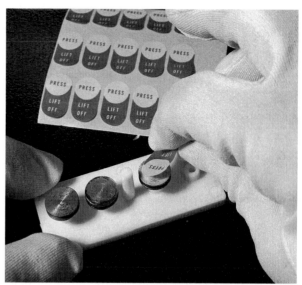

Figure 3-13 Adhesive transfer tabs are used to deposit a small amount of adhesive onto SEM stubs. The adhesive is adequate to hold most small specimens on the stub.

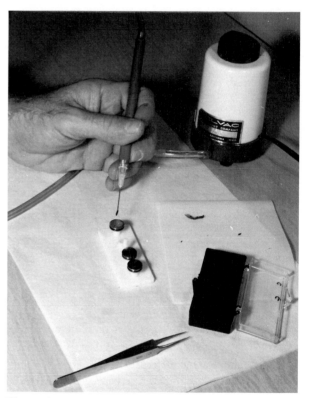

Figure 3-14 Vacuum tweezers used to pick up and deposit small delicate specimens onto SEM stubs.

the vacuum needle shown in Figure 3-14. A small aquarium pump is used to generate a slight vacuum through the hollow needle that is placed onto an expendable area of the specimen. Finger pressure over a hole in the handle causes an increase in the suction so that the specimen is held on the tip of the needle.

No pressure should be exerted on the specimen, but it should be lifted onto the stub or substrate (previously coated with the proper glue or double-stick tape) using the appropriate tool. Once the solvents have thoroughly dried, the specimen stubs may be labeled on the underside with a permanent marker pen, stored in a dust-free desiccator, or coated for conductivity and viewed in the SEM. For more details on specimen mounting methods, see the review article by Murphy (1982). In addition, the excellent collection of articles edited by Murphy and Roomans (1984) covers all of the steps necessary for fixation, drying, mounting, and coating of specimens for SEM.

Specimen Coating for Conductivity

In the standard protocols, after specimens have been mounted onto the specimen stub, they are coated with a thin layer (approximately 20 to 30 nm) of a conductive metal such as gold, platinum, or gold/palladium alloy. Metal coatings usually are applied by sputter coating (see next section) or by thermal evaporation as described in Chapter 5. The metal coatings prevent the buildup of high voltage charges on the specimen by conducting the charge to ground (i.e., the coating is contiguous with the aluminum specimen stub that is connected to the grounded specimen stage of the SEM). In addition, the metal coatings serve as excellent sources of secondary electrons as well as helping to conduct away potentially damaging heat.

Sputter Coating Procedure

The most commonly used method for coating SEM specimens with a thin layer of a metal is the plasma sputtering or, as it is usually termed, the *sputter coating* procedure. Several different designs of instrumentation are possible and have been reviewed by Echlin (1978, 1981). Probably the most commonly used system is the direct current sputtering device similar to the one diagrammed in Figure 3-15. Details of the sputtering chamber and the principle involved in the sputtering process are illustrated in Figure 3-16.

In the sputtering process, after placing the metal stub containing the attached specimen into the spec-

imen chamber, the chamber is evacuated to approximately 0.1 Pa using a rotary vacuum pump. The purpose of evacuation is to remove water and oxygen molecules that might damage the surface of the specimen.

The pumping procedure may be somewhat extended if this is the first time the chamber is being evacuated or if the specimen has any residual moisture present due to adsorbed water vapor or organic molecules from the solvents used to glue the specimen onto the stub. It is best to minimize these molecules by placing the mounted specimen into a drying oven overnight. The temperature of the oven should be as high as possible—but not so high as to damage the specimen. It is very important not to pump the specimen chamber for long periods of time since this will lead to a backstreaming of vacuum pump oil onto the specimen surface (see the discussion of problems associated with vacuum systems in Chapter 6). Normally, if one cannot achieve the desired vacuum in a properly sealed system within 15 to 20 minutes, then either the specimen or the specimen chamber has adsorbed water or organic molecules. Drying the specimen or cleaning the chamber are in order.

It is important to reemphasize that one must remove traces of moisture and oxygen from the specimen chamber, since these molecules will be broken down in the high voltage fields inside the chamber into highly reactive molecules that will oxidize and damage the specimen surface. For this reason, one evacuates the chamber and replaces the atmosphere present with an inert gas such as argon.

After the specimen chamber has achieved the desired vacuum level, an inert gas such as argon is slowly introduced into the chamber. The flow of argon is adjusted so that the vacuum can be maintained at approximately 6 to 7 Pa. A negatively charged, high voltage (1 to 3 kV) field is applied to the area of the chamber termed the *target* so that the argon gas molecules will be ionized into Ar^+ and electrons. The negatively charged target, composed of a heavy metal such as gold or palladium/gold, will be struck with such force by the Ar^+ molecules that some of the metal atoms of the target will be ejected. These atoms are bounced about by the various ions (Ar^+ and electrons) present in the chamber and eventually strike the specimen surface to gradually build up a metallic coating. The fact that the metal particles are knocked about in random paths (rather than in straight lines) is important, since they will strike the specimen at various angles and thereby more uniformly coat the irregularly shaped specimen surfaces. (This is in contrast to the evaporative method of coating specimens with heavy metals de-

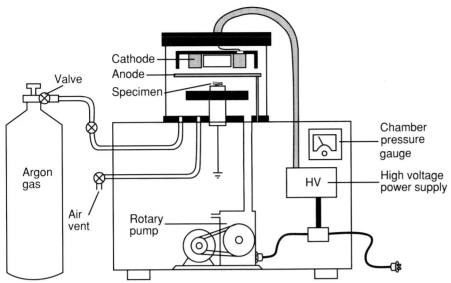

Figure 3-15 Diagram of parts of a commonly used sputter coater. A rotary pump is used to evacuate the specimen chamber to remove atmospheric gases and to permit the introduction of argon gas into the chamber. The application of a high voltage to the target causes the ionization of the argon into Ar$^+$ molecules that strike the cathode (target) and eject metal atoms that then coat the specimen.

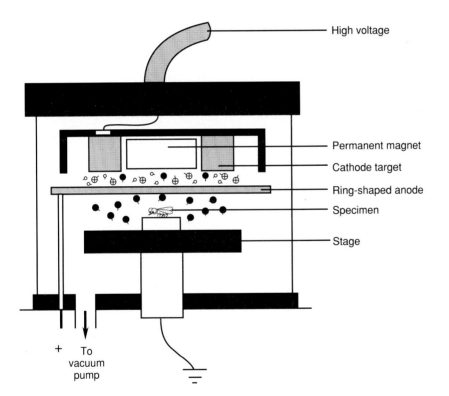

Figure 3-16 Detailed diagram of specimen chamber of a sputter coater. The metal target is struck by ionized Ar$^+$ molecules to cause the ejection of atoms of the metal target (darkened circles) that eventually coat the specimen. A permanent magnet is placed inside of the target to deflect potentially damaging electrons away from the specimen and toward the anode ring. To further protect the specimen from excessive heating, the stage may be cooled with water or by electronic means.

scribed in Chapter 5, where the metal atoms travel in straight lines and are unable to reach areas not in a direct line with the source of the metal.) The sputtering process is continued until a proper thickness of metal coating has accumulated on the specimen surface. The thickness needed depends upon the topography of the specimen: specimens with low topographies need less coating than those with higher topographies. Generally a coating of 15 to 40 nm is adequate for most specimens.

Estimation of Coating Thickness

Some sputter coater systems (see Figure 3-17) have a digital thickness monitor consisting of a crystal of quartz that oscillates at a certain frequency. As metal coating builds up on the crystal that is placed close to the specimens, the oscillation frequency is altered in proportion to the thickness of the coating. A simpler method may involve placing a white piece of paper or a glass coverglass into the chamber and observing the darkening of the ma-

Figure 3-17 Photograph of a commercially available sputter coater. The argon tank is to the right of the sputter coater. This unit is equipped with quartz crystal monitor to give a digital readout of the thickness of metal coating deposited on the specimens.

terial until the desired shade of gray or proper reflectance color is obtained. In any case, for accuracy of measurement, one should establish a series of standards for comparison to recently coated specimens. The thickness of the standards is determined by including a piece of plastic (such as epoxy or methacrylate) in the chamber during the coating process and then cutting an ultrathin section across the coating and measuring the thickness in the TEM. See Chapter 4 for details of ultramicrotomy procedures.

Potential Problems Associated with Sputter Coating

1. **Thermal Damage** may be caused by radiation from the target as well as by the electrons striking the grounded specimen. This may be minimized by (a) deflecting the electrons away from the specimen using a permanent magnet centered in the target, (b) using an anode ring placed above the specimens to attract the electrons, (c) cooling the specimen stage, or (d) pulsing the high voltage—rather than applying the voltage continuously—to prevent heat buildup.
2. **Surface Contamination** may occur due to backstreaming of oil from the rotary pump. This may be minimized by (a) not pumping for extended periods of time on the sealed specimen chamber, (b) maintaining a positive flow of argon gas over the specimen surface and into the vacuum pump, or (c) installing filters or traps in the rotary pump line to the specimen chamber.
3. **Surface Etching** will occur if water vapor, oxygen, or carbon dioxide molecules remain inside of the speci-

men chamber and are ionized in the high voltage field. One must use an inert gas such as argon rather than normal air to generate the ions that strike the target. Water vapor may be detected by noting the color of the glowing plasma generated during the sputtering process. If the color is blue rather than lavender, one must suspect contamination by water vapor and take the appropriate measures (e.g., dry the specimen or clean the specimen chamber). Sometimes it is possible to clean a specimen surface by etching it deliberately. Normally, this is done by reversing the polarity of the instrument so that the specimen is made highly negative and the argon ions strike and erode the specimen rather than the target.

Noncoating Techniques

It is also possible to render biological specimens electrically conductive by methods other than coating by sputtering or thermal evaporation. One limitation of the standard methods of sputtering and thermal evaporation (in contrast to specialized procedures of high resolution sputtering and electron beam evaporation) is the diminished resolution imposed on the specimen surface due to the use of thick coatings of 20 nm or more. In addition, these coatings may not be continuous (leading to localized charging), the specimens may be damaged by heating during the coating process, and the process requires specialized equipment.

Several alternatives for rendering biological specimens electrically conductive have been developed over the years. All of these methods involve the deposition of thin layers of the conductive metals onto the specimen by reducing the metal from aqueous salt solutions of the metal. Since the thickness of the coating layers is in the range of 4 to 20 nm, resolutions of the same order may be expected. These procedures have been reviewed by Murphy (1978, 1980) and Murakami et al. (1983). An example of one method is given in the following box.

Tannic Acid/Osmium Noncoating Technique (Takahashi, 1979)

1. If immersion fixation is used, start at step 4. If perfusion fixation is used, the tissues are initially perfused in an appropriately buffered salt solution to clear the blood from the vessels (see perfusion fixation in Chapter 2). Following this, the tissue is perfused at 4° C for 15 minutes with a freshly prepared mixture of 1.5% glutaraldehyde and 0.5% tannic acid in a buffer system appropriate for the tissues being

studied. (Note: If a phosphate buffer is used, the final solution should be filtered immediately before use.)

2. Perfuse 10 minutes with a freshly prepared mixture of 1.5% glutaraldehyde and 1% tannic acid in the appropriate buffer.

3. Cut tissues into small pieces of 1 mm³ and rinse in buffer for 1 hour. Continue with step 4.

4. Immerse tissue into 2% osmium tetroxide in buffer for 2 hours at 4° C.

5. Rinse in buffer three times (5 minutes each).

6. Immerse in a freshly prepared mixture of 8% glutaraldehyde and 2% tannic acid for 12 hours. (Note: This solution should be changed three times over the 12 hour period. Cacodylate buffer is recommended since phosphate buffers may precipitate during this time period.)

7. Rinse in buffer three times (5 minutes each).

8. Immerse for 2 hours in a 2% solution of osmium tetroxide in cacodylate buffer.

9. Rinse in buffer three times (5 minutes each).

10. Repeat steps 6 through 9.

11. Dehydrate tissue, critical point dry, mount on stub using conductive paint, and examine in SEM.

Specimen Storage

It may be necessary to store the fixed and dried specimens either before or after mounting onto the SEM stub. In either case, certain precautions must be observed to ensure that artifactual changes are not introduced during this period. The specimens are extremely fragile and brittle and may be damaged by contact even with a soft object such as a camel's hair brush. It is best to handle the specimen in an area that is not to be studied in the SEM. Normally, one would store unmounted specimens in small cardboard or plastic boxes (pillboxes) or even petri dishes lined with filter paper or small pieces of aluminum foil (to facilitate picking up the specimens). Specimens that have been mounted onto stubs may be stored in commercially available boxes designed to hold the stubs securely, or one may easily make storage holders by drilling holes in plastic to accommodate the stubs (Figure 3-18).

If plastic containers are used, one must be aware that static charges will often cause unmounted specimens to become displaced, often sticking onto the covering lids. If several different types of specimens are placed in the same container, they may become

Figure 3-18 Various types of containers for storing specimen stubs. Some are commercially available while others may be easily fabricated.

mixed up. Glass or cardboard containers are less likely to develop such static charges. All containers must be dry, clean, and dust-free to prevent contamination of the specimen surfaces.

The specimens must be stored in a dry, dust-free environment. Most laboratories store specimens in a glass or plastic desiccator containing a drying agent of some sort. One should be careful to check that the agent is not exhausted (most have color indicators) and that it is sequestered away from the specimens since most drying agents tend to be powdery on their surface. One may cover the drying agent with lintless lens tissues to prevent the transfer of powdered desiccant onto the specimen. The container should be tightly sealed and kept closed when specimens are not being transferred. Often silicone greases are used to seal the jars, so take care not to contaminate the specimen with these lubricants. Finally, one should avoid storing specimens in the desiccator if they are outgassing solvents (amyl acetate, glues, etc.), since the desiccant and specimen will adsorb these fumes and may be damaged. Often one may detect such fumes by sniffing the jar upon opening. If fumes are detected, another jar should be used or the desiccant replaced.

The dried specimens are quite deliquescent and will absorb water from the environment. This leads to the swelling and shrinking of the specimen as it is moved into or out of the desiccator. Such volume changes may cause fine fracturing of the specimen surface coatings, and delicate features may be damaged. It is best to keep the specimen in the sealed container until mounting and viewing is necessary.

References

Batson, O. V. 1955. Corrosion specimens prepared with a new material. *Anat Record* 121:425.

Becker, R. P., and O. Johari eds. 1979. *Cell surface labeling*. Scanning Electron Microscopy, Inc. (AMF O'Hare, Il) 344 pp.

Boyde, A. 1978. Pros and cons of critical point drying and freeze drying for SEM. *Scanning Electron Microscopy* II:303–14.

Boyde, A., and E. Maconnachie. 1981. Morphological correlations with dimensional change during SEM specimen preparation. *Scanning Electron Microscopy* IV:27–34.

Cohen, A. L. 1979. Critical point drying—principles and procedures. *Scanning Electron Microscopy* II:303–23.

Crankshaw, O. S. 1984. Instruction of replica techniques for scanning electron microscopy. *Scanning Electron Microscopy* IV:1731–7.

Echlin, P. 1978. Coating techniques for scanning electron microscopy and X-ray microanalysis. *Scanning Electron Microscopy* I:109–32.

———. 1981. Recent advances in specimen coating techniques. *Scanning Electron Microscopy* I:79–90.

Falk, R. H. 1980. Preparation of plant tissues for SEM. *Scanning Electron Microscopy* II:79–87.

Haggis, G. H. 1982. Contribution of scanning electron microscopy to viewing internal cell structure. *Scanning Electron Microscopy* II:751–63.

Helinski, E. H., G. H. Bootsma, R. J. McGroarty, G. M. Ovak, E. de Harven, and J. L. Pauly. 1990. Scanning electron microscopic study of immunogold-labeled human leukocytes. *J Electron Microscopy Tech* 14:298–306.

Hodde, K. C., and J. A. Nowell. 1980. SEM of micro-corrosion casts. *Scanning Electron Microscopy* II:89–106.

Kennedy, J. R., R. W. Williams, and J. P. Gray. 1989. Use of Peldri II (a fluorocarbon solid at room temperature) as an alternative to critical point drying for biological tissues. *J Electron Microscopy Tech* 11:117–25.

Maugel, T. K., D. B. Bonar, W. J. Creegan, and E. B. Small. 1980. Specimen preparation techniques for aquatic organisms. *Scanning Electron Microscopy* II:57–77.

Mazia, D., G. Schatten, and W. Sale. 1975. Adhesion of cells to surfaces coated with polylysine. *J Cell Biol* 66:198–200.

Murakami, T., N. Iida, T. Taguchi, O. Ohtani, A. Kikuta, A. Ohtsuka, and T. Itoshima. 1983. Conductive staining of biological specimens for scanning electron microscopy with special reference to ligand osmium impregnation. *Scanning Electron Microscopy* I:235–46.

Murphy, J. A. 1978. Non-coating techniques to render biological specimens conductive. *Scanning Electron Microscopy* II:175–93.

———. 1980. Non-coating techniques to render biological specimens conductive/1980 update. *Scanning Electron Microscopy* I:209–20.

———. 1982. Considerations, materials and procedures for specimen mounting prior to scanning electron microscopic examination. *Scanning Electron Microscopy* II:657–96.

Murphy, J. A. and G. M. Roomans, eds. 1984. *Preparation of biological specimens for scanning electron microscopy*. Scanning Electron Microscopy, Inc. (AMF O'Hare, Il) 344 pp.

Nowell, J. A., and J. B. Pawley. 1980. Preparation of experimental animal tissue for SEM. *Scanning Electron Microscopy* II:1–19.

Pameijer, C. H. 1978. Replica techniques for scanning electron microscopy, a review. *Scanning Electron Microscopy* II:831–36.

Pfefferkorn, G., and A. Boyd. 1974. Review of replica techniques for scanning electron microscopy. *Scanning Electron Microscopy* I:75–82.

Postek, M. T., K. S. Howard, A. Johnson, and K. L. McMichael. 1980. *Scanning electron microscopy: a student's handbook*. Ladd Research Industries, Inc. Burlington, Vt. 305 pp.

Revel, J-P., T. Barnard, G. H. Haggis, and S. A. Bhatt, eds. 1983. *The science of biological specimen preparation for microscopy and microanalysis*. Scanning Electron Microscopy, Inc. (AMF O'Hare, Il) 245 pp.

Scott, E. C. 1982. Replica production for scanning electron microscopy: a test of materials suitable for use in field settings. *J Microsc* 125:337–41.

Shennaway, I. E., D. J. Gee, and S. R. Aparicio. 1983. A new technique for visualization of internal structure by SEM. *J Microsc* 132:243–46.

Takahashi, G. 1979. Conductive staining method. *Cell* 11:114–23.

Tanaka, K. 1989. High resolution electron microscopy of the cell. *Biology of the Cell* 65:89–98.

Watson, L. P., A. E. McKee, and B. R. Merrell. 1984. Preparation of microbiological specimens for scanning electron microscopy. *Scanning Electron Microscopy* II:45–56.

CHAPTER 4

Ultramicrotomy

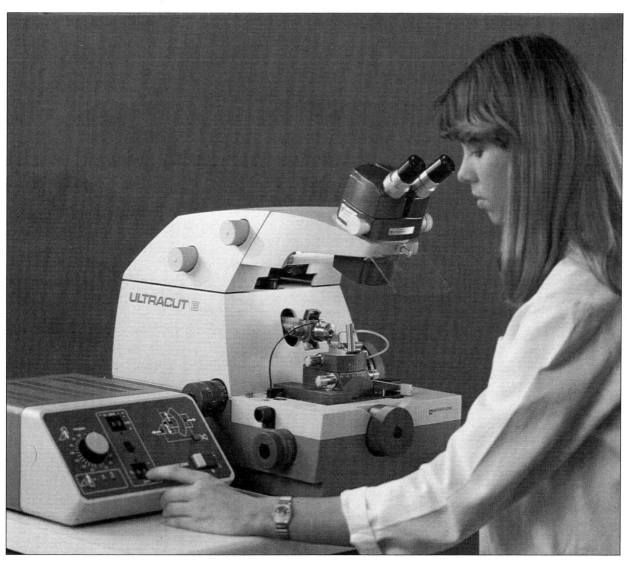

Courtesy of Leica.

Ultramicrotomy is a procedure for cutting specimens into extremely thin slices, or *sections*, for viewing in the transmission electron microscope (TEM). Sections must be very thin because the 50 to 125 kV electrons of the standard electron microscope cannot pass through biological material much thicker than 150 nm. In fact, for best resolution, sections should be from 30 to 60 nm. This is roughly equivalent to splitting a 0.1 mm thick human hair into 2,000 slices along its long axis, or cutting a single red blood cell into 100 slices. Ultramicrotomy is a demonstrably demanding technique that requires much practice and patience from the beginning microtomist, who must pay attention to such details as specimen preparation and embedding, preparation of knives and specimen support grids, and cleanliness of all reagents and utensils. A thorough working knowledge of all equipment involved in the ultramicrotomy process is essential.

When microtomists refer to "ultrathin sections," they mean sections from 50 to 100 nm in thickness, which are suitable for viewing in the standard electron microscope. "Thick" sections in the 0.5 to 2 μm range, are about 10 to 20 times thicker than the thin sections. These thick sections, or *survey sections* as they are often called, are viewed in the light microscope to determine if the right area of the specimen is in position for thin sectioning. It is very common practice to view a thick section in the light microscope first before proceeding with ultramicrotomy or thin sectioning.

To cut thin sections from relatively soft biological specimens, it is first necessary to infiltrate suitably fixed and dehydrated specimens with a liquid plastic that is then hardened or polymerized as described in Chapter 2. The plastic serves as a very hard matrix that provides support to the tissue as the knife passes through the specimen. Another method of hardening the tissue for ultramicrotomy is rapid freezing followed by cutting at temperatures of $-80°$ to $-140°$ C. Known as *cryoultramicrotomy*, this technique will be discussed at the end of the chapter.

The specialized instrument used for cutting sections is called an *ultramicrotome* (see Figure 4-1). Ultramicrotomes advance the specimen in precise, repeatable steps by using either a mechanical or thermal advancement mechanism. In the standard procedure, after the sections have been cut and are floating on the surface of water contained in the trough of the knife, they are picked up on a copper screen mesh, or grid, and stained for contrast using salts of

a heavy metal prior to viewing in the transmission electron microscope.

The sequence of steps in preparing ultrathin sections for examination involves: *trimming or shaping the specimen block, preparing ultramicrotome knives and specimen support grids, cutting sections in the ultramicrotome, picking up the sections onto the specimen grid, and staining them to enhance contrast.* These steps are illustrated in Figure 4-2.

Shaping the Specimen Block

To prepare for ultramicrotomy, it is necessary to shape the plastic specimen block into a small cutting face in order to minimize the stresses imposed on the cutting edge of the knife as the specimen is passed over it. This is usually done in two stages. The specimen block is first *rough trimmed* to exclude excess plastic matrix and to expose the surface of the specimen. A thick section is made from the rough-trimmed block and examined in the light microscope. After locating the areas of interest in the thick section and on the block face, further fine trimming is done to remove unwanted areas of the specimen. The usual goal in fine trimming is to produce a truncated pyramid with its sides sloped at 45 to 60 degrees, as shown in Figure 4-3. A pyramidal shape is most often used because it is a more stable structure than, for instance, one with nonsloped sides. The pyramid should be no larger than 0.5 to 1.0 mm on either of the two parallel sides. Especially when using a glass knife, it is important to reduce these dimensions as much as possible to prevent vibrations generated as the specimen contacts the knife.

Rough Trimming by Hand

Although several instruments are available to shape the block in a precise manner, they are quite expensive (approaching the cost of an ultramicrotome) and may not be justified especially since precision trimming can be done with most ultramicrotomes or even with a single-edged razor blade as follows.

Shaping the Block Using Razor Blades

1. Clamp the specimen block into an appropriate holder or *chuck,* and place the chuck into a stable holder under a dissecting or stereo microscope. Some ultramicrotomes have adapters so that one can use the ultramicrotome optics and accessories for this step.

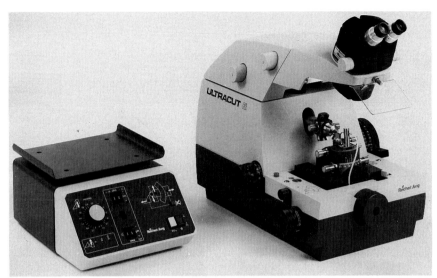

Figure 4-1 Modern ultramicrotome for cutting thin sections of plastic-embedded specimens. (Courtesy of Leica.)

2. Cut into the specimen by making a series of slices with a new, single-edged razor blade that has been cleaned with xylene or acetone to remove oils. The slices should be made parallel to the tabletop and should stop just before the desired area in the specimen is reached (Figure 4-4A). If osmium was used as a fixative, the specimen will be black and readily seen as the trimming takes place. It may be possible even to see the desired area on the flat top surface of the blackened specimen by adjusting the lighting and looking at the cut block face or surface under high magnification in the stereo microscope. The gross structure of the tissues can be seen at the block face due to differential reflections of the various tissue components as the bright light is reflected from the block face surface (Figure 4-5). If it is not possible to select the desired area in this manner, then it will be necessary to cut 0.5 to 2.0 μm thick survey sections as described in the following section. This is the most common way to determine the tissue orientation. The razor cuts should be as thin as possible to prevent removal of large chunks from the specimen block by haphazard fracturing rather than cutting.

3. Begin making a series of razor cuts to a depth of 2 to 3 mm and at a 60 degree angle relative to the desk top along one side of the block (Figure 4-4B). Detach the slices from the block by carefully making one horizontal cut.

4. Rotate the specimen block 180 degrees and begin a second series of 60 degree cuts until the side of the specimen is reached (Figure 4-4C). These cuts will generate a narrow specimen platform on the top of the pyramid. The platform should be wide enough so that no important areas of the specimen are cut off, but not so wide as to cause sectioning difficulties. If in doubt about what to trim, do not trim away any areas until a survey section has been examined in the light microscope.

5. Rotate the specimen block and make a third and fourth series of side cuts at a 60 degree angle to generate a trapezoid-shaped pyramid top (Figure 4-4D and 4-4E).

The roughly trimmed block (Figure 4-5) may be many times larger than the finished block used for thin sectioning; however, it will be fine trimmed once the orientation of the tissue has been determined and important areas located by light microscopy. If the researcher is fortunate enough to be working with homogeneous samples (e.g., clones of cell cultures or cell fractions) and has no need for orientation of sections or selecting areas and retrimming, then the block can be fine trimmed as described in a subsequent section.

Rough Trimming by Machine

Most experienced microtomists prefer to rough trim specimens by hand using razor blades, since it is possible to complete the process in about 5 to 15 minutes. It also is possible to shape the block accurately on the ultramicrotome following the steps for hand trimming (described previously) except that the block is mounted in the ultramicrotome and shaped using a glass knife. Two of the angled sides of the block are produced by shifting the ultramicrotome knife 30 to 40 degrees relative to the initial flat face and approaching the block from the side using the coarse advance mechanism of the ultramicrotome. After the two sloped sides are formed, the specimen chuck is rotated in the specimen arm and the angled knife is used again to remove the third side and, following a third specimen block rotation, the fourth side (Figure 4-4E). A variety of

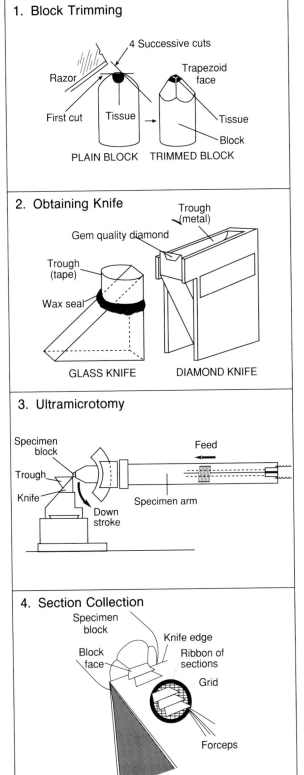

1. Block Trimming

4 Successive cuts
Razor
Trapezoid
face
First cut   Tissue
Tissue
Block
PLAIN BLOCK   TRIMMED BLOCK

Specimen blocks are trimmed to expose the specimen and to form a trapezoidal cutting face. Usually, a single edged razor blade is used for this step which is done by hand while observing the process under a dissecting or stereomicroscope. Automatic trimmers are also available.

2. Obtaining Knife

Trough (metal)
Gem quality diamond
Trough (tape)
Wax seal
GLASS KNIFE   DIAMOND KNIFE

The knife used to cut sections may be made from a special grade of plate glass just prior to use. It is necessary to attach a trough to the glass knife to hold the water onto which the sections will be cut. Alternatively, a diamond knife may be used that has a large metal trough to hold the water.

3. Ultramicrotomy

Specimen block
Feed
Trough
Knife
Down stroke
Specimen arm

The trimmed specimen block is placed into the ultramicrotome and a thick (1–2 μm) section is usually cut and examined in the light microscope to verify that the correct specimen area is in position. Subsequently, ultrathin sections (60–80 nm) may be cut using a mechanically or thermally advanced specimen arm.

4. Section Collection

Specimen block
Knife edge
Block face
Ribbon of sections
Grid
Forceps

Sections that have been cut onto the water surface in the tough are picked up using a mesh or specimen grid usually made of copper. The sections adhere to the grid upon removal from the water and are dried in a dust free environment. The sections may also be stained using special heavy metal salts such as lead citrate and uranyl acetate.

Figure 4-2   Illustration and brief explanation of steps involved in ultramicrotomy process. (Redrawn with permission of J. Murphy.)

Figure 4-3 Specimen block that has been rough trimmed to give general shape to the block. After examination of a thick section taken from the block, the block will be fine trimmed to a considerably smaller shape containing only specimen (shown in black).

Viewed from Side   **Viewed from Above**

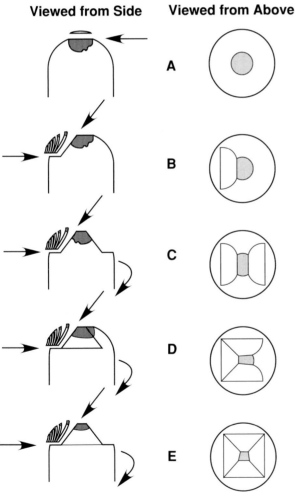

Figure 4-4 Steps in trimming a plastic specimen block. Biological specimen is seen as a darker area in the plastic. Straight arrows indicate razor cuts made in the plastic, while curved arrows indicate turns to be made in the orientation of the block. (Redrawn with permission from *Ultramicrotomy* by N. Reid, 1975.)

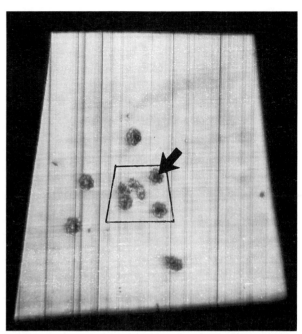

Figure 4-5 Surface view of rough-trimmed specimen block. Gross structure of individual plant cells (arrow) within tissue may be seen in the reflected light. Striations running through the block face are knife marks caused by nicks in the glass knife edge. This block face is approximately 5 mm across at the top. The superimposed trapezoidal shape indicates the shape that the fine trimmed block will assume after excess plastic has been removed using a razor blade.

faces can be made by varying the rotation of the specimen chuck in the arm of the microtome.

Special specimen trimmers may also be used in some research laboratories. One such trimmer (unfortunately, no longer available) is the LKB Pyramitome (Figure 4-6A), which resembles an ultramicrotome except that the Pyramitome can cut only thick sections in the micrometer range. The Pyramitome shapes blocks much more rapidly than the standard ultramicrotome because it has special rotating specimen and knife holders to orient the specimen and knives precisely. In addition, it is possible to cut thick sections on the Pyramitome, evaluate them in the light microscope, and then use a special accessory to superimpose the specimen image in the microscope slide over the block face. This permits very accurate fine trimming of the block face in the Pyramitome.

Another specially designed instrument (still available at this writing) is the Reichert Trimmer (Figure 4-6B), which precisely shapes the block face with a rapidly spinning routerlike tip. Inventive microtomists also have used high speed grinders (available at hardware stores) to remove large portions of

Figure 4-6(A) Specimen trimmer for shaping specimen blocks. The LKB Pyramitome uses glass knives to shape the specimen block. The specimen is advanced several micrometers at each cut. The operator adjusts the orientation of the knives and the specimen to obtain precisely shaped blocks. In addition, the instrument may be used to cut micrometer-thick sections for evaluation by light microscopy.

plastic rapidly and roughly shape the block (Chapter 2). However, use all of these units cautiously since enough heat may be generated during the process to distort or even melt the embedding plastic or, worse yet, to grind away the specimen inadvertently.

CAUTION: The fine dust produced during the grinding process is probably injurious to one's health, so be careful not to inhale the potentially carcinogenic plastic.

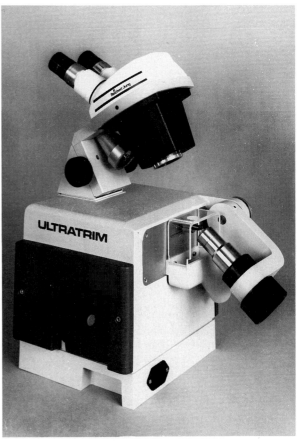

Figure 4-6(B) The Reichert trimmer may be used for shaping the plastic specimen block by means of a rapidly spinning routerlike tip. The specimen is positioned in a clamping chuck, and the specimen is moved across the router to trim the block to the desired shape. (Courtesy of Leica.)

The Mesa Trimming Procedure

In 1966, DeBruijn and McGee-Russell described a special type of trimming that is especially conservative of specimen. This procedure, which can be accomplished only by using machine trimming, is called the *mesa* trim (so named because the shaped block resembles geological features called mesas). After forming the initial flat cut at the top of the block to expose and locate the specimen (Figure 4-7A), the glass knife is moved to the right of the desired specimen area, and a series of cuts are made parallel to the specimen surface to a depth of approximately 15 to 25 μm (Figure 4-7B). Following this, the specimen is rotated 90 degrees, and sections are cut again parallel to the surface to the same depth as the first series of sections (Figure 4-7C). The specimen chuck is now rotated two more times through 90 degrees, and the third and fourth cuts are made on either side of the desired area and to the same depth as the previous two cuts (Figure 4-7D, E). The end product is a small, raised, boxlike mesa with vertical sides cut by the glass knife (Figure 4-7F). After the raised mesa of the specimen has been sec-

tioned down, it is possible to trim another mesa in the same or another area. The advantage of this technique is that a minimum of specimen is removed during the trimming process, making it possible to trim and examine many other areas of the specimen block.

Thick Sectioning

If orientation or a preview of the specimen is desired, as is normally the case, the roughly trimmed block should be mounted into the ultramicrotome and 0.5 to 2.0 μm sections should be cut. Most microtomists use glass or special types of diamond knives to cut such thick sections, which are picked up using the needle from a syringe (Figure 4-8), an eyelash probe, or a wooden spatula made by sharpening an applicator stick to a fine point. The section is then transferred onto a tiny droplet of distilled water on an alcohol- or acid-cleaned microscope

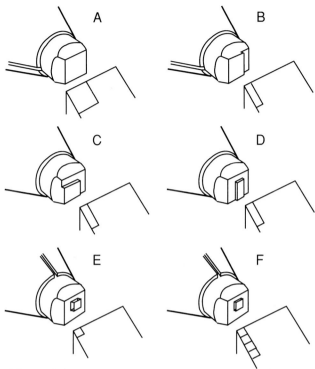

Figure 4-7 Steps in trimming a specimen block using the "mesa" technique. (Redrawn with permission from *Ultramicrotomy* by N. Reid, 1975.)

Figure 4-8 Retrieval of thick section from cutting edge of glass knife using the beveled needle from a tuberculin syringe. The trimmed block, viewed from above, is indicated by an arrow. The section is located just under the block and the syringe point is seen coming into the picture in the lower right.

slide by inverting the needle or wooden applicator so that the section side contacts the water droplet, which draws the section onto the slide. The slide is transferred to a hot plate (70° to 90° C) until the droplet evaporates. After the slide has cooled, a drop

of staining solution (Millipore-filtered 1% toluidine blue dissolved in 1% aqueous sodium borate solution) is placed over the dried section and reheated on the hot plate until the edges of the stain begin to dry. The slide is then gently rinsed in distilled water from a wash bottle and dried by heating it on the hot plate prior to examination in a light microscope. Do not direct the wash water directly on the sections or they may be removed from the slide. Figure 4-9 (small inset) shows a section 1 to 2 μm thick as viewed in a light microscope; the larger print is of the same specimen after ultramicrotomy as viewed in the TEM. In this case, the thick section

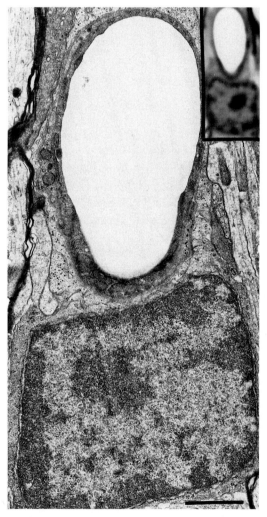

Figure 4-9 (Inset) Thick, 1 to 2 μm section of microglial nerve cell viewed in the light microscope. The large micrograph was obtained after the thick section was reembedded in epoxy plastic, and resectioned in an ultramicrotome to generate the ultrathin section shown. The large clear space is an empty capillary. Marker bar = 1 μm. The electron micrograph is magnified 5 times more than the light micrograph. (Courtesy of J. A. Paterson.)

on the glass slide was covered with epoxy plastic (i.e., reembedded in epoxy), and after removal from the slide, the reembedded *section* was again sectioned in an ultramicrotome to generate the ultrathin section shown.

It may be desirable to collect numerous plastic sections for viewing. If this is the case, one can collect long *ribbons* of connected sections using a scoop readily fashioned from laboratory plastic tubing (Figure 4-10A). Using such scoops, it is possible to mount 200 to 400 sections on a single glass slide (Figure 4-10B). If such collections become routine, it is recommended that an older diamond knife or a knife constructed of industrial diamond be used, since the attached water trough is large enough to accommodate the scoop easily, and a much wider variety of specimens can be cut with the diamond.

Fine Trimming

After the plastic survey sections have been examined by light microscopy and the area of interest located (see Figure 4-5), it is often necessary to retrim the plastic block to exclude areas that are not of interest and to optimize the size of the block face for thin sectioning. This must be done carefully under the stereomicroscope until a cutting block of reasonable size is obtained. These final cuts must be made with a sharp, clean razor blade so that the block sides parallel to the knife edge are quite smooth. It is very important that the two sides parallel to the knife edge be exactly parallel in order to produce straight ribbons of sections. If these two sides deviate from the parallel, curved ribbons that are difficult to handle and fit onto a grid will be produced (Figure 4-11A). In such cases, one will be able to cut only individual sections as shown in Figure 4-11B. A properly trimmed block will permit the microtomist to cut long ribbons of serial sections (Figure 4-12) that are much easier to maintain in the order cut.

Although the trapezoid-shaped block face is most commonly used, other shapes may be more appropriate in certain situations. For example, with elongated specimens such as fibers or rootlets, a block face that accommodates the shape of the specimen is more appropriate (Figure 4-13). In this case, one side may be 3 to 4 times longer than the adjacent side. These types of blocks are best cut in the final thin sectioning process by orienting the shorter side parallel to the knife edge and keeping the shorter dimension under 0.5 mm if possible. It is also possible to trim a block to a square face, the obvious advantage being that one can approach the knife from any of the four equivalent sides. Specimens that are difficult to section can be trimmed to a very small triangular face. In this case, one approaches the knife edge with the point of the triangle rather than one of the three flat sides (Figure 4-14). Although ribbons will not be produced with some of the block shapes described, individual sections of high quality can be organized into groups with an *eyelash probe* (fashioned by cementing an eyelash onto the end of an applicator stick) and the grouped sections picked up on a grid.

HINT: Due to the potential shortage of eyelashes, it is possible to purchase such hairs or to use the fine hair from a dalmatian dog. The inventive microtomist may be able to come up with even more alternatives!

Types of Ultramicrotome Knives

Very early in the development of ultramicrotomy technology, it was determined that metal knives made by meticulously honing standard histology knives or razor blades were too dull to cut ultrathin sections. Such soft and fragile knives rapidly lost their cutting edges when the thin sections required

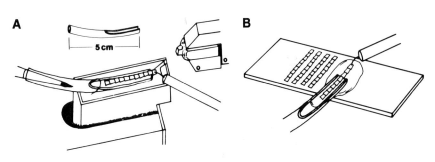

Figure 4-10 (A) Device fashioned from small bore laboratory tubing for picking up long ribbons of sections from the water trough of a diamond knife. (B) Transfer of long ribbons of continuous sections onto a glass slide. After drying onto the slide, the thick sections can be stained and examined in the light microscope. Ultrathin sections may be subsequently cut from the block face. (Courtesy of S. M. Royer and of The Williams and Wilkins Co.)

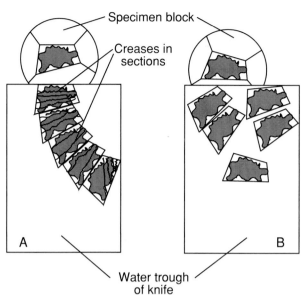

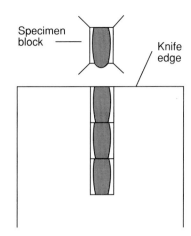

Figure 4-13   Plastic specimen block trimmed to accommodate an elongated specimen.

Figure 4-11   (A) Curved ribbons of sections floating in the knife trough of an ultramicrotome knife. Problems were caused by a lack of parallel sides on top and bottom edges of trimmed block face. (B) It still may be possible to section improperly trimmed blocks by cutting individual sections rather than attempting to allow a ribbon to form.

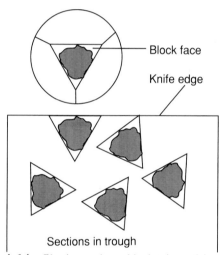

Figure 4-14   Plastic specimen block trimmed in a triangular block face. This shape is used for specimens that may be otherwise difficult to section.

Figure 4-12   Properly trimmed specimen blocks give rise to continuous ribbons of sections. These photomicrographs were taken through the dissecting microscope of the ultramicrotome and represent the view that the microtomist would hope to see during the sectioning procedure.

for transmission electron microscopy were attempted. In 1950, Latta and Hartman discovered that the edge of broken plate glass could be used to cut satisfactory sections. With minor changes, glass is still used extensively in ultramicrotomy for trimming specimen blocks as well as for cutting ultrathin sections. However, certain types of hard specimens such as bone, plants, and thick-walled spores are

difficult to cut even with a good glass knife because the edge dulls too quickly. Fernandez-Moran first indicated that a gem-quality diamond could be used to fabricate a knife for use in the microtomy process. Diamond was a logical material to use, since it is the hardest material known (Table 4-1).

More recently, possibly to take advantage of the relative hardness of sapphire as well as its low cost compared to diamond, the *sapphire microtome knife* was introduced. Such knives cost considerably less than gem-grade diamond knives and are able to cut ultrathin sections of a wide variety of materials. However, since they are rather new to the ultramicrotomy field, their durability and life expectancy are not known. Currently, such knives are approximately one-third the cost of gem-diamond knives, which may average $800 to $1,000 per mm of cut-

Table 4-1   MOHS Hardness Scale

Relative Hardness	Numerical Value	Example of Substance
Soft	1.	Talc
	2.	Gypsum
	3.	Woods Metal
	4.	Soft Iron
	5.	Hard Iron, Tooth, Bone, Soft Glass
	6.	Feldspar, Hard Glass
	7.	Quartz
	8.	Topaz
	9.	Sapphire
Hard	10.	Diamond

Source: *Handbook of Chemistry and Physics*, 59th ed. West Palm Beach, Fl: (CRC Press, 1978–1979) eds. R. C. Weast and M. J. Astle, F-24.

ting edge. They produce ultrathin sections comparable to those produced by good quality glass knives, but probably not as high quality as gem diamonds.

It is also possible to use industrial diamonds for ultramicrotome knives. However, industrial diamonds, in contrast to the near flawless crystal of the gem diamond, contain more imperfections in the crystalline lattice, which result in microscopic flaws in the cutting edge. Although such diamonds are suitable for cutting 0.5 to 1 μm survey sections, they are not recommended for high-quality ultrathin sectioning purposes because knife marks visible only in the electron microscope will result. Industrial-diamond knives are frequently referred to as "histo" knives because the thick sections are more suitable for histological studies conducted at the light microscope level. If the researcher intends to cut orientation or semithick sections on a routine basis, the histo and sapphire knives are recommended because both knives are superior to glass and come with a large water trough attached.

CAUTION: Although gem diamonds and sapphires are very hard materials, they may be damaged if sections thicker than 1 μm are cut. This is especially true if the specimen itself is hard. Always follow the recommendations of the manufacturer in this matter.

Glass Knives

It has long been theorized that, because glass is a supercooled liquid, such knives should be prepared the same day they are used since their sharpness may be diminished by slight molecular flow at the edge. It is now generally accepted that glass knives can be prepared several days in advance without any noticeable effect on their sharpness as long as they are kept in a dry, dust-free environment such as a glass-enclosed desiccator.

Originally, microtomists would purchase 1/4" plate glass from a local glazier and, using a common scoring wheel glass cutter, score and break the glass into 1" wide strips using specially modified glazier's pliers (Figure 4-15). It is still possible to produce high-quality glass knives with such tools, but this technique is both time consuming and difficult for beginning microtomists to master. As a result, several devices are available for routinely producing high-quality knives. The most common instrument is the *Leica Knifemaker* (Figure 4-16). The manufacturer also supplies high-quality glass strips for use in the instrument, thereby eliminating another variable.

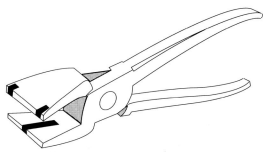

Figure 4-15   Modified glazier's pliers for breaking plate glass. Black areas in jaw are strips of electrical tape used to create raised points to press against the glass that is placed into the pliers. (Courtesy of Leica.)

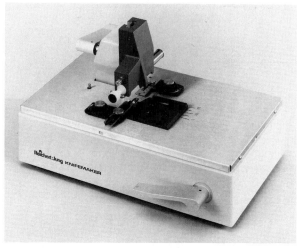

Figure 4-16   A commonly used knifemaker for precisely breaking glass strips into ultramicrotome knives. (Courtesy of Leica.)

During the manufacturing process, glass can be formed into plates either by floating the molten glass on molten tin to produce a very flat, uniform surface or by passing the molten glass through a series of rollers. Leica glass is currently produced by the latter procedure so that the stresses are oriented along the rolling direction. As a consequence, the commercially produced glass strips are scored parallel to these stresses. The unpolished glass is quite uniform in thickness, very hard, and has consistent breaking properties. Such glass may be used in the knifemaker or it may be broken into knives using glazier's tools.

Preparing Glass Knives with a Commercial Knifemaker

The Leica Knifemaker is a precision instrument for securing, scoring, and breaking 1″ glass strips first into squares and then into diagonal knives. If one follows the manufacturer's directions and makes all critical adjustments, this instrument will permit microtomists to produce good quality knives on a regular basis. Problems usually arise when inexperienced individuals attempt to readjust the settings on the knifemaker or mishandle the glass during the cleaning or alignment process. Glass knives are made on this instrument as follows:

1. Wash the glass strips with a standard dishwashing detergent and a lint-free cotton cloth. The strips will be extremely slippery and sharp, so care is in order. Should the glass slip from one's grip, do not attempt to catch it but step away and allow it to fall.
2. Rinse the glass with tap water and dry the strip by wiping gently with paper towels or allow it to dry in a dust-free environment. Again, take care not to cut oneself as the towel is passed over the sharp corners. The glass strips can be wrapped individually in dry paper towels and taped shut for long-term storage.
3. Using cotton or nylon gloves to prevent contaminating the glass, pass a proper length of glass strip into the knifemaker with the factory score line down. Score and break the strip into squares.
4. Place the squares in the machine with the glass oriented as shown in Figure 4-17, and score a diagonal following the manufacturer's directions. Be sure to engage the rubber damping cushion, which absorbs the shock of the break and greatly extends the useful length of the cutting edge. Slowly rotate the breaking knob or lever until a solid-sounding break is heard.

    In addition, it has been demonstrated that placing 1 to 2 drops of distilled water along the diagonal score line of the glass just prior to breaking greatly improves the quality and length of the cutting edge (Slabe, Rasmussen, and Tandler, 1990).
5. Remove the two halves, evaluate them carefully under a stereo microscope, and mount a water trough on the knife as described in "Evaluation of Glass Knives."

Manually Crafted Knives

If one does not own a knifemaker, it is possible to make knives from the commercial glass strips by scoring the strips with a cutting wheel purchased from a hardware store. Glass strips are quite economical and will result in much higher quality knives than can be obtained using glass purchased from a glazier. One principle to follow is the "balanced break" concept in which one applies equal pressure on each side of the score by breaking the glass strip into equal halves in the following manner.

Breaking Glass Knives by Hand

1–2. Same as in section dealing with knifemaker.
3. While wearing cotton or nylon gloves, place the glass strip with the factory score line down on a piece of graph paper ruled in 1″ divisions. Score the strip into 1″ squares using the hand scoring wheel and a ruler. The score should not extend to the very edges, but should stop within 0.5 to 1 mm. A diamond scorer should not be used since it often gouges out rather than scores a uniform line on the glass.
4. Using modified glazier's pliers, place the single raised point of the jaws under the score that falls in the middle of the glass strip. Increase pressure on the pliers until the glass breaks in half (Figure 4-18). Continue breaking the remaining portions of the glass strip into equivalent halves until all strips are broken into 1″ squares.
5. Examine one of the squares and orient it so that the factory score marks are down. It is easy to tell the factory scored side since this ¼″ face will be quite flat compared to the faces broken by hand. Make a diagonal score running to within 1 mm of the corner adjacent to the one good edge (Figure 4-19). Although the ideal knife score should run from corner to corner to give a true 45 degree angle, this is not possible using the hand-scoring method. Instead, the diagonal should be directed slightly toward the good edge indicated in the figure.
6. Grasp the square with the glazier's pliers so that the raised point in the jaw is under the diagonal score. The tip of the pliers should form a right angle relative to the score line. Increase pressure until the glass is broken in two (Figure 4-19). The knives can be stored in a dust-free area until ready for use.

Evaluation of Glass Knives

Carefully examine the two knives produced by bisecting the glass square. Each half should be nearly identical when placed side by side. Pay particular attention to the heel or shelf at the base of each knife (see Figure 4-23B). Knives with good edges usually exhibit very shallow heels, typically between 0.5 amd 1 mm. Generally, one of the knives will be

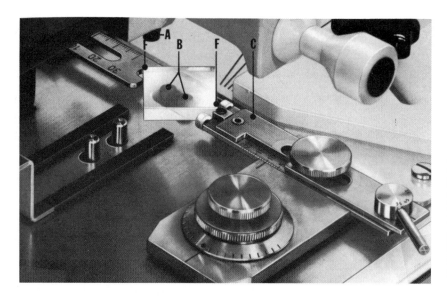

Figure 4-17 Glass squares being placed into knifemaker for diagonal scoring. Two studs (A) in clamping head firmly press down against studs (B) in base to effect the break. Sliding guides or holders (C) hold the square in place so that the score can be made between the two F points. (Courtesy of Leica.)

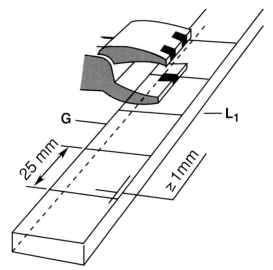

Figure 4-18 Breaking of glass strip into 25 mm squares using glazier's pliers. G indicates the smooth, nonscored portion of the strip where the good part of the knife edge will begin. L1 indicates the serrated edge generated by factory scoring. (Courtesy of Leica.)

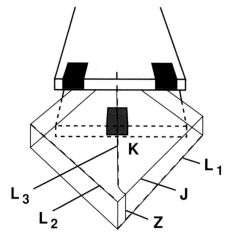

Figure 4-19 Breaking diagonally scored 25 mm glass strips with a glazier's pliers. J is smooth edge of the glass strip. Best part of the knife edge (Z) will begin at intersection with J. L1 is factory score mark. L2 is score used to break 25 mm squares. L3 is diagonal score used to make a good knife (K). The half of the glass knife on the left is termed the counter piece and may also be usable for sectioning purposes. Notice how L3 does not extend to the corner and how the break veers away from the corner. This effectively enlarges the final angle of the knife edge over the angle that was scored onto the glass strip. (Courtesy of Leica.)

of high quality for ultramicrotomy: the one that broke onto the smooth, good edge of the commercial glass strip. The other knife, or *counter piece* as it is sometimes called, will have a much shorter usable cutting edge and can be used for rough trimming or for cutting thick sections. Both knives should be evaluated as follows.

Secure the knife in the microtome or under a stereo microscope with a movable focussed light source. Most modern microtomes provide excellent sources of lighting for evaluation of the knife edge. Move the light source and knife to obtain a very narrow, bright strip of reflected light running along

the very edge of the knife. Some patience is required since the goal is to obtain a thin sliver of light along as much of the edge as possible. Imperfections, nicks, or discontinuities in the edge will show up as dark specks, whiskers, or rough and serrated areas. As you look down on the edge, note the stress line that usually starts in the *left-hand* corner and arcs down toward the heel of the knife (Figures 4-20 and 4-21). Generally, the best part of the knife begins a short distance from the left corner and may

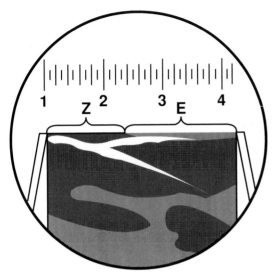

Figure 4-20 Drawing representing top view of glass knife showing stress line curving down towards base of knife. The stress line begins in upper left and marks the point of intersection of J and Z referred to in Figure 4-19. Z is good part of knife edge. E is part of edge containing imperfections or nicks. (Courtesy of Leica.)

Figure 4-22 Plastic box for storing glass knives prior to use. The cover is shown in the raised position with eight glass knives inside of the box.

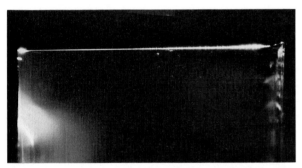

Figure 4-21 View of cutting edge of glass knife showing imperfections in the right half of the cutting edge.

Figure 4-23(A) (Left) One method of attaching a tape trough to a glass knife involves sealing the bottom using either wax or fingernail polish (arrow). (Right) A second type of commercially available plastic or metal trough that has been attached to the glass knife using wax or nail polish for a watertight seal (arrow).

extend from $1/3$ to $1/2$ way across the knife. Only careful examination at the highest magnifications will reveal these desirable areas. Interestingly, the actual angle of glass knives is usually greater by 10 or more degrees than the angle scored (45 degrees) since the fracture of the knife does not follow the score line as it exits the glass square.

A number of knives can be made and stored for several days, but take care not to contact the fragile edge with anything, including fingers. If grease from fingers contacts the area of the water trough, it will be contaminated and it may leak due to a poor seal. Special boxes can be constructed or purchased to store the knives in a dry and dust-free environment (Figure 4-22).

Attachment of Water Trough or Boat

After making the knife and prior to cutting either thick or ultrathin sections, a trough or boat to contain the water used for floating sections is attached to the knife by a variety of means. The most commonly used boat can be fashioned from electrician's, masking, or silver audiovisual slide tape by wrapping a small length around the sloped part of the knife and using paraffin or nail polish to secure a watertight seal at the base (Figure 4-23). These knives should dry for at least 30 minutes prior to use or solvent may dissolve into the flotation fluid and create contaminants that adhere to the sections. Since some of the adhesives and solvents used in formulating the tapes may contaminate the trough, electron microscope suppliers or experienced microtomists should be consulted about the best brands of tapes to use.

More elaborate plastic or metal troughs can be purchased from microtome or knifemaker manufacturers (Figure 4-23A). Such troughs fit the glass tightly to make a good seal and can be rapidly attached to the glass. The major advantage, however, is the rather large area provided for maneuvering various tools to collect many sections. Although they are considered disposable, troughs can be cleaned of wax in a solvent such as xylene and reused many times.

CAUTION: Use xylene in a fume hood since it is toxic and flammable.

A trough for containing only one or two sections can be rapidly prepared by placing a droplet of molten paraffin 4 to 5 mm down from the knife edge. The hydrophobic properties of the cooled wax will repel a water droplet placed above it and provide a small area for floating the sections. Although it is not recommended for quality work, since sections are difficult to retrieve, it is adequate for quickly cutting a survey section or two.

Diamond and Sapphire Knives

Diamond knives are delicate instruments usually costing several thousand dollars each. Consequently, only experienced microtomists are entrusted with their use. The natural gemstones used are usually

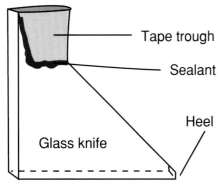

Figure 4-23(B) Drawing of a glass knife showing trough, trough sealant and heel or shelf at base of knife. The best glass knives have a very shallow heel, less than 1 mm in height.

pale yellow, of regular crystal structure, and of the greatest possible purity. Initially, the large stones are cleaved into smaller segments, which are then ground into small slabs on a turntable using diamond dust. The orientation of the crystalline lattice is maintained to within 2 degrees of a predetermined value during the grinding. The grinding process often reduces the stones by 50% of their original weight. The diamond is mounted into a soft metal shaft (Woods metal) and final polishing to a sharp edge is conducted under conditions that manufacturers consider proprietary. The shaft containing the final edge is then mounted in a metal trough or boat and cemented, usually with an epoxy plastic. Typical diamond knives are shown in Figure 4-24.

The final angle of the knives may vary from 40 to 60 degrees. Generally, the smaller-angled knives are used for sectioning softer biological specimens and are capable of cutting thinner sections since the knives are sharper. The larger-angled knives, on the other hand, have sturdier edges and are more suitable for cutting harder specimens such as bone, tooth, or even metals. Such knives are not as sharp and generally will not cut the very thinnest of sections. A reasonable compromise is a 45 to 50 degree angled knife, which has characteristics of both. Such knives are sufficiently sturdy to last many years and are capable of cutting sections on the order of 50 to 60 nm. Diamond knives will cut a greater variety of specimens and can section specimens with varying hardness and containing hard inclusions such as crystals or asbestos fibers. The very best that a diamond knife can achieve is to cut extremely thin, 30 to 50 nm, sections from small block faces and 100 nm sections from block faces that are several millimeters wide.

The lifetime of a diamond knife may vary from a single use to several decades depending on the hardness of the specimen as well as how carefully one handles the instrument. The cutting edge is extremely thin, probably only several molecular layers, and susceptible to damage unless one approaches the cutting edge face on. Any forward or backward pressure, even a touch with a fingertip or filter paper, will damage the edge, so great care must be exercised in cleaning the knife.

REMEMBER: Sections thicker than 1 μm should never be cut using a gem-grade diamond or sapphire knife since the cutting edge may be damaged.

Sapphire knives are constructed from synthetic sapphires in lengths of 4 to 6 mm. Most are supplied

Figure 4-24 (A) Gem grade diamond (arrow) mounted in Woods metal holder and polished to give a 3 mm long cutting edge. This unit is mounted into the large metal trough shown in B and C. (B) Photograph of a diamond knife used in ultramicrotomy. Arrow indicates cutting edge. (C) Photograph of a "histo" diamond knife (left) used for cutting thick plastic sections in the range of 1 μm. Compare the size of the cutting edge and trough to the diamond knife used for ultramicrotomy (right). Industrial grade diamonds are used in the histo knife, whereas gem grade diamonds are used in the standard ultramicrotome knife.

with 45 degree angles and are mounted in an aluminum boat similar to a diamond knife. Like diamond knives, sapphire knives are very convenient since they are always available for sectioning and are similar in performance. They should be considered for use in place of glass knives, possibly as a transition instrument for individuals moving up from glass to diamond knives. They are much more durable than glass and will produce long ribbons of sections; however, they are not recommended for cutting hard specimens such as bone, tooth, or metals. Like those of diamond knives, details of the manufacturing process are closely guarded.

Cleaning Diamond Knives

After sectioning with either diamond or sapphire knives, any adherent sections are removed by wiping with an eyelash probe and rinsing the edge with distilled water from a squirt bottle. Debris lodged on the edge itself can be removed by gently passing a *cleaning stick* soaked in water, a sharpened shaft of Styrofoam, or pithwood (usually supplied with the knife) along the cutting edge (Figure 4-25). Using a toothpick or orangewood stick may harm the edge, so check with the knife manufacturer for the proper types of cleaning sticks to use with a specific knife. No cleaning solutions (acids, bases, harsh detergents, or strong organic solvents such as acetone) should be used, and strict adherence to the manufacturer's recommendations is urged.

Occasionally, the cutting edge becomes difficult to wet with the trough liquid. Overnight submersion of the knife in 1% Tween 20 or mild detergent in distilled water usually solves this problem. If the edge is still difficult to wet, passing a cleaning stick over the edge may help.

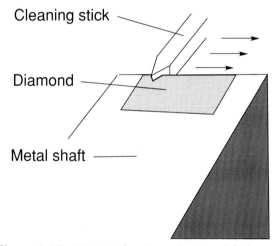

Figure 4-25 Method of passing a cleaning stick along the cutting edge of a diamond knife. Some manufacturers supply the styrofoam rods that can be used.

Wetting the cleaning stick with 50% ethanol may help to remove any stubborn debris. Usually, problems of edge wetability can be traced to allowing debris or sections to dry onto the edge. Therefore, should one leave the knife in the microtome for any length of time, it is advisable to overfill the boat and not allow the level to go below the knife edge. Once the edge has been cleaned, the knife is put back into its special holder while still wet. It is dangerous to use canned or compressed gas to dry the edge since small particles or even organic contaminants such as oil may be impacted onto the edge of the knife.

Diamond knives eventually dull through extended use; however, they can be resharpened usually at 50% to 70% of the cost of a new knife. Some manufacturers guarantee a certain number of re-sharpenings before replacement is necessary. If damage to the edge is too extensive, perhaps due to accidental contact, it may not be possible to resharpen the knife and replacement will be needed. To maintain the integrity of the knife edge, it is highly recommended that one knife be assigned to one person whenever possible, rather than several investigators sharing the knife. Sapphire knives are not resharpened, but replaced when the edges are dulled.

Histo Knives

Histo diamond knives are used to cut sections in the 0.2 to 2 μm thickness range from block faces larger than 1 mm. Histo knives are constructed from industrial diamonds and manufactured by a process different than that used for gem-grade knives. They cost approximately one third that of the higher quality knives. Such 5 to 6 mm knives, while not able to cut ultrathin sections, are cared for in exactly the same way as knives made from natural diamonds.

A special type of glass knife, approximately twice as wide as the standard 6.4 mm knife, can be constructed using a thicker grade of glass available from various EM suppliers and the Leica company. When such knives are manufactured as described previously, the cutting edge is curved, making it difficult to cut large block faces. Fortunately, it is possible to make *RALPH* knives (named after the late Dr. Paul Ralph) with theoretically unlimited lengths and very straight edges. With the RALPH knife, the cutting edge runs along the broad, 1″ flat side of the glass strip rather than along the narrow 6.4 mm thick side (Figure 4-26). The RALPH knife is used most often for cutting paraffin or softer acrylic and epoxy plastic embedded specimens for light microscopy. It will last probably as long as a metal knife. In cutting harder epoxy plastics, however, the knife usually dulls after 20 to 50 cuts. This type of knife is finding frequent use in light microscopy since the edge is much sharper and more durable than the standard metal knife. As with standard glass knives, the RALPH knives can be made by hand or with specially designed instruments. The very best that a RALPH knife can achieve is to cut 10 × 10 mm block faces in the 0.2 to 1.0 μm range.

Trough Fluids

The troughs of all the knives described must be filled with a fluid in order to support the sections as they are cut. Double-distilled water is the most commonly used type of fluid onto which the sections are floated. Occasionally, if the knife edge is difficult to wet using plain distilled water, one can add a drop of Tween 20 or PhotoFlo (Kodak) detergent to each 100 ml of distilled water in order to improve the wetting properties. If sectioning is in progress when wetability problems are encountered, and one does not wish to replace the fluid in the knife, one may dip a dissecting needle or pin into the detergent and touch the tip of the needle to the trough fluid to transfer the proper amount of detergent to the trough. Some researchers may use a trough fluid composed of 1% to 10% ethanol or acetone in distilled water to improve the wetability properties; however, concentrations any higher than 10% are to be avoided since they may damage the epoxy cements used to seal certain diamond knife shanks into the aluminum trough. It is necessary to micropore filter all floatation fluids in order to remove bacteria or other particulates that may be present in distilled water that has been standing for several days.

HINT: Micropore filters should be changed every several weeks since bacteria will grow on and decompose the filter and give rise to filter debris that will contaminate the sections. In addition, one should pass through (and discard) the first 10 to 15 ml of water from the filter to remove any humectants used in the manufacture of the filters.

Grids and Specimen Supports

Grids

A specimen grid is the electron microscope analog of the glass slide used in light microscopy. As the name implies, grids are screens or fine, mesh sup-

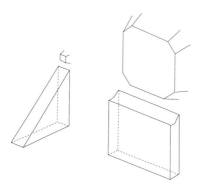

Figure 4-26 Comparison of RALPH knife (right) to standard glass knife. Note relative sizes of specimen blocks.

ports upon which sections and liquid suspensions may be deposited for transport and viewing in the electron microscope. The standard size grid is 3.05 mm in diameter. Most grids have one side that is more brilliant than the other, so microscopists often refer to putting specimens on the dull side of the grid.

Grids were originally made by weaving fine copper wires into a gauze that was then flattened by rollers to remove the undulating surfaces inherent in the weaving process. Discs were then produced with a device similar to a paper punch. Although it is still possible to purchase woven grids, the surfaces are too irregular for quality work. At the present time, most grids are manufactured using an electrolytic process in which metals are deposited onto templates to produce an extremely fine surface free of undulations, burrs, and irregular open spaces.

A wide variety of shapes and configurations is currently available (Figure 4-27) using metals such as gold, platinum, nickel, stainless steel, or rhodium. In fact, just about any nonmagnetic metal can be used to manufacture grids. Nonmagnetic metals are specified since magnetizable metals (such as iron) would interfere with the images formed by the electromagnetic lenses in the microscope. Stainless steel and nickel sometimes exhibit traces of magnetism so that the grids must be demagnetized prior to use (see Chapter 5, Figure 5-6).

Metals are normally used in the fabrication of grids because they will conduct away heat and electrostatic charges in the specimen resulting from bombardment by the electron beam. *Copper* is still considered standard for most purposes. However, the more inert metals may be needed if sections are to be subjected to strong oxidants or acids used in certain cytochemical procedures. If analytical studies are to be conducted on the specimen, then grids constructed of carbon or nylon may be used in place of metals that might interfere with the analysis.

The most popular mesh size is 200 (which means 200 bars per inch), but mesh sizes ranging from 50 to 1,000 are available. Larger meshes provide more unobstructed viewing areas, while the smaller mesh sizes provide better support when viewing extremely thin sections or negatively stained specimens.

In applications requiring a totally unobstructed view of the sections, *single hole* or *slot grids* may be used (Figure 4-28). In these cases, support for the specimen must be provided by a thin membrane of plastic or carbon that bridges the hole or slot. *Support films*, as they are called, are relatively electron transparent.

All grids should be cleaned prior to use by either sonication or swirling in acetone followed by ethanol. After allowing the grids to settle to the bottom of a small beaker and decanting the liquid, the beaker is inverted over a filter paper and the inverted beaker (with the grids stuck to its bottom) and filter paper are placed into a 60 degree oven until dry. The cleaned grids usually fall onto the filter paper when the solvent has evaporated. The grids are separated and stored under dust-free conditions.

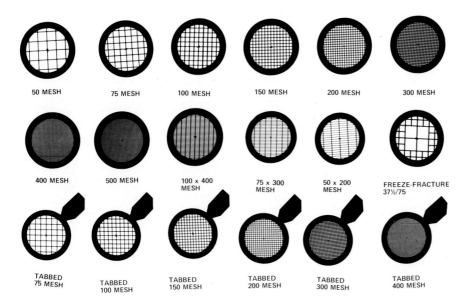

50 MESH   75 MESH   100 MESH   150 MESH   200 MESH   300 MESH

400 MESH   500 MESH   100 x 400 MESH   75 x 300 MESH   50 x 200 MESH   FREEZE-FRACTURE 37½/75

TABBED 75 MESH   TABBED 100 MESH   TABBED 150 MESH   TABBED 200 MESH   TABBED 300 MESH   TABBED 400 MESH

Figure 4-27 Specimen grids are available in a variety of mesh designs. (Courtesy of Ted Pella, Inc.)

Figure 4-28   Single hole and slot grids are used whenever large, unobstructed views of specimens are desired. The large slot is 1.5 mm wide. In addition, "finder-grids" may be used which have coordinates along the sides of the gridwork that are useful for relocating areas that were examined previously.

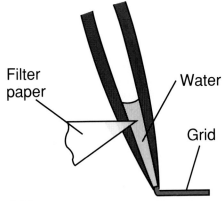

Figure 4-29   Water can be removed from between points of forceps by using a piece of filter paper. For actual photograph, see Figure 5-5.

Handling Grids

Grids are extremely fragile, thin, and difficult to handle with anything but fine-pointed jeweler's forceps. Picking up grids from a very flat surface such as glass or plastic may present a problem, so most grids are placed onto filter or lens papers and retrieved from these surfaces. Often, it is easier to pick up sections if the grid is bent slightly at the very edge. This is done by grasping the grid at the rim and raising the forceps as the grid is kept flat against a hard flat surface such as a glass slide or plastic petri dish. Usually the bend forms a 30 to 40 degree angle. Locking forceps can be made by slipping a small O-ring over the handles of the forceps. By slipping the ring up and down the handle, the points may be opened or locked in the closed position. Grids that have been coated with a plastic film must be very carefully bent to avoid breaking or tearing the film. Pure carbon films are extremely fragile and will probably break if one attempts to bend standard grids containing these types of films. Instead, special grids with tabs or handles (see Figure 4-27) should be used with these support films so that the bend takes place at the handle rather than at the grid edge. Prior to viewing the specimen in the TEM, the tabs are removed with a razor blade or a special detabber or fingernail clipper. Carbon films are best prepared on grids made of nickel since they are more rigid and less likely to be damaged in handling.

When grids are wetted during various procedures, care must be taken to remove the fluid remaining between the points of the forceps before releasing the grid onto filter paper. Otherwise, the specimen grid will be drawn in between the points, often damaging the specimen. Before releasing the grid, wick the fluid from between the points using a small wedge of filter paper (Figure 4-29). Special locking forceps are commercially available, as are anticapillarity forceps that prevent the trapping of water

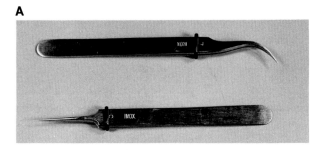

Figure 4-30   (A) Examples of two different types of fine-pointed forceps used to handle electron microscope grids. A grid may be picked up and secured in the forceps by sliding a small rubber O-ring down the shaft towards the points. (B) Closeup photograph of specially designed forceps that lock upon release of finger pressure, thereby securing the grid. One arm of the forceps has a slight bend at the tip to prevent the accumulation of water. These are therefore termed "anti-capillarity forceps."

between the forceps points. Several different types of forceps are shown in Figure 4-30.

Grids may be stored in petri dishes, specimen side *up*, with filter paper liners for short periods of time. Glass dishes are preferable to plastic, since static electricity may cause the grids to jump onto the lid resulting in a loss of orientation. For long-term, safer storage, special grid storage boxes are recommended. Such boxes consist of 20 to 100 numbered slots so that the precise location of all specimens can be recorded until the grids are needed for viewing. Several types of grid storage boxes are shown in Figure 4-31.

Support Films for Grids

Whenever possible, grids without any supporting films should be used. Most plastic sections will withstand the rigors of handling and viewing in the electron microscope without damage; however, very thin sections as well as some of the acrylic plastics are unstable and prone to damage unless some additional support is provided. Usually 20 to 40 nm thin films of plastic and/or carbon are placed onto the grids to provide this extra support. An ideal supporting film should be thin and electron transparent as well as strong.

Plastic Films

Collodion. This plastic is also known as celloidin, pyroxylin, or cellulose nitrate. Collodion is one of the first plastics developed and was originally used as the clear base for motion picture films. Composed of cellulose nitrate, made by dissolving cellulose in nitric acid, collodion is extremely flammable when stored as dry chips or wools. It is used for making support films since it provides adequate support and is relatively easy to make. Collodion is not as stable under the electron beam as are Butvar and Formvar plastics, so it is necessary to use thick 40 to 50 nm films for equivalent support. Although one can purchase the material as solid plastic chips, it is more convenient (and safer) to purchase purified solutions from commercial suppliers.

Preparing Collodion-Coated Grids

1. Make or obtain a 2 to 3% solution of collodion in n-amyl acetate. This should be stored in a dark bottle (ultraviolet light breaks down and weakens collodion) and the bottle placed in a desiccator such as a resealable plastic bag with silica gel to prevent the absorption of moisture.
2. Construct a platform out of stainless steel screen and place it on the bottom of a clean glass petri dish or Buchner funnel that has been equipped with a clamped piece of tubing. The platform should be about 2″ × 3″ and be raised by about ¼″ to ½″ (Figure 4-32).
3. Pour clean distilled water into the container to cover the platform by at least ½″. It is important that the water be free of lint and microorganisms.
4. Place the grids, dull side up, on the top of the platform. The submerged grids must not overlap, but should be placed close together.
5. With a clean, dry, glass Pasteur pipette, take up a small volume of the collodion solution. With the pipette held 1 cm above the water surface, release one drop of the solution onto the water surface. The solution should rapidly spread out to form a uniformly silver layer of plastic after the solvent has evaporated. This first film, which is used to clean the water surface of lint, is removed and discarded using a forceps.
6. A second film is made in the same manner and allowed to dry for 1 to 2 minutes. Examine the film using a fluorescent light held directly over it. One should see an intact, continuous silvery sheet free of lint, tears, and irregular patches of color. Do not proceed until a film that fulfills these criteria is obtained.
7. Lower the water level by draining the Buchner funnel or by aspirating water from the petri dish. Position the collodion membrane over the grids using the tip of the fine forceps as the water is lowered. When the water has been totally drained, remove the platform

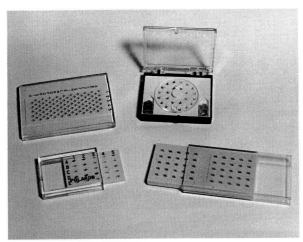

Figure 4-31 Different types of commercially available storage boxes for keeping track of specimen grids.

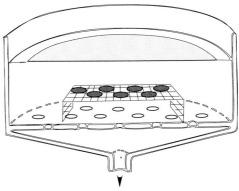

Figure 4-32 Buchner funnel containing platform onto which grids will be placed for coating. The funnel is filled with distilled water and then the platform and grids are arranged under the water prior to floating the appropriate film on the water surface. (Courtesy of L.W. Coggins and IRL Press Limited.)

and place it in a 40° to 50° C dust-free oven for drying. (An alternate method is to lift the screen containing the grids up under the collodion film using a forceps.)

Individual, coated grids may be removed and stored until needed, or a light coating of carbon may be evaporated onto the films to further stabilize them (see Chapter 5). It is important to evaluate the films in the TEM prior to use to ascertain that they are sturdy and free of holes. Films that break or fall apart in the microscope indicate that the collodion content must be increased, while holes indicate the presence of water in the solution. If water is present, the solution should be discarded. The correct thickness of film, indicated by a silver/gold interference color, is important since an overly thick film will degrade both contrast and resolution, while a thin film will drift or break in the microscope. As is the case with all plastic films, the coated grids will be good for several weeks, but should be as freshly prepared as possible for optimum results.

Butvar/Formvar. Butvar (polyvinyl butylate) and Formvar (polyvinyl formate) plastics are much stronger films than collodion and can, therefore, be much thinner (25 to 30 nm) in practice. Unfortunately, since drops of the plastic solution do not readily spread over a water surface, these films cannot be conveniently prepared using the drop technique used with collodion. Instead, a thin layer of the plastic film is stripped from a glass slide onto a water surface.

Preparing Butvar- or Formvar-Coated Grids

1. Clean a standard microscope slide in 95% ethanol by wiping with laboratory tissues. Various brands of slides should be evaluated in this procedure until a suitable one is found.
2. Dip the clean, dry slide into a 0.3 to 0.5% solution of Butvar or Formvar in chloroform (or ethylene dichloride) in a tall glass cylinder. Withdraw and hold the slide in the vapor phase above the liquid for 30 seconds or so. Completely withdraw the slide and allow it to drain vertically on a filter paper. This must be done in a dry, dust-free environment.
3. Place the coated slide on a filter or lens paper and score around the periphery of the slide (3 to 4 mm in from the edge) using a razor blade or fine needle. The object is to break the continuity of the film to facilitate its removal, but not to deeply scratch the glass slide.
4. Slowly lower the slide held at a shallow angle into the dish of water described for making collodion films so that the film is detached from the slide (Figure 4-33).

5. Lower the water level so that the film is deposited on the grids.
   A major problem with this technique is the failure of the plastic film to detach from the glass slide. Numerous methods have been suggested to enhance detachment. Breathing on the coated slide to cause condensation of water vapor immediately before immersing it into the water seems to help, as does precleaning the slides in a mild detergent solution and then wiping the unrinsed slides dry with a lintless cloth just before dipping into the plastic solution. Coating the cleaned, dry slides with the wetting agent Victawet (E. F. Fullam, Inc.) in a vacuum evaporator sometimes helps. To accomplish this, a piece of Victawet about the size of a grain of rice is placed in the wire basket of a vacuum evaporator and slowly heated under the high vacuum. The vaporized Victawet will condense onto the slides. The coated slides are then wiped with a lintless filter paper to remove excess agent just prior to dipping into the plastic solution. It is important to remove the excess Victawet or films perforated with holes will be produced.

An alternate method for coating the slides involves placing the dull side of the grids directly onto the floating plastic film and touching them lightly with the forceps to enhance adherence. The grids are picked up by bringing a clean microscope slide down onto the film and then reversing the slide under water so that the side of the slide with the filmed grids emerges from the water first. This method requires some practice. An easier method involves rolling the floating film and grids up onto a clean glass cylinder or beaker by contacting the cylinder to the edge of the film and rolling the cylinder so that the grids and overlying film go onto it. The plastic films may also be picked up onto Parafilm, filter paper, or stainless steel mesh. After the plastic film has dried, the grids may be carefully removed from the substrate so that the films remain on the grids. It is best to evaluate the filmed grids under a fluorescent light to detect tears, holes, or overlapped areas of plastic film.

Perforated or Holey Films. The *holey film* (Figure 4-34) has many uses in electron microscopy. Not only is it used to support certain types of particulate specimens for high resolution work, but it also makes an excellent standard for checking astigmatism, resolution, and stability of the microscope. These films can be prepared in several different ways.

Preparation of Holey Plastic Films

Bayer and Anderson (1963) described a rather easy method to prepare holey films by dipping a slide into a 0.3% to 0.4% solution of Formvar in ethylene dichloride. Hold the slide in the vapor phase above the liquid for 1

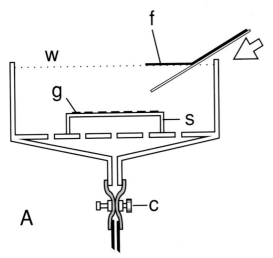

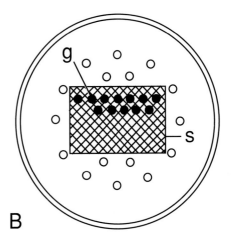

Figure 4-33 (A) Detachment of Forvmar or Butvar film (f) from glass slide and onto surface of water (w) in Buchner funnel. Grids (g) are placed under water onto the wire mesh shelf (s). The water is drained from the apparatus by loosening the clamp (c) so that the film is gently lowered onto the grids on the platform. The platform is removed and al- lowed to dry before removing the coated grids. (B) Top view of Buchner funnel apparatus illustrated in Figure 4-33A, here showing the placement of the grids (g) on the stainless steel screening (s). The open circles represent the drain holes in the support base of the Buchner funnel. (Courtesy of L.W. Coggins and IRL Press Limited.)

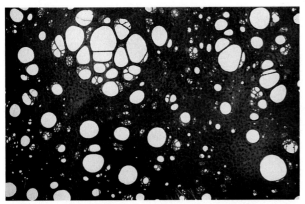

Figure 4-34 Electron micrograph of perforated or holey plastic film that is often used to evaluate the performance of the transmission electron microscope. Such films may also be used to support tiny specimens (crystals, for example) that have been suspended across the holes.

to 2 minutes allowing a very thin layer to form by drainage of the solution to the base of the slide. The slide is rapidly withdrawn from the vapor and immediately breathed upon to condense moisture into the film before it dries in the atmosphere. *IMPORTANT: The film must be thin and the moisture must be condensed into the film before the solvent evaporates. Failure to make satisfactory perforated films is usually due to problems with one or both of these points.* The film is then scored with a razor blade, floated onto a water surface, and grids applied as described in the preceding section.

A second method, described by Harris in 1962, involves suspending a small volume of glycerine in the Formvar solution by vigorous shaking. This generates tiny droplets in the Formvar since the two phases are not miscible. The size of the droplets is determined by the amount of glycerine added. To generate holes of 25, 15, 7 or 5 μm, respectively, 1 ml of glycerine is added to one of the following volumes of the Formvar solution: 8, 16, 32, or 120 ml, respectively.

Dip a Victawet-coated slide into the vigorously shaken Formvar-plus-glycerine suspension immediately after shaking, and withdraw it to drain in the vapor phase of the ethylene dichloride in the tall cylinder. Remove the slide and, after 10 to 15 minutes of drying at room temperature, breathe on the slide (or expose it to the steam rising over boiling water) to help loosen the film and float it onto the water surface. After the grids have been picked up, the dry film may be coated with carbon as described in the sections that follow. The very best perforated films will be made by dissolving away the Formvar backing as described in the section entitled Pure Carbon Films.

Carbon-Coated Plastic Films. It is highly desirable to evaporate carbon (or graphite) onto the plastic films to strengthen them further, since layers of carbon as thin as 2 to 5 nm impart great strength to the plastic. In fact, carbon may be used sometimes to strengthen sections on uncoated grids. Carbon coating may be conducted as follows.

Preparation of Carbon-Coated Plastic Films

1. Place the plastic-coated grids, or grids containing sections, into a vacuum evaporator on a clean glass surface with the plastic film side up and evacuate the chamber to 1×10^{-3} Pa.

2. Slowly heat the carbon rod or braid until it first turns red, and maintain this condition for 10 to 15 seconds to outgas any adsorbed contaminants. Rapidly increase the current to the carbon until evaporation takes place. It is important that the temperature be raised *rapidly* in bursts lasting 5 seconds or so rather than bringing up the current slowly. This will avoid heating and damaging the fragile plastic films. The grids should be placed at least 6″ to 8″ away from the electrodes for the same reason. The thickness of the carbon may be estimated by placing a piece of white porcelain (as from a broken coffee cup) near the grids with a drop of diffusion pump oil on top. When the porcelain turns a very light tan relative to the white area under the oil drop, a thickness of about 4 to 7 nm has been evaporated. Another method for gauging thickness involves placing a white index card with a raised object such as a screw on the card near the specimen. As the carbon is evaporated, the card darkens except in the area shaded by the screw. Hence, one can readily compare uncoated areas to carbonized areas to more readily gauge the thickness. Details of carbon evaporation are given in Chapter 5.

3. After the electrodes have cooled for 3 to 5 min, air is admitted to the vacuum system and the grids are removed.

Pure Carbon Films

Carbon films without any plastic substrates are more desirable supports because they are considerably thinner, less electron dense, and much stronger than plastic alone. Unfortunately, since they are difficult to make, pure carbon films are not used as often as the combination films, except when high stability and resolution are required.

Methods of Preparing Pure Carbon Films

METHOD 1: Place individual Formvar/carbon-coated grids, prepared by the method described previously, onto a stainless steel platform submerged in chloroform. The filmed sides should be facing up. The apparatus is identical to the one described for preparing collodion-coated films (see Figure 4-33) except that it is filled with chloroform rather than water. After several minutes, the Formvar bridging the open spaces on the grid will be dissolved leaving only the carbon. The Formvar sandwiched between the carbon and copper of the grid should remain undissolved and serve as a cement to hold the carbon in place. The length of time (several minutes) for the dissolution of the plastic must be determined by trial and error since too long of an exposure will dissolve all of the Formvar and the carbon will detach. After the chloroform is drained, the stainless steel mesh is placed in a 60 degree oven to permit the chloroform to evaporate. If one uses collodion/carbon grids in this technique, amyl acetate solvent must be used to dissolve the collodion.

DANGER: This procedure must be conducted in a fume hood since chloroform is a potential carcinogen and an anesthetic.

METHOD 2: Another method of preparing carbon supports is coating a Victawet-treated slide (see section on Butvar/Formvar film making) with a layer of carbon in a vacuum evaporator. Instead of using a glass slide, it is also possible to make a substrate for the carbon by delaminating a 1 × 3 inch mica sheet into two thin layers by slipping a clean, single-edged razor blade between the layers and prying them apart. This should be done in a fume hood since mica will shower thousands of fine particles into the air during this process. *CAUTION: Do not breathe the mica particles; they are potentially dangerous.* Transfer the slide or freshly cleaved mica sheet into a vacuum evaporator and coat with a light tan coating of carbon.

Prepare adhesive-coated grids by first placing clean grids (dull side up) on a filter paper and then placing one drop of "grid glue" on top of each of the grids. *Grid glue* is prepared by dissolving the adhesive from 2″ of transparent tape (Scotch Brand) in 10 ml of ethylene dichloride, discarding the tape, and keeping the solution.

Place the adhesive-coated grids, dull side up, on top of the submerged stainless steel rack described in Preparing Collodion-Coated Grids. Slowly lower the carbon-coated slide or mica sheet under the water so that the carbon layer will float free on the water surface. Lower the water level until the carbon comes to rest on top of the grids. Dry the rack in a 60 degree oven overnight.

If one is unsuccessful with this latter method, score the carbonized slide or mica sheet into 3 × 3 mm squares using a clean razor blade, and float off the tiny carbon squares onto a water surface. Lift up each square individually onto an adhesive-coated grid. Blot the excess water by touching the edge of the grid to a filter paper and placing the grids in an oven to dry.

The Ultramicrotome and the Sectioning Process

Development of the Ultramicrotome

The development of the modern ultramicrotome was the major breakthrough that made possible the routine sectioning of tissues for subsequent viewing of cellular ultrastructure in the transmission electron microscope. Prior to the advent of such instruments, only naturally thin specimens such as viruses, bacteria, and edges of whole cells could be examined. Internal cytological details were only surmised based on light microscopy combined with cellular fractionation and biochemical analysis of the fractions. *The ultramicrotome was one of the most important ancillary instruments developed in the field of biological transmission electron microscopy.*

As early as 1939, Von Ardenne attempted to cut sections using a modified histological microtome that cut wedge-shaped sections using metal knives. The outermost edges of the wedge were examined hoping to find sufficiently thin areas. Several years later, improvements by O'Brien and McKinley as well as by Fullam and Gessler led to the development of high-speed or "cyclone knife" microtomes in which the specimen was clamped in a fixed holder and the metal knife spun at 12,500 rpm by an electric motor. As the specimen was advanced into the spinning blade, by means of a fine screw, 0.1 μm or thicker sections were cut and rapidly dispersed into the air in all directions, making collection of sections rather haphazard. Investigators tried even more rapidly spinning blades until it became clear that such high speeds actually hindered their efforts.

A significant advance in the development of the modern ultramicrotome appeared in 1953, when Keith Porter and J. Blum reported the development of two different microtomes and sectioning technologies, the principles of which are used even today. The first was a *thermal advance* ultramicrotome. The plastic-embedded specimen was mounted on the end of a horizontal metal bar and passed over a glass knife. On the return stroke, the specimen avoided bumping the back of the knife by passing to one side, or following a D-shaped path. Thermal advancement of the specimen was achieved by heating the metal bar in which the specimen was mounted with a goose-neck reading lamp. Sjostrand subsequently reported a refinement of this ultramicrotome, which underwent extensive refinement and gave rise to the LKB (now Leica) instrument line of thermal advancement microtomes. In the most modern versions, the specimen arm is advanced by a heating coil while being passed over the knife. The specimen avoids contact with the back of the knife during the vertical return stroke since the knife is retracted 15 μm by a powerful electromagnet. When the specimen reaches the top of its return, the magnet is shut off so that the knife can spring back into position as the specimen is allowed to drop by gravity. The speed of the fall of the specimen arm is controlled by a coil motor so that rates of 0.1 to several mm per second may be achieved. A diagram of the basic features of the microtome are shown in Figure 4-35A.

The second "improved" instrument, the *Porter-Blum MT1*, used a *mechanical advancement* mechanism in which a screw thread advanced a pivot arm that acted as a fulcrum or lever to press the specimen arm forward (Figure 4-35B, C). Since the forward motion of the screw was reduced 1:200 times by the fulcrum, the specimen could be advanced in increments as low as 25 nm. In 1962, the *Sorvall MT2* instrument was introduced, which featured a motorized or hand-driven specimen arm, more precise control of section thickness, and retraction of the specimen arm on the return stroke in order to avoid the back of the knife. Several models later, and following a major redesign of the instrument, an improved microprocessor-controlled version, the *MT-7*, was manufactured by RMC, Inc. (Figure 4-36).

Basic Features of All Ultramicrotomes

Although certain designs and features may vary, all modern ultramicrotomes consist of several basic components that permit: (a) fine advancement of the specimen by mechanical or thermal means; (b) precise orientation of the specimen and knife; (c) coarse mechanical advancement of the knife for approaching specimen and thick sectioning; (d) control of cutting speed; (e) knife avoidance on return stroke; and (f) magnification and illumination of the sectioning process.

The Sectioning Process

The cutting of ultrathin sections is a demanding technique for beginning microtomists. It is important to become thoroughly familiar with the features and details of operation of the particular ultramicrotome being used before attempting to cut sections. Patience, concentration, and a certain degree of manual dexterity are required of all microtomists. It is essential to have adequately embedded and trimmed specimens as well as high quality knives and specimen support grids. The environment surrounding the microtome, as well as all tools and reagents, must be immaculate. Detailed records of the protocols followed must be kept in order to repeat the experiment. One would proceed in the following manner.

General Steps in Ultramicrotomy

Once a specimen block has been trimmed using one of the methods described in the previous sections, the chuck bearing the block is securely clamped into the arm of the ultramicrotome and the following adjustments made.

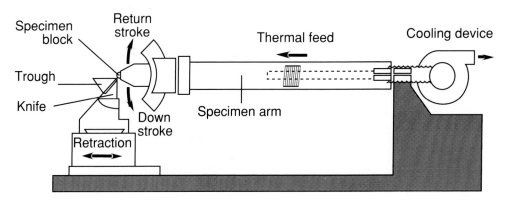

Figure 4-35(A) Diagram showing thermal advancement mechanism used in the LKB (Leica) ultramicrotome. A heating wire wound around the specimen arm is heated to cause the arm to expand. Since the rear of the arm is anchored to a solid base, the expansion of the arm is in the direction of the knife. A motor moves the arm up and down at fixed intervals. The rate at which the arm drops, however, is adjustable by the operator (e.g., the cutting speed). Thickness of the sections is determined by the amount of heating of the specimen arm since the interval between cuts is fixed. In order to avoid bumping the specimen on the back of the knife, the knife is retracted several μm by means of a powerful electromagnet. (Courtesy of Leica.)

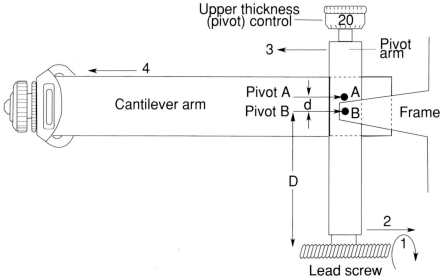

Figure 4-35(B) Diagram showing mechanical advancement mechanism used in the Porter-Blum MT1 ultramicrotome. A fine threaded screw was turned a small amount during each cutting cycle (arrow 1). The turn of the screw moved the pivot arm backwards in the direction indicated by arrow 2. Since the pivot arm was fixed at a pivot point on the frame of the microtome, the backward motion of the pivot arm resulted in a slight amount of forward movement of the cantilever arm (arrows 3 and 4). By adjustment of the upper thickness control, the pivot point was moved either up or down. For instance, if the pivot point was moved to point B, a thicker section would be cut than if the pivot point were at point A. These principles are shown in the simplified diagram shown in Figure 4-35(C). (Courtesy of RMC, Inc.)

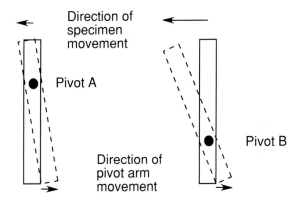

Figure 4-35(C) Conceptual diagram showing the effect of moving the pivot point of the pivot arm from point A (left) to point (B) right. Since the cantilever arm is moved forward by larger increments at point B, thicker sections will be cut compared to point A.

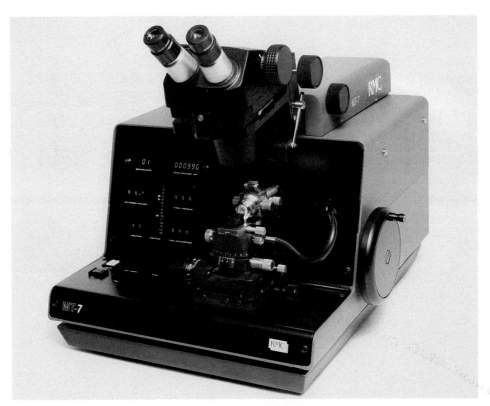

Figure 4-36 The modern version of the original Porter-Blum ultramicrotome. This American-made MT7 incorporates microprocessor control for most of the operator selected functions such as cutting speed, thickness, return stroke speed, etc. This ultramicrotome utilizes a mechanical advancement of the specimen. (Courtesy of RMC, Inc.)

Water Level Adjustment

A knife is mounted in the holder and slightly overfilled with double-distilled water that has been microfiltered using 0.45 μm filters. The edge of the knife should be wetted at this point. If not, use a fine eyelash probe (made by gluing an acetone-cleaned eyelash to the end of an applicator stick) to brush the water onto the edge. While looking down on the water surface, lower the water level using a small syringe until a silver to silver/gray reflection is obtained with the fluorescent light of the microtome (Figure 4-37). It may be necessary to move the light to obtain this reflection. The level should not be so low that the knife edge becomes dry. Never let sections dry down onto the knife edge: they are difficult to remove once dried. If wetability of the knife edge is a problem, then 10% ethanol or acetone may be used in place of distilled water; however, the level must be watched because these fluids evaporate rapidly.

Clearance Angle Adjustment

The top of the knife in its holder is inclined slightly toward the specimen to prevent the specimen block from rubbing the back of the knife during the cutting stroke. A 4 degree angle is used as a starting point, but some experimentation may be necessary as outlined in the section "Methodical Sectioning" (Figure 4-38). Diamond and sapphire knives usually come with a clearance angle recommended by the manufacturer.

Knife Advancement

The knife edge must be adjusted so that it is parallel to the specimen cutting face. Mechanically advance the knife as close as possible to the specimen block face without actually contacting it. This will require the use of both coarse and fine knife movement controls. This is done by looking directly down upon the block and moving the knife stage forward while observing the closing gap between the block and the back of the knife. On some ultramicrotomes, when the microtome light source is adjusted properly, a shadow of the knife is projected onto the block face when viewed through the stereomicroscope. On such instruments, the gap between the knife and the block is easily visualized by the diminishing shadow as the knife is moved forward (Figure 4-39). The knife should also be moved laterally to bring a good cutting edge of the knife opposite the block face.

Figure 4-37(A)   The knife is overfilled with microfiltered distilled water to ensure that the knife edge is wetted along its entire length. A diamond knife is shown clamped into the knife holder of the microtome.

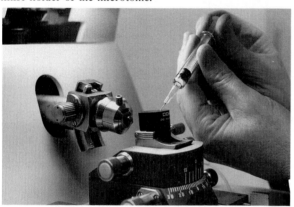

Figure 4-37(B)   The water level is lowered by withdrawing water using a small syringe until a silver-gray reflection of the fluorescent light used to illuminate the boat is seen on the water surface. The knife edge must remain wetted.

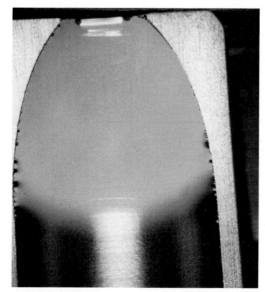

Figure 4-37(C)   Microtomist's view through the stereomicroscope of the microtome showing the silver-gray reflection seen when the water level in the knife is properly adjusted.

Figure 4-38   Knife holder with diamond knife clamped into position. Note the clearance angle adjustment scale (arrow) that is used to measure the angle of the knife inclination relative to the specimen.

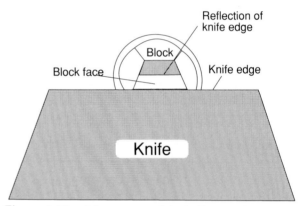

Figure 4-39   Advancement of the knife towards the specimen block may be evaluated in some microtomes by examining the reflection of the knife in the block face of the specimen. It is necessary to adjust the light properly to obtain this view. By studying the reflection, one can not only determine the relative distance of the knife from the specimen, but also use the reflection to align the knife edge parallel to the specimen block face.

Specimen Orientation

The block is oriented in the arm of the ultramicrotome so that the edge that first comes in contact with the knife is parallel to the knife edge and horizontal to the tabletop (Figure 4-40). It is also important to adjust the slant of the specimen cutting face so that it moves parallel to the cutting plane (Figure 4-41). This ensures that the knife will cut a full section rather than a chip from the specimen surface. The best way to achieve this is to establish a reflection of the knife edge in the block face and

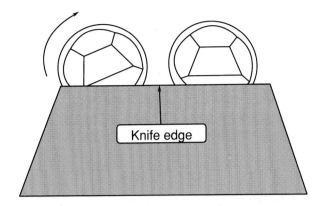

Figure 4-40 Orientation of the specimen block so that the leading edge of the block is parallel to the knife edge is accomplished by rotating the specimen block holder while it is in the ultramicrotome arm. This may be done by using the stereomicroscope of the ultramicrotome to look directly into the block face. Great care must be taken at this step so that the knife is not damaged by moving the block face into the knife edge.

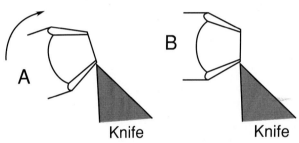

Figure 4-41 Adjustment of the slant of the specimen cutting face. In order to cut full sections, it is necessary to move the specimen block from the position shown in A to the position shown in B. All ultramicrotomes have mechanical adjustments for achieving this reorientation of the specimen block.

to observe the gap between the actual edge and its reflection in the block face. If the gap remains constant as the specimen moves past the knife edge during a downstroke of the microtome arm, then the vertical adjustment is correct. However, if the distance changes, it will be necessary to change the tilt of the specimen in the vertical plane. This is difficult for beginning microtomists; however, if one has just faced and trimmed the block in the microtome, it will be properly oriented. If not, some trial adjustment may be necessary before a full section is cut.

Two major mistakes that may be made during orientation of the specimen include ramming the block face into the knife and not retracting the knife after making adjustments which results in cutting an extremely thick section.

IMPORTANT: The knife should always be moved back away from the block during the specimen orientation procedure to avoid damaging the knife edge.

HINT: Try to keep the vertical orientation adjustment of the microtome always in the same position so that specimen chucks can be removed and returned to exactly the same position each time.

Knife Contact

Using the knife fine-advancement controls, move the knife as close as possible to the block face. This is done under high magnification with good illumination while advancing the knife in increments of 1 μm or less. After an initial thick section is made, the specimen fine-advancement mechanism is activated. If it is essential to collect the very first section, the specimen fine advancement mechanism must be activated several micrometers before initial contact with the block face. This will require patience as the distance is progressively narrowed in steps of 50 to 90 nm.

Ultrathin Sectioning

After setting the thickness control to an appropriate point (usually 80 to 90 nm, initially), ultrathin sections are cut (Figure 4-42) and collected in the trough of the knife. The goal is to produce a uniformly thin section at each pass of the specimen over the knife. One usually starts on the thicker side and adjusts the thickness setting downward until the desired section thickness is obtained. Actual thickness of the sections is determined not so much by the machine settings, but is based on the interference colors generated by shining a fluorescent light onto a water surface to establish a silvery background reflection over the water. An estimate of thickness is given by these colors. Interference colors are generated when white light reflected from the water surface passes through the plastic sections and is refracted so that the emerging light waves interfere with those coming from the water surface.

Most microtomes have an *automatic mode* to move the specimen through its path over the knife and into its return knife-avoidance regime. In older microtomes that lack the automatic feature (e.g., Sorvall MT1), handwheels are turned to move the specimen arm through its path. Microtomes with more precise control of the cutting speed use a motor-driven belt to move the arm. Some models (LKB, for example) control the speed of gravitational drop

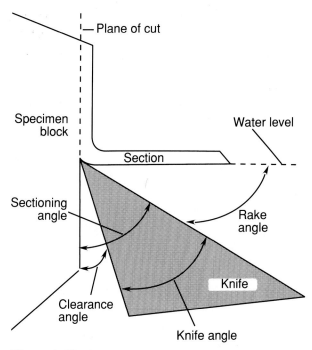

Figure 4-42 The various angles formed by the knife and specimen block must be proper in order to achieve good sections. Compression of the section occurs as it is cut by the knife and floats onto the water surface. The compression can be relieved by slightly heating the sections as described in the text. Thickness of the sections is estimated based on the interference colors on the silvery water surface (see Figure 4-37C). Colorless to grey sections = 30–60 nm, silver/grey to silver = 50–70 nm, gold = 70–90 nm, purple = 100–190 nm, and blue = 200 + nm in thickness.

with a motor connected to the specimen arm by a cord. The *Huxley Microtome* employs an oil-filled piston to control the rate of drop of the specimen arm by adjusting the size of the hole through which the oil exits the piston.

If the *cutting speed* is adjustable, an initial setting would be 0.5 to 1 mm/sec for diamond and sapphire knives and 2 to 3 mm/sec for glass knives. If no problems are encountered, sufficient numbers of sections are collected in the knife trough. An eyelash probe may be used to move the sections in the trough by gently touching the probe to the section (Figure 4-43). Take care not to press too hard against the section, since it has a strong tendency to adhere to the eyelash probe. Should this happen, the section should be removed with a filter paper (not the fingers because body oils will contaminate the probe and trough).

Section Retrieval

The sections should be moved into clusters of 2 to 4 sections with an eyelash probe. Generally, this number of sections will fit conveniently onto the

specimen grid (which has been prepared as described earlier in this chapter). The objective will be to place the sections in the center 2/3 of the grid, since the specimen holder of the microscope will usually obscure the periphery. Although it is possible to pick up sections from overhead by coming directly down onto the sections with a grid, this method is less preferred since the sections will bunch together, overlap, and wrinkle. Instead, grasp a grid by the rim (thick-rimmed grids are desirable) using a fine-pointed forceps and bend up the edge of the grid slightly by pressing the grid against a clean, lintless paper. While still holding the grid in the forceps, pass the naked grid rapidly through an alcohol flame (to diminish repulsion of the water by the grid).

IMPORTANT: Do not use this method with coated grids since the films will be damaged.

Submerge the grid and maneuver it directly under the sections to be collected (Figure 4-44). Slowly lift up the grid parallel to the water surface to collect the sections. The sections will float on the small droplet of water on the grid. Gently contact the underside of the grid and the edge of the filter paper, permitting the water to soak into the paper.

Filmed grids are handled a bit differently to collect sections. Since the plastic and carbon films tend to repel sections, it may be necessary to make them hydrophilic by exposing them to an AC glow discharge in a vacuum evaporator to dissipate the repulsion.

HINT: The same effect may be achieved by storing the filmed grids overnight in the refrigerator or by using a trace of detergent in the fluid in the trough.

In most instances, repulsion is not a problem, but if it occurs then ribbons of connected sections can be retrieved by slowly bringing the grid up under the ribbon at a 45 degree angle relative to the water surface until one of the sections overlaps the edge of the grid. Continue slowly lifting the ribbon of attached sections out of the water until they have been lifted onto the grid.

Sections that are not in a ribbon are sometimes difficult to retrieve onto coated grids. When this difficulty arises, the bent grid is very slowly brought up from under the grouped sections to lift the sections onto the grid. *CAUTION: Take care not to damage the supporting film.* If sections are still being repelled by the grid, then it will be necessary to pick up the group of sections using a 3 mm wide loop

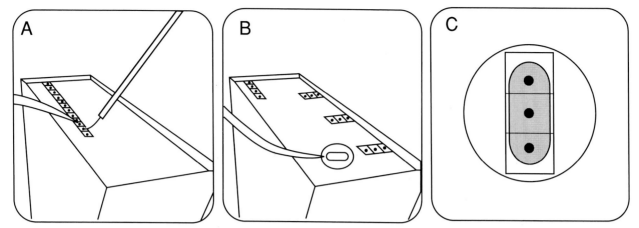

Figure 4-43 Sections can be oriented in the water trough by using an eyelash probe to lightly touch the section and move it as desired. (A) Long ribbons of sections can be broken into smaller ribbons using two probes. (B) Ribbons or individual sections can be moved into groups that will conveniently fit onto a single grid. (C) Slot grid showing three sections along the slotted opening. The area that will be viewable in the TEM is shaded in this figure. (Courtesy of Leica.)

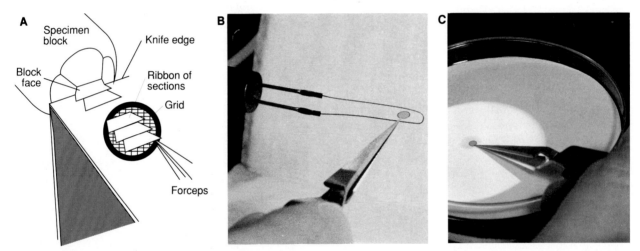

Figure 4-44 (A) One method of retrieving groups of sections is to come up from under the sections with the grid so that the sections are lifted up onto the grid. (B) It is possible to relieve compression of the sections by passing the grid containing the sections close to an electrically heated wire. Care must be taken so that the water on which the sections are floating does not evaporate, or the sections may be damaged by the excessive heat. (C) The grid containing the sections is placed onto a filter paper so that any water remaining from the trough is gently absorbed by the paper. It is best to place the grids directly down onto the filter paper rather than attempting to touch the edge of the wet grid with the paper. Otherwise the sections may be pulled sideways in the direction of the paper. It is useful to first remove any excess water trapped between the arms of the forceps using a filter paper as a wick.

constructed from fine platinum wire. The loop is bent to an angle, submerged, and brought up from under the sections. The sections may now be transferred to a coated grid that has been clamped into a forceps. The grid should have been previously touched to a water surface so that it contains a small drop of water. The loop containing the sections is turned upside down, and the section side is brought directly down onto the water on the grid. Most often, the sections will be transferred onto the grid, which is then allowed to dry while clamped in the

forceps. If one attempts to blot the grid dry, usually the sections will be drawn onto the filter paper.

Perhaps the easiest way to transfer sections from a loop onto a grid involves using a specially designed loop (Figure 4-45) to pick up the sections and deposit them directly onto a dry grid on a filter paper. Proceed as follows:

1. Place the loop above the sections and lower the loop so that the sections are surrounded by the loop. Lift up the loop containing the sections in the water retained in the loop.

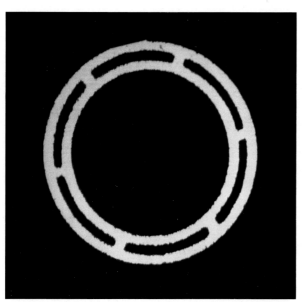

Figure 4-45 Specially constructed loop that facilitates picking up sections and transferring them onto a grid, which is then placed onto a filter paper. (Courtesy of Electron Microscopy Sciences.)

2. Position the loop just above the grid on the filter paper.
3. Lower the loop until the water in the loop just contacts the grid. This will cause the grid to be attracted from the filter paper to the water that lies underneath the floating sections.
4. Continue lowering the loop containing the sections and grid onto the filter paper. This will remove the water and transfer the sections down onto the grid.

It may be necessary to collect a complete series of continuous or *serial sections* as they come off the microtome. In this instance, all sections must be collected or accounted for and grid bars must be eliminated, since they would obstruct a complete view of the cellular details. Usually, when serial sections are taken, it is desirable to trace one or more structures in all of the sections, necessitating the use of single hole or slot grids with a supporting film. Although quite a demanding undertaking, it is possible for the experienced microtomist to collect a complete series of such sections. It will be necessary to use slot or single hole grids with a supporting membrane and to keep careful records of each series of sections as they are made. The specimen block should be trimmed to a shape such as a trapezoid so that the order of sections can be readily ascertained.

A good diamond knife and automatic ultramicrotome with a mechanical advance is desirable if all of the serial sections must be the same thickness.

A mechanical advance is preferred, since it is often necessary to stop microtomy, pick up sections or make other adjustments, and then resume sectioning with a minimum disturbance to the order and continuity of the sections. Thermally advanced units will continue to advance during this time period, so that a much thicker section than desired may be cut upon resumption of sectioning. Nonetheless, it is possible to do serial sectioning on thermally advanced ultramicrotomes by collecting groups of sections in various areas of the trough and retrieving them when the ultramicrotome advance arm is being cooled. Since the arm will have retracted during the collection period, it will be necessary to advance the knife several micrometers in order to resume sectioning. This should be done carefully using micrometer and finer mechanical advancements to avoid cutting thicker than desired sections.

Collecting Serial Sections onto Single Hole or Slot Grids

Rather than attempting to collect the sections directly onto a coated slot grid, it is best to collect them as follows. Grasp an uncoated slot grid in a forceps so that the dull side of the grid is down, and bring the slot or hole directly down on the sections so that they are corralled by the grid (Figure 4-46A). Upon lifting the grid straight up, surface tension will carry the sections in the water that fills the void in the grid. The grid is then placed, sections still on top, onto a plastic film stretched over a 3.5 to 5 mm hole drilled out of a platform made of aluminum stock (Figure 4-46B).

HINTS: (1) To enhance adherence of the grid to the Formvar film, dip the clean grids into a separate 0.5% Formvar solution with a forceps and immediately place the grids on a filter paper until thoroughly dry. (2) The film coated platforms are prepared by first floating the plastic onto a water surface and bringing the platform up from under the film. (3) Care must be taken not to puncture the thin film on the platform with the forceps or edge of the grid. (4) A number of holes can be drilled out of the aluminum so that 25 or more grids can be placed onto one such platform. (5) It is possible either to purchase the ready-made platforms (Structure Probe, Inc.) or to purchase punched stainless steel sheeting (Small Parts, Inc.) that may be cut into the desired sizes of platforms.

The platform is kept warmed to 60° C most of the time (except possibly during the transfer of sections) in order to facilitate spreading of the sections and to prevent wrinkling. After the grids have dried completely, the flat head of a nail or other instrument, about the same size as the grid, is used to push the grids down onto a filter paper placed under the platform (Figure 4-46C). The grids may then be stained and examined in the transmission electron microscope as described in the next two chapters.

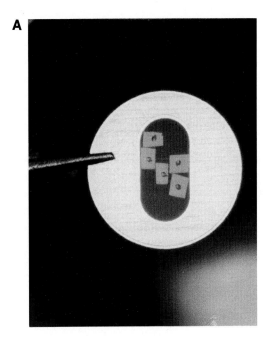

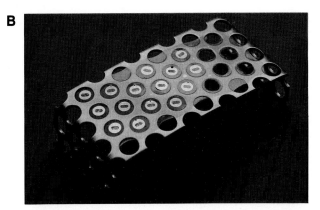

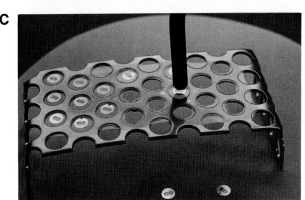

Figure 4-46 (A) Short ribbons of serial sections may be picked up by placing a grid directly down onto the sections so that they are corralled by the grid. Upon lifting up the grid, the sections will be retained in the thin film of water that remains in the slot. It is then possible to stretch the sections by passing the grid close to a heated wire as shown in Figure 4-44(B). (B) The slot grid containing the stretched sections is placed onto a plastic film that extends over the holes in an aluminum platform. The sections remain corralled by the slot grid and will become attached to the plastic film upon evaporation of the water. (C) When completely dried, the slot grids containing the sections on the plastic film are removed by pushing down with a blunt surfaced tool as shown.

Factors Affecting Sectioning

A number of conditions may affect the quality of sections produced. It is unusual when sectioning problems can be attributed to instrument failure, since most microtomes are reliable, sturdy devices that normally last many years before service is needed.

Location of the ultramicrotome is an important consideration since it is more sensitive to vibrations than the modern TEM. The room must be clean and dust-free, quiet and free of distractions, of uniform temperature, without drafts, and must have good lighting (preferably dimmable). Most ultramicrotomes have built-in mechanisms for neutralizing minor vibrations; however, it may be necessary to place the ultramicrotome on a special vibration-damping table or possibly to relocate the microtome in another room if vibrations become excessive.

The *dehydration and embedding* steps may cause problems if the specimen is not adequately dehydrated or if the plastic has not totally infiltrated the cells. Poorly mixed or only *partially polymerized plastic* is a problem sometimes encountered by beginning investigators. Care must be taken when weighing and mixing the resin monomers. The *firmness of the plastic* should match as nearly as possible the hardness of the specimen to minimize the knife's skipping as it passes through areas of different hardness. Normally, one uses harder plastic formulations for diamond knives, while glass knives are better able to section softer blocks.

Proper *trimming* of the block is extremely important to generate parallel, smooth sides that yield straight ribbons of sections rather than individual sections that scatter in the trough and are hard to manipulate and collect. The block face should be as small as possible and consist mostly of specimen. Larger block faces are more difficult to section, even on a diamond knife, so try to maintain the block face at 0.5 to 1.0 mm. When razor blades were used to trim the block, be very careful not to leave small chips of razor blade steel embedded in the block. These will destroy both glass and diamond knives.

The knife must be sharp, securely locked in place in the holder, and free of any defects on the cutting edge. Always check the condition of the knife edge before commencing sectioning. Even diamond knife edges may contain dirt or sections that could affect the quality of sections produced. If multiple users share a diamond knife, it is best to maintain a logbook to keep track of problems with the knife.

Although a 4 degree *clearance angle* will be a good sectioning angle most of the time, it may be necessary to vary this parameter. For example, when sections are being dragged over the knife edge rather than being severed from the block completely or if *chatter* is encountered, then the clearance angle should be adjusted. Chatter will be seen as closely spaced parallel lines running perpendicular to the direction of cut when the sections are viewed in the electron microscope. If one is able to see the striations in sections as they float in the boat or by using a light microscope, then they probably are caused by *low-frequency vibrations* under 1 KHz and are most likely due to vibrations in parts of the ultramicrotome or in the building itself. The latter type of problem may be difficult to solve without purchasing or building vibration damping equipment or even relocating the ultramicrotome.

NOTE: An inexpensive vibration-damping table can be constructed by mounting a heavy platform of slate on top of handballs that have been secured on top of a sturdy table.

Fortunately, the striations may be so widely spaced that they may not present a problem for many areas of the specimen, especially at higher magnifications.

Chatter caused by *high-frequency vibrations* that can be seen only in the transmission electron microscope are nearly always due to embedding problems or the wrong sectioning parameters. When chatter extends over the entire section, then the clearance angle, knife angle, and cutting speed must be varied as described in the next section of this chapter. If only small areas inside the cells are showing chatter, it may not be possible to readily correct the problem.

The *cutting speed*, or rate at which the specimen is passed over the knife edge, is another important parameter affecting the quality of sections. Normally, one uses 1 mm/sec with diamond knives,

while 2 mm/sec seems better suited to glass knives. However, the hardness of the specimen, angle and sharpness of the knife, as well as thickness of the section all should be considered when selecting this speed. Although it is thought that harder specimen blocks should be cut at slower speeds (1 mm/sec or less), some specimens such as bone and tooth may section better at faster speeds (5 to 10 mm/sec). Faster speeds will cause greater compression of the sections so that it may be necessary to apply heat or organic vapors such as xylene or chloroform over the surface of the sections so they can return to nearly normal size. To avoid organic fumes, use an electrically heated wire loop (Figure 4-44B). Such loops can be purchased from several EM supply houses or made from a flashlight by removing the bulb and soldering a nichrome wire into the leads of a broken bulb. Sections are stretched by placing the grid containing the sections on the droplet of water between the wire loop and observing the stretching process under a dissecting microscope. Do not overheat the sections and do not allow the water to completely dry out. Once stretched, the moist grid is placed onto a filter paper to dry.

Methodical Sectioning

Manufacturers of diamond knives have often predetermined such parameters as knife angle, optimum clearance angle, and cutting speed. However, they may need to be determined empirically if problems occur when using glass, and sometimes even diamond, knives. It is necessary to use a methodical approach such as outlined in the flow diagram devised by the LKB (Leica) company as follows:

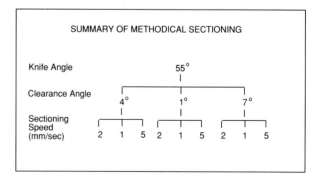

Start with a knife of a known angle and try cutting the specimen using a 4-degree clearance angle and cutting speeds of 2, 1, and 5 mm/sec. If high-quality sections are not produced, the clearance angle is changed to 1 degree and the same three speeds again attempted. Should this also prove unproductive, then a 7 degree clearance angle is used with the three sectioning speeds. If sections are still not obtained, make another glass knife with a different knife angle and repeat the process. Try knife angles of 55, 60, and 65 degrees in this test.

A Guide to Sectioning Problems and Causes

Failure to cut any sections: (a) microtome at end of fine advance, (b) dull knife, (c) specimen block too soft, (d) knife or block not secure, (e) negative clearance angle, (f) wet block face, (g) vibrations, (h) temperature fluctuations nearby.

Thickness variations from one entire section to another: (a) dull knife, (b) feed set too low, (c) bumping of instrument, (d) microtome problem, (e) drafts and temperature variations, (f) loose component: knife, specimen, mechanical advancement lock, (g) block too large or soft, (h) wrong knife angle or cutting speed.

Thickness variations within same section: (a) inhomogeneous specimen: areas of hard/soft, (b) large areas of plain plastic around specimen, (c) low frequency vibrations, (d) dull portion of knife.

Wrinkled sections: (a) block face too large, (b) soft block, (c) dirty or dull knife, (d) clearance angle too great, (e) water level too low, (f) cutting speed too fast, (g) knife loose or angle wrong.

Compressed sections: (a) block too soft, (b) inadequate expansion: try organic vapors or heat, (c) cutting speed too fast.

Chatter: (a) high-frequency vibrations during sectioning process: try different cutting speed, knife, and clearance angles, (b) block too tall with small base, (c) dull knife or soft block, (d) block or knife slightly loose, (e) knife too high in holder.

Specimen block lifts section on return stroke: (a) water level too high, (b) block face dirty, wet, or hydrophilic in nature due to nature of specimen, such as sperm cells, or that of the embedding plastic, (c) clearance angle too small, (d) dirty knife or wet back, (e) static electricity on block face.

Block face gets wet: (a) see a–e, above, (b) block face too large, (c) cutting speed too slow.

Sections are dragged over knife edge: (a) cutting speed too slow, (b) knife sitting too low in holder, (c) water level too high, (d) clearance or knife angle too low.

Sections have holes: (a) bubbles in resin, (b) incomplete infiltration, (c) hard objects in specimen.

Specimen drops out of block: (a) poor infiltration, (b) block too soft.

Sections have striations perpendicular to knife: (a) nick in knife edge, (b) dirt on knife edge, (c) knife damaged by hard material embedded in block.

Sections do not form ribbons: (a) block edges not parallel to edge of knife or each other, (b) water level wrong, (c) cutting speed too slow, (d) static electricity on block.

Ribbon of sections is curved: (a) leading and trailing edges of block not parallel, (b) compression on one side of section.

Sections hard to see: (a) illumination angle wrong, (b) water level wrong, (c) viewing angle wrong.

Sections hard to move in boat: (a) flotation fluid is contaminated and should be changed (replace with clean fluid containing trace of detergent or touch needle dipped into detergent into the water trough of knife).

Sections stick to eyelash probe: (a) dirty eyelash or roughened end of eyelash, (b) bearing down on section too much with eyelash.

Sections move away from grid: (a) dirty grid: clean in acid/alcohol or flame naked grids in alcohol lamp.

Cryoultramicrotomy

Cryoultramicrotomy is a procedure for cutting ultrathin frozen sections of biological specimens. Cryoultramicrotomy appears to offer several advantages over conventional ultramicrotomy techniques. One major advantage was originally thought to be *speed*, since it should be possible to freeze and section a specimen in 1 to 2 hours. This would offer the pathologist excellent diagnostic capabilities similar to those currently obtained using cryostat sectioning of surgical biopsies. To date, however, such uses for cryoultramicrotomy have not been realized on a routine basis probably because methods to obtain conventionally embedded and sectioned specimens in 4 to 5 hours have been developed (see Chapter 2).

A second and more practical reason for using cryoultramicrotomy is either for the localization of antigens using immunocytochemical procedures (Chapter 9) or for the elemental analysis of diffusible substances or ions by X-ray analysis (Chapter 15). In the latter case, it appears that cryoultramicrotomy offers probably the only way that one can localize diffusible substances.

For immunocytochemical procedures, it is best to first try a conventional approach by fixing in aldehyde fixatives (1% glutaraldehyde/2% formaldehyde) followed by rapid dehydrating in ethanol and embedding in an acrylic resin (Lowicryl or L.R. White resin) that is polymerized at relatively low temperatures ($-20°$ to $40°$ C, respectively). The plastic-embedded specimens can be sectioned on a conventional ultramicrotome and stained using the immunocytochemical procedures outlined in Chapter 9. If this should fail after repeated attempts, then one may consider resorting to cryoultramicrotomy. Realize, however, that this may involve a considerable investment of time and finances (the cryokit attachment for most ultramicrotomes can cost as much as an ultramicrotome alone) without any assurances of success. A conservative approach would be to contact a researcher currently using the technique and to collaborate on the research.

The technique for cryoultramicrotomy was first reported in the scientific literature in 1951 by Fernandez-Moran; however, specifics of the procedure were not made available in that publication. In 1964, W. Bernhard and M. T. Nancy described a procedure that gave good ultrastructural details of biological tissues. In this procedure, the specimen was fixed in 2.5% glutaraldehyde and, after rinsing in buffer, transferred into a 10% to 20% solution of gelatin to provide a supporting matrix. Following a dehydration in 30% glycerol, the specimen was placed onto a metal specimen carrier and frozen in liquid nitrogen. After a quick trimming of the frozen specimen block with a chilled razor blade, sections were cut either into a trough containing 50% DMSO (as a antifreeze agent for the trough fluid) or onto a dry knife at −50° C. If a dry knife was used, the dry frozen sections were transferred onto a Formvar-coated grid and flattened onto the grid using slight pressure. This work was accomplished by placing a conventional ultramicrotome inside of a cryostat or cabinet that had been chilled from −50° to −70° C. While such an arrangement affords a stable temperature control, it is not possible to achieve temperatures lower than −70° C in a cryostat. This is an important point, since ice crystal damage occurs if the temperature rises above −70° C. The use of a trough fluid for the collection of cryosections was eventually abandoned due to the thickening of antifreeze agents at such low temperatures.

In 1972, many significant contributions to the field of cryoultramicrotomy were published independently by Christensen and Paavola, Dollhopf, and Persson. These researchers reported the development of kits or attachments for the Porter-Blum, Reichert, and LKB ultramicrotomes, respectively. Instead of placing the conventional ultramicrotome inside of a cryostat, an insulated chamber was fashioned around the specimen and knife. Since this area was cooled with liquid nitrogen, it was possible to regulate the temperature down to −170° C. Some of the kits even permitted independent control of the temperature of the knife and the specimen. Once these cryokits were made available commercially, researchers were able to refine the specimen preparation techniques, and many technical papers were published during the mid-1970s and 1980s. Examples of these attachments installed on two ultramicrotomes are shown in Figure 4-47. A diagram illustrating the design of the cryo-chamber of the RMC instrument is shown in Figure 4-48. Out of these research and development efforts, two major procedures for obtaining ultrathin frozen sections on dry knives are used most often.

Dry Knife, Dry Retrieval Method

In his 1971 article, A. K. Christensen described not only the development of an apparatus for cutting ultrathin frozen sections, but he also detailed the first readily usable procedure for cutting and collecting ultrathin frozen sections. Photographs of the first commercially available cryo unit for the Porter-Blum MT2 ultramicrotome are shown in Figure 4-49. The Christensen procedure was used to produce dry cryosections of unfixed tissues. The reason for cryosectioning unfixed specimens and picking them up under dry conditions (as opposed to the sucrose droplet retrieval method of Tokuyasu, which is described subsequently) is to prevent the loss of diffusible substances. See Christensen's original paper for more details. The Christensen procedure is illustrated in Figure 4-50 and summarized below.

1. Fresh tissue is cut to a small size, placed onto a copper specimen stub, and rapidly frozen by pressing the tissue against a polished copper surface that has been precooled in liquid nitrogen. This technique, known as slam freezing, can also be used for the Tokuyasu procedure (described subsequently). Slam freezing gives better preservation than that obtained by placing the copper stub into a chilled liquid (such as liquid nitrogen or Freon).

   An alternative method for freezing specimens is to use a specially designed pair of pliers with flat copper jaws (Figure 4-51). The pliers are chilled in liquid nitrogen, and the specimen is rapidly squeezed between the two jaws. After freezing is completed, pieces of the tissue are cut out while submerged in liquid nitrogen and mounted into a special vise-type specimen holder for the ultramicrotome (Figure 4-52).

2. Transfer the specimen holder into the cryochamber of the cryoultramicrotome. A small container for transporting the specimen holder under liquid nitrogen can be made by cutting down the top half of a Styrofoam coffee cup.

3. Rapidly transfer the frozen specimen holder onto the precooled specimen arm of the ultramicrotome. This must be accomplished rapidly in order to avoid any temperature changes that would permit ice crystals to form inside the tissue. Generally, specimen tem-

A

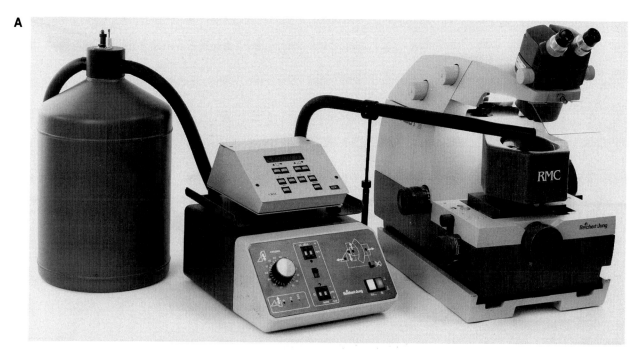

B

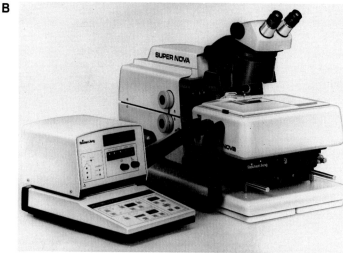

Figure 4-47 Cryoultramicrotomes for cutting ultrathin frozen sections. All ultramicrotome manufacturers have accessories for cryoultramicrotomy. (A) A universal cryo-attachment that fits all commonly used brands of ultramicrotomes is now available. The unit, sold by RMC, Inc., is shown installed on a Reichert-Jung ultramicrotome. Reichert-Jung and RMC also sell different units designed specifically for their own brands of ultramicrotome. (Courtesy of RMC, Inc.) (B) The LKB Cryo Nova ultramicrotome set up for cutting ultrathin frozen sections. (Courtesy of Leica.)

peratures range from −80° to −120° C for cryoultramicrotomy.

4. Advance the chilled knife carefully towards the specimen until it nearly contacts it. Use the fine mechanical advancement of the microtome until the first section is cut. Take care not to cut a section thicker than 1 μm; otherwise the block face may be shattered and damaged. In addition, only the uppermost 5 μm or so of tissue will have undergone vitrification, and ice damage may become more pronounced as one proceeds deeper into the tissue block.

NOTES: Some ultramicrotomes permit one to adjust the temperature of the knife independently of the specimen. It may be necessary to experiment with various knife temperatures until a proper combination is obtained. A good starting point is to set the knife at the same temperature as the specimen. Either glass or dia-mond knives may be used. Glass knives appear to last longer at the colder temperatures, so diamond knives may not be needed. As in conventional sectioning, the glass knife must be of high quality as determined by careful examination just prior to use. Modern knife makers can produce such high quality knives for cryoultramicrotomy. Plain knives, without attached troughs, are used since sections will be cut onto a dry knife.

5. Begin cutting ultrathin sections using the automatic advance mechanism of the ultramicrotome. A cutting speed of 1 mm/sec, or less, is a good starting point. A fast return cycle is recommended in order to minimize thermal changes to the specimen block.

6. As the dry sections are cut, they will collect along the knife edge. These sections may have a clear to slightly translucent appearance and may exhibit interference colors similar to plastic sections floating

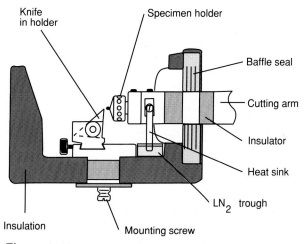

Figure 4-48 Diagram of interior of cryoultramicrotome for cutting ultrathin frozen sections. (Courtesy of RMC, Inc.)

on a water surface. These colors, however, cannot be used to estimate the thickness. Use an eyelash probe to move the sections down the knife so that they will not be in the way of the next section. Although one may occasionally obtain ribbons of sections, normally, individual sections are formed. Continue cutting sections as long as possible. If static electricity becomes a problem, a static eliminator (gun or radioactive strip) can be employed.

7. With an eyelash probe, transfer the sections onto a Formvar-coated grid. This is facilitated by using a small platform containing a supply of chilled grids positioned near the knife edge. The sections are moved onto the grid, and the grid is placed onto a flat, chilled copper surface inside the cryochamber.

8. A copper rod with a polished flat end that is approximately the same size as the grid is used to flatten the sections onto the grid by gently pressing down onto the sections.

9. At this point, the sections may be freeze-dried by simply leaving them on a cold copper surface inside of the cryochamber, or they may be transferred into a freeze-drying apparatus. Alternatively, the frozen and hydrated sections may be transferred into the microscope and examined while still in the frozen state by means of a special cryotransfer apparatus and cold stage that maintains the specimen at liquid nitrogen temperatures while inside the TEM. Unfortunately, the morphological features in frozen hydrated sections are rather poor compared to dried sections. In addition, such sections are more prone to damage by the electron beam and are of little use when analyzing for elements that emit low energy X rays (Na or Mg, for instance).

10. The sections may be examined for diffusible ions by X-ray microanalysis or electron energy loss spectroscopy (see Chapter 15).

Dry Knife, Wet Retrieval Method

This method was originally described by Tokuyasu in 1973 and is the predominant method for producing cryosections for use in cytochemical and immunocytochemical procedures. The chief advantages are that the technique is relatively simple, reproducible, and yields high-quality sections with good preservation of enzymatic and immunological reactivities. The steps for this procedure are as follows:

1. The tissues are cut into appropriately sized pieces (or centrifuged into a compact pellet if individual cells are used). This is usually done rapidly at temperatures of 4° to 5° C in order to minimize ultrastructural changes.

2. If a fixative is desired, the tissues are placed into the chilled fixative for an appropriate amount of time. Normally, one uses only an aldehyde fixative such as 1% to 2% glutaraldehyde or a mixture of 0.5% to 2% glutaraldehyde and 1% to 4% formaldehyde for 15 to 30 minutes. Other aldehyde fixative combinations and times may be more appropriate, depending on the investigation.

3. Rinse in buffer to remove unreacted fixative. The type of buffer, as well as its tonicity, will depend upon the tissue. In addition, it is recommended that 0.1 M glycine be incorporated into the buffer in order to inactivate any residual aldehyde groups that may be associated with tissue components. This will help to minimize nonspecific staining.

4. Place the specimen into 2.1 to 2.3 M sucrose (or 1.8 M sucrose in 15–20% polyvinyl pyrollidone) in phosphate buffered saline (0.1 M phosphate buffer containing 0.9% NaCl) for 15 to 60 minutes. The time for infusion of the sucrose into the tissue depends on the size of the tissue.

5. Mount the specimen pieces onto metal holders (usually copper stubs or pins). It is particularly useful to shape the specimen pieces at this time (rather than after they have been frozen) so that they have a wide base in contact with the copper pin and so that they have a pointed tip. This is similar to trimming plastic specimen blocks into the pyramid shape.

6. Quick-freeze the mounted specimen by contacting it to a chilled metal surface or by plunging the copper holder into liquid nitrogen. If plunge freezing is used, a quenchant such as isopentane or Freon or even nitrogen slush is needed to enhance the freezing rate. The specimen must be moved or shaken while submerged in the coolant in order to achieve rapid freezing. The goal is to *vitrify* the remaining water inside the cells (i.e., to freeze the water so rapidly that it forms solid amorphous water rather than ice crystals). The specimens can be stored indefinitely in liquid nitrogen.

7. Transfer the frozen specimen (while still submerged in liquid nitrogen) into the cryochamber of the ultramicrotome and proceed to cut ultrathin frozen sec-

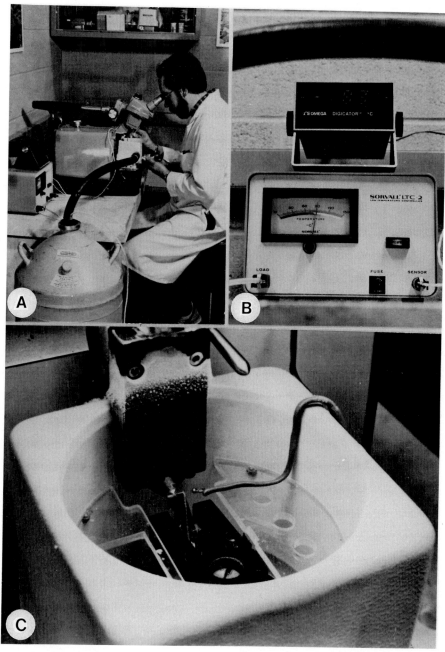

Figure 4-49 The first commercially available version of the Christensen cryoultramicrotomy kit for use on the Porter-Blum MT2B ultramicrotome. (A) Overall view of the ultramicrotome with attached cryo chamber (white area in center of top left photograph). The large tank in the foreground contains liquid nitrogen that is converted to cold gas and transported to the cryo chamber for cooling. (B) Low temperature controller for setting and maintaining the temperature of the cryo chamber. The digital thermometer on the top of the controller is for sensing the actual temperature inside the cryo chamber. (C) Closeup view of the cryo chamber. The wire extending down into the chamber and approaching the knife edge is the temperature sensor for the digital thermometer. Inside the cryo chamber one can see the glass knife and a specimen being cut by the knife. Note the formation of frost on the top part of the specimen arm. The constant flowing of cold, dry nitrogen gas out of the cryo chamber prevents frost from forming inside the chamber. Plexiglass shelves surround the knife and permit the storage of chilled grids and other tools (eyelash probes, rods for flattening sections, etc.). (Courtesy of A. K. Christensen and Wiley-Liss Publishers.)

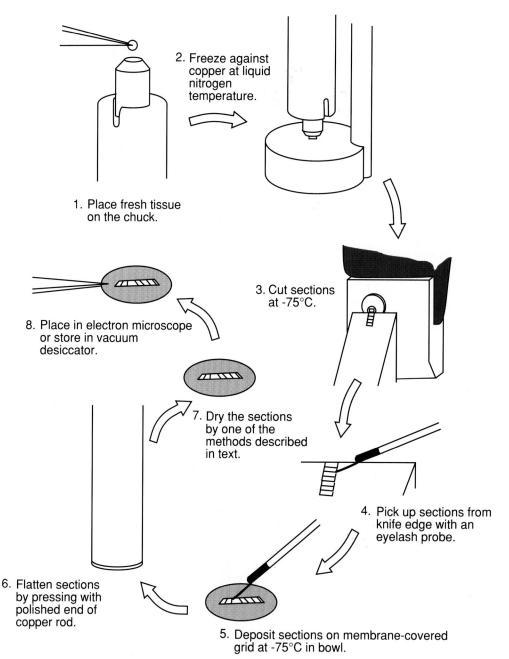

2. Freeze against copper at liquid nitrogen temperature.

1. Place fresh tissue on the chuck.

3. Cut sections at -75°C.

8. Place in electron microscope or store in vacuum desiccator.

7. Dry the sections by one of the methods described in text.

4. Pick up sections from knife edge with an eyelash probe.

6. Flatten sections by pressing with polished end of copper rod.

5. Deposit sections on membrane-covered grid at -75°C in bowl.

Figure 4-50   Summary of Christensen technique for cutting ultrathin frozen sections on a dry knife and transferring the dry sections onto a coated grid. The sections could then be freeze-dried by leaving them in the dry atmosphere of the cryo chamber. (Redrawn with permission of A. K. Christensen.)

tions (Figure 4-53A) as described in steps 4–6 of the Christensen procedure.

8. Move the sections into groups that will fit conveniently onto a grid using a chilled eyelash probe (Figure 4-53B). Pick up the sections using a 1 to 2 mm wide platinum loop that has been dipped into 2.3 M sucrose in phosphate buffered saline so that a droplet of the sucrose solution remains in the loop (Figure 4-53C). Approach the group of sections with the droplet and

contact the sections to the underside of the droplet. Take care to allow the droplet to cool but not to freeze in the cryochamber, otherwise the sections may not attach to the droplet. Do not contact the knife with the droplet or it may freeze onto the knife.

9. Withdraw the droplet containing the sections outside of the cryochamber. The droplet containing the sections will probably have frozen by this time. Allow several seconds for the droplet to thaw—as evidenced

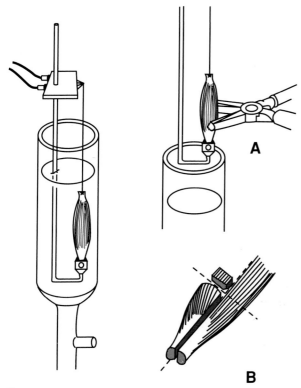

Figure 4-51 Specially designed pliers are used for freezing the specimen. The specimen, muscle tissue, is pressed between the liquid nitrogen cooled jaws of the pliers (A). Upon freezing, small portions of the frozen tissue are excised (B) into smaller pieces that may be further trimmed and clamped into a special vise holder as shown in Figure 4-52. (Courtesy of D. Parsons and Electron Microscopy Society of America.)

Figure 4-53 (A) Ultrathin frozen sections shown collecting on a glass knife. The specimen is the large white object in the center of the photograph. (B) Manipulation of the frozen sections is accomplished by a chilled eyelash probe. (C) The sections are picked up on the underside of a drop of 2.3 M sucrose suspended in the platinum wire loop. (Courtesy of A. K. Christensen, T. E. Komorowski, and Wiley-Liss Publishers.)

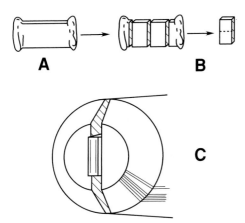

Figure 4-52 Special vice-type holder for clamping specimens that were frozen in the special pliers shown in Figure 4-51(B). The frozen tissue, shown in A, is first trimmed to a smaller piece as shown in B and clamped into the special vice-type holder shown in C. (Courtesy of D. Parsons and Electron Microscopy Society of America.)

by a transition from milky white to clear. Touch the underside of the droplet (containing the sections) onto a Formvar- and carbon-coated grid (Figure 4-54) so that the sections will be transferred onto the grid.

10. Immediately pick up the grid (absolutely avoid drying of the sections), and float the grid (sections down) onto a droplet of phosphate buffered saline containing 1% bovine serum albumin or 2% gelatin.

11. Proceed to stain the sections using the appropriate cytochemical or immunocytochemical procedure (see following examples).

If cytochemical procedures are not needed, then the sections can be stained by floating the grids on four changes of distilled water (to remove the phosphate buffer) followed by flotation on uranyl acetate as described by Tokuyasu in his 1989 paper. After a quick rinse in distilled water, the grid can be placed onto a filter paper to dry. For an example of results obtainable using this procedure see Figure 4-55.

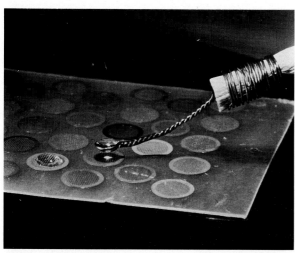

Figure 4-54 Transfer of the thawed, cryoultramicrotomed sections onto Formvar-coated grids is accomplished by touching the underside of the sucrose droplet (section side) to the grid. (Courtesy of A. K. Christensen, T. E. Komorowski, and Wiley-Liss Publishers.)

Example of the Use of Frozen Sections in an Immunocytochemical Localization Procedure

Bastholm, L., M. H. Nielsen, and L.-I. Larsson. 1987. Simultaneous demonstration of two antigens in ultrathin cyrosections by a novel application of an immunogold staining method using primary antibodies from the same species. *Histochemistry* 87:229–231.

PROCEDURES: In this study two different antigens, human pituitary growth hormone and synthetic human ACTH, were localized using two different rabbit antisera against each of the hormones. After cryosectioning mouse pituitary glands using the Tokuyasu procedure, the sections were reacted with the rabbit antiserum against human growth hormone followed by gold-labeled goat anti-rabbit antibodies. The gold-labeled antibodies were used at a high concentration so that all of the rabbit antibodies were totally saturated. After treating the sections with paraformaldehyde fumes to stabilize the rabbit and goat antibodies, the sections were treated with a glycine-containing buffer to inactivate residual aldehyde groups, and reacted with the rabbit antibody against ACTH. Following a staining with gold-labeled goat anti-rabbit antibodies (in this case the gold particles were of a different size than was the first gold label), the sections were rinsed and ultimately stained in uranyl acetate.

RESULTS: The researchers were able to discriminate against the two hormones since two differently sized gold probes were used (see Figure 4-55). Since a rabbit antiserum was used to localize two different antigens, it was very important that the first set of rabbit antibodies be totally saturated and stabilized (hence the use of excess gold anti-rabbit antibodies followed by formaldehyde linking of the antigen-antibody complex) before the second gold labeled anti-rabbit stain was used. The electron micrographs shown in Figure 4-55 reveal that the two hormones are localized in the granules of different cells. One cell type is responsible for the production of the growth hormone (GH, on the micrograph), while a different cell was responsible for the production of the ACTH (A, on the micrograph). Note that the gold particles are of sufficiently different sizes (15 nm versus 5 nm) that the different antigenic sites are clearly distinguishable. The researchers found that either size of gold probe could be used with either antigen and that the order of the application of the different sizes did not matter in this instance.

For further reading on immunocytochemical applications, see Chapter 9 and some of the selected references at the end of this chapter. The combination of cryoultramicrotomy and the use of double labeling represents a challenging yet powerful technique for the localization of multiple antigenic sites. Before embarking on cryoultramicrotomy, however, one should first attempt the localization using conventionally embedded specimens as described in Chapters 2 and 9.

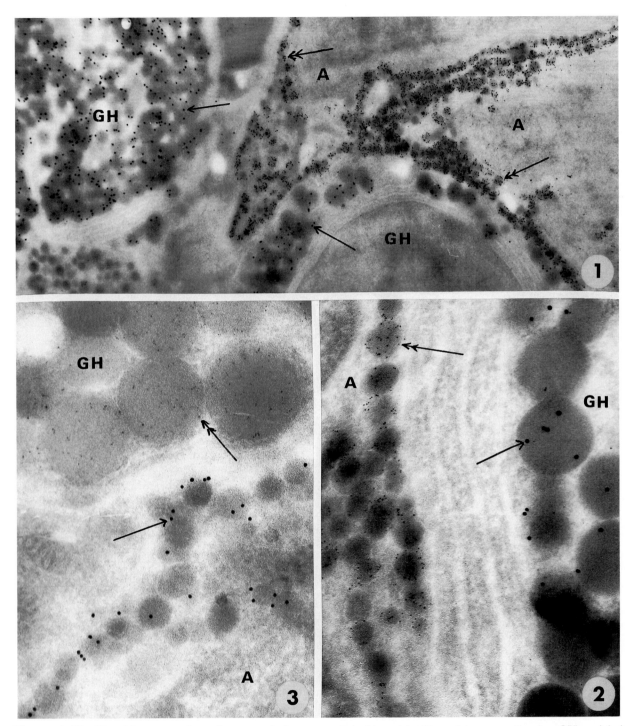

Figure 4-55 Ultrathin cryosection of mouse pituitary showing the localization of growth hormone (GH) in the cells to the left and lower middle areas of the micrograph. ACTH hormone (A) is localized in the cell in the top middle portion of the micrograph. Note the difference in sizes of the gold particles (A particles are smaller than GH particles). The probes used were gold labeled goat anti-rabbit antiserums. (Courtesy of L. Bastholm, M. H. Nielsen, L.-I. Larsson, and Springer-Verlag Publishers.)

References

Ultramicrotomy

Abad, A. 1988. A study of wrinkling on single-hole coated grids using TEM and SEM. *J Electron Microsc Tech* 8:217–22.

Bayer, M. E., and T. F. Anderson. 1963. The preparation of holey films for electron microscopy. *Experientia* 19:1–3.

Coggins, L.W. 1987. Preparation of nucleic acids for electron microscopy. In *Electron microscopy in molecular biology: a practical approach.* J. Sommerville and U. Scheer, eds. (IRL Press, Oxford, England) Part of The Practical Approach Series.

DeBruijn, W. C., and S. M. McGee-Russell. 1966. Bridging a gap in pathology and histology. *J Royal Microscopical Society* 85:77–90.

Fernandez-Moran, H. 1953. A diamond knife for ultrathin sectioning. *Exptl Cell Res* 5:255–6.

Fullam, E. F., and A. E. Gessler. 1946. A high speed microtome for the electron microscope. *Rev Scient Instrum* 17:23–31.

Harris, W. J. 1962. Holey films for electron microscopy. *Nature* 196:499–500.

Latta, H., and J. F. Hartmann. 1950. Use of a glass edge in thin sectioning for electron microscopy. *Proc Soc Exp Biol Med* 74:436–9.

Lindner, M., and P. Richards. 1978. Long-edged glass knives ('Ralph Knives'): their use and the prospects for histology. *Science Tools* 25:61–7.

O'Brien, H. C., and G. M. McKinley. 1943. New microtome and sectioning method for electron microscopy. *Science* 98:455–6.

Porter, K. R., and J. Blum. 1953. A study in microtomy for electron microscopy. *Anat Rec* 117:685–709.

Royer, S. M. 1988. A simple method for collecting and mounting ribboned serial sections of epoxy embedded specimens. *Stain Technol* 63:23–6.

Slabe, T. J., S. T. Rasmussen, and B. Tandler. 1990. A simple method for improving glass knives. *J Electron Microsc Tech* 15:316–7.

Von Ardenne, M. 1939. Die Keilschnittmethode, ein Weg zur Herstellung von Microtomschnitten mit weniger als 10^{-3} mm. Starke fur electronenmikroskopische Zwecke Z. Wissensch. Mikroskopie 56:8–15.

Cryoultramicrotomy

Appleton, T. C. 1974. A cryostat approach to ultrathin 'dry' frozen sections for electron microscopy: a morphological and x-ray analytical study. *J Microsc* 100:49–74.

Bastholm, L., M. H. Nielsen, and L.-I. Larsson. 1987. Simultaneous demonstration of two antigens in ultrathin cryosections by a novel application of an immunogold staining method using primary antibodies from the same species. *Histochemistry* 87: 229–31.

Bernhard, W., and M. T. Nancy. 1964. Coupes a congelation ultrafines de tissu inclus dan la gelatine. *J Microscopie* 3:579–88.

Christensen, A. K. 1971. Frozen thin sections of fresh tissue for electron microscopy, with a description of pancreas and liver. *J Cell Biol* 51:772–804.

Christensen, A. K., and T. E. Komorowski. 1985. The preparation of ultrathin frozen sections for immunocytochemistry at the electron microscope level. *J Electron Microsc Tech* 2:497–507.

Dollhopf, F. L., G. Lechner, K. Neumann, and H. Sitte. 1972. The cryoultramicrotome Reichert. *J Microscopie* 13:152–3.

Fernandez-Moran, H. 1951. Application of the ultrathin freezing-sectioning technique to the study of cell structures with the electron microscope. *Ark Fys* 4:471–83.

Griffiths, G., A. McDowell, R. Back, and J. Dubochet. 1984. On the production of cryosections for immunocytochemistry. *J Ultrastructure Res* 89:65–84.

Parsons, D., D. J. Bellotto, W. W. Schulz, M. Buja, and H. K. Hagler. 1984. Towards routine cryoultramicrotomy. *Bull of Electron Microsc Soc of Am* 14(2):49–60.

Persson, A. 1972. Equipment for cryo-ultramicrotomy. The LKB Cryokit. *J Microsc* 13:162.

Roberts, I. M. 1975. Tungsten coating: a method of improving glass microtome knives for cutting ultrathin frozen sections. *J Microsc* 103:113–9.

Saubermann, A. J. 1986. Comparison of analytical methods for x-ray analysis of cryosectioned biological tissues. *Bull Electron Microsc Soc of Am* 16: 65–9.

Somlyo, A. P., M. Bond, and A. V. Somlyo. 1985. Calcium content of mitochondria and endoplasmic reticulum in liver frozen rapidly in vivo. *Nature* 314:622–5.

Tokuyasu, K. T. 1973. A technique for ultracryotomy of cell suspensions and tissues. *J Cell Biol* 57:551–65.

———. 1989. Use of poly(vinylpyrrolidone) and poly(vinyl alcohol) for cryoultramicrotomy. *Histochemical Journal* 21:163–71.

Zierold, K. 1986. Preparation of cryosections for biological microanalysis. In *The science of biological specimen preparation*. M. Muller, R. P. Becker, A. Boyde, and J. J. Wolosewick, eds. Scanning Electron Microscopy, Inc. (AMF O'Hare, IL 60666) pp 119–27.

Reference Books

Glauert, A. M., and R. Phillips. 1965. The preparation of thin sections. In *Techniques for electron microscopy*, 2d ed. D. H. Kay, ed., Oxford: Blackwell Scientific Publications, pp 213–53.

Hayat, M. A. 1970. *Principles and techniques of electron microscopy: biological applications*, Vol. 1. New York: Van Nostrand Reinhold Co., pp 183–237.

Reid, N. 1975. *Ultramicrotomy*. A. M. Glauert, ed. New York: Elsevier/North Holland Biomedical Press.

Robards, A. W., and U. B. Sleytr. 1985. *Low temperature methods in biological electron microscopy*. A. M. Glauert, ed. New York: Elsevier/North Holland Biomedical Press.

Specimen Staining and Contrast Methods for Transmission Electron Microscopy

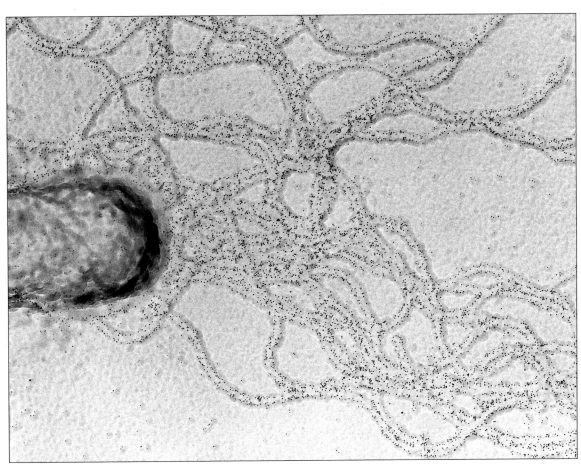

Courtesy of M. Yamaguchi.

In the light microscope, the "white" light that illuminates the specimen usually consists of a spectrum of colors. These colors become readily visible when the light strikes a specimen that has been stained with various dyes. Biological stains absorb certain colors in the spectrum, but transmit or reflect others to the eye. For example, in a specimen that has been stained with two commonly used biological stains, safranin and fast green, certain parts of the cell will appear red (safranin will stain acidic components of the cell such as DNA in the nucleus), while others will stain green (fast green stains basic components of the cytoplasm).

Color does not exist in any electron microscope, since the illumination source is of a single wavelength determined by the accelerating voltage. In contrast, the light microscope uses white light composed of a spectrum of wavelengths. In fact, as explained in Chapter 6, a single wavelength is necessary to minimize the phenomenon termed *chromatic aberration*, which will degrade the resolution of the microscope. Since the illumination used in the electron microscope is of a single wavelength beyond the range of sensitivity of the human eye, one must examine images on viewing screens coated with phosphorescent materials. Such phosphors convert the kinetic energy of the short wavelength electrons into longer wavelength (green to yellow-green) light that can be perceived by the eye. The images generated on such screens consist of various brightness levels ranging from very bright to quite dark so that the electron microscopist sees contrasting shades on a yellow-green screen. The darker shades are associated with areas of the specimen that have greater density, whereas brighter areas of the specimen have less density. Dense areas in the specimen deflect beam electrons to such an angle that they are subtracted from the total image to generate the various tonal ranges or contrast (discussed further in Chapter 6). Since unstained biological tissues have little density differences—even compared to the surrounding environment—contrast is usually quite low.

It is important to increase the contrast of most biological specimens by reacting various cellular components with heavy metals. For instance, the fixative osmium tetroxide reacts with many organic molecules of the tissue and is reduced to metallic osmium. Such osmicated molecules are thereby increased in density and will appear darker on the viewing screen. Only occasionally will this impart sufficient contrast to render an adequate image. In the majority of cases, it is necessary to further increase overall contrast by using other heavy metals. The use of salts of heavy metals (such as lead citrate and uranyl acetate) further increases contrast of osmicated tissues when the metal ions of the salt react with various cellular components. Heavy metal salts that are used to increase contrast levels are therefore termed *stains*. This chapter covers the application of several of the more commonly used stains, as well as the use of various metal shadowing techniques, to enhance contrast of specimens for viewing by TEM. Techniques for enhancing the contrast of SEM specimens are covered in Chapter 3.

Positive Staining

An electron stain is said to exhibit *positive contrast* when it increases the density of a biological *structure* as opposed to the background. In positive staining, the heavy metal salts attach to various organelles or macromolecules within the specimen to increase their density and thereby increase contrast differentially. This differs from the *negative contrast* situation where the background area surrounding the specimen is made dense by a heavy metal salt so that the specimen appears lighter in contrast to the darkly stained background. Figure 5-1 illustrates the positive staining of ultrathin, sectioned bacterial cells, while Figure 5-2 demonstrates negatively stained whole intact bacterial cells.

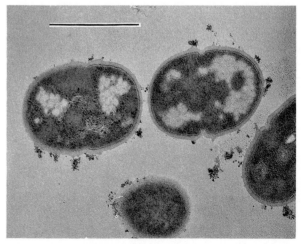

Figure 5-1 Ultrathin section through the gram positive bacterium *Streptococcus mutans*. The cells were positively stained with uranyl acetate and lead citrate. Marker bar = 0.5 μm.

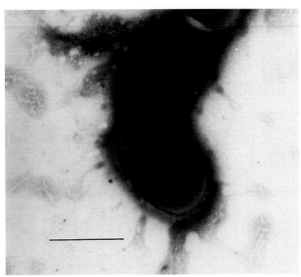

Figure 5-2 Negatively stained cells of *Streptococcus mutans*. In this method, whole cells were mixed with negative stain (2% phosphotungstic acid) and deposited onto Formvar/carbon-coated grids. Marker bar = 0.5 μm.

The Physical Basis of Contrast

The heavy metal salts used as stains in the electron microscope consist of ions of a high atomic number with a large number of protons and electrons that scatter the beam electrons. When beam electrons encounter the atomic *nucleus* of a heavy metal ion, they will be deflected with minimal energy loss (elastic scattering) at such a wide angle that they will not enter the imaging lenses. This subtractive action (see Figure 6-33) gives rise to the various tonalities or shades evident on the screen. This contrast, which is dependent on the elimination of a number of electrons, is termed *amplitude contrast*. The electrons of the heavy metal stain may also deflect the beam electrons to give rise to amplitude contrast.

A low-angle deflection of the beam electrons may also occur so that the beam electrons lose some energy (inelastic scattering), but are still able to enter the imaging lenses. In thin specimens (<60 nm), these lower energy electrons have shorter focal lengths than do higher energy electrons and give rise to *phase contrast*, which may be seen as a light halo surrounding a dense object. In thick specimens, too many different energy levels of electrons enter the imaging lenses, so that resolution is degraded due to chromatic aberration (see Chapter 6).

The two most commonly used positive stains are uranyl acetate (MW = 422) and lead citrate (MW = 1054). The exact mode of action of these stains is not completely understood. It is known that uranyl ions react strongly with phosphate and amino groups so that nucleic acids and certain proteins are highly stained. With lead stains, it is thought that lead ions bind to negatively charged components such as hydroxyl groups and osmium-reacted areas. Phosphate groups may also be involved in this phenomenon, since the use of phosphate buffers often enhances overall staining with lead ions.

Uranyl and lead stains are termed *general or nonspecific stains* when used in the routine manner since they will stain many different cellular components. This contrasts to their use in certain *cytochemical* and *immunocytochemical* techniques where prior treatment of the tissues may render macromolecules reactive with specific stains. For a discussion of cytochemistry and immunocytochemistry, consult Chapters 9 and 10.

Preembedding, Positive Staining with Uranyl Salts

Reaction of the tissues with uranyl stains may take place either before or after embedding in plastic. There are certain advantages to preembedding, or *en bloc*, uranyl staining. In fact, membrane preservation is so enhanced that many cytologists consider uranyl salts to have fixative properties.

Adjustment of the uranyl stain to a pH of 5.2 also leads to the fine structural preservation of DNA filaments as well as cell junctions, mitochondria, myofibrils, nucleoproteins, and phospholipids. Depending on the specimen, ultrastructural preservation is so much better when uranyl ions are used following aldehyde/osmium fixation, that it should be considered for use prior to embedding.

Preembedding Staining Procedure (Karnovsky, 1967)

1. After fixation of the tissues in an aldehyde and osmium fixative (the exact concentrations, buffers, times, etc., vary with the specimen type—see Chapter 2), the specimen is rinsed three times for 15 minutes each in 0.05M sodium hydrogen maleate-NaOH buffer adjusted to pH 5.2. (The maleate buffer is prepared by dissolving 0.58 g maleic acid in 100 ml double-distilled water and adjusting the pH with 0.05 M NaOH [0.2 g NaOH in 100 ml distilled water]).
   NOTE: This rinse must be extended even longer and with more exchanges of fresh maleate buffer if a phosphate buffer was used for the earlier fixations, since phosphate ions will cause precipitation of the uranyl ions in the tissues.
2. Immerse specimens in the freshly prepared uranyl stain consisting of 0.5% to 2% uranyl acetate in the maleate buffer usually for 2 hours at 4° C.
   NOTE: Uranyl salts are photolabile and must be protected from bright light. The stain is good for about one week

and should be centrifuged or micropore-filtered prior to use. The final pH of the stain may vary from 4.2 to 5.2, depending on the concentration of uranyl acetate used. The stain is formulated by adding the desired amount of uranyl salt to pH 6.0 maleate buffer. If 0.5% uranyl acetate is added, the pH drops to 5.2, while 2% solutions will drop to pH 4.2. Precipitation may result if one attempts to titrate the pH. The safest way is to prepare a 4% aqueous uranyl acetate solution and slowly add this to the maleate buffer until the desired final concentration is obtained. If uranyl acetate solutions rise higher than pH 6.0, precipitation will result.
CAUTION: Salts of uranium are toxic and may contain radioactive isotopes. Take precautions to avoid breathing the dust when weighing the powder. It is highly recommended to do all weighing in an enclosed glove box or a fume hood, or to wear a dust mask or filter. The scale and area where the weighing was done must be cleaned with a wet paper towel. Dispose of all uranyl salts and solutions as recommended by the local pollution control authorities. Radioactive uranyl salts are low-level radiation emitters; a radiological control office should be consulted for recommended disposal procedures.

3. Transfer specimen through dehydration series and embed in appropriate plastic prior to sectioning.
   NOTE: The dehydration should be as expeditious as possible since uranyl stains are slowly removed by organic solvents. Prolonged storage in alcohols, acetone, or propylene oxide may remove most of the stain. Small tissue blocks are highly recommended in order to speed the dehydration steps.

Some workers feel that solutions of 0.25% to 2% uranyl acetate in double-distilled water are as effective as the buffered uranyl solutions, so some experimentation is in order to suit the stain to one's own system. It is possible to refrigerate specimens overnight in the more dilute aqueous solutions of uranyl acetate.

Even if tissues have been stained prior to embedding, it is still possible (and advisable) to stain the sections with uranyl salts followed by lead stain as described in the following sections.

Postembedding Staining with Uranyl Salts

Staining Trimmed Specimen Blocks
Most uranyl staining is carried out on sectioned material that has been embedded in plastic. However, it is possible to stain a trimmed plastic specimen block *before* sectioning. Only lead staining will be needed after the sections are cut.

Staining Trimmed Specimen Blocks in Uranyl Acetate

1. Trim specimen block so that specimen is exposed on as many surfaces as possible.
2. Immerse specimen block in a small container of 2% uranyl acetate dissolved in 95% ethanol. Tightly seal the container and place in a 60° C oven for 12 to 24 hours.
   NOTE: Phosphotungstic acid may be used in place of uranyl acetate.
3. Rinse the specimen block quickly in several changes of ethanol, allow the block to dry, and proceed to cut ultrathin sections that may then be stained with lead.

Depending on the type of plastic used, the staining extends to a depth of 10 to 15 μm so that many sections can be cut that will not need subsequent uranyl staining. This method is particularly useful for thick sections to be viewed in a high voltage transmission electron microscope (Chapter 16).

Staining Ultrathin Sections with Uranyl Salts

Aqueous Solutions. One of the more common methods of staining thin sections is with 2% aqueous solutions of uranyl acetate. Usually, staining is accomplished in a simple device constructed from a petri dish, a sheet of dental wax, and a piece of filter paper (Figure 5-3). The filter paper is placed in the bottom of the petri dish, and a small amount of distilled water is added to soak the paper. A piece of dental wax (or Parafilm) is placed over the wet filter

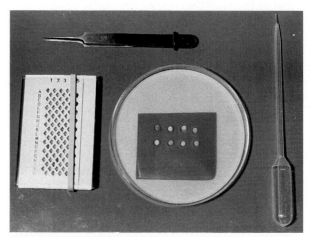

Figure 5-3 Setup required for staining of thin sections using aqueous uranyl acetate. Fine pointed forceps (top) are used to manipulate the grids on the stain droplets on the rectangular piece of dental wax in the petri dish. A storage and filing system for the grids is shown on the left while a plastic pipette for depositing the stain droplets on the dental wax is shown to the right in the photograph.

paper, and drops of the stain are placed on the wax surface. Individual grids are placed onto the droplets to float, section side down, and the petri dish lid is installed.

The moist atmosphere created by the damp filter paper prevents the stain from evaporating, so that staining times of 15 minutes to 18 hours at room temperature can be safely carried out. It is possible to accelerate the staining process by increasing the temperature to 40° to 60° C. If elevated temperatures are used, one must take care not to melt the dental wax or paraffin sheeting. As with all uranyl stains, the solution should be protected from the light by placing the apparatus inside a drawer or simply by covering it with aluminum foil. In addition, the stain should be centrifuged (5,000 × g for 20 minutes) or forced through a syringe microfilter prior to use to prevent precipitates from attaching to grids. Cloudy solutions should be discarded. It is recommended to use freshly prepared solutions of uranyl acetate and not to use solutions that are more than 1 to 2 weeks old.

After staining has been accomplished, the grids are individually removed using a jeweler's forceps and rinsed with drops of distilled water from a plastic squeeze bottle held within 1 cm of the grid. Grids must be rinsed gently so that the supporting films are not broken or the sections lost. Alternatively, the grids can be washed by floating them, section side down, on large droplets of distilled water on dental wax or on distilled water contained in small beakers. One good method involves gently dunking the grids into several beakers of distilled water using 10 to 25 gentle dunks per beaker (Figure 5-4). After rinsing, the grids may be stained immediately with lead stains or dried and stored until lead staining is desired.

If one wishes to store the grids, it is essential to remove the small amount of water remaining between the prongs of the forceps before placing the grid onto a clean surface. If this water remains in the forceps, the grid will be pulled between the prongs when the forceps is spread apart. To prevent this, a wedge-shaped piece of filter paper is placed between the prongs to soak up the water prior to releasing the grid (Figure 5-5). *NOTE: It is important to minimize the carryover of the various staining reagents from one staining solution to another by using the filter paper wedge between the prongs of the forceps.* Normally, one would store the grid on a clean filter paper in a petri dish or in a special grid storage box (Figure 5-3, left).

Figure 5-4 Series of beakers of distilled water used for rinsing stained grids by dipping the grids into each of the beakers.

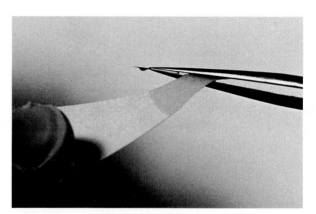

Figure 5-5 A small wedge of filter paper may be used to remove liquids trapped between prongs of forceps that hold grid. This will prevent contamination of grid surface with the liquid.

Helpful Suggestions when Staining

If the grid still adheres to the forceps when one attempts to transfer it onto a filter paper, it is helpful to place a drop of distilled water on the filter paper before placing the grid (section side up) on the moistened spot. The points of the forceps should be kept clean and free of burrs to prevent the adherence of the grid. Forceps points can be cleaned with filter paper and deburred by pulling a fingernail file between the closed forceps prongs. When working with nickle grids, it may be necessary to demagnetize both the forceps and grid by passing them through a demagnetizing loop (Figure 5-6).

Another method of staining using aqueous solutions of uranyl salts is by submerging the grids in

Figure 5-6   Certain types of grids (nickle) as well as the stainless steel forceps often must be demagnetized by passage through a demagnetizing loop. Otherwise, grids will be attracted to the forceps and difficult to handle.

Figure 5-7   Sections can be stained by placing the edge of the grid into heat-softened dental wax and applying a drop of uranyl acetate to completely cover the grid. Caution is necessary not to carry over any wax onto the grid and subsequently into the microscope.

the stain. One way of doing this is to stand the grids on edge by pressing them lightly into heat-softened dental wax and then placing a drop of stain to completely cover the grid (Figure 5-7) as described by Springer (1974). When staining is completed, the

grid is removed by forceps and rinsed by dipping in several changes of distilled water. The advantage of this technique is that uncoated sections are stained on both sides. Take care not to damage the fragile grid or to carry over wax into the microscope. A better way of immersion staining is to use one of the holders available for multiple staining (Figure 5-12) or special glass containers (Figure 5-8).

Uranyl acetate is the most commonly used aqueous uranyl salt; however, 7.5% solutions of magnesium uranyl acetate in triple-distilled water give comparable, or possibly even better, results depending upon the specimen (Frasca and Parks, 1965). They are used in the same manner as previously described for uranyl acetate.

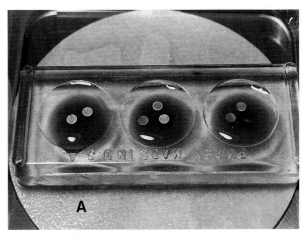

Figure 5-8(A)   Molten dental wax was poured into the depressions in this glass spot plate to form a flat platform. The grids were submerged in the stain and arrayed along the wax surface.

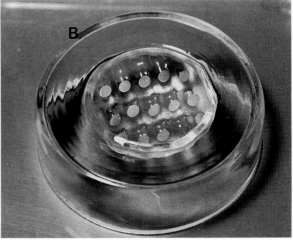

Figure 5-8(B)   Specially designed glass dish with individual pockets to accommodate individual grids that are submerged in the positive stain. (Courtesy of Steve Schmitt.)

Alcoholic Solutions. Some types of tightly cross-linked resins (Spurr's epoxy resin, for example) may not be adequately stained using aqueous solutions. To enhance penetration, it is possible to formulate a stain with the heavy metal salt dissolved in an alcohol such as ethanol or methanol.

To prevent excessive evaporation of the alcohol, a chamber is fabricated as described for aqueous uranyl staining (Figure 5-3) except that the filter paper is moistened with 100% alcohol. Although it is possible to stain the sections by floating the grids on drops of stain on a wax base, care must be taken since the small drops tend to evaporate rapidly and cause precipitation of the uranyl salts. Safer methods for staining using alcoholic solutions are illustrated in Figures 5-7, 5-8, and 5-12.

Staining Methods for Alcoholic Solutions of Uranyl Salts

1. Ethanolic Solutions of Uranium Salts (Epstein and Holt, 1963).

Probably the most commonly used alcohol-based staining solutions range in uranyl acetate concentration from 2% to 4% (weight/volume) to a saturated solution of uranyl acetate in 50% ethanol. *NOTE: Saturated solutions should be prepared at least one day prior to use since uranyl salts dissolve slowly. The stain should be centrifuged immediately before use (5,000 × g for 15 to 20 minutes) in a tightly sealed tube to prevent evaporation.* Staining times of 15 to 30 minutes are used at room temperature for sections of most epoxy-embedded specimens, but it may be necessary to increase the staining time to 1 to 2 hours or longer, and possibly to elevate the temperature to 40° to 60° C, if Spurr's epoxy resin (*CAUTION: Carcinogenic!*) is used. If extended times and temperatures are used, then the grids should be submerged in the stain rather than being floated on small droplets.

After the grids have been stained, they are removed individually using a jeweler's forceps and quickly rinsed in several changes of 50% ethanol. The safest method involves dunking the grids into several small beakers of 50% alcohol, taking care to minimize the carryover of alcohol from one container to the other by using a filter paper to blot the area between the prongs of the forceps. It is very important to transfer the grids quickly into the alcohols to minimize evaporation of the alcohol, which will cause a precipitation of the uranium salts on the sections. After the last rinse, the grids are placed on filter papers to dry and then a second, usually lead, stain may be applied.

If one intends to apply a second stain (usually aqueous solutions of lead salts) to the grids immediately after the last rinse in 50% ethanol, then the grids should be dipped into a beaker of 25% ethanol followed by several changes of double-distilled water. The hydrated sections may then be stained using aqueous solutions.

2. Absolute Methanolic Solutions of Uranyl Salts (Stempak and Ward, 1964).

In this method, a 25% solution of uranyl acetate is prepared in absolute methanol. *NOTE: Absolute methanol is prepared by placing 100% methanol over a drying agent such as Molecular Sieves for several weeks to remove traces of water. Methanol may dissolve certain plastic substrates such as collodion, so supported sections cannot be used in this procedure. In addition, some types of embedding plastics (methacrylates, for instance) may be weakened or dissolved by methanol. Caution is in order with this method.* The grids are submerged in the solution (they will not float on concentrated alcohols) and stained for 10 to 15 minutes.

After staining, the grids are removed using a fine forceps and gently but rapidly dipped 10 to 25 times each in 3 or 4 changes of absolute methanol. The prongs of the forceps should be drained with filter paper after each rinse to minimize carryover of uranyl stain. Great care must be taken to transfer the grid as rapidly as possible to prevent drying, which will cause a precipitation of the uranyl salts onto the surface of the sections (Figure 5-9). After the final rinse, the prongs are blotted and the grid gently placed on a filter paper and allowed to dry completely before staining with aqueous lead stains.

An alternate method involves the use of 2% uranyl acetate dissolved in absolute methanol containing 1% dimethylsulfoxide (DMSO). The DMSO is thought to enhance penetration of the stain into the plastic.

WARNING: This solution must be handled very carefully since DMSO will permit the rapid penetration of uranium ions through the skin. Gloves must be worn, and the solution should be handled in a fume hood.

Postembedding Lead Staining

Reynolds' Lead Citrate

After the specimen has been contrasted by uranyl staining, it is customary to apply a second stain, usually lead citrate dissolved in double-distilled water. The most commonly used lead citrate stain is that described by Reynolds in 1963. Although time consuming to prepare, the stain is stable for many months if kept tightly sealed and protected from carbon dioxide.

Preparation and Use of Reynolds' Lead Citrate Stain

1. In a scrupulously clean, 50 ml volumetric flask, combine 1.33 g lead nitrate, 1.76 g sodium citrate, and 30 ml of CO_2-free double-distilled water (see next section). Shake the solution vigorously for several minutes and then 5 or 6 times over a 30-minute period. This will generate a milky white suspension of lead stain with no large particles. If one sees large chunks, con-

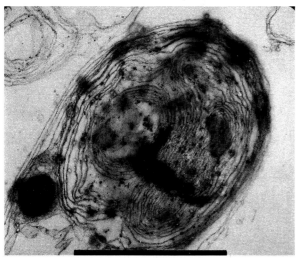

Figure 5-9(A) Grids contaminated with uranyl acetate salts that have precipitated due to slow transfer of the grids from one rinse to another.

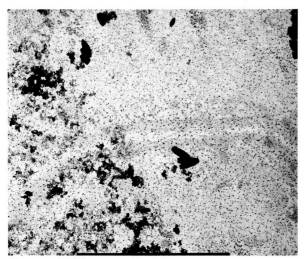

Figure 5-9(B) Lead precipitate usually appears more granular than does the uranyl acetate precipitate.

tinue shaking until the particles are dissociated, or start over.

2. Add 5 to 7 ml of 1N NaOH freshly prepared in CO_2-free water and swirl to mix the two solutions. The milky solution should turn clear. If not, add a few more drops of NaOH up to a total of 8 ml. If the solution still does not turn clear, something is wrong and the stain should be discarded.

3. Withdraw a small amount of clear stain and check the pH using a microelectrode. The pH should be 12.0 +/ −0.1. If the pH is too low, add more NaOH to the clear solution in the volumetric flask in step 2. If the pH is above 12.1, start over, this time adding a smaller amount of 1N NaOH to the milky suspension generated in step 1.

NOTES: A proper pH is extremely important with this stain. If the pH varies by more than 0.1 unit from pH

12.0, poor staining or precipitation will occur. Failure to verify the pH is the major reason for failure of this highly reliable stain. It is not possible to adjust the pH using acidic solutions, since precipitation will occur. A second reason for failure of this stain involves not using CO_2-free water.

CAUTION: Lead salts are extremely toxic. Exercise care in weighing the powders: wear gloves, use a fume hood or filter over face and mouth, wash hands after using, and clean work area thoroughly. Dispose of all lead salts and stains in a manner outlined by pollution control authorities.

4. After the pH has been verified, add CO_2-free water to the volumetric flask to bring the solution to a final volume of 50 ml. To prevent entry of CO_2 and possible spoiling of the staining solution, the volumetric should be tightly stoppered with a plastic (rather than ground glass) stopper.

5. Grids may be stained by flotation on the concentrated stain for 30 seconds to 15 minutes. Methacrylate sections must be stained for shorter times than epoxy plastics. In fact, it may be necessary to dilute the stain 1:10 to 1:100 using 0.01N NaOH to prevent over-staining of some acrylic plastics.

Unfortunately, lead stains are easily precipitated upon contact with CO_2. In order to prevent this, it is necessary to protect the staining solution during storage and especially during the staining process. The latter is easily accomplished by constructing a staining apparatus similar to that used for staining with uranyl acetate. In this case, the filter paper is wetted with 1N NaOH and several pellets of NaOH are placed on the filter paper and around the periphery of the dental wax (Figure 5-10). The NaOH will rapidly scavenge any CO_2 that may waft into the chamber.

Preparing CO_2-Free Water

All double-distilled water used in lead citrate staining should be free of CO_2. Such water may be conveniently prepared by autoclaving screw-capped bottles or flasks of water and then sealing them immediately upon removal from the autoclave (while still very hot). Alternatively, the flasks of water may be boiled 10 minutes and then sealed while hot. When opened, such flasks should emit a hissing sound, indicating that they were properly sealed. Opened flasks of water rapidly absorb CO_2 and should not be used after 10 to 15 minutes. They may, however, be reheated, resealed, and then cooled to room temperature prior to use.

Reynolds' lead citrate stain has a rather long shelf life (up to 6 months) if sealed tightly. Even when slight turbidity is noted, the stain is often still

Figure 5-10 Setup for staining grids with lead citrate. Special precautions are necessary to prevent carbon dioxide precipitation of the lead salts. Sodium hydroxide pellets are normally used to scavenge any carbon dioxide.

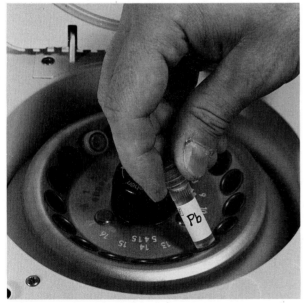

Figure 5-11 Lead stains may be centrifuged before use by placing a small volume in a microcentrifuge tube and spinning the sealed tube at high speeds (15,000 × g) for 5 minutes. Conventional centrifugation or filtration of the stain is also acceptable.

useable after centrifugation (5,000 × g for 10 minutes) or passage through a microfilter. One particularly convenient method of clarification involves placing 1 ml of stain in a polypropylene, sealable microtube (Figure 5-11), and centrifuging the sample at high speeds for 4 or 5 minutes. The sealed tube is impervious to CO_2 for many hours and may

readily be recentrifuged as needed. If microfiltration is used, a few milliliters of stain or 0.01N NaOH must first be run through the filter to remove any wetting agents used in the manufacture of the filter.

Staining in lead citrate is accomplished in the protected environment of the petri dish apparatus by first floating the grid, section side down, on a large drop of CO_2-free water for several seconds. The grid is then removed so that a drop of water remains over the sections, and the grid is transferred to a freshly deposited drop of lead citrate stain. The stain should not be used after standing exposed in the petri dish for longer than 15 to 30 minutes and it should not be reused. After use, the stain droplets should be discarded in an appropriately labeled container for removal by a pollution control authority.

After 30 seconds to 15 minutes, the grid is removed from the droplet of lead citrate so that a large drop remains covering the sections, and the grid is immediately dipped gently 25 times in a small beaker of freshly prepared 0.01 NaOH. After this, the grid is dipped for the same number of times in CO_2-free double-distilled water and rinsed in several changes of plain double-distilled water. The prongs of the forceps must be blotted with filter paper at each transfer step to prevent carryover of stain. After the final rinse, the grid is placed on a filter paper and allowed to dry prior to examination in the transmission electron microscope.

Other Methods for Lead Citrate Staining

Many other methods for lead staining are described in the literature and appear to work as well as Reynolds' lead citrate stain. Each method has its own advantages and problems, so that some trial and error should be expected if attempting some other method. The staining procedures are similar to those for the Reynolds' method and staining times of 2 to 15 minutes are to be expected, depending on the specimen.

It is possible to purchase lead citrate from a number of chemical companies, rather than making it by combining lead nitrate and sodium citrate as in the Reynolds' method. In a popular method described by Venable and Coggeshall (1965), 0.01 to 0.04 g of lead citrate is added to 10 ml of double-distilled water in a sealable centrifuge tube. After adding 0.1 ml of 10 N NaOH, the tube is capped and shaken vigorously until the lead citrate goes into solution. Prior to use, the sealed tube is centrifuged and used in the same manner as described for Reynolds' lead citrate.

In another method described by Fahmy (1967), one pellet of NaOH (0.1 to 0.2 g) is placed into 50 ml of autoclaved, CO_2-free water in a sealable centrifuge tube.

After adding 0.25 g of lead citrate, the sealed container is shaken until the stain is dissolved. This solution has a long storage life, but should be centrifuged before use.

Staining Many Grids

Until experienced in using the various staining methods described, researchers must be careful to prevent ruining valuable sections. At every step, it is possible to introduce contamination and precipitation on the sections (Figure 5-9) or possibly even to wash away the sections in the various solutions. Consequently, the conservative approach is to stain only some of the grids at one time. As confidence in the methodology is gained, then more grids can be stained at one time. When stains are freshly prepared, it is very important to test the stains on some expendible sections in order to quality-control the batch of stain. In the same manner if the stain is thought to be too old, caution dictates that it should be discarded or tested prior to use.

Normally, one can expect to stain 5 to 10 grids per session so that the manual methods described above are quite adequate. When larger numbers of grids (over 100) must be stained on a regular basis, it becomes necessary to resort to multiple or automated grid staining devices. Such systems must be thoroughly evaluated prior to use on valuable specimens or many hours of work may be lost. Some typical multiple-staining devices are shown in Figure 5-12. In addition, programmable machines are available that automate the entire staining process.

Negative Staining

In positive staining, the heavy metal ions react with macromolecules resulting in an increase in the density and contrast of the molecules. By comparison, in negative staining the macromolecule itself is usually unstained but is instead surrounded by the dense stain (Figure 5-2). As a result, the specimen appears in negative contrast (lighter in tone against a dark background). Negative stains are not used on sectioned materials, but are used to contrast whole, intact biological structures (viruses, bacteria, cellular organelles, etc.) that have been deposited on a supporting plastic or carbon film. The principle of this procedure is illustrated in Figure 5-13.

Because whole specimens are deposited on a grid, several conditions must be satisfied in order to

achieve optimal results. A firm, structureless substrate is essential to support the specimen. Collodion or Formvar substrates stabilized with carbon are satisfactory (see Chapter 4). Negatively stained specimens are usually examined at high magnifications, so that a double condenser lens and an anticontaminator in the specimen area are needed to minimize specimen damage. A minimum of beam current must strike the fragile specimen, so the bias must be adjusted to achieve minimal, yet adequate, illumination. Very often, the specimens have not been fixed, but are simply surrounded by the dried stain. In spite of this, the resolutions obtainable with the negative staining method are often better than with sectioned materials since thick plastic embedments are missing. Unlike fixatives, which may chemically combine with the specimen, negative stains enrobe fine structures in a firm, amorphous supporting matrix.

The negative staining technique is quite simple, rapid, and requires a minimum of experience and equipment. It may be used to evaluate cell fractionations (ribosomes, vesicles, tubules, etc.) and even to provide a rapid clinical diagnosis with certain viral infections.

Commonly Used Negative Stains

As was the case with positive stains, negative stains are salts of heavy metals such as uranium, tungsten, and molybdenum. Usually, 1% or 2% aqueous solutions are used, and the specimens can be viewed within minutes of staining. Some of the more popular negative stains are listed in Table 5-1. Uranyl acetate and phosphotungstic acid are the two most commonly used negative stains.

Preparation of Three Commonly Used Negative Stains

Uranyl acetate is dissolved with stirring in double-distilled water to give a 1% or 2% (w/v) final concentration. The solution should be prepared one day in advance since uranyl acetate dissolves slowly in water. Uranyl solutions should be crystal clear, protected from light, and discarded if turbid. Contact with phosphate buffers must be avoided, since they will precipitate uranyl salts. Cacodylate and tris buffers are compatible with uranyl acetate.

Phosphotungstic acid (PTA) is prepared as a 1% or 2% aqueous solution, and the pH is adjusted to 5.0 to 7.0 using 1N KOH. Different pH conditions will affect the staining, so some experimentation is in order with each specimen. When the pH is adjusted above 7.0, the stain becomes unstable and is more prone to precipitation. Below pH 6.0, the stain is stable for many weeks if refrig-

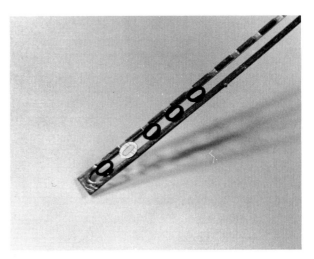

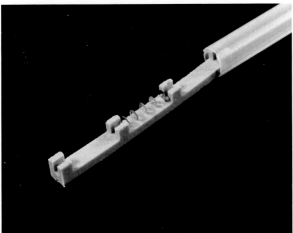

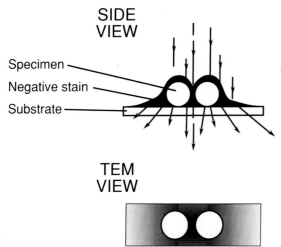

Figure 5-12 Devices for holding multiple grids in position during the staining process. After placing the grids into the various holders, the devices are placed into stain-filled receptacles for staining. Pictured from top left are the holders designed by Hiraoka, Synaptek GridStik™, and Giammara.

Figure 5-13 Principle of negative staining showing side view of specimen surrounded by stain. The electron-dense stain fills in depressions in the specimen and is held tightly around the specimen by surface tension forces.

erated. To improve the spreadability of the stain (to overcome hydrophobic forces on the plastic/carbon substrate), one can add several components to the PTA: several drops of fetal bovine serum, 3 or 4 drops of a 10% solution of soluble starch, or bovine serum albumin to a final concentration of 0.01% (w/v). For osmotic balance, 0.4% sucrose may be added to PTA; however, this will decompose in the electron beam and contaminate the microscope unless anticontaminators are in place over the specimen. Phosphate buffers are compatible with PTA.

Ammonium molybdate, 1% or 2% in double-distilled water, is a useful negative stain when examining certain enzyme subunits, membranes, or cell fractions. The contrast obtained with this stain is not as great as with uranyl acetate, but is comparable to PTA. Ammonium molybdate has a finer background than uranyl acetate and so may permit better resolution of extremely fine details. It does not appear to be as stable as PTA, but has a shelf life similar to uranyl acetate. Compatibility with various buffer systems and osmotic agents should be evaluated prior to use.

Table 5-1 Some Commonly Used Negative Stains

Salt	Preparation	Uses	Reference
Ammonium molybdate	1–3% aqueous	Membranes, enzyme subunits, cell fractions	Muscatello & Horne, 1968
Phosphotungstic acid (PTA)	0.5–2% aqueous, adjust pH to 5.5-8.0 with 1 M KOH	Viruses, bacteria, cell fractions, frozen sections, macromolecules (DNA, actin, enzymes, etc.)	Valentine & Horne, 1962; Horne, 1967
Uranyl acetate	0.5–2% aqueous	Same as above	Van Bruggen et al., 1960
Uranyl magnesium acetate	1% aqueous	Same as above	Valentine & Horne, 1962; Horne, 1967
Uranyl oxylate	12 mM uranyl oxylate + 12 mM oxalic acid. Mix equal parts and titrate to pH 6.5–6.8 with ammonium hydroxide	Small macromolecules	Mellama et al., 1967
Uranyl formate	0.5–2% aqueous solution adjust pH to 4.5–5.2 with ammonium hydroxide	Same as above	Leberman, 1965

Negative Staining Procedures

Drop Method

In this procedure, a Formvar/carbon-coated grid is clamped into a locking forceps, and a drop of the sample is placed on the grid (Figure 5-14). After waiting 1 minute to permit adsorption of the specimen onto the substrate, a drop of negative stain is placed onto the grid and the excess is removed by touching a filter paper to the edge of the grid. After drying for 15 to 30 minutes, the sample can be examined in the electron microscope. It is also possible to mix the sample with an equal portion of stain (1 drop of each) on a Parafilm sheet, to apply the sample/ stain mixture onto the grid, and then to blot and examine the grid as usual. Still another variation involves simply mixing equal portions of sample and stain and then using a pipette or clean platinum loop to transfer the sample/ stain mixture onto coated grids arrayed on a piece of filter paper (Figure 5-15).

Flotation Method

A coated grid is floated on a droplet of sample on Parafilm or dental wax for 1 minute to permit adsorption of the

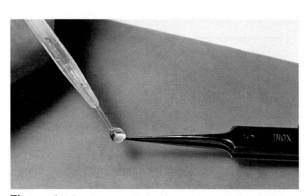

Figure 5-14 One method of negative staining involves deposition of the specimen onto a coated grid that has been locked into a forceps (shown). After allowing the specimen to adsorb to the substrate, a drop of negative stain is applied to the grid.

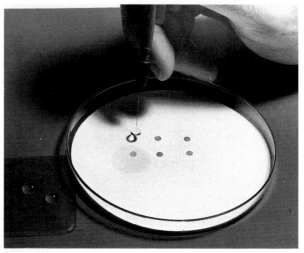

Figure 5-15 Negative staining can also be accomplished by mixing equal volumes of specimen and negative stain and depositing the mixture on a grid using either a loop (as shown) or a pipette.

specimen. The grid is then transferred onto a nearby drop of negative stain for 30 seconds, blotted with a filter paper, and then air-dried for 30 minutes. Alternatively, the sample can be mixed with stain and the grid floated on the mixture.

Spray Method

This technique was popular a number of years ago, but is seldom used except in special circumstances today. Several milliliters of sample is mixed with an equal volume of negative stain and aerosolized in a special nebulizer/sprayer. Tiny microdroplets of specimen/stain then settle onto a number of grids to generate a fine background stain. This technique is dangerous when pathogenic agents are involved—not to mention the possibility of inhaling toxic negative stains unless properly vented hoods are used. Either of the previous methods should give comparable results, so this method probably is not necessary in most cases.

Metal Shadowing Techniques

Besides the negative staining technique, it is possible to enhance contrast and reveal topographic features by spraying particles of vaporized metal onto whole specimens. This technique is used extensively to reveal the fracture surfaces of specimens that are processed for freeze-fracture and -etching (see Chapter 14), but it can also be used to advantage in other studies.

Metal shadowing is commonly used to enhance structural details of small macromolecules such as DNA (Figure 1-5), ribosomes, cell walls, and other cell fractions. The technique requires special vacuum evaporators and skill in the preparation of specimens so as not to introduce artifacts.

In practice, a specimen is placed in a vacuum evaporator (Figures 5-16 and 5-19) and pumped to a vacuum of 10^{-4} Pa or better. Platinum or some other heavy metal (see Table 5-2) is melted and evaporated from a heated electrode (Figure 5-17) so that it is vaporized into a monoatomic state. The metal molecules travel in straight lines until they strike a solid object where they will condense into tiny particles about 2.5 nm in size. Specimen surfaces that face the platinum electrode will receive a heavier deposit of platinum than will areas shielded from the platinum source. Heavily coated areas will be electron dense and appear dark on the viewing screen in contrast to uncoated areas that will be light in appearance (Figure 5-18).

Figure 5-16 Photograph of a vacuum evaporator used for deposition of thin films of carbon and other metals for contrast purposes. (Courtesy of Denton Vacuum.)

Metal Evaporation Procedures

Depending on the material involved, as well as the resolution desired, the shadowing material may be evaporated by several different methods summarized in the following paragraphs.

Vacuum System

A vacuum evaporator (Figure 5-19) is used to achieve the high vacuums needed for the metal shadowing procedure. Most systems consist of rotary and diffusion pumps, although some modern units have dry systems using turbomolecular or cryopumps (see Chapter 6). Whatever system is used, it is very important that the interior of the specimen chamber is clean and that backstreaming from rotary and diffusion pumps be minimized using cryo or other trapping methods. To minimize cleaning of the internal components of the vacuum evaporator's specimen chamber, all evaporative metal sources can be placed inside a special housing equipped with a small aperture from which the metal particles may exit.

The metal source and specimen should be adjustable relative to each other so that distances can be varied and the angle of the specimen relative to the metal source can be adjusted. Such adjustments are particularly important if one wishes to calculate the height of a specimen (see Expression 5-2: How

Table 5-2   Materials for Shadowcasting

Material	M.P. °C	Granularity	Evaporation Method
Carbon	3650	Extremely fine	Carbon rods, braids, electron beam
Chromium	1900	Fine	Coiled filament (basket) molybdenum boat
Gold	1064	Coarse	Tungsten filament, carbon rod
Gold/Palladium 60/40	1465	Coarse	Tungsten filament, carbon rod
Niobium	2470	Extremely fine	Electron beam
Platinum	1774	Extremely fine	Carbon rod, tungsten filament, electron beam
Platinum/Iridium	NA	Extremely fine	Carbon rod, tungsten filament, electron beam
Tungsten	3382	Extremely fine	Electron beam, resistance heating of filament
Tantalum	2996	Extremely fine	Electron beam
Tungsten/Tantalum	3030	Extremely fine	Electron beam

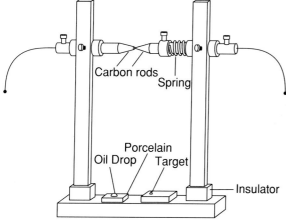

Figure 5-17   Platinum can be melted and evaporated from a heated electrode as shown.

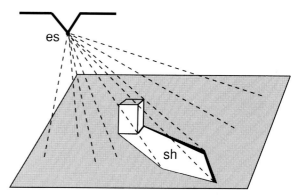

Figure 5-18   Principle of shadowing technique. Heavy metal particles travel in straight lines from the evaporating source (es) and accumulate along areas of the specimen that face the source. Other areas (sh) will receive less coating and appear lighter (less dense) in the TEM.

To Calculate Height Of Specimen). Ideally, the metal source should have a shutter apparatus to shield the specimen from heat until the metal begins to evaporate, at which point the shutter is opened to permit passage of the metal particles through an aperture. An *aperture* is needed to help generate a "point source" of evaporating metal that will greatly enhance the sharpness and contrast of shadows created by the system.

Specimens can be either statically or rotary shadowed. In *static shadowing*, the specimen is placed under vacuum and the metal is allowed to strike the immobile specimen. *Rotary shadowing*, which is commonly used to contrast DNA and other small macromolecules, is carried out using a mo-

torized stage to rotate the specimen during the evaporation process. Such rotated specimens will be shadowed from all directions to give details in areas that might otherwise have been sheltered from the metal particles.

Methods for Metal and Carbon Evaporation

The method that one will use in the evaporation procedure is determined largely by the material to be evaporated as well as the resolution desired.

Electron Beam Evaporation. In one variation of this procedure, a tungsten filament is coiled around (but not in contact with) a high melting point metal such as tantalum. Normally, the tantalum wire is wound around a peg of

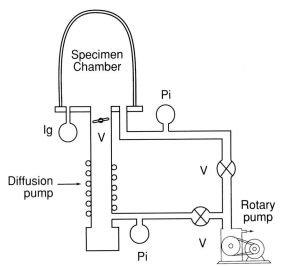

Figure 5-19 Schematic of high vacuum evaporator showing the vacuum system, valves (v) involved in switching vacuum pumps and specimen chamber where evaporation is conducted. Ionization gauge (Ig) and Pirani gauge (Pi) for reading vacuum levels are also shown.

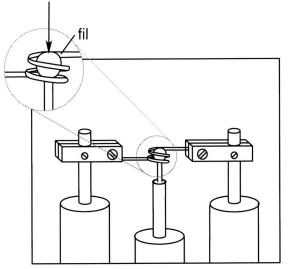

Figure 5-20 Schematic of electron beam evaporator. Electrons generated by the coiled tungsten filament (fil) bombard the tantalum target (arrow) to generate heat and cause the vaporization of tantalum atoms, which then coat the specimen.

tungsten that is then grounded to form an anode (Figure 5-20). After the coiled tungsten filament has been heated, several thousand volts of power is applied to the coil to cause the ejection of electrons from the coil. *NOTE: This is similar to the process used to effect the emission of electrons from the filament in the gun of the electron microscope.* These accelerated electrons strike the tantalum metal on the anode causing it to melt and to be evaporated into the vacuum apparatus. Several different metals can be vaporized using this method (see Table 5-2). A great deal of heat may be generated in this process, so it may be nec-

essary to cool the electrodes. Although the electrodes needed to accomplish this procedure are more expensive than the standard resistance heated electrodes, electron beam evaporation generates some of the highest resolution metal films that can be used in TEM and high resolution SEM studies.

Sputtering (Cathodic Etching). This procedure, which is used for high resolution studies, is based on the principles of the sputter coater employed in the preparation of specimens for SEM studies (Chapter 3). A noble gas, such as argon, is ionized in the presence of a high voltage field, and the positive ions are used to bombard a metal target (Figure 5-21). Electromagnetic lenses are used to focus the ions onto the target that ejects molecules of metal due to the impact of the argon ions. The metal particles settle around the specimen to generate fine metal films with resolutions similar to the electron beam evaporation process. Since these devices do not generate as much potentially damaging heat as the other methods of making thin metal films, it is thought that they eventually will be favored for high resolution studies.

Resistance Heated Electrodes. This is the oldest and still the most extensively used method—primarily because the apparatus needed for this procedure is less expensive than the electron beam evaporator or sputter ion devices. This procedure involves heating an electrode (either carbon or a high melting point metal) with an electrical current so that the material to be evaporated is melted and then vol-

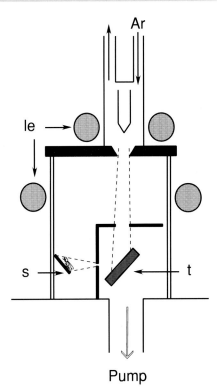

Figure 5-21 Diagram of sputtering (cathodic etching) unit where ionized argon (Ar) atoms are focussed by electromagnetic lenses (le) onto a target (t) to cause the ejection of molecules of metal from the target onto the specimen (s).

atilized from the electrode while in the high vacuum evaporator. Several different materials can be vaporized in this apparatus using various electrode configurations:

(a) Carbon Rods. Carbon is easily evaporated by sharpening spectroscopic-grade soft graphite rods (the hard rods are very difficult to evaporate), such that the end that is heated is shaped into one of the configurations shown in Figure 5-22. The opposite carbon rod provides support for the evaporated rod. A slight amount of tension is established by springs to maintain contact between the two rods (Figure 5-23).

After achieving a vacuum in the 10^{-4} Pa (10^{-6} Torr) range, the carbon rods are heated until the carbon is volatilized. First, one gradually heats the carbon rod until it glows a dull red. After waiting 10 to 20 seconds to purge any adsorbed gases from the carbon rods, the rods are rapidly raised to bright white heat (a variable DC transformer with 30 to 50A of power is recommended) for a few seconds and then lowered to red. After waiting a few seconds for the carbon to settle, the thickness is determined and the "burst of power" procedure repeated as many times as necessary to achieve the desired thickness of carbon. Slow evaporation of carbon is not recommended since excessive heat is generated that may damage the specimen.

NOTE: The thickness of the deposited carbon layer can be roughly gauged by placing a piece of glazed white porcelain (a portion of a broken coffee cup) near the specimen. A small drop of diffusion pump oil placed on the porcelain will prevent the carbon from reaching the porcelain and will show up as a comparatively white zone contrasted to the adjoining area, which will darken from the deposited carbon. As an approximation of carbon thickness, a barely visible, light tan color indicates a thickness of 5 nm, while a light chocolate color indicates about 10 nm.

To avoid the use of diffusion pump oil, some researchers place in the specimen chamber a small piece of white index card containing a raised object, such as a small screw, on top of the card. During the evaporation process, the screw prevents some areas of the card from receiving carbon. These uncoated, shielded areas serve as a relative gauge of thickness when compared to the areas of the card that received a darker coating of carbon. For more accurate measurements, digital resistance and quartz crystal monitors are available.

Due to its high melting point (3650° C), carbon electrodes can serve as a heat source to vaporize lower melting point metals. For instance, platinum and gold/palladium wires may be evaporated from heated carbon electrodes. Often carbon is evaporated along with the metals. This simultaneous evaporation is advantageous since carbon tends to "wet" specimen surfaces and thereby permits the metals to form thinner, more continuous films. To evaporate the metals, a proper length of acetone-cleaned metal wire is tightly wound around the tip of a pointed carbon rod that is then heated under high vacuum.

(b) Carbon Braid. Some EM supply houses sell braided carbon fibers in 3 or 4 foot lengths. This type of carbon is easier to evaporate and is suitable for making filmed carbon grids or for general coating purposes. To use it, a short length of braid is cut with a scissors and placed into special clamp-type electrodes (Figure 5-24). Although less amperage is needed to evaporate this type of carbon, enough heat may still be evolved to damage delicate specimens.

(c) Tungsten Filaments. One may make these filaments from tungsten wire purchased from EM or scientific supply houses. A V-shaped, downward pointing filament is desirable since it maintains the molten metal near the tip to help form a point source. It is also possible to purchase multistranded tungsten filaments preformed into the V-shape (Figure 5-25A). These filaments are highly reliable, may be reused several times, and are less likely to break if overheated. One disadvantage is that they require more power to melt the metal (some vacuum evaporators may not be able to achieve the power levels needed) and more heat will be generated. It is also possible to purchase "tornado-shaped" tungsten filaments (Figure 5-25B) to hold chunks of metal or metal filings and powders (chromium chips, for example).

(d) Molybdenum Boats. These trough-shaped strips of metal (Figure 5-26) are also used to evaporate powders, metal filings, and chips of metal. They require a great deal of power, generating high heat that may affect delicate specimens—so caution is in order. They are useful for cleaning certain components of electron microscopes such as aperture strips. Since the metal tends to flow out into the trough as it melts, it is more difficult to achieve a point source of evaporation so that shadows may be less sharp with this method of evaporation.

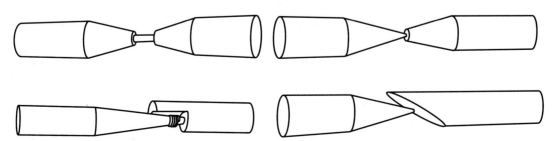

Figure 5-22 Carbon rods can be sharpened into various shapes to form the electrodes that will be used to vaporize carbon or other metals wrapped around the carbon electrodes.

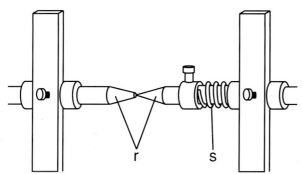

Figure 5-23 Carbon rods (r) are held snugly against each other by using spring tensioned (s) holders.

Figure 5-24 Carbon braid can be used instead of carbon rods as a source of carbon coating. A short length of carbon braid is clamped into the special holder as shown.

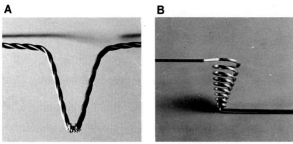

Figure 5-25 (A) Multistranded tungsten filament with platinum wire wound around it. Heating the tungsten filament will melt and vaporize the platinum. (B) Spiral basket of tungsten wire used to melt chunks of metals such as chromium for evaporation.

Expression 5-1: Estimating Amount of Metal Needed

It is useful to know approximately how much of the metal is needed to achieve a particular thickness of evaporated film. The following formula will help in this estimation:

$$M = \frac{4 \pi r^2 t d}{\sin \alpha}$$

where M = weight of metal in grams
         r = distance from source to specimen in cm

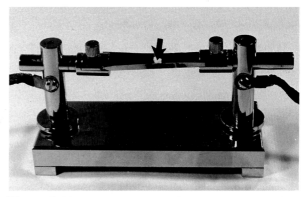

Figure 5-26 Molybdenum trough clamped into electrode holder. Chunks of metal (arrow) may be placed into the trough for evaporation.

---

        t = thickness of deposit in Angstroms
        d = density of metal in gm/cm²
        α = angle of shadowing

Some Applications of Metal Shadowing and Negative Staining

Making Height Measurements Using Metal Shadowing

Metal shadowing can be used not only to reveal morphological features present in a specimen, but it may be used to measure the height of a specimen based on basic geometric principles. Figure 5-27 illustrates the relationship of the specimen to the evaporating electrodes. If one knows several parameters, it is possible to calculate the height of a particle or specimen based on measurement of the length of its shadow.

Expression 5-2: How to Calculate Height of Shadowed Specimens

When small particles or organelles have been deposited onto a coated grid, it is possible to determine the width by directly measuring the image on the negative. The height of the specimen can be calculated using the following equation.

$$H = \frac{b}{c} (l)$$

where H = height of specimen
        b = height to filament from level of specimen
        c = distance from point directly under filament to specimen
        l = length of shadow

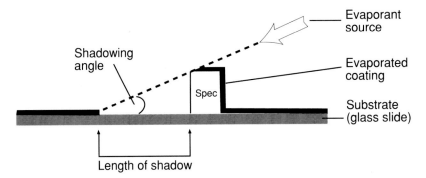

Figure 5-27 Principle of metal shadowing procedure showing deposition of metal along side facing filament source. The specimen physically prevents the vaporized metal from reaching certain areas of the substrate. These "shaded" areas will show up as electron dense, a white shadow will be generated.

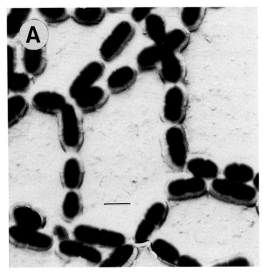

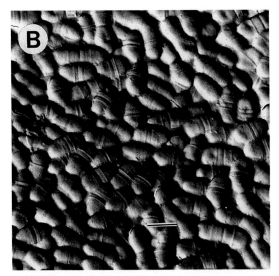

Figure 5-28 Platinum/carbon-coated bacterial cells. (A) *Streptococcus mutans* cells are still present in the preparation so that they obstruct fine details of the coating.

(B) The bacterial cells were dissolved using chromic acid. Only the platinum/carbon replica remains. Marker bars = 0.5 μm.

For example, suppose that the filament is situated 6 cm above the level of the specimen and that the distance measured from a point directly under the filament to the specimen is 12 cm. If one measured the length of the shadow and determined it to be l = 10 nm, then:

The height of the particle, H = 6/12 × 10 nm = 5.0 nm.

In this example, it may be noted that the shadow is twice as long as the specimen height.

One may recognize the term b/c as being equivalent to the geometric equation for calculating the tangent of an angle formed by a line connecting the filament to the edge of the shadow. Consequently, on some vacuum evaporators equipped with a specimen stage that is graduated in degrees, one needs only to measure the length of the shadow and multiply this by the tangent of the angle to determine the height of the particle.

Replication of Biological Surfaces for Transmission Electron Microscopy

The SEM has made possible the investigation of the topographic features of a variety of biological structures. If a SEM is not available or if one needs better resolution than can be obtained in a particular SEM, it is possible to make a replica of a biological surface and to examine the replica in the TEM. (Replication methods for viewing biological specimens in the SEM are covered in Chapter 3, Replication Procedures). The resolution of this technique is approximately 2 to 3 nm; however, specimens with high relief cannot be replicated easily. Several different methodologies are available, depending on the specimen and the nature of the information desired. Only the more commonly used methods will be covered in this book. For more possibilities, consult the reference book by Willison and Rowe (1980).

Platinum-Shadowed, Single-Stage Carbon Replicas. These types of replicas offer the greatest resolution and are to be considered before other methods, if the specimen is suitable. This is the same method used in preparing replicas in freeze-fracture and freeze-etching, except that the specimen is not frozen when the replica is made. The specimen must be well-fixed, dried, and capable of withstanding the conditions inside a vacuum evaporator.

Preparing Single-Stage Replicas

1. Fix and dry the specimen in a manner consistent with the preservation of the details that are to be studied. In some specimens (insects, plant materials such as leaves and seeds, or other hard specimens), this may mean simply allowing the samples to air-dry. Other samples (mammalian, bacterial cells, or other soft tissues) must undergo fixation, dehydration, and drying (critical point or freeze-drying) prior to replication. This technique works best with specimens that have little relief and can be dissolved using appropriate solvents (e.g., the sample should be expendable).
2. Place specimen inside vacuum evaporator and obtain vacuums in the 10^{-4} Pa range.
3. Evaporate platinum onto the specimen surface at a 45° angle followed by carbon evaporated at a 90° angle. The thickness of the two evaporated materials must

be determined by trial and error. Overly thick coatings will lack detail, while thinner ones will be fragile and easily broken during the processing steps.

4. Remove sample and, if possible, score the surface into 2×2 mm squares that will fit conveniently onto a TEM specimen grid. Razor blades or scalpels are normally used in this step.
5. Float off the replica squares from the surface of the specimen by slowly submerging the specimen into a container of distilled water. If the replica does not strip, try treating the specimen with 10% hydrofluoric acid to assist in the separation.
   NOTE: Difficult to strip coatings may sometimes be removed by pressing Scotch Brand tape onto the coated surface and pulling back the tape. The tape is then placed in chloroform to dissolve the adhesive and to free the replicas into the chloroform. After several changes in chloroform, the small replica squares are then placed into the proper organic solvent, as in step 6.
6. Scoop up the replica squares using 30 mesh stainless steel screening (Small Parts, Inc., Florida) and transfer the replicas into an appropriate solvent to dissolve the organic materials: (a) full strength chlorine bleach (sodium hypochlorite) for several hours to overnight, or (b) 50% chromic acid (prepared by dissolving 5 gm sodium dichromate in 100 ml of 50% sulfuric acid). Either of these two solvents should remove any biological materials trapped in the replica; however, other solvents or enzymes may be needed initially to dissolve some types of organic materials.
7. After the organic material has dissolved (as determined by examining one of the processed replica squares in the TEM), scoop up the replica squares with the screening and transfer the replicas through at least four changes of double-distilled water.
8. Pick up individual replica squares by either: (a) lifting the replicas onto uncoated 200 mesh copper grids, or (b) picking up the replicas with a loop and transferring them onto a Formvar-coated grid. Experience is needed in both of these procedures since the replicas will tend to roll and fold over onto themselves.
9. Blot the grids, allow them to dry, and examine them in the TEM.

Figure 5-28 shows a platinum/carbon single-stage replica made of some bacterial cells. Figure 5-28A shows the presence of the bacterial cells in the replica (prior to dissolution in chromic acid) while Figure 5-28B shows only the replica remaining after dissolution of the organic material. Figure 5-29 shows bacterial cells viewed in the SEM.

Two-Stage, Negative Replicas. In bulk specimens that cannot be sacrificed, it may be possible to make a replica of the surface using the two-stage, negative replica procedure. The resolution is not as good

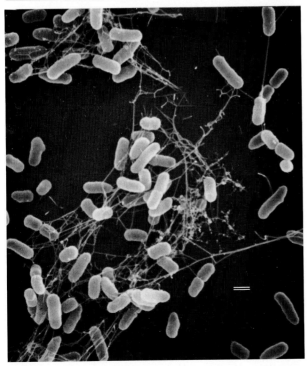

Figure 5-29 Some bacterial cells as in Figure 5-28 only viewed in the SEM. Marker bar = 0.5 μm.

Two-Stage Replication

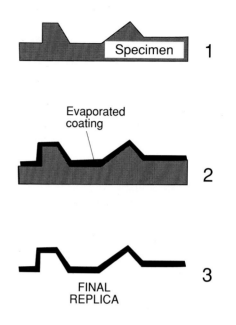

Single-Stage Replication

Figure 5-30 Comparison of the single stage and two-stage replication procedures. As may be expected by examination of the steps involved, the single-stage procedure yields replicas with better resolution.

with this technique, however, since one first makes a plastic replica of the biological surface and then coats the plastic with platinum/carbon. A schematic comparison of the single- and two-stage replication methods is given in Figure 5-30.

Preparing Two-Stage, Negative Replicas

1. Onto a suitable specimen (fixed, dried, etc.) apply a thin layer of 1% collodion in amyl acetate. When dry, apply several more layers of 2% collodion.

2. With a blade, score the surface into squares and strip the plastic film from the surface as described in step 5 of the single-stage replication procedure. If adhesive tapes are used to assist stripping of the collodion replica, avoid chloroform solvents since they will dissolve the plastic replica.

3. Pick up the plastic replica, place it on an uncoated grid, and blot it dry.

4. Place the grids in the vacuum evaporator with the side that was in contact with the specimen surface facing up.

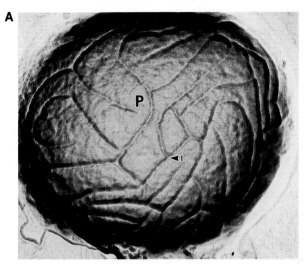

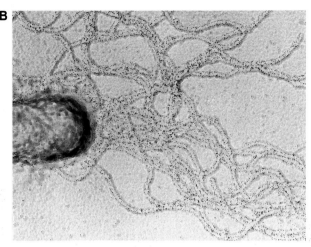

Figure 5-31 (A) Replica of the surface structure in a yeast cell showing invaginations (I) of the plasma membrane (P). The replica was produced using the plasma polymerization method of Tanaka, Sekiguchi and Kuroda (1978). (Courtesy of T. Hirano). (B) Replica of surface of the bacterium, *Pro-* *teus vulgaris,* prepared by plasma polymerization method. This replica includes colloidal gold particles that localize antigens on the flagella. Notice how the bacterial cell (C) is not labeled, indicating that these antigens are restricted to the flagella. (Courtesy of M. Yamaguchi.)

5. Prepare the platinum and carbon coatings as described in step 3 of the single-step replication procedure.
6. Gently lower the grids into chloroform to dissolve the collodion (as evidenced by the platinum/carbon replica becoming free of the grid).
7. Scoop up the replica onto the grid and allow the chloroform to evaporate prior to placing the grid in the TEM.

Extraction Replicas. This procedure is useful to physically extract insoluble components from the biological specimen for further study in the analytical electron microscope. It would be useful, for example, in the extraction and identification of carbon particles, asbestos, talc, etc., from lung or other tissues. Briefly, the tissues are covered with a layer of 5% to 10% polyvinyl alcohol (PVA) in water. After drying, the thin film is stripped from the specimen (pulling with it any particles in the tissues). After coating with carbon, the PVA is dissolved in water, leaving behind the carbon replica containing the particles, which are examined in the electron microscope. For more detail, see the reference by Henderson (1975).

Plasma Polymerization Replicas. Tanaka and colleagues (1978) described a novel procedure for producing a high resolution replica by polymerizing naphthalene gas onto the surface of a specimen. The naphthalene gas is introduced into an evacuated bell jar and exposed to a high voltage discharge for 10

seconds. This polymerizes the gas into a thin film that clings to the surface of the specimen. A replica may then be produced by dissolving the biological material using a 1% solution of sodium hypochlorite for 5 to 10 minutes. The replicas may also be produced from the surfaces of frozen-hydrated or dried specimens, and replicas may be made from cells that have been labeled with colloidal gold probes (Figure 5-31). (See Chapter 9 and reference by Yamaguchi and Kondo, 1989.)

Visualizing Macromolecules
The imaging of the nucleic acids, DNA and RNA, is routinely done using either negative staining or metal shadowing techniques. Although better contrast may be obtained using metal shadowing, resolution will generally be better with negative staining techniques.

Rotary metal shadowing is preferred to static shadowing of nucleic acid molecules. In this procedure, negatively charged nucleic acids are usually suspended in a stabilizing *hyperphase* solution (consisting of ammonium acetate, EDTA [ethylenediaminetetraacetate], cytochrome c, and about 0.1 to 0.5 μg nucleic acid/100 μl of total solution). The hyperphase is allowed to flow down a scrupulously clean microscope slide onto either a distilled water surface or the surface of a dilute solution of salts (the *hypophase*). The molecules spread over the surface of the hypophase, and the cytochrome c holds

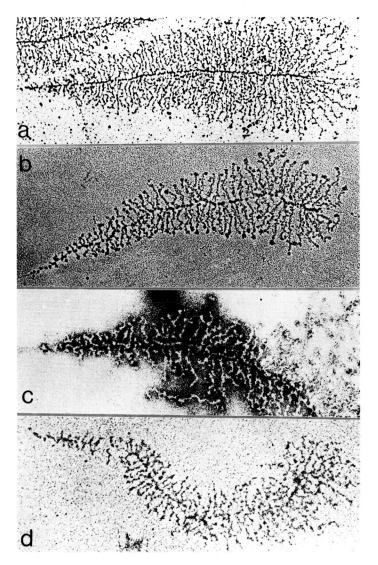

Figure 5-32 Different modes for visualization of spread amphibian genes in the act of transcription. The central dark line is DNA while the fibrils jutting out from the DNA are pre-RNA transcriptional units. (a) The most frequently used staining method for spread chromatin is phosphotungstic acid in 70% ethanol. (b) Phosphotungstic-acid stained preparations may be enhanced for contrast by additional shadow-casting using a heavy metal. (c) When phosphotungstic acid at a neutral pH is used, contrast reversal may be encountered. (d) In some instances, if the spread nucleic acids are picked up on extremely thin carbon films, the genes may be visualized without any additional heavy metal staining. Marker bars = 0.5 μm. (Courtesy of M. F. Trendelenburg and IRL Press.)

the nucleic acids firmly in a monomolecular film of denatured protein.

DNA Spreading and Shadowing Technique

The hyperphase solution containing the nucleic acid is allowed to spread over the surface of the hypophase (0.25 M ammonium acetate, for instance) that has been dusted with talcum powder or powdered graphite. The nucleic acid/cytochrome c molecules will push the talc aside so that cleared areas containing specimen will be outlined by the talc/graphite. Collodion-coated grids are gently touched to the clear zones of nucleic acid to permit adsorption of the nucleic acid/cytochrome c and then blotted gently by a filter paper touched to the edge of the grid. The grids can then be floated on a stain consisting of 50 mM uranyl acetate and 50 mM HCl in 95% ethanol. After rinsing in 95% ethanol, the grid is placed on a filter paper and allowed to air-dry. The dried grid is then placed on the stage of a vacuum evaporator, pumped to high vacuum, and rotated as a shadowing metal is evaporated. Figure 1-5 shows some DNA molecules that have been prepared using this technique. For more procedural detail, see the book edited by Sommerville and Scheer (1987).

Negative staining is conveniently used to reveal fine structural details of nucleic acids, enzyme subunits, and a number of other macromolecules (Figure 5-32). In the simplest approach, the macromolecule-containing sample is mixed with the negative stain and deposited on a collodion- or Formvar/carbon-coated grid as described previously in this chapter. After blotting with a filter paper, the grid is allowed to dry completely prior to examination in the electron microscope. A minimum of technical skills and equipment is needed with this procedure.

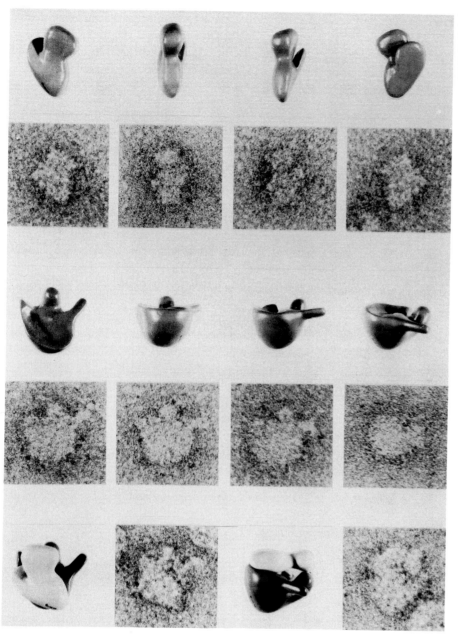

Figure 5-33 Three dimensional model of ribosome that was constructed by examining many different views of negatively stained ribosomes. (Courtesy of J. Lake.)

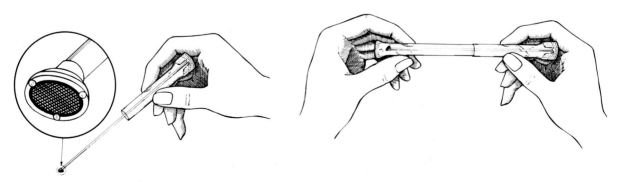

Figure 5-34 Diagram of portable device for collecting viral specimens from patients. (Courtesy of Ted Pella, Inc.)

Imaging Subcellular Components by Negative Staining

When cells have been fractionated into subcellular components by a variety of procedures, it may be necessary to examine the components to determine purity of the preparation or possibly to study the ultrastructure of a specific constituent.

The structure of the ribosome, for instance, was elucidated largely by examination of thousands of negatively stained ribosomal particles to generate models that represented three-dimensional views of the organelle. Figure 5-33 shows a panel of such views and a model that was generated from them by Dr. James Lake.

It is possible to view any subcellular component (membrane fractions, miscellaneous cytoplasmic filaments such as actin and myosin, mitochondria, etc.) using this procedure. The only limitation is that the structure not be so thick that it impedes the passage of electrons.

Revealing Viruses Using Negative Staining

It is possible to use negative staining in clinical situations as well as in research settings. Virologists have used the negative staining technique for over thirty years to study viral preparations in the laboratory, but the procedure of examining clinical isolates for the presence of virus is only occasionally used today—despite the potential for rapid diagnosis. In the laboratory, one method involves mixing a drop of viral preparation with a drop of negative stain and floating a coated grid on the surface of the mixture for 1 minute. After blotting and drying, the grid is examined in the electron microscope. Figures 19-129 to 19-131 shows some virus images obtained using this procedure.

A slightly different approach is needed when clinical specimens are involved. Some infectious samples (biopsy tissues, feces, urine, pus, etc.) may require clarification and concentration by high-speed centrifugation in order to increase the number of viral particles, whereas others (vesicle fluid, for example) may be examined directly. It may prove beneficial to fashion a portable sampling kit as described in Figure 5-34 for use in a clinical setting. Such devices are small, easily used, and mailable if an electron microscope is not available on site. For more information regarding the use of electron microscopy in viral diagnosis, consult the reference by Hsiung and Fong (1982).

References

Bozzola, J. J. 1987. Clinical sampling device for rapid viral diagnosis by transmission electron microscopy. *J Electron Microsc Tech* 5:243–8.

Epstein, M. A., and S. J. Holt. 1963. The localization by electron microscoppy of HeLa cell surface enzymes splitting adenosine triphosphate. *J Cell Biol* 19:325–6.

Fahmy, A. 1967. An extemporaneous lead citrate stain for electron microscopy. *Proc 25th Annu EMSA Meeting*, pp 148–9.

Frasca, J. M, and V. R. Parks. 1965. A routine technique for double-staining ultrathin sections using uranyl and lead salts. *J Cell Biol* 25 (No. 1, Pt. 1): 157–61.

Hayat, M. A. 1972. *Principles and techniques of electron microscopy: biological applications*, Vol. 2. New York: Van Nostrand Reinhold Co.

Henderson, W. J., et al. 1975. Analysis of particles in stomach tumors from Japanese males. *Environ Res* 9:240–9.

Horne, R. W. 1967. Electron microscopy of isolated virus particles and their components. In *Methods in virology*, Vol. 5. K. Maramorosch, and H. Koprowski, eds. New York: Academic Press, pp 521–74.

Hsiung, G. D. and C. K. Y. Fong. 1982. *Diagnostic virology illustrated by light and electron microscopy*. New York: Yale University Press.

Karnovsky, M. J. 1967. The ultrastructural basis of capillary permeability studied with peroxidase as a tracer. *J Cell Biol* 35:213–36.

Leberman, R. 1965. Use of uranyl formate as a negative stain. *J Molecular Biol* 13:606.

Lewis, P. R. and D. P. Knight. 1977. *Practical methods in electron microscopy: staining methods for sectioned material*. A. M. Glauert, ed. Amsterdam, The Netherlands: Elsevier/North-Holland Publishing Co.

Mellama, J. E., E. F. J. Van Bruggen, and M. Gruber. 1967. An assessment of negative staining in the electron microscopy of low molecular weight proteins. *Biochim Biophys Acta* 140:180–2.

Muscatello, U., and R. W. Horne. 1968. Effect of the tonicity of some negative-staining solutions on the elementary structure of membrane-bound systems. *J Ultrastructure Res* 25:73–9.

Reynolds, E. S. 1963. The use of lead citrate at high pH as an electron-opaque stain in electron microscopy. *J Cell Biol* 17:208–12.

Sommerville, J. and U. Scheer. eds. 1987. *Electron microscopy in molecular biology: A practical approach.* Oxford: IRL Press.

Springer, M. 1974. A simple holder for efficient mass staining of thin sections for electron microscopy. *Stain Tech.* 49:43–6.

Stempak, J. G., and R.T. Ward. 1964. An improved staining method for electron microscopy. *J Cell Biol* 22:697–701.

Tanaka, A., Y. Sekiguchi, and S. Kuroda. 1978. *J Electron Microsc.* 27:378–81.

Trendelenburg, M. F. and F. Puvion-Dutilleul. 1987. Visualizing active genes. In: *Electron microscopy in molecular biology: A practical approach.* J.

Sommerville and U. Scheer, eds.: IRL Press, Oxford, England.

Valentine, R. C., and R. W. Horne. 1962. An assessment of negative staining techniques for revealing ultrastructure. In: *The interpretation of ultrastructure.* R. J. C. Harris, ed. New York: Academic Press, pp 263–78.

Van Bruggen, E. F. J., E. H. Wiebenger, and M. Gruber. 1960. Negative-staining electron microscopy of proteins at pH values below their isoelectric points. Its application to hemocyanin. *Biochim Biophy Acta* 42:171–2.

Venable, J. H., and R. Coggeshall. 1965. A simplified lead citrate stain for use in electron microscopy. *J Cell Biol* 25(No. 2, Pt. 2):407–8.

Willison, J. H. M., and A. J. Rowe. 1980. *Practical methods in electron microscopy, Vol. 8: Replica, shadowing and freeze-etching techniques.* A. M. Glauert, ed. Amsterdam, The Netherlands: Elsevier/North-Holland Publishing Co.

Yamaguchi, M., and I. Kondo. 1989. Immunoelectron microscopy of *Proteus vulgaris* by the plasma polymerization metal-extraction replica method: differential staining of flagellar (H) and somatic (O) antigens by colloidal golds. *J Electron Microsc* 5: 382–8.

The Transmission Electron Microscope

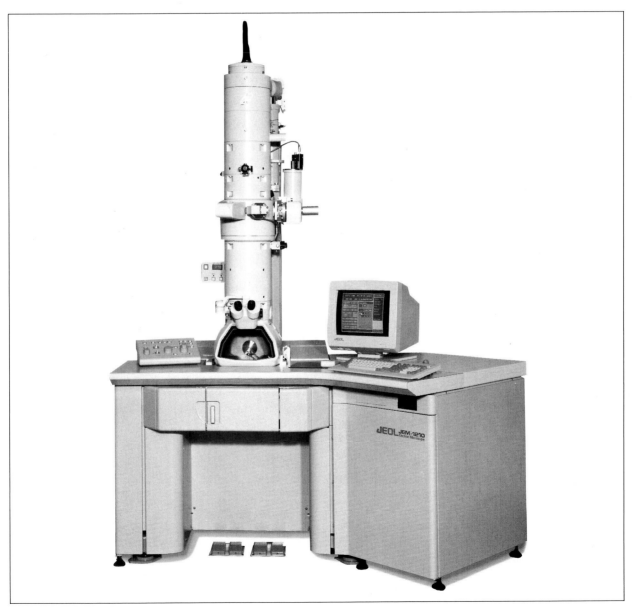

Courtesy of JEOL.

A transmission electron microscope, or TEM, has magnification and resolution capabilities that are over a thousand times beyond that offered by the light microscope. It is an instrument that is used to reveal the *ultrastructure* of plant and animal cells as well as viruses and may provide an image of the very macromolecules that make up these biological entities. The TEM is a complex viewing system equipped with a set of electromagnetic lenses used to control the imaging electrons in order to generate the extremely fine structural details that are usually recorded on photographic film. Since the illuminating electrons pass *through* the specimens, the information is said to be a *transmitted* image. The modern TEM can achieve magnifications of one million times with resolutions of 0.1 nm.

The concepts of magnification and resolution may be understood as one views the letters on this printed page. If one moves away from the page, the distinct image of various letters slowly blurs into one indistinct object, as if the letters had merged. Conversely, if one moves closer to the page, two phenomena occur: the letters become larger and sufficient detail becomes apparent to distinguish the individual letters again. We have, in fact, "zoomed in" on the letters to both magnify and resolve finer detail. In a strict sense, *magnification* is a measure of the increase in the diameter of a structure, and *resolution* is the ability to discriminate two closely placed structures that might otherwise appear as one and to see more details within the objects.

Higher magnification does not necessarily mean better resolution or vice versa. It is possible to enlarge an object without being able to see or resolve any more detail. For example, advertisements in comic or adventure books sell various "microscopes" with magnifications of over a thousand times. While such toys are able to magnify objects as claimed, the magnified images produced are very inferior because the overextended or *empty magnification* lacks resolution. Both magnification and resolution are necessary to produce a quality image.

Small objects are normally brought close to the eye in an attempt to magnify and resolve finer detail. However, the human eye cannot resolve two points that are any closer than 0.1 to 0.2 mm, no matter how close to the eye the object is placed. A simple hand lens may be used to project both a magnifed and resolved image of an object on the retina of the eye so that the object can be seen in greater detail. The lenses of the electron microscope follow the same optical principles of magnification and resolution to permit one to observe the ultrastructure of biological specimens. Unlike glass lenses, which use visible light to form magnified and resolved images on the retina, the lenses of the electron microscope use shorter wavelengths of electromagnetic radiation that are not directly visible to the human eye. How the electron microscope uses such electromagnetic lenses to generate visible images on the human retina is the subject of this chapter.

Visible Light, Electrons, and Lenses

Electromagnetic Radiation and the Diffraction Phenomenon

Visible light represents a very small segment of a spectrum of waves making up the family known as *electromagnetic radiation* (Table 6-1). All of these waves travel at the speed of light and differ from each other only in the distance from the top of one wave to the next (Figure 6-1). Starting at the long wavelength end of the spectrum and proceeding toward the shorter, there are the following waves: radio, infrared or heat, visible light, ultraviolet, X rays, gamma rays, and cosmic rays. Radio waves may have wavelengths ranging from several miles to several millimeters, while cosmic rays have wavelengths measured in femtometers.

Electromagnetic waves radiate from a source. For radio waves, the source is the transmitter, while a possible source of light waves may be a tungsten filament in a light bulb. The radiating waves emanate in ever-widening circles from the source until they come in contact with a solid object. When the waves strike the solid object, another series of waves is radiated from the edge of the object. The result is a new source of waves that merges with the original waves so that the light now appears to bend around the corner (Figure 6-2). This phenomenon is called *diffraction*.

Both the wavelike nature of electromagnetic radiation and the diffraction phenomenon may be readily demonstrated by observing the action of waves in a bowl of water. A drop of water allowed to fall into the bowl will generate a series of waves that radiate from the point where the drop entered the water. When the waves encounter a solid object such as a wall with an opening or aperture, another series of waves will be generated from the free edges

Table 6-1 Electromagnetic Radiations

Radiation Type	Usual Source	Usual Detector	Wavelength Ranges (Meters)
Radio	Electric Circuits	Electric Circuits	
AM			545–188
FM			3.40–2.79
Television			5.55–0.34
Radar			3×10^{-1}–3×10^{-3}
Infrared (heat)	Hot Objects	Thermometers Thermocouples Nerve Cells	3×10^{-3}–8×10^{-7}
Visible Light	Electric Arcs, Hot Objects	Photocell, Photographic Film, Eye	8×10^{-7}–4×10^{-7}
Ultraviolet	Electric Arcs	Photocell, Photographic Film	4×10^{-7}–1×10^{-9}
X Rays	Impact of electrons on metal target	Photographic Film, Ionization Chamber, Geiger Counter	1×10^{-9}–1×10^{-11}
Gamma Rays	Radioactive Nuclei	Ionization Chamber, Geiger Counter	$< 1 \times 10^{-11}$

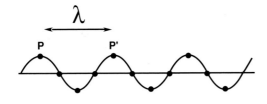

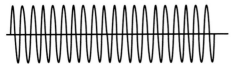

Figure 6-1 Wavelength (λ, top) is measured from the top of wave P to the top of wave P′. A short wavelength (bottom) has a smaller distance between waves.

of the wall, creating the diffraction phenomenon (Figure 6-3).

When the diffraction phenomenon occurs with electromagnetic radiations, the diffracted waves *interfere* with the initial illuminating waves' front. This results in an *unsharp image of the edge of the object being irradiated.* The edge appears to have a series of bands or fringes, called *Fresnel fringes* (named after the French physicist Augustin Fresnel and pronounced *fre-nell*) running parallel to the

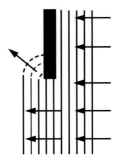

Figure 6-2 Diffraction phenomenon demonstrated by a series of parallel waves that strike the edge of a solid object. From the edge, a new series of waves (dashed lines) are generated that merge with the original front.

edge. Unless magnified, these bands will not be seen individually but will appear to blend together to give the impression of a fuzzy rather than a distinct edge. When diffraction occurs around a solid object, a series of concentric halos or fringes appear around the object. The net result is that the object appears slightly larger and indistinct (Figure 6-4).

Effect of Diffraction on Resolution

Resolution is degraded due to the diffraction phenomenon. Classically, loss of resolution has been illustrated by viewing a series of pinholes made in

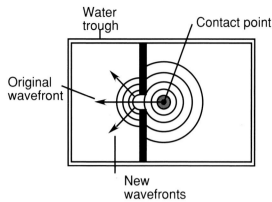

Figure 6-3 Demonstration of the diffraction phenomenon in a trough of water. A drop of water striking the water surface gives rise to waves that generate two more series of waves when they contact the solid walls extending into the water. These two new wavefronts merge with the original front that continues through the opening, or aperture, in the walls.

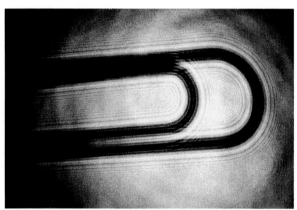

Figure 6-4 Diffraction of light using a solid object illuminated by a point source of light. The paper clip was illuminated by a laser and the shadow image recorded on a piece of photographic film. The diffracted waves interfere with the original waves, giving rise to the series of bands or Fresnel fringes around the object.

a thin metal foil. If the foil is held up to a strong light source and viewed with a lens, sharp points of light are not seen. Instead, one sees a bright central area surrounded by a series of Fresnel fringes originating from the edges of the foil (Figure 6-5). In fact, even if one could generate a true point-source of light (as opposed to looking at light through holes), the Fresnel fringes would still be generated since the waves would diffract from the edges of the lens or from any *apertures* in the lens system.

Apertures are simply holes of various sizes that may be used to control the amount of light passing through a lens. Most 35 mm cameras, for instance, have adjustable apertures that control the exposure by varying the amount of light striking the photographic film. The f/stop markings on the camera refer to the ratio of the diameter of the aperture to the focal length of the lens. Thus, a 5 mm aperture opening in a 55 mm lens has an f/stop of f/11. The amount of illumination passing through a lens is inversely related to f/stop: smaller numbered f/stops allow more light to pass. The f/stop of a transmission electron microscope lens with a focal length of 2 mm and a 50 μm aperture would be f/40 (i.e., 50 μm/2,000 μm).

Figure 6-6 is a photograph of some backlighted pinholes viewed in a light microscope. These ringed patterns are termed *Airy discs* after the 19th century astronomer Sir George Airy, who first described this pattern. The ringed patterns, or *circles of confusion*, tend to increase the apparent size of the holes so as to cause an overlap in some places. If the holes are too close, the overlap would obscure the distinct nature of the holes. Thus, *resolution is decreased by the diffraction phenomenon.*

To determine resolving power, it is important to know the radius of the Airy disc. The radius of the Airy disc as measured to the first dark ring is expressed by Equation 6-1:

Equation 6-1: Radius of an Airy disc

$$r = \frac{0.612\,\lambda}{n(\sin\,\alpha)}$$

where 0.612 = a constant
λ = wavelength of illumination
n = refractive index
α = aperture angle of the lens

The constant, 0.612, is based on an equation derived from the observation of self-luminous points such as stars. The use of this constant in microscopy is problematical, since one deals with objects that generate diffraction effects when the illumination strikes them.

The term *refractive index*, (*n* in the equation) is a measure of the optical density of a medium. Light passing through a medium with a high refractive index is slowed down. When light passes through two media with dissimilar refractive indices, it will bend at the interface of the two rather than continue in a straight line. If one wishes light to continue in a straight line, the refractive indices of all of the media through which it passes must be similar. Some typical refractive indices are: air = 1.000, water = 1.333, glass = 1.5–1.6, standard immersion oil = 1.515. A vacuum is considered to have a refractive index of one.

Aperture angle (α in the equation) refers to the half angle of the illumination a lens can accept. The larger this acceptance angle, the more information will enter the lens. Consequently, as shown in Figure 6-7, a lens with a large aperture angle will accept more information from

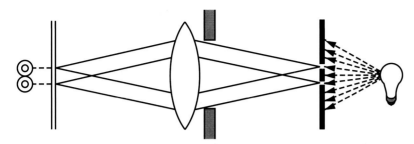

Figure 6-5 Degradation of resolution due to diffraction. Two pinholes held in front of a light source are viewed in a lens and the image is projected onto a flat surface. Instead of two sharp, bright spots, one sees two spots surrounded by diffuse rings (left). These enlarged, indistinct spots are caused by Airy discs. The thick cross-hatched barrier in front of the lens is an aperture.

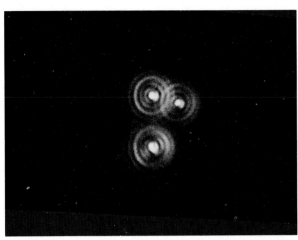

Figure 6-6 Airy discs generated by viewing three pinholes in a light microscope. A thin film of palladium/gold was deposited onto a glass slide, and the slide was examined for naturally occurring pinholes in the film. Magnification of micrograph is 1,000×.

the object being viewed. In Figure 6-8, we see that the use of immersion oil in the light microscope increases the aperture angle and allows more information to effectively enter the lens, thereby increasing resolution.

The term *numerical aperture* or N.A. is equivalent to $n(sin\ \alpha)$ and is often engraved on the barrel of light microscope objective lenses. The higher this value, the greater is the information-gathering ability (and resolution) of the glass lens. The highest N.A. currently obtainable in glass lenses is around 1.5.

As early as 1896, Rayleigh established the criterion that if Airy discs are placed in such close proximity that their first dark rings contact each other, overlap will occur to the extent that they can no longer be resolved as two distinct units. Therefore, it is apparent that *the equation for the radius of the Airy disc is the equation for resolving power.* We can formally define *resolving power* as the minimum distance that two objects can be placed apart and still be seen as separate entities. Consequently, the shorter the distance, the better (or higher) is the resolving power of the system.

In order to obtain the best resolving power in an optical system, it is necessary to use the shortest possible wavelength and the largest aperture angle. For example, with a light microscope one would specify violet light (wavelength = 380 nm, from Table 6-2), immersion oil with a high refractive index, and a lens with a wide acceptance angle to obtain a resolving power of approximately 0.2 μm, as shown in Equation 6.2.

Equation 6-2: Calculation of Optimal Resolving Power of Light Microscope

$$r = \frac{0.612 \times 380\ nm}{1.5 \times 0.9} \qquad r = 172\ nm$$

Note: 0.9 is the value of the sine of a 64° angle, representing half the 128° acceptance angle of a lens (a typical value for a glass lens).

Thus, the theoretical best resolving power of the light microscope is 0.172 μm. The figure usually quoted is 0.2 μm.

In order to achieve the widest possible acceptance angle with any lens, the focal length of the lens must be as short as possible (Figure 6-7). This explains why the high resolution, oil-immersion objective lens on the light microscope is so close to the specimen (Figure 6-8).

Electrons, Waves, and Resolution

Physicists have demonstrated that, besides being discrete particles having a negative charge and a mass of 9.1×10^{-23} kg, *electrons also have wave properties.*

Table 6-2 Wavelengths of Visible Light

Color	Wavelength in nm
red	760–630
orange	630–590
yellow	590–560
green	560–490
blue	490–450
violet	450–380

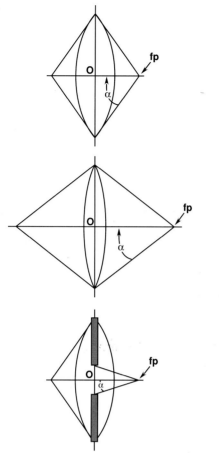

Figure 6-7 As aperture angle (α) increases, resolution improves. A high resolution lens (top) is a strong, short focal length lens with a large aperture angle formed when the lens is brought close to the specimen. Lower resolution lenses (middle) are generally weaker, with longer focal lengths and smaller aperture angles. An aperture placed into a high resolution lens (bottom) may diminish resolution, but contrast will be improved. Note how the aperture has decreased the aperture angle of the lens. α = aperture angle or half the angle of light leaving the lens, O = center of lens, fp = focal point where rays converge. The focal length of lens is expressed as the length of the line from point O to fp.

In fact, the wavelength (λ) of an electron is expressed by the equation of the French physicist de Broglie as follows.

Equation 6-3: de Broglie Equation for Wavelength of an Electron

$$\lambda = h/mv$$ where h = Planck's constant (6.626 × 10^{-23} ergs/sec)

m = mass of the electron

v = electron velocity

After appropriate substitutions associating kinetic energy to mass, velocity, and accelerating voltage, the equation may be expressed:

$$\lambda = \frac{1.23}{\sqrt{V}} \text{ nm}$$ where V = accelerating voltage

Therefore, if one were operating a transmission electron microscope at an accelerating voltage of 60 kV, the wavelength of the electron would be 0.005 nm, and the resolving power of the system—after substitution of these values into Equation 6-2—should be approximately 0.003 nm. In fact, the actual resolution of a modern high resolution transmission electron microscope is closer to 0.1 nm. The reason we are not able to achieve the nearly 100-fold better resolution of 0.003 nm is due to the extremely narrow aperture angles (about 1,000 times smaller than that of the light microscope) needed by the electron microscope lenses to overcome a major resolution-limiting phenomenon called spherical aberration. In addition, the *diffraction phenomenon* as well as chromatic aberration and astigmatism (to be discussed later) all degrade the resolution capabilities of the TEM. To appreciate these problems, it is necessary to understand how lenses function.

General Design of Lenses

A lens may be thought of as a device that refracts or bends electromagnetic radiations to converge at a certain distance from the lens. Since it is impossible to represent waves adequately when drawing lens diagrams, one normally uses straight lines to represent rays of light (Figure 6-9A).

By convention, most drawings show only the outermost rays entering the lens. Be aware that there are numerous other rays between the peripheral ones that are not being represented in most drawings (Figure 6-9B shows only four rays).

It is important to note that both light and electron waves behave similarly as they pass through lenses and so both observe the same optical principles.

In studying Figure 6-10, it must be realized that the object being imaged is composed of many points, of which we are labeling only two, A and B or A' and B'. Furthermore, each point radiates information that passes through the entire curvature of the lens. The focal point (F) at which the rays that run parallel to the optical axis (line WXY in Figure 6-10) converge or *cross over* defines the *focal length* of the lens (i.e., the distance between points X and F = focal

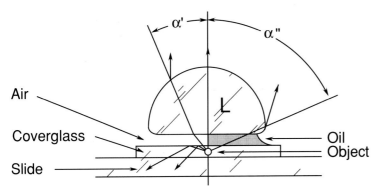

Figure 6-8 Refraction or bending of light to different degrees as it passes through media of different densities. On the left side of this combined drawing, one sees the situation when light travels through an object (O), glass slide/coverglass, and through air into an imaging lens (L). On the right side, oil has been added to fill the gap between the lens and the slide/coverglass. A lens brought close to a specimen on a glass slide will accept more information if oil (with a similar refractive index as the lens and slide) is used in the gap between the lens and slide (compare α' angle without oil to α'' angle with oil). Since more information enters the oil-immersion lens (at a wider α angle), greater resolution is possible.

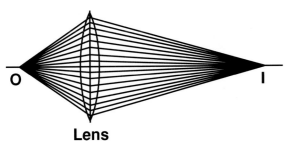

Figure 6-9(A) Diagram illustrating how lenses and illuminating rays are drawn. Object O radiates information that enters a lens and is imaged at I.

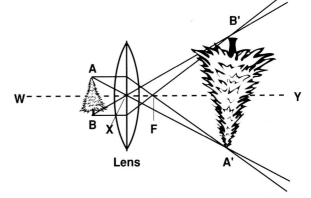

Figure 6-10 Image forming lens. An object, AB, is shown magnified to A'B'. Dashed line WXY marks the optical axis of the lens. The enlargement or magnification of the final image may be calculated by dividing the distance between points A' and B' by the distance between A and B. Focal length of this lens is the distance from point X to F (the focal point).

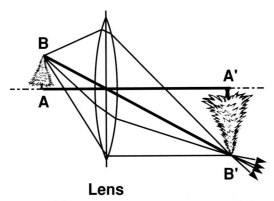

Figure 6-9(B) A single object point B is shown imaged at B' after passing through the lens. The reader should realize that each object point radiates many rays that are not shown in this drawing. Only four rays are shown.

length of lens). The type of lens shown is a double-convex or converging lens, the principle type of lens in all electron and most light microscopes. The point of entry of the electron into the lens field (central or peripheral) will determine the extent to which the ray will be refracted. Rays that pass far away from the exact center of the lens, or the optical axis, will be bent or refracted to a great degree, whereas those rays that pass along the optical axis will not be refracted at all. The manner that the lenses of an

electron microscope refract electrons is discussed in the next section.

Design of Electromagnetic Lenses

Since electrons are particles with such small mass that they will be stopped even by gas molecules present in the air, glass lenses are of no value in an electron microscope. However, *since electrons have a charge, they can be affected by magnetic fields.* For example, an electron accelerated through a vacuum will follow a helical path when it passes through a magnetic field generated by a coil of wire with a direct current (DC) running through it (see Figure 6-11). Such simple electromagnetic coils are termed *solenoids.*

Suppose one illuminates a specimen (the arrow shown in Figure 6-12) with a beam of electrons such that some of the electrons that interact with a spec-

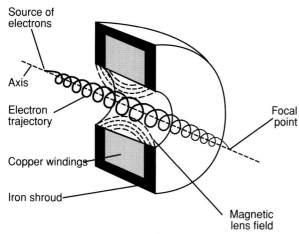

Figure 6-11 Single electron passing through electromagnetic lens. Instead of traveling in a straight line along the axis of the lens, the electron is forced by the magnetic field to follow a helical trajectory that will converge at a defined focal point after it emerges from the lens. Therefore, electromagnets, which are DC powered, behave similar to converging glass lenses.

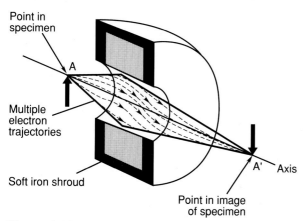

Figure 6-12 A group of electrons originating from point A in the specimen plane pass through an electromagnetic lens to all be focused at an appropriate point (A') in the image plane. The specimen is represented as the heavy arrow in this drawing. The electromagnetic lens behaves as a thin biconvex glass lens as shown in Figure 6-9.

imen point (A in Figure 6-12) are transmitted through the specimen and enter the electromagnetic lens. Depending on their precise trajectories as they enter the magnetic field, they will assume various helical paths as they speed through the lens. After leaving the lens, the electrons will focus at point A' to generate an image point of the specimen. The distance from the center of the lens to where the electrons converge at A' represents the focal length of the electromagnetic lens.

It is possible to change the focal length of an electromagnetic lens by changing the amount of DC

current running through the coil of wire. This relationship is expressed in Equation 6-4:

Equation 6-4: Focal Length of Electromagnetic Lens

$$f = K \left(V / i^2 \right)$$

where K = constant based on number of turns in lens coil wire and geometry of lens
V = accelerating voltage
i = milliamps of current put through coil

As the accelerating voltage of the electron is increased, the focal length is also increased since the electrons pass much more rapidly through the lens and assume looser helical routes. An increase in current put through the lens coil, however, results in a shorter focal length by forcing the electrons to assume tighter helical trajectories.

Being able to change the focal length of a lens is of practical importance, because this is how one can focus an image formed by a lens as well as change the magnification. In the light microscope, where the glass lenses are of a fixed focal length, focussing is done by physically moving the specimen into the proper plane of focus for each objective lens or vice versa. Similarly, magnifications are changed by removing an objective lens of one fixed focal length and replacing it with another. Obviously, the electromagnetic lenses of the electron microscope are advantageous because they permit one to change focal lengths (e.g., change focus and magnification) by varying the current running through the lens coil without having to move the specimen or physically change lenses.

The efficiency of the electromagnetic lens can be greatly improved by concentrating the magnetic field strength close to the path of the electrons. This is accomplished by shrouding the coil on top, bottom, and side with a soft-iron casing so that the magnetism will run through the shroud (Figure 6-13A). The strength of the lens is thereby increased. (The term "soft-iron" refers not only to the hardness of the metal, but also indicates that the iron is magnetized only when the electromagnetic field is conducted through it.)

The strength of the lens can be further increased by concentrating the magnetism to an even smaller area inside the lens bore by means of a liner termed a *polepiece* (so named because it sits in the north-south poles of the magnet). The cylindrical polepiece (Figure 6-13A and B) consists of upper and lower cores of soft iron held apart by a nonmagnetic

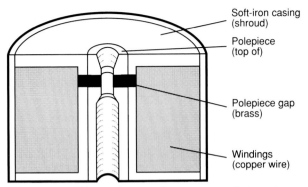

Soft-iron casing (shroud)

Polepiece (top of)

Polepiece gap (brass)

Windings (copper wire)

Figure 6-13(A)   Diagram of electromagnetic lens showing soft-iron casing (shroud) and soft-iron polepiece that slips down inside bore of lens.

brass spacer. The magnetic field is now concentrated between the top and bottom (north and south) iron components of the polepiece. These north and south cores of the polepiece are bored much smaller than the polepiece liner and must be as symmetrical as is mechanically possible in order to achieve high resolution. In practice, they are rarely perfect and may possess a number of defects that may degrade resolving power.

Defects in Lenses

A number of imperfections in lenses may reduce resolution. *Astigmatism* results when a lens field is not symmetrical in strength, but is stronger in one plane (north and south, for example) and weaker in another (east and west) (Figure 6-14). A point would not be imaged as such, but would appear elliptical in shape; a cross would be imaged with either the vertical or horizontal arm, but not both, in focus at one time.

Some *causes of astigmatism* are an imperfectly ground polepiece bore, nonhomogeneous blending of the polepiece metals, and dirt on parts of the column such as polepieces, apertures, and specimen holders. Since it is impossible to fabricate and maintain a lens with a perfectly symmetrical lens field, it is necessary to correct astigmatism by applying a correcting field of the appropriate strength in the proper direction to counteract the asymmetry. Such a device is called a *stigmator* and can be found in the condenser and objective lenses of the electron microscope (Figure 6-35B).

Astigmatism in a glass lens could be corrected by regrinding the curvature of the lens so that the strength is symmetrical, or by imposing another lens

Figure 6-13(B)   Photograph of a polepiece that fits into the electromagnetic lens coil shown in the background. The soft iron casing of the lens coil is removed to reveal the wire windings around the spool. In the polepiece (bottom photo), the north pole is arbitrarily on top, followed by a nonmagnetic brass spacer that holds north and south poles apart.

field of the appropriate strength over one of the aberrant fields of the original lens—as is done with correcting eyeglasses.

Chromatic aberration results when electromagnetic radiations of different energies converge at different focal planes. With a glass lens, shorter wavelength radiations are slowed down and refracted more than are longer wavelengths of light. Effectively, the shorter, more energetic wavelengths of light come to a shorter focal point than do the longer wavelengths (Figure 6-15). In an electromagnetic lens, the reverse is true: *shorter wavelength, more*

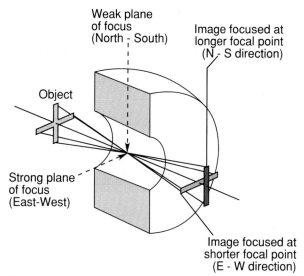

Figure 6-14 Astigmatism in a lens. Since the lens field is asymmetrically weaker in the north/south plane, objects oriented along the north/south axis will focus at a longer distance. By contrast, due to a stronger east/west lens field, objects oriented east/west will come to focus at a shorter distance from the lens. The effect is that only some portions of the image (either north/south or east/west) will be in focus at one time. Obviously, resolution will be degraded since the image will be focused in only one plane.

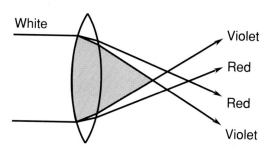

Figure 6-15 Chromatic aberration in a glass lens. Different wavelengths do not come to focus at the same point. Note how the violet part of the spectrum (stippled) focuses at a shorter distance than does the red part of the spectrum. This results in an enlarged, unsharp point rather than a smaller, focused one. Resolution of the point will be degraded.

energetic electrons have a longer focal point than do the longer wavelength electrons. In both cases, however, chromatic aberration results in the enlargement of the focal point (similar to the Airy disk phenomenon caused by diffraction effects) with a consequential loss of resolution. Chromatic aberration can be corrected by using a monochromatic source of electromagnetic radiation. With glass lenses, one would use a monochromatic light (possibly by using a shorter wavelength blue filter). In an electromag-

netic lens, one would insure that the electrons were of the same energy level by carefully stabilizing the accelerating voltage and having a good vacuum to minimize the energy loss of the electrons as they passed through the column. Thicker specimens give rise to a spectrum of electrons with varied energy levels and consequently worsen chromatic aberration (Figure 6-16A). Thin specimens are therefore essential for high resolution studies.

Chromatic change in magnification occurs when thick specimens are viewed at low magnifications using a low accelerating voltage. The *image appears to be sharp in the center, but becomes progressively out of focus as one moves toward the periphery* (Figure 6-16B). This is because the lower energy electrons are imaged at a different plane than the higher energy electrons. The effect is maximal at the periphery of the image, since these electrons are closer to the lens coils and, thus, are more affected by the magnetic field. This problem may be minimized by using thinner specimens, higher accelerating voltages, higher magnifications, and by correcting any other distortions that may be present in the lens.

Spherical aberration is due to the geometry of both glass and electromagnetic lenses such that rays passing through the periphery of the lens are refracted more than rays passing along the axis. Unfortunately, the various rays do not come to a common focal point, resulting in an enlarged, unsharp point (Figure 6-17A). At some distance, however, one should encounter the sharpest possible point

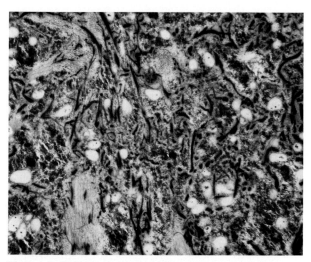

Figure 6-16(A) Chromatic aberration in an overly thick section is evidenced by an image that is blurred overall due to degraded resolution.

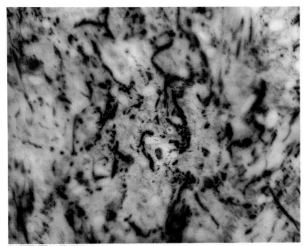

Figure 6-16(B) Chromatic change of magnification occurs when an overly thick specimen is viewed at low magnifications with a low accelerating voltage. Only the central part of the image is sharp since the effect is maximal at the periphery.

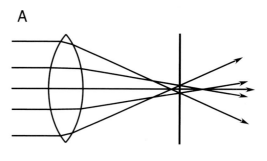

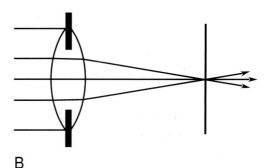

Figure 6-17 (A) Spherical aberration in a lens. Peripheral rays are refracted more than central rays, so that all rays do not converge to a common, small focal point. Instead, an enlarged, diffuse spot like the Airy disc will be generated. The vertical line indicates the one point where the point will be smallest, i.e., having the smallest circle of confusion. (B) Correction of spherical aberration with an aperture (here shown inside the lens) to cut out peripheral rays and thereby permit remaining rays to focus at a common small imaging point. Resolution will be improved since individual image points in the specimen will be smaller.

that would constitute the *circle of minimum confusion* (i.e., the smallest Airy disc) and the practical focal point of the lens. *Spherical aberration may be reduced by using an aperture to eliminate some of the peripheral rays* (Figure 6-17B). Although apertures must be used in the electron microscope to reduce spherical aberration as much as possible, they decrease the aperture angle and thereby prevent the electron microscope from achieving the ultimate resolving power specified in the equation for resolution (Equation 6-1). In addition, the worsening of resolution as a result of using a longer focal length lens is shown in Equation 6-5.

Equation 6-5: Limit of Resolution, d_s, Imposed by Spherical Aberration

$$d_s = k_s \bullet f \bullet \alpha_o^3$$

where k = a constant related to lens characteristics
 f = focal length of lens
 α = aperture angle of lens, normally the objective lens

From Equation 6-5, we see why a short focal length lens combined with a smaller aperture (to generate a smaller aperture angle) will help to reduce the degradation of resolution caused by spherical aberration. Consequently, smaller apertures are generally more desirable than larger ones to improve resolution—in spite of the theoretical advantage offered by large apertures as indicated in Equation 6-1.

Certain types of image *distortions* may arise when spherical aberration occurs in the final imaging

(or projector) lenses of the transmission electron microscope. Since peripheral electrons are refracted to a greater extent than central rays, the image formed by these electrons will be at a greater magnification and in a different focal plane than the image generated from more centrally positioned electrons. A grid of lines would not be imaged as square, but would assume the shape of a sunken pillow, hence the name *pincushion distortion* (Figure 6-18B). This type of distortion may occur when attempting to operate the transmission electron microscope at excessively low magnifications. Another type of imperfection, *barrel distortion*, occurs when one attempts to use an electromagnetic lens in a demagnifying mode rather than the normal magnifying mode. In this case, the central part of the image is magnified more than the periphery so that the gridwork assumes a swollen or barrel shape (Figure 6-18C).

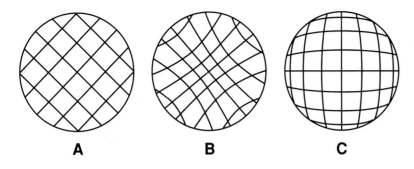

Figure 6-18 Distortions in a lens. (A) Normal image of grid pattern. (B) Image with pincushion distortion. (C) Image with barrel distortion.

Fortunately, it is possible to neutralize one type of distortion with the other. For instance, if one projector lens is displaying excessive pincushion distortion, it is possible to operate another projector lens in the demagnifying mode to introduce an opposing barrel distortion. The lens systems of modern electron microscopes are designed to automatically counterbalance the various types of distortions throughout a wide magnification range. In older microscopes, however, one must take care not to introduce these distortions in the lower magnification range.

As one begins to use the electron microscope, a curious phenomenon called *image rotation* will be noticed as one changes magnification. This occurs because the electrons follow a spiral path through the lenses, and the spiral shifts as the strength of the lens is varied. Image rotation not only rotates the image on the viewing screen, but it also exaggerates the effects of distortion. It is possible to minimize or eliminate image rotation entirely by ensuring that a series of lenses have opposing rotations rather than all having rotations in the same direction. This is accomplished by running the lens current through the coil in the opposite direction (i.e., reversing polarity) and is a principle utilized in some of the newer transmission electron microscopes.

Magnification

Besides forming images with good resolution, the lenses of the electron microscope are able to further magnify these images. Magnification refers to the degree of enlargement of the diameter of a final image compared to the original. As will be discussed later, there are at least three magnifying lenses in an electron microscope: the *objective, intermediate,* and *projector lenses.* The final magnification is calculated as the product of the individual magnifying powers of all of the lenses in the system or:

Equation 6-6: Calculation of Total Magnification, M_T, of the TEM

$$M_T = M_O \times M_I \times M_P$$

where M_T = total magnification or mag
M_O = mag of objective lens
M_I = mag of intermediate lens
M_P = mag of projector lens(es)

For example, if the transmission electron microscope is operating in the high magnification mode, typical values for the respective lenses might be: $200 \times 50 \times 20 = 200,000\times$. If one were to operate the microscope in the low magnification mode, perhaps the values would be: $50 \times 0.5 \times 50 = 1,250\times$.

Intermediate magnifications may be produced by varying the current to the various lenses. Sometimes, it is desirable to view as much of the specimen as possible in order to evaluate quickly the quality of the preparation or to locate a particular portion of the specimen. In this case, an extremely low magnification is obtained by placing the microscope in the *scan magnification* mode, which can be accomplished by shutting off the objective lens and using the next lens (the intermediate lens) as the imaging lens as follows: $1 \times 0.5 \times 100 = 50\times$.

Although it is theoretically possible to increase the magnification indefinitely, the quality of the image magnified is dependent on the resolving power of the lenses in the system. Consequently, the term *useful magnification* is used to define the maximum magnification that should be used for a particular optical system. It is defined by the formula:

Equation 6-7: Useful Magnification

$$\text{Useful Magnification} = \frac{\text{resolution of the human eye}}{\text{resolution of the lens system}}$$

In the case of the light microscope, a typical value would be $1,000\times$, because the resolving power of the human eye is about 0.2 mm, while the resolving power of the light microscope is approximately 0.2 μm. An electron microscope with a resolving power of 0.2 nm could be expected to have a top magnification of approximately 1,000,000×, or a thousand times greater than the light microscope. In practice, due to the diminished illumination at such high magnifications, microscopists would probably take the micrograph at a magnification of 250,000× and photographically enlarge the negative to the needed magnification. However, only rarely do biologists need such high magnifications.

Nearly all modern electron microscopes have digital displays that give the approximate total magnification when the lenses are automatically changed as one varies the magnification. Older microscopes usually have a gauge that may either read the current to one lens, or they may have a magnification knob with a series of click stops that may be correlated to a particular magnification. However, all microscopes (including light microscopes) must be calibrated in order to determine more accurately the total magnification, since a number of variables may cause the magnification to vary by as much as 20% to 30% over a short period of time. Even modern instruments with direct reading digital displays are only guaranteed to be accurate to +/− 5% to 10% of the stated values. The procedure for magnification calibration is discussed later in this chapter.

Design of the Transmission Electron Microscope

Comparison of Light Microscope to Transmission Electron Microscope

The transmission electron microscope is similar in many ways to the compound light microscope. For instance, in both microscopes, electromagnetic radiations originating from a tungsten filament are converged onto a thin specimen by means of a condenser lens system. The illumination transmitted through the specimen is focused into an image and magnified first by an objective lens and then further magnified by a series of intermediate and projector lenses until the final image is viewed (Figure 6-19). Both kinds of microscope may record images using a silver-based photographic emulsion since it is sensitive to both types of radiations.

Of course, the lenses of light microscopes are composed of glass or quartz rather than the electromagnetic solenoids used in electron microscopes.

Electron microscopes require an elaborate vacuum system to remove interfering air molecules that would impede the flow of electrons down the column. Such high vacuums are necessary from the point of origin of the electrons (the filament) up to, and usually including, the photographic film. Since the specimen is also subjected to these high vacuums, living specimens would be rapidly dehydrated if placed directly into the electron microscope. Nonetheless, it is possible to view rapidly frozen, hydrated, thin specimens by using cryostages that maintain the frozen state of cellular water even under bombardment by the electron beam.

Basic Systems Making Up a Transmission Electron Microscope

The transmission electron microscope (Figures 6-20A, B and C) is made up of a number of different systems that are integrated to form one functional unit capable of orienting and imaging extremely thin specimens. The *illuminating system* consists of the electron gun and condenser lenses that give rise to and control the amount of radiation striking the specimen. A *specimen manipulation system* composed of the specimen stage, specimen holders, and related hardware is necessary for orienting the thin specimen outside and inside the microscope. The *imaging system* includes the objective, intermediate, and projector lenses that are involved in forming, focusing, and magnifying the image on the viewing screen as well as the *camera* that is used to record the image. A *vacuum system* is necessary to remove interfering air molecules from the column of the electron microscope. In the descriptions that follow, the systems will be considered from the top of the microscope to the bottom. See Tables 6-3 and 6-4.

Illuminating System
This system is situated at the top of the microscope column and consists of the electron gun (composed of the filament, shield, and anode) and the condenser lenses.

Electron Gun. Within the electron gun (Figure 6-21), the *filament* serves as the source of electrons. The standard filament, or *cathode* (Figure 6-22), is composed of a V-shaped tungsten wire approximately 0.1 mm in diameter (about the thickness of a human hair). Being a metal, tungsten contains positive ions and free electrons that are strongly attracted to the positive ions. Fortunately, it is possible to entice the outermost orbital, or valence, electrons

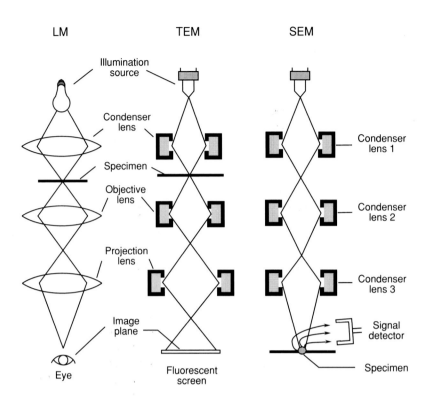

Figure 6-19 Comparison of light microscope (LM) to transmission (TEM) and scanning (SEM) electron microscopes.

out of the tungsten by first applying a high voltage to the filament and then heating the metal by running a small amount of DC electrical current through the filament while operating within a vacuum. In other words, a certain amount of energy must be put into the system to cause the electrons to leave the filament. The amount of energy necessary to bring about electron emission is termed the *work function* of the metal. Although tungsten has a relatively high work function, it has an excellent yield of electrons just below its rather high melting point of 3,653° K.

In practical terms, one first applies a fixed amount of negative high voltage (typically 50, 75, or 100 kV) and then slowly increases the amount of direct current running through the filament to heat it to achieve the emission of electrons (*thermionic emission*). As one applies more heat to the filament, the yield of electrons increases until the filament begins to melt and evaporate in the high vacuum of the microscope (see Figure 6-23). At some optimal temperature, the gun achieves good electron emission as well as an acceptable filament life: this is termed the *saturation point* (discussed later). By examining Figure 6-23, it becomes apparent that a good electron yield is achieved at 2,600° K where the filament life is around 100 hours. Oversaturating the gun by heating the filament even 200° K beyond 2,600° K results in a dra-

matic decrease in filament life. The average life of a filament ranges from 25 hours in older electron microscopes to over 200 hours in microscopes with good vacuum systems and scrupulously maintained gun areas. The *major causes of premature filament failure* are: oversaturation, high voltage discharge caused by dirt in the gun region, poor vacuum, and air leaks or outgassing from contaminants in the gun chamber. The two controls on the panel of the microscope that are used to initiate the flow of electrons are usually labeled *accelerating voltage* (or possibly kV, HV, or HT) and *emission* or *saturation* (or sometimes *filament*).

If one induces the emission of electrons from a filament as described above, this will result in the emanation of electrons in all directions. Without a mechanism for guiding them, most of the electrons would not enter the illuminating system. A second part of the electron gun, the *shield* (also called Wehnelt cylinder, bias shield, or grid cap), is involved in assuring that the majority of the electrons go in the proper direction. The shield is a caplike structure that covers the filament and is maintained at a slightly higher negative voltage potential than the filament. Because it is several hundred volts more negative than the 50 to 100 kV electrons, the shield surrounds the electrons with a repulsive field that is breachable only through a 2 to 3 mm aperture

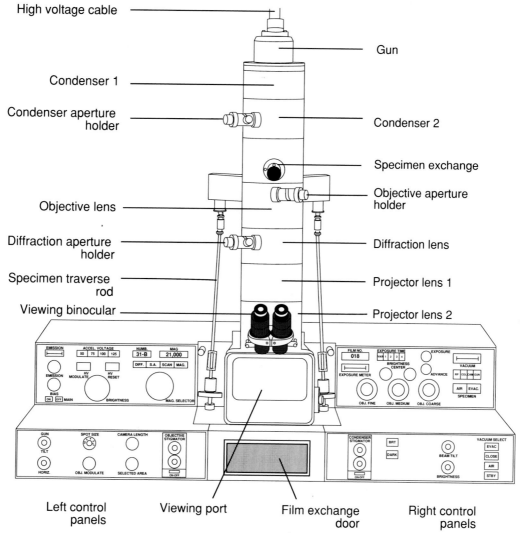

Figure 6-20(A) Diagram of a modern transmission electron microscope with major components labeled. The function of each of the components is listed in Tables 6-3 and 6-4.

directly in front of the filament tip. Electrons exit the shield aperture and are drawn toward an apertured disc, or anode, the third part of the electron gun (Figure 6-21). The anode is connected to ground so that the highly negative electrons are strongly attracted to it. In fact, the highly attractive pull of the anode in combination with the negative surface of the shield act as an electrostatic lens to generate a crossover image of the electron source near the anode (see Figure 6-24). Therefore, it may come as a surprise that the first "lens" in an electron microscope is not an electromagnetic one at all.

NOTE: Just as electromagentic lenses have two forces or poles, electrostatic lenses have positively and negatively charged surfaces to attract or repel and, thereby, focus electrons. The term *crossover* refers to the point where the electrons converge and cross over each other's paths.

Variable Self-Biased Gun. It was stated earlier that the high voltage shield is slightly more negative than the filament. This difference in negative potential, or *bias*, is established by connecting the shield directly to the negative high voltage line while placing a variable resistor in the high voltage line to the filament (Figure 6-21A). By varying the value of this resistor, the filament may be made less negative than the shield (usually by 100 to 200 volts). The greater the value of the variable bias resistor, the more negative the shield will become relative to the filament. As the shield becomes more negative, fewer electrons will be able to pass through the shield aperture since they are now repulsed to a greater degree by

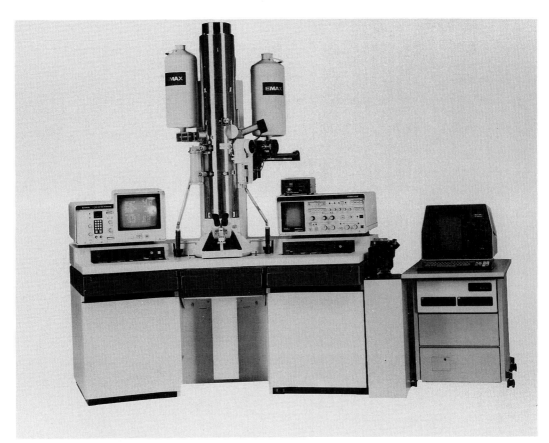

Figure 6-20(B)   Photograph of a modern transmission electron microscope equipped for doing x-ray analysis (see Chapter 15). (Courtesy of Hitachi Scientific Instruments.)

the shield. The overall effect of the *variable bias*, therefore, is to regulate the escape of electrons through the shield aperture.

In addition to high voltage, one applies a certain amount of direct current to the filament in order to heat up the filament and enhance electron emission. As this current passes through the variable bias resistor, a certain amount of voltage is generated and applied to the shield in order to make it more negative. Therefore, as one continues to increase the heating current to the filament, the numbers of electrons coming off the filament will increase. But since the shield is becoming progressively negative, the total number of electrons actually passing through the shield aperture does not increase significantly. The so-called *saturation point* of the gun is the point where the number of electrons emitted from the gun no longer increases as the filament is heated. The gun is, therefore, said to be self-biasing, since it throttles back on electron emission as the heat is increased. *It is important that the operator realize that increasing the heat of the filament beyond the saturation point will not increase the brightness of the*

gun but will considerably shorten the filament life. On the other hand, *undersaturation* of the filament may lead to instabilities in the illumination of the specimen and cause problems if analytical procedures (such as X-ray analysis) are to be attempted. The arrangement for controlling electron emission in modern electron microscopes is termed the *variable self-biased gun.*

Controlling the Amount of Illumination Striking the Specimen. It is possible to make practical use of the variable bias to regulate the amount of illumination that strikes the specimen. For example, when operating at high magnifications with small condenser spot sizes, it may be necessary to alter the bias to effect greater gun emissions. Of course, the filament life will be shortened, but this may be necessary in order to critically view and focus the specimen.

Moving the filament closer to the shield aperture will permit more electrons to pass through to the condenser lenses. However, if the filament is placed too close to the aperture, the bias control by the shield will be lost, and the emission will become excessive. Filaments placed too far away from the

COMPONENTS OF TYPICAL TRANSMISSION ELECTRON MICROSCOPE COLUMN

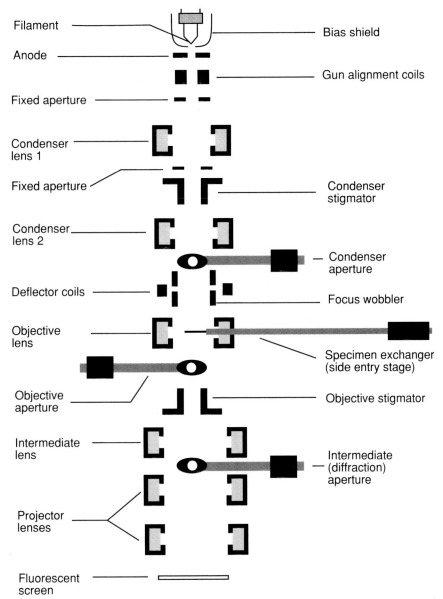

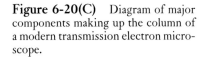

Figure 6-20(C) Diagram of major components making up the column of a modern transmission electron microscope.

aperture, on the other hand, may never yield sufficient numbers of electrons from the gun. Therefore, careful placement of the filament relative to the shield aperture is very important and should be in accordance with the manufacturer's specifications.

The distance of the anode from the filament and shield is also important. As one moves the anode closer to the filament, more electrons will be extracted from the gun. This becomes a consideration when using lower accelerating voltages where it may be necessary to move the anode closer to assist in the extraction of the lower energy electrons. Some

electron microscopes have an external adjustment screw that will mechanically adjust the height of the anode, while other models have a pneumatically actuated "anode lifter" that changes in response to the kilovolt selected by the operator. Older microscopes may have no such adjustment.

Other Gun Designs. The majority of transmission electron microscopes use the gun design described in the previous section. However, some notable variations are also available, depending on the needs of the researcher. The filament shape may be altered

Table 6-3   Column Components*

Component	Synonyms	Function of Components
Illumination System		
Electron Gun	Gun, Source	Generates electrons and provides first coherent crossover of electron beam
Condenser Lens 1	C1, Spot Size	Determines smallest illumination spot size on specimen (see Spot Size in Table 6-4)
Condenser Lens 2	C2, Brightness	Varies amount of illumination on specimen—in combination with C1 (see Brightness in Table 6-4)
Condenser Aperture	C2 Aperture	Reduces spherical aberration, helps control amount of illumination striking specimen
Specimen Manipulation System		
Specimen Exchanger	Specimen Air Lock	Chamber and mechanism for inserting specimen holder
Specimen Stage	Stage	Mechanism for moving specimen inside column of microscope
Imaging System		
Objective Lens	—	Forms, magnifies, and focuses first image (see Focus in Table 6-4)
Objective Aperture	—	Controls contrast and spherical aberration
Intermediate Lens	Diffraction Lens	Normally used to help magnify image from objective lens and to focus diffraction pattern
Intermediate Aperture	Diffraction Aperture, Field Limiting Aperture	Selects area to be diffracted
Projector Lens 1	P1	Helps magnify image, possibly used in some diffraction work
Projector Lens 2	P2	Same as P1
Observation and Camera Systems		
Viewing Chamber	—	Contains viewing screen for final image
Binocular Microscope	Focusing Scope	Magnifies image on viewing screen for accurate focusing
Camera	—	Contains film for recording image

* Not included are the mechanical adjustment screws for centering and tilting various lenses or apertures.

as illustrated in Figure 6-25b, where the tip was first flattened and then sharpened to a point. It is also possible to purchase a pointed filament made by welding a single crystal of tungsten onto the curved tip of a standard filament (Figure 6-25c). Both types of *pointed filaments* have a considerably shorter lifetime than do standard filaments. However, since the initial gun crossover image is much smaller and the beam is highly coherent, they are necessary for high resolution studies where beam damage may be a consideration (e.g., viewing crystalline lattice planes).

Besides being made of tungsten, filaments may also be constructed of *lanthanum hexaboride*, which has a lower work function. Typically, these filaments operate at temperatures 1,000° K lower than tung-

sten and have a brightness several times greater than a standard tungsten source. The lifetime of such filaments ranges from 700 to 1,000 hours. This type of filament may be made from a single LaB_6 crystal with one end having a point measuring only several micrometers across (Figure 6-26). LaB_6 filaments are coming into use slowly, since they are considerably more expensive than tungsten filaments and are extremely chemically reactive when hot. For the latter reason, vacuums greater than 10^{-5} Pa are essential (see section on vacuum systems), and special filament mounts must be constructed from such nonreactive elements as rhenium or vitreous carbon. LaB_6 filaments are useful when small beam crossover sizes containing large numbers of electrons are nec-

Table 6-4   Components on Control Panels*

Component	Synonyms	Functions of Component
Filament	Emission	Effects emission of electrons upon heating
Bias	—	Adjusts voltage differential between filament and shield to regulate yield of electrons
High Voltage Reset	HV, kV Reset	Activates high voltage to gun
High Voltage Select	HV, kV Select	Selects amount of high voltage applied to gun
Magnification Control	MAG	Controls final magnification of image by activating combinations of imaging lenses
Brightness	C2	Controls current to second condenser lens
Gun Tilt	—	Electronically tilts electron beam beneath gun
Gun Horizontal	—	Electronically translates electron beam beneath gun
Spot Size	C1	Controls final illumination spot size on specimen
Objective Stigmator	OBJ STIG	Corrects astigmatism in objective lens
Focus Wobbler	Focus Aid	Helps focus accurately at low magnifictions
Exposure Meter	—	Monitors illumination for accurate exposures
Vacuum Meter	VAC	Monitors vacuum levels in various parts of scope
Focusing Control	Focus—fine, medium, coarse	Controls current to objective lens for accurate focusing of image
Brightness Center	C2 Centration	Translates entire illumination system onto screen center
Condenser Stigmator	COND STIG	Corrects astigmatism in condenser lenses
Bright/Dark	—	Selects brightfield or darkfield operating mode
Main	—	Main power switch to console
Main Evac	EVAC	Main switch to vacuum system
HV Wobbler	HV Modulate	Wobbles high voltage to locate voltage center for alignment
Objective Wobbler	OBJ MODUL	Wobbles current to objective lens for alignment

* *Not included* are less frequently used controls as lens current switches or meters, current normalizers, selector switches for various magnification modes, diffraction controls, darkfield centration and stigmation controls, film counters and switches, vacuum status indicators, camera calibration switches, current and voltage stabilizer switches.

essary—as in high magnification/resolution studies, for elemental analysis, or in high resolution scanning electron microscopy.

A totally different gun, nearly a thousand times brighter than the standard gun, may also be used under certain conditions. In the *field emission gun*, the filament is a single crystal of tungsten with its atomic crystalline lattice precisely oriented to maximize electron emission. Electrons are not generated by thermionic emission (heating), but are actually drawn out of the tungsten crystal by a series of positive high voltage anodes that act as electrostatic lenses to focus the gun crossover to a spot size of 10 nm (Figure 6-27). A major disadvantage of the field emission gun is the ultrahigh vacuum required (greater than 10^{-8} Pa) and the extreme susceptibility of the filament to contaminants. Field emission guns are very useful in high resolution scanning and scan-

ning transmission electron microscopes and are now being incorporated into conventional transmission electron microscopes.

Condenser Lenses. This second major part of the illuminating system gathers the electrons of the first crossover image from the gun and *focuses electrons onto the specimen.* Modern transmission electron microscopes have two condenser lenses, unlike the first microscopes that had only one. The first condenser lens (designated C1) is a demagnifying lens that decreases the size of the 50 μm gun crossover to generate a range of spot sizes from 20 μm down to 1 μm. The second condenser lens (C2), on the other hand, enlarges the C1 spot. The overall effect of both lenses is to control precisely the amount of electron irradiation or illumination striking the specimen. *The operating principle for using C1 and C2 is to generate a spot on the specimen of the proper*

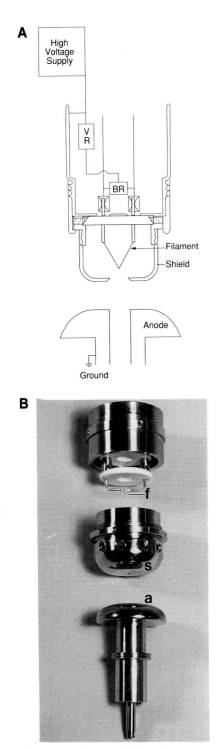

Figure 6-21 (A) Diagram of an electron gun showing filament, shield, and anode. The shield is connected directly to the high voltage, whereas the high voltage leading to the filament has a variable resistor (VR) to vary the amount of high voltage. The output from the variable resistor is then passed through two balancing resistors (BR) which are attached to the filament. (Modified from a drawing provided by Hitachi Scientific Instruments). (B) Actual electron gun from TEM showing filament (f), shield (s), and anode (a). Compare to line drawing in 6-21(A).

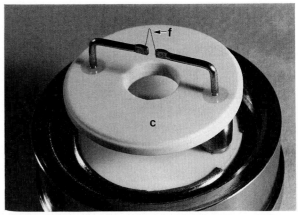

Figure 6-22 Standard V-shaped tungsten filament (f) used in most electron microscopes. The filament is spot welded to the larger supporting arms which pass through the ceramic (c) insulator and plug into the electrical leads of the gun.

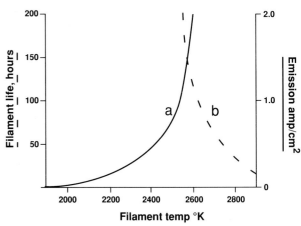

Figure 6-23 The importance of proper saturation of the filament is shown in these two curves. (a) Relationship between electron emission from a filament as a function of the temperature of the filament. The optimal temperature is 2,600° K. (b) When one exceeds 2,600° K filament temperature, the filament life drops dramatically.

size to irradiate only the area being examined. Therefore, at higher magnifications smaller spot sizes should be focused on the specimen (Figure 6-28B), while larger spots may be used at lower magnifications (Figure 6-28A). Spot sizes are controlled to minimize beam damage to parts of the specimen not being viewed.

Suppose one is working at a magnification of 50,000×. At this high magnification, the C1 lens should be highly energized to demagnify the 50 μm illumination spot from the gun down to 1 to 2 μm. Next, the C2 lens should be used to adjust the size of the C1 illumination spot to cover only the specimen area being viewed. Since the average viewing

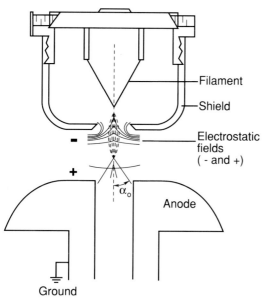

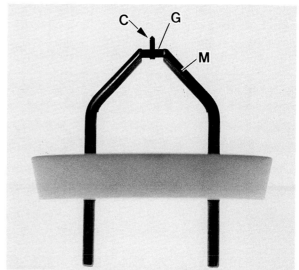

Figure 6-24 The self-biased electron gun. The shield (Wehnelt cylinder) is slightly more negative than the filament to control the release of electrons from the gun. A variable bias resistor (see Figure 6-21A) regulates the degree of negativity of the filament. The anode serves as a positive attracting force and as an electrostatic lens (in combination with the shield) to help focus the electrons into a crossover spot approximately 50 μm across.

Figure 6-26 Lanthanum hexaboride cathode. The crystal (C) is held in place by means of pyrolytic graphite (G) blocks with compressive force generated by molybenum (M) alloy posts designed to withstand extremely high temperatures. (Cathode provided by FEI Company)

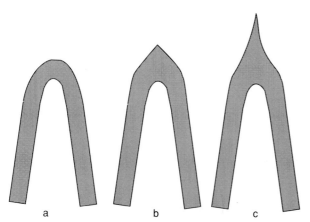

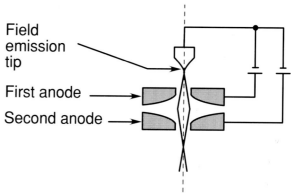

Figure 6-25 Drawing of three different types of filaments tips. (a) Standard V-shaped filament tip. (b) Standard filament tip that was flattened and then sharpened to a fine point. (c) Filament tip where a crystal of tungsten was spot-welded onto the curved end.

Figure 6-27 The field emission gun. Electrons are extracted from a single crystal of tungsten by a series of anodes that are several thousands volts positive. It is not necessary to heat this type of filament.

screen is about 100 mm across, a 2 μm spot of illumination enlarged 50,000× would just cover the screen (2 μm × 50,000 = 100 mm). Therefore, the C2 lens should also be highly energized to generate a 2 μm spot on the specimen. At a magnification of 10,000×, it is possible to keep C1 highly energized but to use C2 to magnify or spread out the 2 μm

spot an additional 5× to just cover the 100 mm screen. However, the illumination will be about 5 times dimmer.

If one studies Figure 6-28, it is apparent why smaller spot sizes are necessarily dimmer. If C1 is highly energized in order to generate a small spot (Figure 6-28B), the focal length is made so short and the aperture angle so great that many electrons are refracted to such an extent that they do not enter C2. On the other hand, if C1 is weakened to generate a larger spot, the focal length is longer and the aperture angle is smaller so that effectively all electrons may now enter C2 (Figure 6-28A). Therefore,

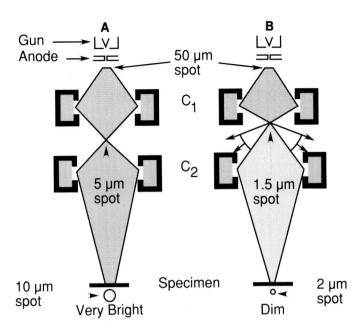

Figure 6-28   The condenser lens system. (A) In this mode, the 50 μm gun crossover is reduced to 5 μm by condenser lens 1, C1, and then slightly enlarged by condenser lens 2, C2, to yield a 10 μm spot on the specimen that is five times brighter than the initial gun crossover. (B) At higher magnifications, the 50 μm gun crossover is reduced to 1.5 μm by a highly energized C1. This refracts the peripheral electrons to such a great angle that they cannot enter C2 and are therefore lost. After C2 slightly enlarges the C1 spot, the resulting 2 μm spot is rather dim.

as the C1 spot is made progressively smaller, overall illumination tends to diminish. This poses a problem at higher magnifications where very small spot sizes are needed. Illumination may become so dim that microscopists must allow 30 to 60 minutes for their eyes to adapt to working under such dark conditions. However, it may be possible to increase the illumination on the specimen using the techniques described previously in this chapter section entitled Controlling the Amount of Illumination Striking the Specimen.

Apertures in Condenser Lenses. Depending on the design of the transmission electron microscope, one or both condenser lenses may have apertures of variable sizes. Generally, the C1 aperture is an internal aperture of a fixed size, while the C2 aperture is variable by inserting into the electron beam pathway apertures of different sizes attached to the end of a shaft. A popular method is to use a molybdenum foil strip containing 3 or 4 holes of 500, 300, 200, and 100 μm in diameter (see Figure 6-29). The various sizes may be inserted and centered using external adjustment screws during the alignment procedure (described later in this chapter). Larger condenser apertures permit most of the electrons to pass through the lens and, therefore, yield a brighter spot on the specimen. Smaller apertures cut out more peripheral electrons and, hence, reduce the illumination on the specimen. However, since spherical aberration is concomitantly reduced, greater resolution is possible using smaller condenser apertures. The operational principle to remember is *larger con-*

denser apertures give more illumination but with more spherical aberration.

Specimen Manipulation System

Most biological specimens are mounted on a copper meshwork or *grid*. For additional backing, a thin membrane or film may be used to support the specimen over the open areas of the screen. Thin specimens are viewable in the "windows" formed by the lattice of the grid (see Chapter 4 for a detailed description of grids). Grids are placed into a specimen holder and, after insertion into an air lock, the chamber is evacuated and the specimen holder is inserted into the stage of the microscope. (In very old microscopes, no air locks were provided, so it was necessary to admit air to the entire column in order to insert a specimen. Such changes would take 5 to 10 minutes versus 30 or so seconds in modern airlocked microscopes.)

The *specimen stage* is a micromanipulator for moving the specimen in x and y directions in increments as small as 10 nm, the width of a cell membrane. Depending on the design of the specimen holder and stage, it may also be possible to tilt and rotate the specimen inside the column of the electron microscope. Some of the newer microprocessor-controlled TEMs have automated stage controls that permit motorized and precise movement of the specimen. It is possible to program the stage controller to scan specimens and even to systematically record images as, for example, when one wishes to make a montage of the numerous individual electron micrographs (see Chapter 8 for more information on montages). An important feature of such com-

Figure 6-29 Variable aperture holder from a TEM. The rod contains a molybdenum strip (m) with apertures of various sizes. Positioning screws (s) permit the precise alignment of the apertures in the electron beam. An O-ring seal (o) permits the aperture to be sealed off inside the vacuum of the microscope column.

puter-controlled stages is the ability to memorize specified coordinates and to be able to return to these locations on command.

Top Entry Stage. One type of *specimen holder* is a brass cartridge with a long cylinder that enters the objective lens as the holder is placed in the stage on top of the objective lens (Figure 6-30). The grid sits on the end of the cylinder and is held firmly in place by a tight-fitting sleeve that slips over the cylinder.

Firm contact between grid and specimen holder is essential in order to dissipate the buildup of heat and static charges resulting from bombardment by the electron

beam. To optimize this dissipation, insert the copper mesh facing the electron beam (i.e., specimen down). However, be careful that the specimen and supporting membrane are well adhered to the copper grid, or they may detach and fall onto the microscope stage or objective lens, necessitating time-consuming disassembly of the microscope.

After the grid has been inserted in the specimen holder, the cartridge is placed in an air lock where the air is removed by the vacuum system. The air lock is opened to the high vacuum of the microscope and the specimen holder transferred onto the movable stage of the microscope. It is again important that the contact between the specimen holder and brass stage be firm in order to dissipate any static charges. For instance, a dirty specimen stage or holder will prevent such contact and may lead to thermal or electrostatic drift in the specimen.

Some top entry holders are designed to tilt mechanically several degrees, and some stages have gearing that will enable the specimen to be rotated 360 degrees. It is possible to vary the length of the cartridge nosepiece and construct holders suited to special purposes. For instance, short nosepiece holders are useful for high contrast and low magnification applications (longer focal length objective lens), while longer nosepieces are suitable for high resolution studies (short focal length objective lens). Although it is possible to insert only one cartridge holder into the stage, some microscopes have air locks that accommodate several cartridges so that it is not necessary to break vacuum in order to insert another specimen (Figure 6-31).

Figure 6-30 (left) Short, top-entry grid holder for high contrast, low-magnification work. Resolution is not as good with this type of grid holder since the specimen is placed higher in the objective lens, necessitating a longer focal length of the lens. (right) Standard top-entry specimen grid holder for high resolution work. The specimen grid is placed on the end of the grid holder shaft and held in place with a sleeve that is slipped over the shaft. The holding sleeve is shown in place and indicated by an arrow.

Side Entry Stage. In this type of stage, the specimen grid is introduced into the microscope stage by en-

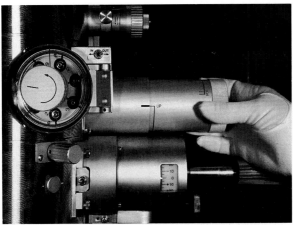

Figure 6-31 Multiple grid holder specimen airlock. In this design, it is possible to load six grid holders into the airlock, evacuate the sealed chamber, select the holder, and insert it into the column using the exchanger arm shown.

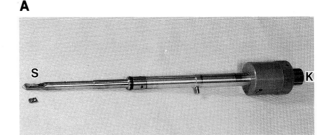

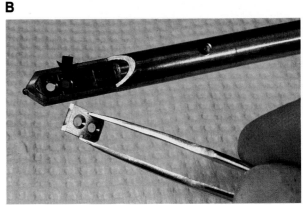

Figure 6-32 (A) Side entry, multiple specimen grid holder. k = specimen selection knob for positioning proper grid, s = specimen holder area where grids are inserted and held. (B) Closeup view of side entry, multiple grid holder showing a specimen grid in the countersunk depression on the right (arrow). The grids are held in place by means of the plate shown in the prongs of the forceps. The plate is positioned over the grids and held in place by a spring loaded latch.

tering through the side of the objective lens polepiece. The specimen holder resembles an aperture holder consisting of a rod with a flat plate on one end that has one or more recessed areas for holding grids (Figure 6-32).

The grid is placed into the flanged recess and held in position with a clip or plate (Figure 6-32B). The specimen rod may then be evacuated in the air lock and the rod inserted into the microscope stage. With some types of rods, the end bearing the specimen grids may be inserted into the stage and detached while the carrier rod remains in the air lock. Other types of rods are inserted into the specimen stage and remain in the stage during the viewing process. Since the rod enters through the side of the polepiece, it is necessary to design the polepiece with a large enough gap to permit entry and allow for tilting of the specimen by as much as 65 degrees from the horizontal. Side entry stages provide much more versatile manipulation of the specimen. Besides the standard x and y horizontal movements, the specimen holder may permit tilting, rotation, a second axis of tilt (double-tilt stage), and special modifications as described in the next paragraph. Since it is also necessary to accurately set the specimen in the correct focal plane of the objective lens, a z-axis or vertical movement is always provided. Modern side entry stages offer high resolution capabilities nearly comparable to top entry stages and permit more versatility for specimen manipulation and orientation for analytical purposes. For these reasons, the side entry stage is currently favored over the top entry stage in the latest generation of TEMs.

Special Stages. It is possible to manipulate the specimen in the electron microscope in a number of ways using special specimen stages or holders. For instance, the specimen may be subjected to stretching and compression in a *tensile stage*, and heating or cooling in specially mod-

ified *thermal stages*. Of particular interest to biologists is the *cold stage*, since it permits the examination of rapidly frozen specimens (such as live virus preparations) that are still hydrated and have not been exposed to chemical fixation or staining. Besides examination of fluid specimens, it is also possible to study ultrathin frozen, hydrated sections of unprocessed biological materials for elemental analysis. Although specimen preparatory techniques are still being refined, cold stages offer tremendous potential when combined with the analytical capabilities of the TEM.

Imaging System

This part of the microscope includes the objective, intermediate, and projector lenses. It is involved in the generation of the image and the magnification and projection of the final image onto a viewing screen or camera system of the microscope.

Objective Lens. By far, this is the single most important lens in the transmission electron microscope, since it forms the initial image that is further magnified by the other imaging lenses. In order to achieve such high resolutions, the lens must be highly energized to obtain the short, 1 to 2 mm focal lengths necessary. The lens must be free of astigmatism and have minimal aberrations. This means that the polepieces must be constructed from homogeneously blended metals, be as symmetrical as possible, and contain devices, or stigmators, for correcting astigmatism. The objective lens is used primarily to focus and initially magnify the image, whereas other lenses are used to magnify this image further. Of all the lenses used in the magnification of an image, the objective lens is the least variable so that it can maintain the very short focal lengths necessary for high resolution and is convenient to focus (i.e., if the strength of the objective lens were varied over a wide range, refocusing would require major adjustments of the lens current to the lens). Currently, as magnifications are changed, the adjustments to the objective lens needed to bring the image into focus are not excessive.

Because any fluctuations in either lens current or high voltage would affect the focus of the objective lens, both must be made extremely stable. Since contamination may introduce astigmatism into any lens system, some way of minimizing contamination in the objective lens is needed. Such devices, called *anticontaminators*, are now essential for high quality work.

Images are formed in the objective lens by a "subtractive" action. Depending on specimen thickness and density of various parts of the specimen, some electrons (inelastically scattered ones) will pass through the specimen and into subsequent lenses with a loss of some energy. Other electrons may be deflected upon contact with parts of the specimen and rendered unable to enter the objective and other imaging lenses. Still other electrons may lose all of their energy upon impact with the specimen and are likewise lost. Most of the electrons that enter the objective lens are ultimately projected onto the phosphorescent viewing screen to cause a certain level of brightness. The more electrons passing through any one point on the specimen, the brighter the image generated.

Regardless of how an electron is lost, this loss is evidenced on the screen as a darker region. Hence, areas of high density/thickness will appear darker than areas of less density/thickness resulting in various "gray levels" on the screen. Since biological specimens have inherently small density differences between the various parts of the cells, it is necessary to enhance these differences by reacting high density metals (osmium, lead, uranium, etc.) with specific subcellular structures. These heavy metals are introduced during the specimen preparation processes of fixation and staining (see Chapters 2 and 5).

Apertures in Objective Lens. As will be illustrated in subsequent chapters, obtaining a thin specimen with good contrast is not always easily done. *The function of the objective aperture is primarily to enhance contrast* by trapping more of the peripherally deflected electrons (Figure 6-33). Apertures of various sizes may be positioned in the polepiece gap just under the specimen. Arranged on a similar positioning rod as the condenser apertures, these apertures are much smaller in size (70, 50, 30, and 20 μm, for example) and more prone to contamination. Small objective apertures give increased contrast, although at the expense of overall illumination.

Photographers, as well as electron microscopists, make use of apertures not only to control the amount of illumination entering a lens, but further to control the depth of field. It is well known that "stopping down" or decreasing the size of the aperture results in bringing more of the foreground and background into focus, whereas wide open apertures result in only a narrow zone being in focus. *Depth of field, therefore, refers to the depth in the specimen plane that is in focus.* As is demonstrated in Equation 6-8, *smaller apertures increase the depth in the specimen that is in focus.*

Equation 6-8: Depth of Field

$$D_{fi} = \frac{\lambda}{\sin \alpha^2}$$

where λ = wavelength of radiation
α = aperture angle

If we are using an accelerating voltage of 60 kV, the wavelength of the electron is 0.005 nm. A large 200 μm aperture would generate an aperture angle of illumination in the objective lens of approximately 10^{-2} radians. Upon substitution in the equation, we obtain a depth of field of approximately 50 nm. Now, if a 100 μm aperture is used, the aperture angle is approximately 10^{-3} radians, which yields a 100 times greater depth of field of 5 μm. Consequently, when using smaller apertures in both the objective and condenser lenses to generate narrow aperture angles, the entire depth of the specimen is in focus. This is in contrast to the light microscope, where larger aperture angles result in rather narrow depths of field, making it necessary to focus through the various levels to view the entire depth in the specimen. Depth of field and depth of focus are illustrated in Figure 6-36, later in this chapter.

Apertures are usually constructed from high melting point metals such as molybdenum or platinum. Although they may be configured as single

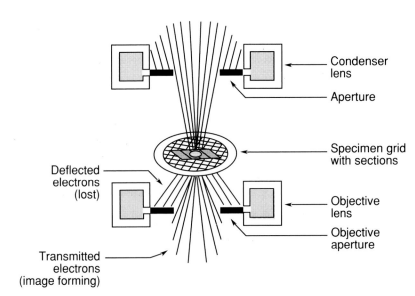

Condenser lens

Aperture

Specimen grid with sections

Deflected electrons (lost)

Objective lens

Objective aperture

Transmitted electrons (image forming)

Figure 6-33 Objective aperture located between upper and lower parts of polepiece, just under the specimen. The major function of the aperture is to help remove peripherally deflected electrons to enhance image contrast. In addition to the specimen and objective aperture, a chilled anticontaminator blade (see Figure 6-34) may also be inserted just above the specimen (or sometimes above and below the specimen) to prevent contaminants from condensing on specimen.

discs with a central hole, they are more commonly fabricated from thin foils cut into a strip. Individual holes are precisely drilled through the metal and scrutinized for burrs that would affect the symmetry of the field. Using the external controls on the TEM column, a single hole may be selected and positioned symmetrically around the axis of the electron beam.

It is possible to purchase standard strips containing holes commonly used in a particular instrument, or one can order customized aperture strips with specific types and thickness of metal and specified hole diameters. For optimum performance, apertures must be cleaned periodically, the frequency depending on the types of specimens and general cleanliness of the microscope itself. Molybdenum apertures are cleaned by placing the strip into a flat holder of tungsten or molybdenum and heating the strip in a vacuum evaporator (to avoid oxidation). After reaching a cherry red color, most contamination will be evaporated and removed by the vacuum system. However, it may be necessary to repeat this process several times. One should not exceed the cherry red color, since the molybdenum may weld onto the holder. Platinum strips may be cleaned by passage through a propane gas flame followed by immersion in hydrofluoric acid and ammonium hydroxide. Aperture strips are easier to clean if contamination has not built up over time.

A *thin foil aperture* may also be used in the objective or condenser lenses. Such apertures are made of an extremely thin layer of gold and are "self-cleaning." In practice, one must regularly run the focused electron beam over the rim of the aperture in order to bake off the contaminants. When such foils can no longer be cleaned using such a procedure, they must be replaced.

Anticontaminators in Specimen Area. Anticontaminators are found in close proximity to the specimen, aperture, and polepiece of the objective lens. They are essentially metal surfaces that are chilled with liquid nitrogen from a reservoir outside the column of the microscope. Most contaminants originating from the specimen or the microscope will condense onto the extremely cold anticontaminator and be removed from the system.

Anticontaminators are sometimes called *cold fingers*. Some anticontaminators may resemble an aperture holder, except that the brass plate is much thicker and the aperture much larger (Figure 6-34). Other anticontaminators are ring-shaped and encircle the specimen. Since anticontam-

Figure 6-34 Specimen anticontaminator or cold finger. The large container (c) is filled with liquid nitrogen to chill the cold finger blade (b) that is located just above and below the specimen. An O-ring seals the apparatus from the atmosphere. (Courtesy of Gatan, Inc.)

inators must fit into the same cramped space that also accommodates the specimen and objective aperture, they must be designed very carefully. Anticontaminators must be polished clean periodically to remove condensed materials and then carefully positioned to avoid contact with the specimen holder or objective aperture.

Astigmatism Correction. Stigmators are located beneath not only the objective but also the condenser lenses. They function to correct the radial lens asymmetries that prevent one from focusing the image in all directions and generating circular illumination spots. Since an astigmatic lens is stronger in one direction (north-south, for instance) than another, one creates a compensating field of equivalent strength in the opposite direction (east-west). Two parameters must be considered: direction of the astigmatism (*azimuth*) and strength of the astigmatism (*amplitude*). One must be able to adjust both variables to suit the particular situation.

Older stigmators were composed of pairs of magnetic slugs that could be mechanically rotated into position to compensate for astigmatism. Newer microscopes use primarily electromagnetic stigmators since they are less expensive to build, easier to use, and somewhat more precise in their correction. Electromagnetic stigmators may consist of eight tiny electromagnets encircling the lens field. By varying the strength and polarity of various sets of magnets, one can control both amplitude and azimuth in order to generate a symmetrical magnetic field (Figure 6-35). When stigmators become dirty, they will no longer effectively compensate for astigmatism and must be withdrawn from the microscope and cleaned.

Intermediate (Diffraction) Lens.

As one proceeds down the column, this lens immediately follows and is constructed similarly to the objective lens. In simpler microscopes, magnification is altered by varying the current to this lens, while in more sophisticated microscopes the preferred method is to use combinations of several lenses to allow a wider, distortion-free magnification range. The major function of this lens is to assist in the magnification of the image from the objective lens. At very low magnifications, the objective lens is shut off and the intermediate lens used in its place to generate the primary image. Although the image produced by the very long focal length intermediate lens is poor com-

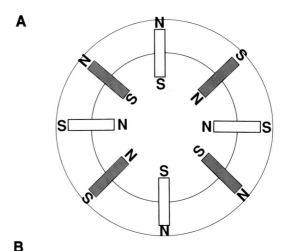

A

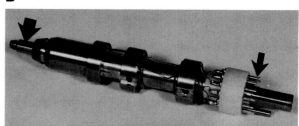

B

Figure 6-35 (A) Conceptual drawing of electromagnetic stigmator showing orientation of eight electromagnets around lens axis. Strength and direction are controlled by adjusting appropriate combinations of magnets to generate a symmetrical field. The stigmator is located under the condenser and the objective lens polepieces. (B) Actual stigmator apparatus taken from an electron microscope. Large arrow indicates one of the eight electromagnetic iron slugs oriented around the central axis. The entire apparatus fits up into the bore of the objective lens so that the area indicated in the large arrow is positioned just under the specimen. The smaller arrow points out individual electrical contacts through which current flows to energize the electromagnets. The closeup photograph shows some of the electromagnets that are positioned near the specimen.

pared to that generated by using all three lenses, it is adequate for low magnification work. The intermediate lens may be equipped with an aperture that

is used when operating the microscope in the diffraction mode (Chapter 15).

Projector Lens. Most modern transmission electron microscopes have two projector lenses (P1 and P2) that follow the intermediate lens. Both P1 and P2 are used to further magnify images from the intermediate or diffraction lens. Except for very high magnifications, only three of the four imaging lenses are normally energized at any one time, and various triplet combinations are used to achieve the magnification range desired. In a microscope with four imaging lenses, the first projector lens can also be used as a diffraction lens, and it may be possible to insert a specimen into a specially modified holder located either between P1 and P2 or below P2 for specialized diffraction studies. As with intermediate lenses, projector lenses suffer from distortions that have less effect on resolution than do aberrations occurring in the objective lens.

Projector lenses are said to have great *depth of focus*, meaning that the final image remains in focus for a long distance along the optical axis. This is determined by the equation:

Equation 6-9: Depth of Focus

$$D_{fo} = \frac{M^2 \cdot RP}{\alpha}$$

where M = total magnification
RP = resolving power of instrument being used
α = aperture angle established by objective lens

At a magnification of 100,000×, in an instrument with resolving power of 0.2 nm and having an aperture angle of 10^{-2} radians, the depth of focus of the projector lens may be calculated to be 200 meters. This becomes important when one realizes that the photographic film is not in the same plane as the viewing screen. For the same reason, it is possible to locate multiple image recording devices at various points beyond the projector lens, since they will all be in focus. However, the magnification will be different (further planes will be at a higher magnification). The relationship between depth of field and depth of focus relative to aperture angle is shown in Figure 6-36.

Viewing System and Camera. The final image is projected onto a viewing screen coated with a phosphorescent zinc-activated cadmium sulfide powder attached to the screen with a binder such as cellulose

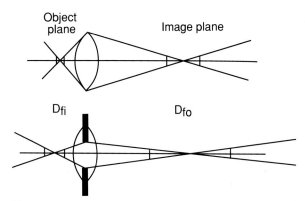

Figure 6-36 Depth of field (D_{fi}) occurs in the object plane, while depth of focus (D_{fo}) refers to the depth in the image plane that is in focus. In the bottom figure, note that an aperture increases both the depth of field and depth of focus.

nitrate. Most electron microscopes provide for an inclination of the viewing screen so that the image may be conveniently examined either with the unaided eye or with a *stereomicroscope*. With the stereomicroscope, although the image may appear to be grainy due to the 100 μm sized phosphorescent particles making up the screen, it is necessary to view a magnified image in order to focus accurately. Some microscopes may provide a second, smaller screen that is brought into position for focusing. In this case, the main screen remains horizontal, except during exposure of the film. All viewing screens will have areas marked to indicate where to position the image so that it will be properly situated on the film.

Pre-evacuated films are placed in an air lock (*camera chamber*) under the viewing screen and the chamber evacuated to high vacuum. The chamber is then opened to the column to permit exposure of the film. It is necessary to have a supply of pre-evacuated films for insertion into the microscope, since it may take several hours to remove residual moisture or other contaminating gases from the film. Cameras may hold from one to fifty individual 3¼" × 4" films, or one may use a variety of roll cameras loaded with 35 mm or larger films. Most films used in electron microscopy are orthochromatic and thus may be handled using an appropriate safelight (see Chapter 8).

All modern electron microscopes provide for metering the exposure and have a shutter to time the exposure accurately. Since the screen is coated with the same type of material as are camera light meters, it is possible to monitor the current generated by the electron beam striking the screen.

Some electron microscopes provide for an integrated reading when the image is viewed on the

large main screen and for a localized or "spot" reading when a smaller viewing screen is used for imaging. Normally, integrated readings are adequate; however, if the image has extremes of contrast, the spot meter is very valuable to insure proper exposure. It is usually necessary to calibrate the exposure meter when different film is being used or a new microscope is installed. This is done by taking a series of micrographs with the meter set for different sensitivity levels. After the developed negatives are evaluated, the proper setting may be recorded and the meter adjusted.

A shutter is provided to time the exposure so that the proper negative density (as determined by the previous calibration) may be obtained. Most electron microscopes have timers that vary from a fraction of a second to "hold" positions in which a timer may be used for very long manual exposures. Electron micrographs are exposed for 0.5 to 2 seconds in order to record all density levels and to minimize image shift or *drift* (i.e., slow movement of the image after exposure to the beam). Once the time has been selected, the illumination level is adjusted with the C1 and C2 lens controls until the exposure meter reaches the calibration point. The film is then advanced under the viewing screen, and the screen is moved to permit electrons to pass onto the film. As one begins to raise the viewing screen, the beam is blocked by the shutter until the screen is totally raised. The shutter is then opened for the proper interval, after which the beam is again blocked until the screen is repositioned.

Vacuum System

The vacuum system is a major assemblage of pumps, switches, and valves that are involved in evacuating air primarily from the pathway of the electron beam, but also from other selected areas of the microscope. It may be thought of as a plumbing system designed to connect the vacuum pumps to selected areas (specimen, camera, gun, column) that need evacuation. Although earlier electron microscopes had vacuum systems that were more manual in operation, the standard automated vacuum plant in present-day electron microscopes is little changed from its predecessors. Besides the obvious advantage of convenience, the major advantage of current systems is the inclusion of extensive safeguards against contamination, misvalving, and power and water failure (although these safeguards were sometimes incorporated even in the earliest of microscopes). The price for such conveniences is not only monetary,

but includes the added burden of more complex maintenance problems.

A vacuum is needed in the electron microscope not only to increase the *mean free path* of an electron (i.e., the distance an electron must travel without encountering an interfering gas molecule), but also to prevent high voltage discharges between the filament/shield and the anode. Such discharges are one of the major causes of filament failure. In addition, all filaments (especially field emission) are extremely sensitive to oxidation and must be protected. A fourth reason why a vacuum is needed involves the removal of contaminating gases (especially water vapor and organics present in laboratories) that are broken down under high energy electron bombardment and generate corrosive radicals that combine with the specimen to destroy fine structure.

Vacuum Terminology

Since the terminology used to describe vacuum situations may be confusing even to the practicing electron microscopist, it is important to be familiar with certain basic concepts and phrases. Confusion arises since vacuum is expressed in units of pressure: a high vacuum is low pressure. In addition, a general acceptance of a standardized nomenclature (*bar* versus *torr* versus *Pascal*) has met with resistance. It is not uncommon to see all three terms used even today in the literature. For that reason, one should be aware of the relationship between the various terms to fully understand the discussion.

If one takes a long tube sealed on one end, fills it with mercury, and inverts the tube in a bowl, some of the mercury will flow out into the bowl. At sea level, the height of the column of mercury remaining in the tube, measured from top to fluid level in the bowl, would be 760 mm. The 760 mm of mercury is supported in the glass tube by the pressure of the atmosphere pressing down into the bowl. This simple barometer was developed in the mid-1600s by Evangelista Torricelli and is the basis for many pressure measurements. If the atmospheric pressure is lowered, the mercury begins to flow out into the bowl, and the height of the column becomes lower. At some point, only 1 mm of mercury will be supported. This 1 mm of mercury, supported by 1/760 atmosphere, is equivalent to one *torr* (named in Torricelli's honor) and would be the atmospheric pressure one might encounter 28 miles above sea level. One torr = 1.32×10^{-3} atmosphere.

The international standard for pressure measurement is the metric unit termed the *Pascal* (named in honor of the French philosopher and scientist Blaise Pascal, who lived from 1623 to 1662), and is abbreviated *Pa*. One Pascal equals 9.92×10^{-6} atmosphere. Since there are 133 Pa per torr:

$$1 \text{ Pa} = 1/133 \text{ torr}$$
$$= 7.52 \times 10^{-3} \text{ torr}$$

To convert a pressure stated in torr into Pascal, multiply by 133; to convert a figure stated in Pascal into torr, divide by 133. Since these figures may not be easily remembered, a crude approximation may be obtained if we add 2 to the torr exponent to obtain Pa. So, a vacuum of 10^{-7} torr is very roughly equivalent to 10^{-5} Pa (actually 1.33×10^{-5} Pa).

The relationship between the three major pressure measurement terms is given in Table 6-5.

Vacuum ranges for the TEM vary from the so-called "high vacuums" of 10^{-1} to 10^{-4} Pa, to "very high vacuums" in the 10^{-4} to 10^{-7} Pa range, and "ultrahigh vacuums" in the 10^{-7} to 10^{-9} Pa range. It may be of interest to note that even in electron microscopes operating at a high vacuum of 10^{-4} Pa, there are still over 10^{10} air molecules per cubic centimeter. However, the mean free path of an electron under such conditions is over 50 meters.

Rotary and Diffusion Pumps. The majority of electron microscopes achieve operating vacuum conditions using two types of pumps: *rotary pump* and *diffusion pump*. The rotary pump is used first to lower the pressure into the 10^0 to 10^{-1} Pa range (*rough pumping*), and then the diffusion pump is used to achieve higher vacuums in the 10^{-3} to 10^{-4} Pa range. These pumps use different principles to achieve their ultimate vacuums, and certain limitations in each should be realized.

Rotary Pump. This *mechanical pump* is also sometimes called a *rough pump* or *forepump* since it is often used to establish a rough starting vacuum before the diffusion pump is used. A typical *rotary pump* is shown in Figure 6-37. A rotating cylinder with sliding spring-mounted vanes is mounted off center inside a larger cylindrical space. As the rotor turns clockwise from position A to B, an enlarged space is created with a lowered pressure that draws air into B. (A similar phenomenon occurs when one releases a squeezed rubber bulb of an eyedropper pipette to draw fluid or air into the bulb.) As the rotor continues turning, space B is sealed off and becomes smaller as the rotor moves the contents into space C. The gas in space C is compressed to the extent that a spring-loaded valve is forced open and the air is exhausted from

Table 6-5   Relationships Between Commonly Used Units of Pressure

	Pascal	Torr	Millibar
1 Torr			
= 1 mm Hg	133	1	1.33
1 Millibar	100	0.75	1
1 Atmosphere	1.01×10^5	760	1.01×10^3

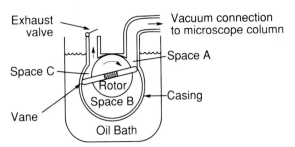

Figure 6-37(A)   Rotary pump. Illustration of principle of operation of pump module. As the rotor turns, space A becomes enlarged, creating a vacuum that sucks air into the space. When the rotor rotates further and seals off the space by means of the spring-loaded vane, a large volume of air has been removed from the TEM and into the pump. Upon further rotation of the rotor, the sealed space designated B becomes smaller and the air is compressed. Eventually, the compressed air is moved over to space C by the rotor and a spring-loaded valve opens to exhaust the air to the outside of the system. The oil is used to lubricate the moving rotor and vanes and to carry frictional heat to the outside of the case.

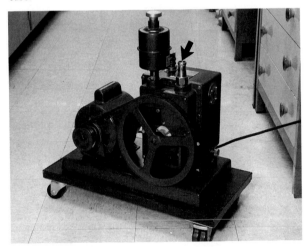

Figure 6-37(B)   Photo of a belt-driven, two-stage rotary pump. The electric motor driving the pump is to the left. The cylindrical cannister on top of the pump module is a high-efficiency filter to remove oil fumes from the exhaust of the pump. The port through which the vacuum is applied is indicated by an arrow.

Figure 6-37(C)   Photograph of a direct-drive, two stage rotary pump found on most modern electron microscopes.

the system (similar to squeezing a rubber bulb filled with fluid). The pump has thereby moved some of the air from inside the closed microscope system to the outside. It is possible to link two such systems in series (exhaust of one into the inlet of a second pump) to achieve a two-stage pump, the most commonly used type of rotary pump. Two-stage pumps do not necessarily pump faster, but do achieve better vacuums.

Since friction is a problem as the vanes rub along the inside of the larger cylinder, it is necessary to lubricate the system. This is accomplished by immersing the pump module in an oil bath so that some oil is constantly coating the two rubbing surfaces. The oil serves not only to lubricate the system and provide a vacuum seal between the two surfaces, but also to transfer any frictional heat to the outside of the pump casing. Being mechanical pumps, the seal between the rotor and casing will, after extensive use, wear so that good vacuums can no longer be obtained. In addition, the oil absorbs moisture and other laboratory solvents, and the lubrication and pumping efficiency falls off. Depending on the pumping loads and atmospheric moisture, the rotary pump oil must be replaced at regular intervals (1 to 4 times a year). *One must be aware that used oils are carcinogenic, so gloves must be worn and proper disposal procedures must be followed* (some service stations will take the oil). Since all rotary pumps release a fine mist of oil out of the exhaust port, it is very important that the exhaust be vented to the outside or, at the very least, fitted with efficient oil mist traps. The major causes of rotary pump demise are failure to maintain the proper oil level and infrequent oil changes.

Diffusion Pump. In contrast to the rotary pump, diffusion pumps contain no moving parts but are able to complete the pumping operation started by the rotary pump. Diffusion pumps were developed nearly 70 years ago and still form an important component in most vacuum systems. The basic design, shown in Figure 6-38, consists of a series of stacked towers with umbrella-like caps with highly polished surfaces. The towers are placed inside of a water cooled cylindrical enclosure that is open on the end connected to the microscope and has an electric heater and side-mounted outlet at the opposite end. Special oils in the bottom of the pump housing are boiled, and the pressurized vapor is forced up the towers to be deflected downwards by the umbrellas or jet assemblies. Since the oil molecules are traveling at supersonic speeds, any air molecules that have diffused into the pump will be knocked to the bottom of the diffusion pump chamber where they will be removed by a rotary pump connected to the side port. *Diffusion pumps cannot function unless they are "backed" by or connected to operating rotary pumps to remove the accumulated air molecules.* The hot oil vapors strike the sides of the pump housing, condense on the water-cooled surface, and flow to the bottom of the pump to be boiled and recycled. In an effort to slow down the creep of diffusion pump oil into the column, water- or liquid nitrogen-chilled baffle plates are often installed at the mouth of the diffusion pump to trap contaminants. Such baffles not only provide a cleaner vacuum environ-

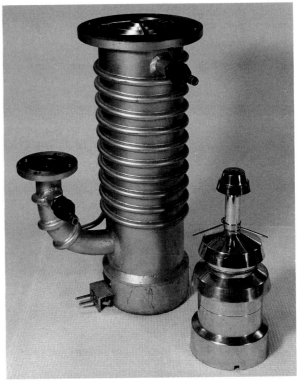

Figure 6-38(A) Diffusion pump. Photograph of exterior of a diffusion pump showing cooling coils surrounding body. Stacked, umbrella-like caps (lower, right) fit down into the body of the pump.

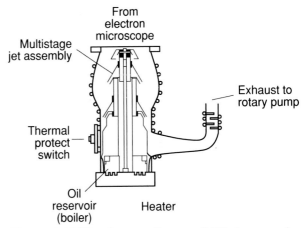

Figure 6-38(B) Cutaway diagram of diffusion pump interior.

ment, but also act as cryopumps to improve the pumping efficiency of the diffusion pumps.

The number of umbrella towers determines the number of stages in a diffusion pump, with each stage capable of lowering the pressure by nearly one Pascal. Therefore, a three-stage diffusion pump should improve the vacuum by a factor of 10^{-3} Pa. If a vacuum system had been previously rough pumped to a vacuum of 10^{-1} Pa, the final vacuum would be in the 10^{-4} Pa range.

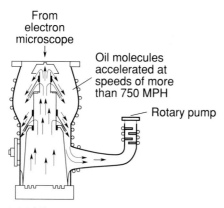

Figure 6-38(C) Operating principle of pump: supersonic oil particles drive air molecules to base of pump to be withdrawn by a rotary pump (not shown) attached to the side arm. (B and C, courtesy of Varian Associates—Vacuum Products Division.)

Different types of oils, with very *low vapor pressures* (i.e., not easily vaporized), are used in diffusion pumps. Silicone-based oils (Dow Corning 704 or 705) may be used in some vacuum systems. However, since they are very difficult contaminants to remove, polyphenyl ethers (Santovac 5, Convalex 10) or perfluoropolyethers (Fomblin Y VAC) are preferred in clean systems. One simply cannot replace one oil with another, however, since the boiler temperatures and vapor pressures of the various oils are different and must be matched to the specific system.

Diffusion pumps require little maintenance and may function for many years without problems. However, should they be exposed to pressures greater than 1 Pa, some oil may be blown into the rotary pump. When the diffusion pump oil level is sufficiently lowered, pumping efficiency will be diminished and the boiler may overheat and burn out the heater. Therefore, the oil level should be checked if pumping efficiency drops significantly and no other causes can be found. The pump should be unbolted, drained of oil, cleaned with the proper solvent, dried, and refilled with the precise amount of proper oil. The Viton, heat resistant, sealing gaskets should be replaced, if deformed.

Reading Vacuum Levels. The level of vacuum may be measured using several different monitors. Vacuums generated by the rotary pump are evaluated using either a thermocouple or Pirani gauge.

The *thermocouple gauge* (Figure 6-39) consists of a filament that is heated with a constant amount of current. The temperature of the wire is measured directly by an attached thermocouple (thermometer). Large numbers of air molecules conduct away heat and lower the temperature of the wire. Instead of reading temperature, the meter scale is replaced with one expressed in vacuum

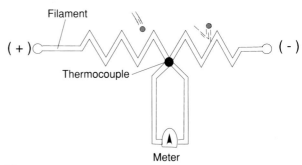

Figure 6-39 Principle of operation of thermocouple gauge. A thermocouple senses the temperature of a heated wire. As air molecules are removed, the temperature of the wire rises so that temperature may be correlated to vacuum level. (Courtesy of Varian Associates—Vacuum Products Division.)

units. As air molecules are removed due to an improved vacuum, the wire gets hotter and the needle reads better vacuum conditions rather than higher temperatures. Therefore, in this gauge, the amount of voltage fed to the wire is the constant, while the temperature is the variable that is measured on the surrogate scale.

The *Pirani gauge* consists of a filament that is maintained at a constant level of electrical resistance by running the proper voltage through it. The resistance is established by heating the filament to a certain temperature. Air molecules conduct heat away from the filament, causing a decrease in temperature and resistance. A Wheatstone bridge circuit senses the resistance drop and applies more voltage to the filament to maintain the resistance level or temperature. A voltmeter is used to sense the amount of voltage being put into the circuit. As air molecules are removed from a system, the filament loses less of its heat and less voltage is needed to maintain the resistance. The voltmeter scale may be replaced with a scale calibrated in vacuum units rather than volts, so that one reads the vacuum directly. Lower voltage is read as better vacuum. Therefore, in the Pirani gauge, the temperature and resistance are maintained at a constant level, while the voltage is the variable that is measured (and converted into vacuum units).

As the pressure decreases to 10^{-1} Pa or less, the Pirani and thermocouple gauges are no longer usable. Instead, a cold cathode gauge or an ionization gauge may be used.

The *cold cathode gauge* (Figure 6-40A) consists of cathodes and an anode or collector wire that are maintained at several thousand volts relative to each other. The high voltage field ionizes gas molecules present in the gauge and causes the electrons to travel to the anode. The electrons collide with gas molecules and ionize them. A powerful external magnet increases the sensitivity of the meter by causing the electrons to assume long spiral trajectories, rather than straight lines. The positive ions travel to the negatively charged cathode and the resulting ion current is measured and expressed in vacuum units rather than amperage. This gauge is used in the 10^0 to 10^{-6} Pa ranges.

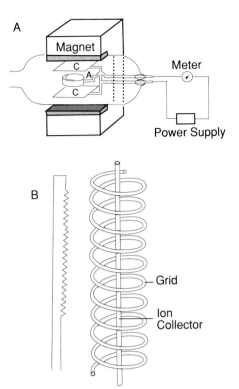

Figure 6-40 (A) Diagram of cold cathode vacuum gauge. Air molecules are ionized between the high voltage cathodes (c) and anode (a), and the flow of ions is measured as current by a meter. The magnet increases the path lengths of the electrons to increase ionization and boost sensitivity of the gauge. (B) Diagram of ionization gauge. A heated filament (left) generates electrons that travel towards the positively charged grid. As they travel to the grid, the electrons ionize any gas molecules present and generate gas ions that are collected by the central wire. (Both figures courtesy of Varian Associates—Vacuum Products Division.)

For wider measurements of vacuums in the 10^{-2} to 10^{-12} Pa ranges, an *ionization gauge* (Figure 6-40B) is desirable. This gauge consists of a hot filament that generates electrons that travel towards a positively charged coil of wire called a grid. In their travels to the grid, the electrons ionize air molecules and cause a flow of positively charged gas molecules that are attracted to a collector. The ion current collected is proportional to the pressure in the chamber.

Total Vacuum Systems. The principles behind a complete vacuum system are more readily understood by viewing a simplified diagram as shown in Figure 6-41. The system consists of the rotary and diffusion pumps, switching valves, connecting pipes, air locks, and vacuum gauges. In all vacuum systems, certain operational principles or rules must be followed: (a) sealed chambers must be evacuated first with a rotary pump followed by a diffusion pump;

(b) diffusion pumps must be backed by a rotary pump; (c) air may be admitted to chambers only if closed off from all pumps; (d) rough pumps and diffusion pumps must not pump on the same sealed chamber at the same time. Descriptions of several operational situations follow.

Turning On System. When one first turns on the electron microscope, there is an initial period of rough pumping of the column area while the diffusion pumps are warming up. Cooling water is run through the diffusion pump cooling coils while the oil is brought to a boil. The diffusion pumps are sealed from the column but are being backed by a rotary pump. This usually takes 15 to 20 minutes.

Column Pump Down Sequence. After the column area (gun chamber, column proper, camera chamber) has reached 10^0 to 10^{-1} Pa, the rotary pumps are closed off from the chamber and the diffusion pump is opened to the column. After pumping for a period of time (perhaps 20 to 30 minutes), the microscope has reached the proper operating vacuum so that the filament may be activated for viewing.

Specimen Insertion. A specimen grid is loaded into the holder, inserted into the sealed specimen air lock, and rough pumped into the proper range. The rough pump is disconnected and the specimen inserted into the high vacuum environment inside the microscope for examination.

Specimen Removal. The specimen air lock is rough pumped, disconnected from the rough pump, and the specimen retrieval system is used to move the specimen back into the sealed air lock. Air is admitted to the air lock and the specimen removed.

Removal of Exposed Films. After electron micrographs have been taken, the camera chamber is sealed off from all vacuum pumps and air is admitted to the camera chamber. After removal of exposed films and replenishment with unexposed, pre-pumped films, the sealed-off camera chamber is first pumped by rotary pump and then properly connected to the diffusion pump and the column.

Shut Down Sequence. After shutting off the lenses and high voltage, all air-locked chambers are evacuated and sealed. Power to the diffusion pump heater is shut off, but cooling water and rotary pump backing is maintained for 15 minutes to permit the oil to condense into the bottom of the diffusion pump boiler. The backing pump is sealed from the diffusion pump and power to the backing pump is shut

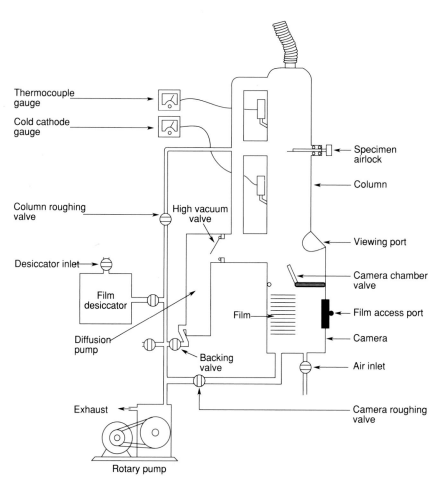

Figure 6-41 Simplified diagram of a vacuum system in a transmission electron microscope.

Thermocouple gauge

Cold cathode gauge

Specimen airlock

Column

Column roughing valve

High vacuum valve

Viewing port

Desiccator inlet

Camera chamber valve

Film desiccator

Film

Film access port

Diffusion pump

Camera

Backing valve

Air inlet

Exhaust

Camera roughing valve

Rotary pump

off. Air may be bled into the vacuum line between pumps and sealed. Cooling water is shut off.

Figure 6-42 is a diagram of an actual vacuum system from a manually valved electron microscope and shows a complicated system of lines interconnected with several switching valves. However, even in this system, the basic rules outlined at the beginning of this section are applicable.

Vacuum Problems and Safety Features

A number of accidental situations may disable a vacuum system and possibly contaminate the column of the electron microscope. Descriptions of the more common types of accidents follow.

If the diffusion pump is connected to the column and the flow of cooling water is interrupted, the diffusion pump oil will not condense and the vapors may drift in the reverse direction to contaminate the column. This is usually prevented either by installing an overtemperature cutoff switch on the diffusion pump or a flow detector in the water line that will shut off power to the heater if water flow stops. The use of liquid nitrogen-chilled baffles or traps (anticontaminators) placed above the diffusion pumps will also lessen or prevent the contamination.

Another problem may result when the backing rotary pump is disabled, perhaps due to a broken drive belt or a power outage. If the diffusion pump is connected to the column when this occurs, then the lower pressure of the column will draw the oil from the rotary pump through the diffusion pump and into the column creating a major disaster requiring extensive cleaning. Most automated vacuum systems will detect the deterioration of vacuum in the column and close all valves. Since the valves are pneumatically activated from a pressurized tank, no electrical power is needed. In a manually valved system, it is poor practice to leave an unattended microscope in high vacuum. In manual systems, it is possible to build a battery-powered warning bell that will be activated by a loss of vacuum in either the column or the backing line of the rotary pump.

Misvalving of the microscope may also lead to messy situations. If one opens the diffusion pump to a column containing too much air, the rush of air into the diffusion pump may blow the diffusion pump oil out into the rotary pump or possibly back up into the lower part of the column. Worst of all, if the column is connected to high vacuum and one improperly opens the bleed valve to vent the lines between the rotary pump and diffusion pump (as may be last done on system shut down), air will rush through the bottom of the diffusion pump and blow va-

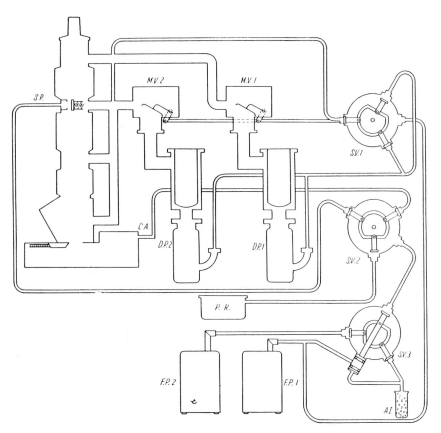

Figure 6-42 Diagram of vacuum system in older, manually valved transmission electron microscope. Newer microscopes use pneumatically actuated valves that are opened and closed automatically as vacuum is needed in various parts of the microscope. Nonetheless, the basics of the systems are very similar in terms of numbers of pumps, valves, and vacuum plumbing needed. (Courtesy of Hitachi Scientific Instruments.)

porized diffusion pump oil into the column with great force. Probably the diffusion pump towers will be displaced and damaged by impact with the cooling baffles at the top of the diffusion pump. Obviously, a major cleaning of the column, as well as disassembly and cleaning of the vacuum system, is needed. The only safety measure to prevent misvalving of a manually operated vacuum system is a thorough understanding of the proper valving sequences of the vacuum system.

Other Types of Vacuum Pumps

Besides the rotary and diffusion pumps, several other specialized vacuum pumps may be more suited to particular electron microscope applications as, for example, when ultraclean vacuums are needed. The more commonly used types of specialized pumps are:

Turbomolecular pumps (Figure 6-43) are oil-less systems that consist of a series of rapidly spinning rotors with blades or vanes inclined at an angle. Between the rotors are a series of static blades (stators) slanted in the opposite direction. Individual rotors and stators are arranged alternately in 8 to 12 layers or stages. The rotors spin at 20,000 to 50,000 rpm and strike gas molecules with such impact that they are driven down to the next stage when the next rotor knocks them deeper into the pump with increasing momentum. The stators encourage

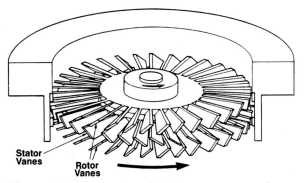

Figure 6-43 Illustration of moving rotor vanes and nonmoving stator vanes of a turbomolecular pump. (Courtesy of Varian Associates—Vacuum Products Division.)

the downward movement due to their inclination. Upon reaching the bottom of the pump, the gas molecules are removed usually by a mechanical rotary pump. Unlike diffusion pumps, turbo pumps do not require water cooling and may pump a sealed chamber at atmospheric pressures. The 10^{-7} to 10^{-9} Pa attainable vacuums are better than diffusion pumps, although the rate will be slower. Turbo pumps are more expensive and require more frequent maintenance than diffusion pumps, but are much cleaner and less susceptible to misvalving problems. These pumps are used more often in scanning rather than transmission electron microscopes, since vibration may be a problem in the latter case.

Entrainment pumps (ion, cryogenic) work on a principle totally different than other pumps that move gas molecules by impact (diffusion, turbomolecular) or positive displacement (rotary) methods. These pumps entrain or hold gas molecules onto chemically reactive or extremely cold surfaces rather than remove the molecules totally from the system. They are used in applications requiring extremely clean, ultrahigh vacuums in the 10^{-9} Pa or better range as, for instance, in lanthanum hexaboride and field emission electron guns. Having no moving parts, these pumps are less prone to breakdown but must be regenerated or purged of the entrained molecules periodically. A major disadvantage of these types of pumps is the differential pumping speeds for various molecular species. For instance, cryogenic pumps remove water molecules quite efficiently but other gas species as nitrogen or methane may be less efficiently entrained by the pump. It may be necessary to combine two different types of entrainment pumps to cover the range of molecules likely to be encountered in any one research situation. Cryosorption pumps entrain molecules on a cooled surface by means of van der Waals forces. The pumping mechanism is based on cryocondensation and cryotrapping onto such cooled surfaces as molecular sieves, porous silver, and activated charcoal. Figure 6-44 shows the design of one system that is chilled with liquid helium from

a recirculating refrigeration loop. Note that different gas molecules are trapped at various chilled surfaces inside the pump module.

Ion pumps, such as the sputter ion pump, operate on the same principle as the cold cathode ionization gauge used in measuring vacuum status. High voltage cathodes generate a flow of electrons toward high voltage anodes. The electrons collide with and ionize gas molecules causing the positive ions to flow to the cathode while the electrons continue their collision path toward the anode. The gas molecules collide with the surface of the cathode and are chemically held to the surface by a reactive coating of titanium or zirconium.

Preparing the Transmission Electron Microscope for Use

Alignment Theory

After the microscope has reached high vacuum status, one must prepare the electron optics for optimal results. This usually takes place at the beginning of the microscopy session or after maintenance has been performed on any of the systems associated with image formation. Although the details of this process, called *alignment,* may vary significantly from one transmission electron microscope to another, certain basic conditions must be satisfied in order to obtain high quality micrographs. This section will explain and outline a general procedure for fulfilling these conditions. More detailed protocols are furnished by the various microscope manufacturers.

The major reasons for aligning any optical system are convenience of operation and improved resolution. For instance, when increasing the magnification, it is very annoying to have the image and even the illumination move off the viewing screen. These phenomena are caused by misaligned imaging and illumination systems, respectively. *In alignment, the optical axes of the illuminating and imaging lenses are lined up along a common axis; the various apertures are centered relative to this axis, and any astigmatism present in the condenser and objective lenses is corrected.* The corrections may involve certain mechanical and electronic adjustments that may take anywhere from 15 minutes for a routine "touch up" alignment to several hours if major systems have been moved, for example after cleaning the components (lens polepieces, stigmators, etc.) inside the TEM column.

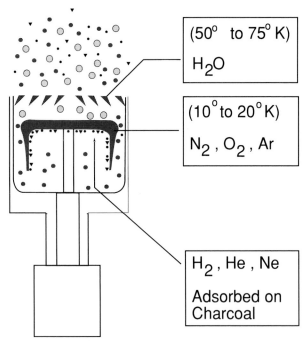

Figure 6-44   Cryosorption pump schematic diagram. Various molecular species are cryocondensed or adsorbed onto reactive cold surfaces. Liquid helium travels inside of the supporting stalk and chills the pump surfaces down to within 10 degrees of absolute zero. Drawing indicates various temperatures inside the pump as well as where various gases are adsorbed. (Courtesy of APD Cryogenics, Inc.)

Since it is impossible to fabricate a perfectly symmetrical electromagnetic lens, the optical axes of the various lenses seldom correspond to the physical center of the lens but rather correspond to either a voltage center or current center. The voltage and current centers are axes around which the image is seen to expand or rotate as either the high voltage or current to a lens is changed, respectively. One way to find the optical axis would be to vary or "wobble the filament high voltage" (i.e., vary it by several hundred volts) and observe the rotation of the image around some point. The appropriate lens would then be shifted so that the rotation point is conveniently centered on the viewing screen. This is called *voltage centering* and is commonly used in the alignment procedure. A second way to locate the optical axis of a lens, *current centering*, is to vary the current to that lens and observe the rotation of the image around the current center. The lens is then moved so that this rotation is centered on the viewing screen. The current and voltage centers will not coincide exactly, but will be very close to each other in a well-constructed lens.

The optical axis is placed on the screen center because any electrical disturbances are minimal at the optical axis. For instance, any high voltage fluctuations will cause the image to expand and rotate around the screen center. If the important part of the image is centered on the screen, then it will show the least effect, whereas the movements are more exaggerated toward the edge of the field (see Figure 6-45). Similarly, any current instabilities or image drift will be minimal at the optical axis, which was placed at the screen center.

The movements necessary to situate the optical axis of an electromagnetic lens along the common alignment axis may involve a horizontal x and y movement, called *translation*, and an inclination along a vertical axis, called *tilt*. Some lenses have provisions for both types of movements, others may have only a translation capability, while some may be factory aligned and not conveniently moved at all. There has been a tendency in newer microscopes to factory align major lenses mechanically and to encourage users to trim this alignment electronically using electromagnetic deflection coils. In these systems, alignment is much faster and easier to accomplish.

Alignment Practice

1. **Optional steps (when replacing burnt-out filament).** Routine alignment normally would begin as described in the next paragraph. However, when the

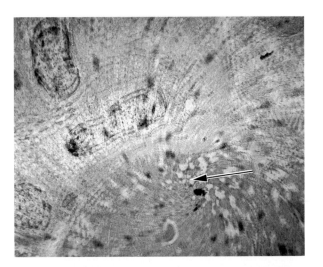

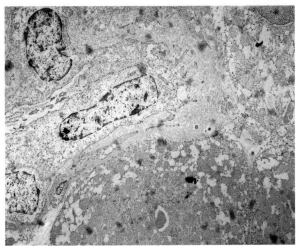

Figure 6-45 (top) Instabilities in a lens are minimal at the optical axis (arrow). Notice how the image is less smeared at the optical axis compared to the region that is off axis. This image was obtained by wobbling the current to the objective lens. (bottom) Normal micrograph without instabilities in the objective lens current.

filament burns out, it will be necessary to remove the cathode cartridge, clean the gun, and install a new filament. After setting the new filament in the cathode cartridge and replacing the shield, the filament height and centration are carefully adjusted relative to the shield aperture. The cathode cartridge is then attached to the high voltage assembly, the system sealed, and the microscope evacuated. During the pumpdown, one should withdraw the smaller objective and diffraction apertures from the column, while the condenser apertures may remain in position if they were previously centered. One may now proceed to the next step and begin alignment of the column.

2. **Alignment begins at the top of the microscope column with the illumination system.**
   If the microscope was very much misaligned (perhaps

in an attempt to locate the illumination after the filament failed), then it may be necessary to switch off some of the imaging lenses in order to find the beam. Otherwise, the lenses are left on and set to the recommended positions for alignment.

3. **Filament emission and rough saturation.**
   The high voltage is applied to the filament and the filament heated to the approximate *saturation point* by watching the emission meter until it no longer continues to rise. If no reading is obtained on the meter, or if the reading is excessive, then the bias to the shield should be changed until the recommended emission is obtained (usually 15 to 25 μA). If this fails, then the filament distance from the shield aperture should be rechecked and the procedure repeated.

4. **Gun and condenser lens 2 (C2) alignment.**
   Vary the C2 lens current or brightness controls until some illumination is seen on the screen center. If possible, decrease the spot size using C2 control and maintain the central location on the screen by translating the condenser lens. Maximize the illumination by translating the gun, but take care not to burn the screen with excessive beam irradiation by spreading the beam, as needed, using the C2 control. Slowly desaturate the gun using filament emission control and focus the filament image using C2. Translate the gun relative to the entire illumination system until a symmetrical image of the undersaturated filament halo is obtained at the center of the screen (Figure 6-46).

5. **C2 aperture centration.**
   Start with crossover at center of screen. Check the centration of the C2 aperture by taking the C2 control through the crossover point and back. If the illumination spot expands and contracts symmetrically at the screen center, then the aperture is centered. If the illumnation sweeps off the screen, the two aperture centration controls should be adjusted until the focal point stays centered.

6. **Condenser lens 1 alignment.**
   Since the illumination spot size may be varied to suit the magnification, the C1 lens must also be aligned. This is accomplished by bringing C2 to crossover at the screen center and varying the C1 lens control

through the various spot sizes (1 through 10 μm, for example) and observing the positioning of the spots on the screen. Translate the C1 lens so that the different spots remain centered.

7. **Condenser lens stigmation.**
   Astigmatism in the condenser lenses results in an illuminating spot that is not circular and generates an asymmetrical aperture angle that degrades resolution. One method of astigmatism correction is to focus the image of the undersaturated filament on the screen using C2 and then sharpen up the image of the filament using the condenser stigmator (Figure 6-47).

8. **Imaging lens alignment.**
   a. *Objective lens.* Depending on the microscope, the imaging lenses may have various combinations of translate and tilt adjustments. In some instruments, the objective lens may have one or both movements, while in other microscopes this lens is stationary and other lenses are moved relative to it. If the objective lens is alignable, this is usually accomplished by wobbling the current through its lens coil (or by reversing the polarity of the lens) and observing the motion of a focused image like a holey film (see Chapter 4). The objective lens is adjusted until the rotation of the image occurs about the screen center.

   b. *Diffraction lens.* The intermediate (diffraction) lens may be centered by bringing it to the crossover or diffraction point, reversing or varying the current through the lens, and translating the lens until the focused bright spots remain close to screen center. This very bright diffraction or crossover spot is sometimes called a *caustic figure*, since it may actually burn a mark on the viewing screen. If the microscope is to be used for diffraction purposes, then the diffraction spot must also be centered at the various camera lengths to be used.

   c. *Projector lens.* The projector lens is usually centered by varying the current to the projector lens and translating the system until the image rotates about the center of the viewing screen. In some

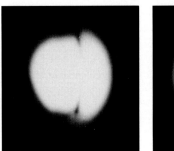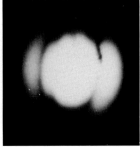

Figure 6-46 Nonsymmetrical filament image (left) prior to alignment. A nearly symmetrical filament image (right).

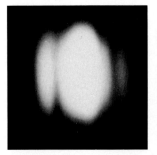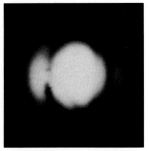

Figure 6-47 Astigmatism in the condenser lens is evidenced by a smearing of the image of the undersaturated filament in the north-south direction (left panel). Stigmated condenser shows a more symmetrical undersaturated filament image.

microscopes with two projector lenses, the projector alignment is accomplished in two stages. After turning off the first projector lens (P1), the second projector (P2) lens is first aligned by observing the two caustic figures generated by reversing polarity of the intermediate lens and moving the most peripheral caustic figure to screen center using the P2 controls. The P1 lens is then turned on and centered in the same way.

9. **Specimen height adjustment.**
On specimen stages with adjustable specimen height settings (z-axis controls), it is important to place the specimen in the proper focal plane of the objective lens. In this location, optimal resolution will be obtained and the image will remain centered if the specimen is tilted. To accomplish this, center a distinctive part of the test specimen and focus it. Tilt the stage +/− 10° from the horizontal and observe if the focused image shifts off center. If so, return the image to center using the z-axis control (and possibly other stage centration controls, depending on design of the TEM). This setting is critical not only for optimal resolutions, but also for convenience of operation if tilting is to be done. In addition, a misplaced z-axis will also affect accuracy of the magnification settings.

10. **Illumination system tilt.**
Most electron microscopes provide a tilt of the illuminating system that must also be accomplished after the other lens alignments. This is an important correction since the illumination may be inclined at a wrong angle to properly enter the optical axis of the objective lens. Voltage centering is achieved by first moving the focused image of a distinctive object such as a holey film to the center of the screen using the specimen traverse controls and activating the high voltage wobbler device to vary the high voltage. The image will rotate and expand around some point that should be returned to the center of the viewing screen using the illumination system tilt controls. When properly tilted, the image will rotate and expand around the screen center as the high voltage is wobbled (Figure 6-48).

11. **Objective aperture centration.**
With the specimen in place, vary the lens current to the intermediate (diffraction) to form the caustic figure. A diffraction aperture is inserted and centered using the aperture centering controls. The aperture is properly adjusted when the bright image stays centered as the the diffraction lens current is varied. The diffraction image is sharpened; the objective aperture is inserted and moved until its outline is centered around the bright central spot (Figure 6-49). The diffraction aperture may then be removed.

12. **Objective lens stigmation.**
A holey film is inserted into the microscope and focused at high magnification (at least twice the magnification that one will be using). As the image of the hole is focused, a bright line, or Fresnel fringe, will be seen around the hole. Focusing is carefully ac-

complished and the symmetry of the white fringe around the hole is observed. In this overfocused position, the white fringe will be readily seen against the background structure of the film. The stigmator is adjusted until the white overfocused fringe is symmetrically arranged around the edge of the hole (Figure 6-50). Note that the background details become sharper and more contrasted when astigmatism is corrected. In fact, this is another way to correct for astigmatism. Namely, a carbon or other finely structured film is focused as sharply as possible at a very high magnification. The stigmator controls are then used to fine focus the image to maximize contrast (Figure 6-51).

Major Operational Modes of the Transmission Electron Microscope

During the alignment procedure, one should be aware that the conventional transmission electron microscope may be set up for operation in several different operational modes. Depending on the design of the microscope, this may involve relatively few or many mutually exclusive adjustments. In addition, certain specimen preparation techniques may be utilized to further enhance these operational modes.

High Contrast

A constant problem with biological specimens is their low contrast. In the high contrast mode, the instrument is adjusted to give contrast at the expense of high resolution. As a result, this mode is generally used at magnifications under 50,000×. The conditions that may be changed to enhance contrast are summarized below.

How to Obtain High Contrast

1. **The focal length of the objective lens is increased.**
This necessitates using shorter specimen holder cartridges (Figure 6-30, left) in a top entry stage to position the specimen higher above the objective lens. In a side entry stage, adjustment of the z-axis or specimen positioning may also be needed if a special holder is not provided. It may be recalled that longer focal lengths result in narrower aperture angles, a worsening of chromatic aberration, and a loss of resolution.

2. **Lower accelerating voltages are used.** The resulting lower energy electrons are more readily affected by differences in specimen density and thickness, and contrast will be thereby increased. Unfortunately, this interaction with the specimen generates a population of imaging electrons with a wide range of energies, resulting in an increase in chromatic aberration. Lower

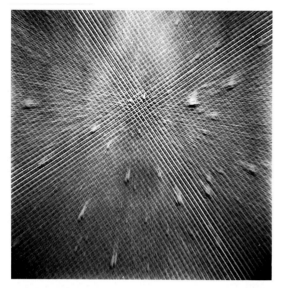

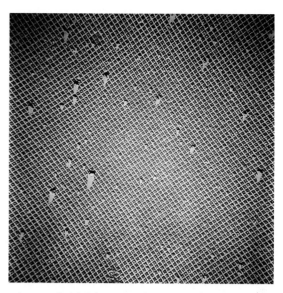

Figure 6-48 Expansion of the image around the voltage center of an electromagnetic lens (left). This image was obtained by modulating the high voltage and recording the image shifting during the modulation. The more distinct micrograph (right) shows the image after centering of the voltage and termination of the modulation. (Courtesy of S. Schmitt.)

Figure 6-49 Off-centered objective aperture (left) compared to the image of a properly centered objective aperture (right). (Courtesy of Scott Pelok.)

accelerating voltages are also more damaging to the specimen, since the electrons are slowed down more and transfer more energy to the specimen, resulting in excessive heating. Lower energy electrons are more susceptible to poor vacuum conditions, with the exacerbation of chromatic aberration. Clean, high vacuums are needed to minimize electron energy losses, and the microscope itself should be clean, since these electrons are more easily affected by astigmatism. Lastly, it will be recalled that lower energy electrons have longer wavelengths, so that the resolving power will be degraded.

3. **Smaller objective apertures should be utilized.** These apertures will remove more of the peripherally deflected electrons from the specimen, so that the subtractive image from the objective lens will be accentuated in contrast (i.e., the signal-to-noise ratio is increased). Small apertures are more prone to astigmatism problems, making clean vacuums and specimen anticontaminators essential.

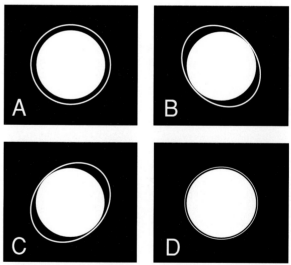

Figure 6-50 Correction of astigmatism (A) Establish overfocused fringe. (B) Change focus until direction of astigmatism becomes apparent. (C) Place the stigmator direction or azimuth so that it is 90 degrees to the astigmatism. (D) Adjust strength or amplitude to generate a symmetrical field.

4. **Photographic procedures may be employed.** Most images generated in the transmission electron microscope are enhanced for contrast using photographic techniques. During exposure of the electron micrograph, the sensitivity of the exposure meter may be adjusted to slightly overexpose the film. Underdevelopment will then enhance the contrast range in the final negative. Details will necessarily be lost in the

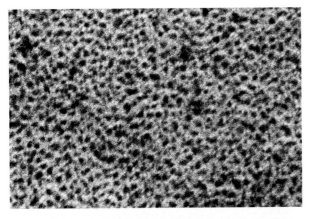

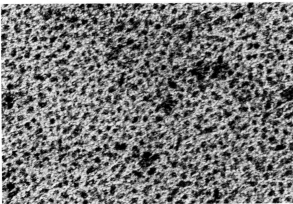

Figure 6-51 Alternate method for correcting astigmatism using Pt/Ir or carbon evaporated film. (A) Focus image as best as possible using objective lens. (B) Sharpen focus and contrast using stigmator controls.

intermediate density ranges. Of course, during the printing of the negative, one may use higher contrast photographic papers (see Chapter 8).

5. **The specimen may be prepared to enhance contrast.** Standard fixation and staining techniques will increase density by depositing the heavy metals along various organelles. Certain embedding media (polyethylene glycol) that may be dissolved or etched away will help boost contrast, or one may utilize stained, frozen sections without any embedding media. The easiest approach is simply to cut thicker sections; however, the resulting chromatic aberration and superimposition of structure will degrade resolution.

High Resolution

Most of the conditions used to achieve high resolution in the electron microscope are the opposite conditions discussed above for the high contrast mode. Since contrast will be lacking in these specimens, efforts should be made to boost contrast using appropriate specimen preparation and darkroom techniques, as described in the previous section.

How to Obtain High Resolution

1. **The objective lens should be adjusted to give the shortest possible focal length** and the proper specimen holders used. In some systems, this is simply a matter of pressing a single button, whereas, in certain microscopes several lens currents must be changed concomitantly. Perhaps it may even be necessary to insert a different polepiece in the objective lens.

2. **Adjustments to the gun, such as the use of higher accelerating voltages,** will result in higher resolution for the reasons already mentioned in the discussion on high contrast. Chromatic aberration may be further lessened by using LaB_6 or field emission guns since the energy spread of electrons generated from such guns is considerably narrower. (The energy spread for tungsten = 2 eV, LaB_6 = 1 eV, and field emission = 0.2-0.5 eV). In an electron microscope equipped with a conventional gun, a pointed tungsten filament will generate a more coherent, point source of electrons with better resolution capabilities.

3. **Use apertures of appropriate size.** For most specimens, objective lens apertures of 20 to 50 μm will give adequate contrast and diminish spherical aberration, to a significant degree. Smaller apertures may be used, but they must be kept clean since dirt will have a more pronounced effect on astigmatism. Small condenser lens apertures will diminish spherical aberration, but this will be at the expense of overall illumination. The illumination levels may be improved by altering the bias to effect greater gun emissions; however, this may thermally damage the specimen.

4. **Specimen preparation techniques** may also enhance the resolution capability. Extremely thin sections, for instance, will diminish chromatic aberration. Whenever possible, no supporting substrates should be used on the grid. To achieve adequate support, this may require the use of holey films with a larger than normal number of holes (holey nets, see Chapter 4). The areas viewed are limited to those over the holes.

5. **Miscellaneous conditions** such as shorter viewing and exposure times will minimize contamination, drift, and specimen damage, and help to preserve fine structural details. Some of the newest microscopes have special accessories for minimal electron dose observation of the specimen and may even utilize electronic image intensifiers to enchance the brightness and contrast of the image. Anticontaminators over the diffusion pumps and specimen area will diminish contamination and resolution loss. High magnifications will be necessary, so careful adjustment of the illuminating system is important. It may take nearly an hour for the eyes to totally adapt to the low light levels, and this adaption will be lost if one must leave the microscope room. Alignment must be well done and *stigmation must be checked periodically* during the viewing session. The circuitry of the microscope should be stabilized by allowing the lens currents and high voltage to warm

up for 1 to 2 hours before use. Bent specimen grids should be avoided since they may place the specimen in an improper focal plane for optimum resolution. In addition, they prevent accurate magnification determination and are more prone to drift since the support films are often detached.

Darkfield

In the normal operating mode of the transmission electron microscope, the unscattered rays of the beam are combined with some of the deflected electrons to form a brightfield image. As more of the deflected or scattered electrons are eliminated using smaller objective lens apertures, contrast will increase. If one moves the objective aperture off axis, as shown in Figure 6-52, left, the unscattered electrons are now eliminated while more of the scattered electrons enter the aperture. This is a crude form of darkfield illumination. Unfortunately, the off-axis electrons have more aberrations and the image is of poor quality.

Higher resolution darkfield may be obtained by tilting the illumination system so that the beam strikes the specimen at an angle. If the objective aperture is left normally centered, it will now accept only the scattered, on-axis electrons and the image will be of high quality (Figure 6-52, right). Most microscopes now have a dual set of beam tilt controls that will permit one to adjust the tilt for either brightfield or darkfield operation. After alignment of the tilt for brightfield followed by a darkfield alignment, one may rapidly shift from one mode to the other with the flip of a switch. Both sets of controls also provide for separate stigmation controls to correct for any astigmatism introduced by the tilting of the beam to large angles.

The darkfield mode can be used to enhance contrast in certain types of unstained specimens (thin frozen sections) or in negatively stained specimens. An example of a darkfield image is shown in Figure 6-53.

Diffraction

In specimens that contain crystals of unknown composition, the diffraction technique may be used to measure the spacing of the atomic crystalline lattice and determine the composition of the crystal, since different crystals have unique spacings of their lattices. *The diffraction phenomenon is based on the reflection or diffraction of the electron beam to certain angles by a crystalline lattice.* Instead of focusing a conventional image of the crystal on the viewing screen using the objective lens, one uses the intermediate or diffraction lens to generate a diffraction pattern on the screen. Since the crystalline lattice reflects the electron source to form bright spots on the viewing screen (similar to the mirrored rotating

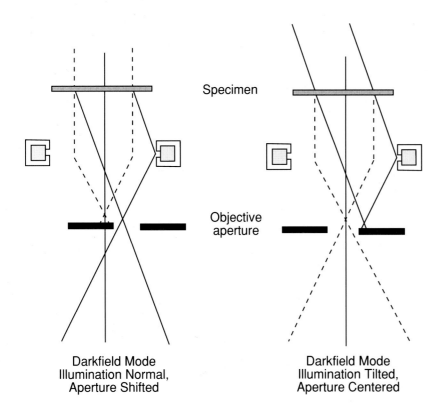

Specimen

Objective aperture

Darkfield Mode
Illumination Normal,
Aperture Shifted

Darkfield Mode
Illumination Tilted,
Aperture Centered

Figure 6-52 Schematic diagram showing two ways of setting up microscope for darkfield imaging: (left) displacement of objective aperture off-axis; (right) tilt of illumination system into on-axis objective aperture.

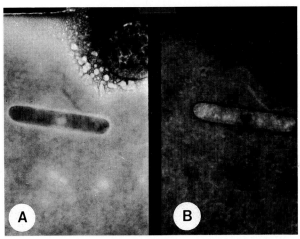

Figure 6-53 (A) Darkfield image obtained by tilting illumination system. (B) Same specimen viewed in standard brightfield mode. Specimen is a gram negative bacterium, *Halobacterium volcanii*, that grows in high salt concentrations in the Dead Sea. (Specimen provided by Larry Kepple.)

sphere sometimes used in ballrooms to reflect a light source onto the walls), the image will consist of a central, bright electron source surrounded by a series of spots, which are the reflections. The distance of these spots from the bright central spot is inversely proportional to the spacing of the crystalline lattice. A crystal with small lattice spacings will diffract the central beam to greater angles to give spots that are spaced far from the central spot. This is unfortunate for biologists, since organic crystals, such as protein, with large lattice spacings will diffract the beam so little that the spots will be crowded around the central bright spot and engulfed by its brilliance. With organic crystals, the specialized technique of high dispersion electron diffraction must be used.

Diffraction Practices

In practice, the crystal is located using either the brightfield or darkfield imaging modes. After placing the TEM into the selected-area diffraction mode (a pushbutton on some microscopes), a diffraction aperture of the appropriate size is inserted and used to select the area to be diffracted. After focusing on the edges of this aperture with the intermediate lens, the image is carefully focused with the objective lens for the last time. Following a slight defocus of C2 to provide a coherent source of illumination (and to prevent burning the viewing screen), the bright central image of the source is focused on the screen using the intermediate lens. Withdrawal of the objective aperture will reveal the diffraction pattern on the screen. Focus may be sharpened using the intermediate lens only. In order to cut down on the glare from the bright central spot, a physical beam stopper is inserted to cover it. Ex-

posures are usually made for 30 seconds to several minutes in the manual mode since the illumination levels will be very low. Single crystals will generate separate spots while polycrystalline specimens will produce so many spots around the central point that they will blend to form a series of concentric rings (Figure 6-54). Some biological applications of diffraction may be to confirm that a crystal present in human lung tissue is a form of asbestos, or to identify an unknown crystal in a plant or bacterial cell. See also Chapter 15 and the reference sources at the end of this chapter.

Checking Performance

Alignment

It is possible to check and make minor corrections to the alignment of the electron microscope in less than 10 minutes, as follows:

Check that the C2 aperture is centered by varying the current to the C2 lens and observing that the illumination stays centered. Insert a holey film and set the microscope at a low magnification. Bring the illumination to the smallest possible crossover spot size and then increase to the top magnification. Both the illumination spot and the image should stay close to the center of the screen. If either move off screen, then adjustments will be needed. Now, repeat this process using the largest spot selectable by C1 and make corrections if the illumination does not stay centered. Wobble the high voltage at a high magnification and observe that the focused image rotates around the center of the screen. Vary the current to the objective lens and observe similarly. Go to the diffraction mode with the intermediate

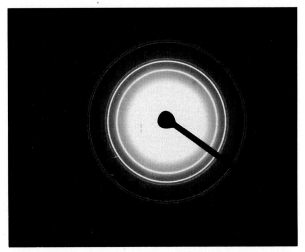

Figure 6-54 Diffraction pattern obtained from polycrystalline specimen showing characteristic ring pattern.

lens and observe that the caustic pattern is centered. Check centration of the objective aperture while in the diffraction mode. Finally, stigmate the microscope if necessary.

Electrical Stability

The microscopist should be aware that a recently turned on microscope will be somewhat unstable until the high voltage and lens circuits have warmed up for perhaps 30 to 60 minutes. If imaging problems are still encountered and instabilities in the microscope are suspected, several areas may be checked as follows:

Problems with the filament may be checked by undersaturating the filament and carefully observing its focused image on the screen. Changes in the size or brightness of the halo may indicate problems with the vacuum, bias controls, contamination in the gun area, or it may indicate that a weak filament is about to burn out. Instabilities in the high voltage, diffraction, or projector lenses may be detected by observing the caustic image. If the image changes in sharpness or moves, then instabilities in these circuits should be pursued further. High voltage or objective lens instabilities may be detected by observing an overfocused Fresnel fringe in a holey film. Should the fringe change in any way, high voltage and objective lens circuits should be checked. All electron microscopes have built-in gauges or test points in circuit boards that may be monitored using appropriate testing equipment. However, unless one has a thorough understanding of the process and an appreciation of the dangers involved, this is best left to trained personnel. At least one will have diagnosed the problem as a microscope—not a specimen—problem.

Image Drift

Gradual shifting of the image, usually in one direction, is a common and annoying problem, encountered especially when support films are not used.

Drift might be caused by heating of the specimen due to excessive beam irradiation, in which case a smaller spot size, condenser aperture, or less filament emission may be tried. If the section or plastic substrate is not firmly attached to the grid, it will move as the grid heats up—the amount of movement is related to the beam intensity and the area illuminated. This may be confirmed by examining a test specimen known to be thermally stable. Contamination in the area above the specimen may lead to charging and shifting of the image as the static is discharged. This should be suspected if the image

shifts as the C2 illumination level is changed. Moderate drift of the specimen on the screen may be seen through the viewing binoculars. An easy way to test a specimen for drift is to place a recognizable specimen structure next to a fixed point on the viewing screen and observe the movement of the specimen over an interval 2 to 3 times greater than the exposure time for the negative. Drift may be documented by taking an exposure of an overfocused hole, waiting for 1 to 2 minutes with the specimen still being irradiated, and then taking a second exposure on the film. This double-exposed film is then developed and the distance traveled by the hole may be converted into nanometers traveled per second (Figure 6-55). For example, if one is hoping to resolve 1 nm and the exposure time is 4 seconds, then the specimen should not drift any more than 1 nm per 4 seconds. At a magnification of 500,000×, the drift should not exceed 5 mm over 4 seconds. (See discussion on magnification calibration in subsequent section of this chapter.)

Contamination

Deposition of contaminants onto the specimen that is being subjected to electron bombardment will degrade resolution. One should be prepared to quantitate the rate of contamination in order to determine when it has become unacceptable for the resolution level needed.

The contamination rate may be quantitated by double exposing a hole in a test specimen as described in the previous paragraph. The decrease in

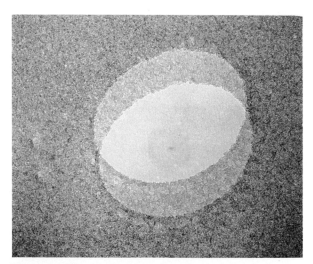

Figure 6-55 Drift in an electron microscope. Two exposures are made on the same negative with an interval of 1 to 2 minutes between exposures. The rate of drift may be calculated based on the distance moved over the intervening time period.

the diameter of the hole, by deposition of contamination along the rim, is an excellent indicator of the rate of contamination. A 2 to 3 minute exposure should indicate a contamination rate of less than 0.01 nm per second with all anticontaminators in operation. Without a specimen anticontaminator, the rates may be ten times higher even in a clean microscope. It should be realized that a recently cleaned microscope column is actually quite contaminated with organic vapors used in the cleaning procedure, so that the microscope should be allowed to remain in high vacuum for one full day in order to remove these molecules. The anticontaminators on the diffusion pumps should be filled with liquid nitrogen, but the specimen anticontaminator should not be chilled during that time, since it will quickly load up with contaminants that may later be released onto the specimen.

Magnification

It is necessary to calibrate the magnification settings of all electron microscopes, since different mechanical and electronic alterations will result in significant variations. Even in the most modern of microscopes, some manufacturers warrant the figure displayed to be within only +/– 5% to 10% of the actual value.

Although it is possible to calibrate the entire magnification range in the microscope, this may not be necessary if one uses only certain settings. The frequency of calibration varies with the individual need for accuracy, as well as with the servicing intervals for major systems in the microscope. Once or twice a year may be sufficient for routine work where high accuracy is not required. In critical operations, calibration should take place during each use of the microscope or, alternately, one may include a standard of known size on the same grid as the specimen. A number of situations may cause a change in previously determined magnification figures.

After *cleaning* operations involving the polepieces, specimen stage, or specimen holders, magnifications should be recalibrated. *Servicing* of electronics may affect magnifications if lens current or reference circuits were repaired.

A major cause of erroneous figures arises from lens *hysteresis* (also called *remanence*), or the residual magnetism left in the soft-iron polepieces even after the electromagnetic field strength has been changed. This may be minimized by starting the magnification calibration series at a low magnification and taking

micrographs at increasingly higher magnifications. Similarly, the low- to high-magnification scheme should be followed when viewing and recording images. One should also verify that the calibration is accurate at different accelerating voltages.

The placement of the *specimen height* in the objective lens is very critical. Since the position of the grid must be within 40 μm for accurate replication of a particular magnification setting, the grid must be perfectly flat and the z-axis (specimen height) must be accurately set. This is a significant consideration with a side-entry stage since the z-axis is adjustable over a very wide range. If the stage is tilted, the x and y traversement will result in a change of specimen height. Bent grids should be avoided, and the specimen should always be on the same side of the grid relative to the beam.

Magnification Calibration

Calibration may be accomplished by first acquiring accurate standards with established sizes and properties. More than one standard will be needed to cover the entire magnification range.

For instance, from the lowest magnifications up to perhaps 40,000 to 50,000×, one may use commercially available cross ruled *diffraction grating replicas* with 2,160 lines/mm (Figure 6-56). The procedure followed is to focus carefully on the granularity of the grating and take an electron micrograph at the desired settings. After development, the negatives are placed on an illuminated view box and distances measured in millimeters. Measurements should be made from the same relative positions in each line (e.g., middle of the dark band of one line to a similar position on the second line). One then counts the number of spaces included in the specific millimeter distance measured. These values are substituted into the following equation:

Equation 6-10: Magnification Calculation from Diffraction Grating Replica

$$M = \frac{2.16\,(A)}{B} \times 10^3$$

where M = magnification
A = distance in mm between lines on electron micrograph
B = number of spaces between lines

Several *sources of error* exist with the diffraction grating method. The grating is a platinum/carbon replica of a standard optical grating and may have undergone some distortion during the mounting on the grid, thereby affecting the accuracy of measurements. Generally "waffle" gratings are preferred to the parallel line gratings, since

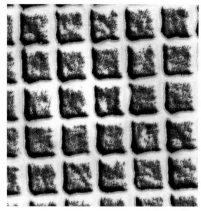

Figure 6-56 Magnification calibration standard. This series of micrographs are of a standard diffraction grating containing 2,160 lines/mm. The magnifications were calculated to be 10,000×; 21,000×; and 30,000×, respectively.

the crossed lines permit measurements in both directions to increase the accuracy. In order to have an accuracy of 2%, it is necessary to include at least 10 spaces in the distance measured. In practical terms, this means that gratings are accurate at the 2% level only up to a magnification of 25,000×. However, if a 5% level of accuracy is acceptable, then they may be used up to 50,000 to 60,000×. At the higher magnifications, in order to minimize the error involved in counting only a fragment of a space, one measures a small piece of debris at a known accurate magnification and then remeasures this same object at the unknown magnification. The amount of enlargement between the two is the factor by which the known setting is multiplied. An object measuring 5 mm at 20,000× and now measuring 12.5 mm is 2.5 × larger, so that the new magnification is 2.5 × 20,000 = 50,000×.

Above 50,000×, one must resort to *organometallic crystals* with established lattice spacings. A good standard is beef liver catalase prepared by placing a drop of the catalase suspension on a coated grid for 5 to 10 seconds and then negatively staining the specimen by floating the grid on a drop of 2% aqueous solution of ammonium molybdate or potassium phosphotungstate (see Chapter 5). The largest lattice, beef catalase, has spacings of 8.8 nm. Alternative standards might include bacteriophage particles that may be obtained from colleagues or purchased from the American Type Culture Collection.

Resolution

Most electron microscopes have a guaranteed optimum resolution figure that was verified by the manufacturer usually upon installation of the microscope. As long as one obtains satisfactory micrographs, it may be mistakenly assumed that the resolving power has not changed. One should be aware of several methods to verify this, since the degradation of resolving power may be so insidious as to go unnoticed until the quality of work is

brought into question, perhaps to the embarrassment of the microscopist.

The *point-to-point method* is the most readily accepted method since it graphically demonstrates the ability of the microscope to visualize two fine points separated by a specific distance. As a test specimen, one may use either a thin carbon film alone or with a thin film of platinum-irridium alloy evaporated onto the carbon. A series of micrographs are made at the top magnification needed to demonstrate the resolution. Even if all conditions are perfect, it is very difficult to obtain satisfactory results from a single micrograph, since precise focusing is essential. Consequently, a *through-focus series* is made by first focusing the image as best as possible and then backing off the finest focus knob counterclockwise by two click stops. Micrographs are then rapidly taken, advancing the focus clockwise for each exposure until a total of five shots have been made. The five micrographs are later examined on a light box and several separated points are located on the two films that are closest to focus. The smallest distances between two points are located on both negatives and converted into actual nanometer distances based on an accurate knowledge of the magnification (Figure 6-57). It is important to locate the points on two films, since electron noise ("snowy" background) may give an impression of two points that do not physically exist. Incidentally, *a through-focus series may be necessary in order to obtain the most accurate focus in instances where it is critical for optimum resolution.*

The *lattice test* is based on demonstrating a crystalline lattice of known spacings. Simply stated, if one sees the lattice structure in a graphitized carbon particle or a single crystal gold foil (Figure 6-58),

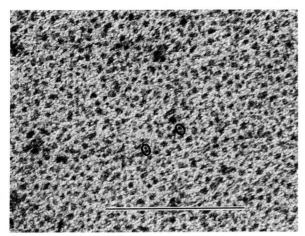

Figure 6-57 Resolution standard, evaporated Pt/Ir film showing a resolution of 0.4 nm at points circled. Magnification bar = 50 nm.

Figure 6-58 Resolution standard, gold foil. The lattice spacings show a resolution better than 0.204 nm. Final magnification of print is 2.4 million times.

then distances of 0.34 nm and 0.20 nm, respectively, are being resolved. There are several problems with taking such figures as being strictly accurate, since astigmatism and electron noise may artificially enhance the resolving capabilities supposedly being demonstrated. Most microscope manufacturers will provide two figures for resolving power, for example, lattice = 0.14 nm and point-to-point = 0.30 nm.

The *Fresnel fringe method* for calibrating resolution is undoubtedly the most convenient method since the test specimen, a holey film, is readily available to all microscopists. After correcting for astigmatism, the fringe is focused as accurately as possible and a series of micrographs is taken by varying the focus slightly between the various exposures.

The films are examined for that negative that shows the finest fringe by being slightly overfocused. The fringe width is then evaluated by measuring the distance from the center of the overfocus fringe (e.g., the dark line on the negative) to the center of the white space just inside of the dark line (Figure 6-59B). This distance in nonometers is the resolving power of the instrument. Of course, one must have a good magnification calibration in order to determine the fringe width accurately. Since this method is based on accurate stigmation and focusing of the microscope, an error of 20% is not unusual. However, with experience this simple method may prove adequate for all but the most precise situations.

Levels of Usage of the Transmission Electron Microscope

The preceeding chapter was directed toward giving the investigator a sound background in the fundamentals of the transmission electron microscope. For serious, committed electron microscopists maintaining their own equipment, this level of detail—and even much more—is a necessity. However, there are several other levels of experience that might be appropriate for individuals using an electron microscope.

The experienced user and maintainer. This is an individual who has responsibility for the operation and maintenance of the microscope. Considerable experience in the theoretical basis of microscopy, electronics, and a mechanical inclination are necessary. If the microscope fails, and even new ones do fail, this individual must have the capability to repair it. Rarely do electron microscope users have full maintenance capability. If the microscope breaks down or if specialized replacement parts are needed, then service representatives may be called upon.

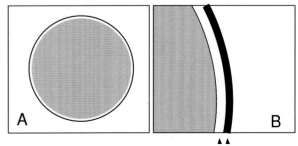

Figure 6-59 (A) Convenient resolution standard, Fresnel fringe method. The fringe width is measured, based on known magnification, and the finest fringe width measured is equal to resolution. (B) Enlargement showing distance to be measured indicated by arrowheads.

The experienced user with a maintenance contract. Such researchers are involved in the routine operation of the microscope and the recording of images. The level of familiarity with the instrument may include routine alignment and perhaps routine servicing (cleaning of apertures and specimen anticontaminators, replacment of burnt-out filaments, calibration of magnification and resolving power, etc.). Major repairs and routine cleaning of sensitive microscope parts (polepieces, specimen stages, column liners, etc.) may be left to service engineers as part of an annual service contract. Such yearly contracts may vary from 5% to 10% of the original cost of the instrument. In situations where productivity cannot be interrupted by a researcher becoming involved in microscope maintenance, such contracts are probably quite cost effective. In addition, the contracts may include the replacement of extremely expensive parts (i.e., high voltage tanks, vacuum system components, electronic circuit boards, stages and specimen holders, etc.).

The assisted viewer. Individuals who have only a temporary need for electron microscopy may rely on more experienced personnel to assist them in specimen preparation and the viewing of specimens in the electron microscope. Often, researchers in need of electron microscopy will contact a microscopist at their own or perhaps at nearby institutions to arrange for the needed services. In this case, the researcher may simply view the specimens while a trained individual oversees the operation of the microscope. With time, and a continued need for electron microscopy, the researcher may opt to become more personally involved in the operational procedures. However, a more efficient use of the researcher's time may still involve assistance by an experienced microscopist.

A user contracting for service. Occasionally a researcher may need electron microscopy as a small part of a study, perhaps to confirm some data obtained by other techniques. The researcher may have little experience in image interpretation and no experience or interest in learning even the basics of microscope operation. Such individuals usually contact *consultants* who are experienced in the area of research in which the investigator is involved and who have access to an electron microscope. The researcher may send specimens to the consultant, who then studies the specimens in the electron microscope, records images, and provides the researcher with photographic prints and perhaps even an analysis of the micrographs. Such consultants generally command premium fees since they are experienced in both electron microscopy as well as the interpretation of the images.

Shared Facilities

Due to the high cost of acquiring and maintaining an electron microscope and associated support equipment, shared facilities are becoming more common than individual researchers having such units. A shared or centralized facility may involve only one electron microscope serving several researchers with similar research interests or it may be quite extensive with many electron microscopes equipped for various specialized needs. Users of such facilities may be operating at different levels of experience, so that trained microscopists and technicians may be involved in various service and training activities. Central facilities are usually financially supported to various extents by a central administration and/or by a system of external grants and contracts. Often there is an established fee structure for the various levels of usage of the facility.

References

Agar, A. W., R. H. Alderson, and D. Chescoe. 1974. Principles and practice of electron microscope operation. In *Practical methods in electron microscopy*, Vol. 2. A. M. Glauert, ed. Amsterdam, The Netherlands: Elsevier/North Holland Biomedical Press.

Beeston, B. E. P., R. W. Horne, and R. Markham. 1972. Electron diffraction and optical diffraction techniques. In *Practical methods in electron microscopy*, Vol. 1 of *Practical methods in electron microscopy* Series, A. M. Glauert ed. Amsterdam, The Netherlands: Elsevier/North Holland Biomedical Press.

Goldstein, J. I., D. E. Newberry, P. Echlin, D. C. Joy, C. Fiori, and E. Lifshin. 1981. *Scanning electron microscopy and x-ray microanalysis. A textbook for biologists, materials scientists and geologists.* New York: Plenum Press.

Meek, G. A. 1976. *Practical electron microscopy for biologists.* New York: John Wiley and Sons.

Misell, D. L., and E. B. Brown. 1987. *Electron microscopy for biologists.* New York: John Wiley and Sons.

Misell, D. L., and E. B. Brown. 1987. Electron diffraction: an introduction for biologists, Vol. 12 of *Practical methods in electron microscopy* Series, A. M. Glauert, ed. Amsterdam, The Netherlands: Elsevier/North Holland Biomedical Press.

O'Hanlon, J. F. 1980. *A user's guide to vacuum technology.* New York: John Wiley and Sons.

Varian Vacuum Products Division. 1986. *Basic vacuum practice.* Palo Alto: Varian Associates Vacuum Products Division.

Wischnitzer, S. 1981. *Introduction to electron microscopy.* New York: Pergamon Press.

Basic Systems of the SEM

Electron Optical and Beam Management Systems

Condenser Lenses 1 and 2

Final Condenser Lens

Stigmator Apparatus

Apertures in the Final Condenser Lens

Resolution in the SEM

Interaction of Electron Beam with Specimen

Elastic Scattering

Inelastic Scattering

Specimen Manipulation

Electron Detector, Signal Processing, and Recording Systems

Signal Versus Noise

Secondary Electron Detector

Signal Processing

Image Recording

Three Dimensionality of the SEM Image

Stereo Imaging with the SEM

Generating Two Micrographs with Separate Views

Merging Two Micrographs with Separate Views to Generate a Stereo Image

Major Operational Modes of the SEM

High Resolution

Great Depth of Field

Imaging Other Types of Specimen Signals

Backscattered Electrons

Backscattered Electron Detection

Important Considerations When Using Backscattered Imaging with Biological Specimens

Cathodoluminescence

Using Secondary and Backscattered Electrons in Biological Studies

Specialized Instrumentation for Observing Unfixed Tissues

Observation of Frozen Specimens

Observation of Fresh Specimens

The Scanning Electron Microscope

Courtesy of AMRAY Inc.

About the same time the first transmission electron microscope (TEM) was nearing completion in the 1930s, a prototype scanning electron microscope (SEM) was constructed by Knoll and von Ardenne, in Germany. Following several refinements made by Zworykin at the RCA laboratories in the United States, as well as improvements made by McMullan and Oatley at Cambridge University in England, a commercial SEM became available in 1963. A later version shown in Figure 7-1A, the Cambridge instrument had resolving powers of about 20 to 50 nm, useful magnifications of 20 to 50,000×, and a depth of field that was 300 times greater than the light microscope.

A recent model SEM with capabilities for x-ray analysis (Chapter 15) is shown in Figure 7-1B. This instrument resolves 3.5 nm with magnifications up to 200,000×.

The SEM received much notice by the popular press due to the readily recognizable, greatly enlarged, three-dimensional images of insects, flowers, pollen, etc. Even now it is not unusual to find popular books dedicated to the beautiful, often surre-

alistic images observed with these instruments (Figure 7-2). In contrast to the TEM, that is used to view *thin* slices of biological specimens, the SEM can be used by biologists to study the *three-dimensional* features of individual cells and even whole organisms. Some SEMs permit one to insert specimens 3″ to 5″ in size into the specimen chamber and view them with a depth of field of several millimeters. In addition, the SEM is a standard instrument in the electronics industry where it is often situated in the assembly line area to control the quality of microcircuitry.

Even though the condenser and objective lenses of the SEM are constructed similarly to those of the TEM (Chapter 6), these electromagnetic lenses do not form an image of the specimen according to the optical principles used by the more conventional light and transmission electron microscopes. Instead, the lenses of the SEM are used to generate a demagnified, focused spot of electrons that is scanned over the surface of an electrically conductive specimen (Figure 7-3). As these impinging electrons strike the specimen, they give rise to a variety of signals (see Chapter 15), including low energy *secondary electrons* from the uppermost layers of the specimen. Some of the secondary electrons are collected, processed, and eventually translated as a series of *pixels* (picture elements) on a cathode ray tube or monitor. For each point where the electron beam strikes the specimen and generates secondary electrons, a corresponding pixel is displayed on the viewing monitor. The brightness of the pixel is directly proportional to the number of secondary electrons generated from the specimen surface. Since the electron beam is scanned rapidly over the specimen, the numerous, minute points appear to blend into a continuous-tone image composed of many density levels or shades of gray. The shading is similar to an ordinary black and white photograph in which light and dark areas give the impression of depth.

Basic Systems of the SEM

The SEM can be subdivided into a number of component systems that carry out various functions. Among these systems, an *electron optical system* is involved in the focusing and control of the electron beam. A *specimen stage* is needed so that the specimen may be inserted and situated relative to the beam. A *secondary electron detector* is used to collect

Figure 7-1(A) One of the first commercially produced SEMs, the Cambridge Mark II Stereoscan. (Courtesy of Leica.)

Figure 7-1(B) Photograph of a modern Cambridge SEM equipped to do x-ray analysis. (Courtesy of Leica.)

the electrons and to generate a signal that is processed by *electronics* and ultimately displayed on viewing and recording monitors. A *vacuum system* is necessary to remove air molecules that might impede the passage of the high energy electrons down the column—as well as to permit the low energy secondary electrons to travel to the detector.

Certain components of the SEM are identical to those found in the TEM. For instance, the construction of the electron gun, electromagnetic lenses, and vacuum systems are so similar that it is necessary to read these sections in Chapter 6 prior to reading this chapter. This chapter will deal only with those components and functions that are unique to the SEM. Likewise, since some of the techniques for biological specimen preparation are quite different from those used in the TEM (Chapter 2), the SEM procedures were described in Chapter 3. An overall schematic representation of the SEM is shown in Figure 7-3.

Electron Optical and Beam Management Systems

This system of electromagnetic lenses, deflection coils, and stigmators is involved in the control and refinement of the electron beam after it leaves the electron gun and before it strikes the specimen. As in the TEM (Chapter 6), most SEMs use a V-shaped, tungsten filament that is heated to effect the thermionic emission of electrons, which are accelerated in the direction of the anode due to the application of negative high voltage. After leaving the bias shield and forming an initial focused spot of electrons of approximately 50 μm in diameter, a series of two or three condenser lenses are used to successively demagnify this spot sometimes down to 2 nm or less. As will be seen later, *small spot sizes are essential for the better resolutions required at high magnifications.*

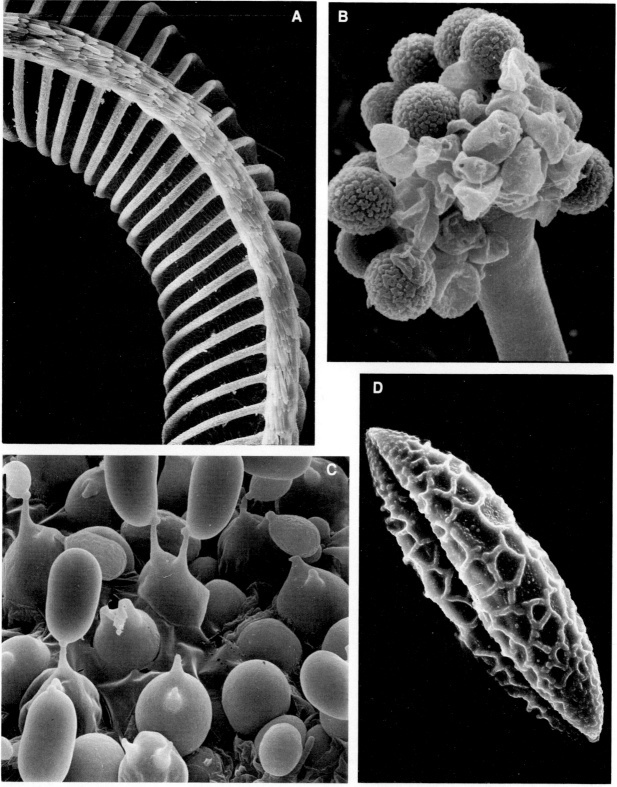

Figure 7-2 (A) Antenna of moth, 50×. (Courtesy of S. Schmitt) (B) Fungal fruiting body on soybean leaf, 1,700×. (C) Spore structures on underside of mushroom, 2,500×. (Courtesy of A. Krajec) (D) Day lilly pollen grain, 750×. (Courtesy of S. Schmitt)

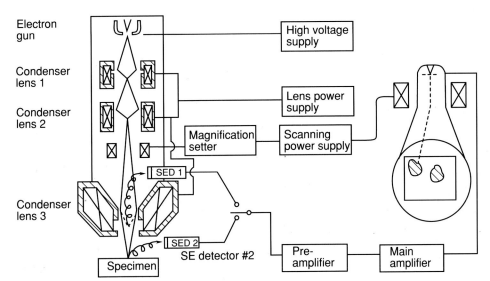

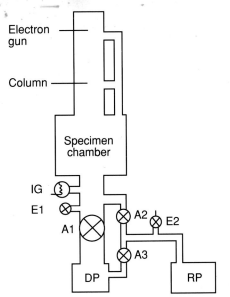

Figure 7-3(B) Schematic diagram of vacuum system. A1-3 = valves, DP = diffusion pump, E1, 2 = air admittance valves, RP = rotary pump, IG = ionization gauge. (Courtesy of Hitachi Scientific Instruments.)

Condenser Lenses 1 and 2

The number of condenser lenses may vary from two to three, depending on the resolving powers needed. The *first condenser lens, C1*, begins the demagnification (i.e., decrease in size) of the 50 μm focused spot of electrons formed in the area of the electron gun.

As the amount of current running through the first condenser lens (C1) is increased, the focal length of the lens becomes progressively shorter and the focused spot of electrons becomes smaller. It is

apparent from studying Figure 6-28 (Chapter 6) that a short focal length C1 lens subsequently causes such a wide divergence of the electrons leaving this lens that many electrons are not able to enter the next condenser lens (C2) and are lost from the electron beam. *The overall effect of increasing the strength of C1 is to decrease the spot size, but with a loss of beam electrons: Resolution improves, but the overall signal (number of secondary electrons) coming from the specimen will be weaker since fewer beam electrons strike the specimen.*

Apertures are placed in the lenses to help decrease the spot size and to reduce spherical aberration (Chapter 6) by excluding the more peripheral electrons. Each of the condenser lenses behaves in a similar manner and possess apertures, some of which may be either fixed in size and placement in the column or which may be variable and adjustable using controls on the column of the SEM.

Final Condenser Lens

The *final condenser lens*, often inappropriately called the objective lens, is the strongest lens in the SEM and does the final demagnification of the spot. Whereas one of the main functions of the first and second condenser lenses is to govern the *beam current* by regulating the number of electrons entering the next lens in the series, the final or third lens is used primarily to fine-tune the spot size without a loss of beam electrons. As will be discussed later, *the final lens is used to focus the image seen on the monitor or cathode ray tube (CRT).*

The final lens usually contains two sets of *deflection coils* and a stigmator. The deflection coils are

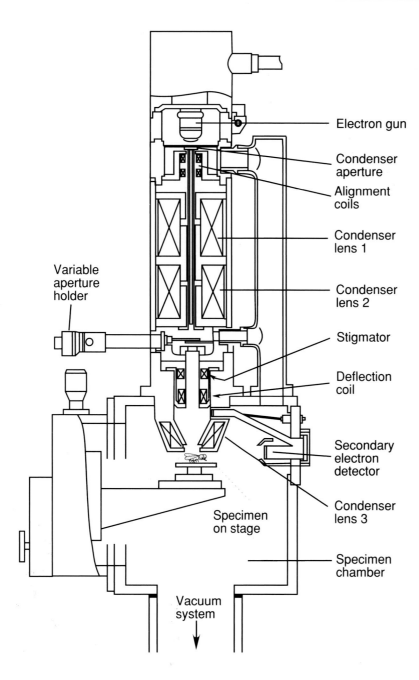

Electron gun

Condenser
aperture

Alignment
coils

Condenser
lens 1

Condenser
lens 2

Variable
aperture
holder

Stigmator

Deflection
coil

Secondary
electron
detector

Condenser
lens 3

Specimen
on stage

Specimen
chamber

Vacuum
system

Figure 7-3(C) More detailed diagram of column of standard SEM showing major components. (Courtesy of Hitachi Scientific Instruments.)

connected to a *scan generator* to raster the electron spot across the specimen. Rastering not only moves the spot in a straight line across the specimen, but also moves the spot down the specimen as well (i.e., possesses both x and y movements) (Figure 7-4). A *change of magnification* is achieved by varying the length that the beam is scanned over the specimen. For instance, if the electron probe is scanned over a 10 mm distance on the specimen, and displayed on the monitor at a final length of 10 cm, this represents a magnification of 10×. Going to a smaller scan length of 1 μm would give a final magnification of 100,000× on the 10 cm viewing screen. Therefore:

$$\text{Magnification} = \frac{\text{length displayed on screen}}{\text{length scanned on specimen}}$$

Dual Magnification Mode

A useful feature present on some SEMs is the *dual magnification* mode that permits the simultaneous viewing of two different magnifications either on two separate viewing screens (some SEMs have more than one) or on a single screen that is split into two areas. This is accomplished by sending alternate scans taken at different mag-

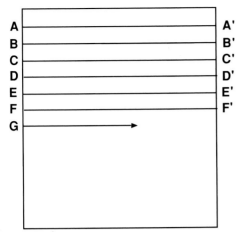

Figure 7-4   The focused beam of electrons is scanned in a raster pattern over the specimen surface. The first scan is from A to A', with the beam moving down and then scanning line B to B', etc. (Redrawn from Postek, et al, 1980.)

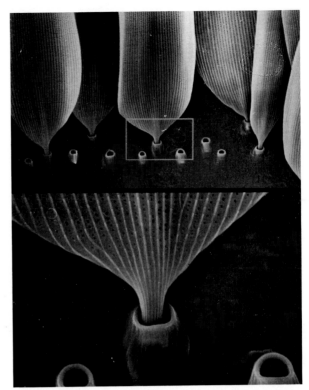

Figure 7-5(B)   Scanning electron micrograph of scales on butterfly wing taken using the dual magnification function. The small box in the top half of the micrograph indicates the area enlarged in the bottom half of the micrograph. Both images are viewed simultaneously in the SEM. (Courtesy of S. Schmitt.)

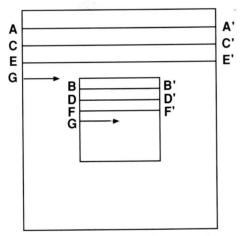

Figure 7-5(A)   In the dual magnification mode, one set of lines (A to A', C to C', etc.) is scanned a particular length over the specimen, while alternating lines (B to B', D to D', etc.) are scanned a shorter distance. (Redrawn from Postek, et al, 1980.)

nifications to different display screens. For instance, the first scan across the specimen may be at 100×, while the next scan would be at a magnification of 1,000×. If all of the odd-numbered scans are then displayed on the 100× area, while the alternate even-numbered scans are displayed on the 1,000× area of the screen, one sees the two different magnifications at the same time (Figure 7-5). This feature is useful for rapidly locating an area of interest at a low magnification while still being able to scrutinize detail at a higher magnification.

Stigmator Apparatus. The *stigmator* (*stigma* means "mark" or "spot" in Latin) is a device that is used to control any distortions in the roundness of the spot formed by the electron probe that is scanned

over the specimen. Since the pixel on the viewing monitor is round, it is important that secondary electrons from the corresponding point on the specimen also emanate from a round spot. For instance, if the electron beam spot is not round but elliptical, then extraneous information (i.e., secondary electrons) is being put into the round spot on the viewing monitor (one is attempting to "fit an elliptical peg into a round hole"). Beam spots that are not round will generate an image on the viewing monitor that is smeared in one direction. This phenomenon, called *astigmatism*, is one of the major reasons for loss of resolution in the SEM. The major cause of astigmatism is contamination on one of the apertures (usually the final aperture), since this causes a distortion of the symmetrical electromagnetic field.

Stigmators usually consist of 6 to 8 small electromagnetic coils inside the lens bore of the final condenser lens. As described in Chapter 6, stigmators are able to correct the asymmetrical distortions to the electromagnetic field by introducing an opposing field of appropriate strength (amplitude) in the proper direction (azimuth) so as to counterbalance the offending field. This is usually accomplished manually; however, the newest SEMs have *autostigmation* modules that permit the rapid and precise correction of astigmatism, which saves time and assures accurate stigmation.

Apertures in the Final Condenser Lens. The final lens usually has externally adjustable apertures that may be readily varied in size. Normally, final apertures on the order of 50 to 70 μm are used to generate smaller, less energetic spots for secondary electron generation and imaging. Larger apertures (200 μm or so) are used to generate larger spots with greater numbers of electrons. These large spots contain a great deal of energy and may damage fragile specimens. They are used primarily to generate X rays for elemental analysis rather than for imaging purposes (Chapter 15).

Apertures are important parts of electron microscopes because they not only affect spot size and beam current, but also affect depth of field and help diminish spherical aberration.

Apertures and Depth of Field

Apertures can be used to control the *depth of field* in the specimen. Depth of field refers to the depth in the specimen that appears to be in focus. As discussed in Chapter 6, depth of field is expressed as:

Equation 7-1: Depth of Field

$$D_{fi} = \frac{\lambda}{NA^2}$$

where λ = wavelength of illumination
        NA = numerical aperture

Variation in the depth of field as a result of changes in aperture size is shown in Table 7-1. As illustrated in Figures 6-36 and 7-6, smaller apertures generate narrower

---

Table 7-1   Effect of Aperture Size on Depth of Field at Various Magnifications and at a 10 mm Working Distance

	Depth of Field		
Mag	100 μm Aperture	200 μm Aperture	600 μm Aperture
10×	4 mm	2 mm	670 μm
50×	800 μm	400 μm	133 μm
100×	400 μm	200 μm	67 μm
100,000×	0.4 μm	0.2 μm	0.067 μm

Reference: Goldstein, et al. 1981. *Scanning Electron Microscopy and X-ray Analysis: A Text for Biologists, Materials Scientists, and Geologists.* (New York: Plenum Publishing Corp.), 134.

---

beams with smaller aperture angles. The diameter of the beam (i.e., spot size) varies less along the length of such narrow beams compared to wider angled beams. Consequently, as the beam is scanned along the contours of the specimen, the spot size varies less at the various levels, so that the specimen will appear sharply in focus at the various levels.

NOTE: As the working distance increases, the aperture angle becomes narrower (see Figure 7-7). Consequently, the depth of field figures given in Table 7-1 are valid only at a working distance of 10 mm.

Depth of field is also affected by the distance the specimen is situated from the final condenser lens, the so-called *working distance.* From Figure 7-7 it can be seen that the aperture angle decreases as one increases the working distance. Consequently, the depth of field will increase as one increases the working distance. As with smaller apertures, this increase in depth of field is at the expense of resolution, since the numerical aperture decreases and since long focal length lenses are more susceptible to chromatic aberration (see Chapter 6). The relationship between working distance and image quality is summarized in Table 7-2.

Resolution in the SEM. As with any magnifying instrument, *resolution* in the SEM refers to the ability of the instrument to image two closely placed objects as two entities rather than a single object. The size of the final spot is related to the resolution of the SEM: *Smaller beam spot sizes permit better resolution.* The final lens is used to focus the size of the illuminating beam spot to match the magnification used. Since the secondary electrons arising from the beam spot striking the specimen are summarized and displayed as a spot of a fixed size (usually around 100 μm) on the viewing monitor, the diameter of the beam spot must not exceed a certain size as defined by the following equation:

Equation 7-2: Maximum Allowable Spot Size of Beam

$$\text{Maximum Spot Size} = \frac{100 \ \mu m}{\text{Magnification}}$$

For example, if one is working at a magnification of 10×, the beam spot size must not exceed 10 μm. A magnification of 100,000× would require a beam spot size of 1 nm or less. If the beam spot size goes beyond this size, secondary electrons are generated from areas outside of what is being summarized on the monitor pixels. This results in an unsharp image, since extraneous information is present in the dis-

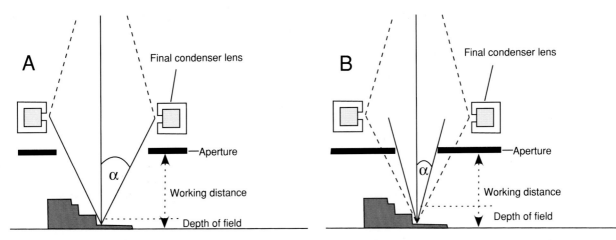

Figure 7-6 Depth of field (the depth that is in focus in the specimen) is increased by using smaller apertures as shown on right.

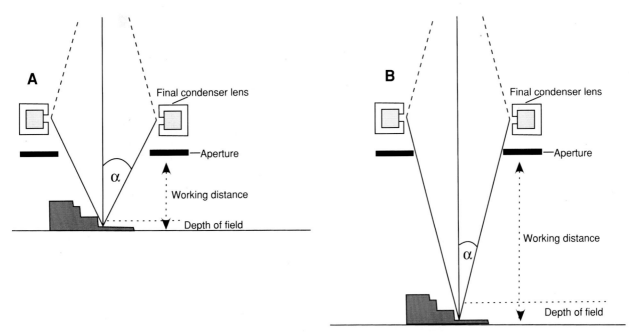

Figure 7-7 When the working distance is increased as shown in B, this decreases the aperture angle alpha so that the depth of field is also increased.

Table 7-2 Working Distance and Quality of Image

Working Distance (mm)	5	10	20	35
Resolution	Best ——————→ Worst			
Depth of Field	Shallow ——————→ Deep			
Signal Strength	Strong ——————→ Weak			

play spot. The final lens may be used to decrease the beam spot and thereby "focus" the image. Obviously, the final lens can reduce the beam spot only so much, so that one must rely on the other condenser lenses to further assist in the demagnification. Therefore, if one cannot achieve a satisfactory focus using the final lens, it may be necessary to increase the strength of the other condenser lenses to achieve smaller spot sizes. The relationship between beam spot size and image quality is summarized in Table 7-3.

Table 7-3   Beam Spot Size and Quality of Image

Beam Spot Diameter (nm)	1	100	500
Resolution	Best ⟶ Worst		
Signal Strength	Weak ⟶ Strong		

Table 7-4   Aperture Size and Quality of Image

Aperture Diameter (μm)	30	200	400	600
Resolution	Best ⟶ Worst			
Depth of Field	Deep ⟶ Shallow			
Signal Strength	Weak ⟶ Strong			

From the above discussions, it is apparent that smaller apertures not only give better resolutions due to decreased spherical aberration and spot size, but they also increase the depth of field. Unfortunately, these gains are at the expense of a decreased signal coming from the specimen since fewer electrons are present in the impinging beam. The relationship of aperture size to image quality is given in Table 7-4.

Interaction of Electron Beam with Specimen
When the accelerated 15 to 25 KV electrons of the SEM strike the specimen, they give rise to a number of emanations (Figure 15-1, Chapter 15). Depending on the speed of the electrons, as well as the density of the specimen, the beam may penetrate to a variable depth in the specimen. The initial point of entry may be at a 2 nm diameter spot; however, the electrons scatter randomly throughout the specimen until their energy is dissipated by interaction with atoms of the specimen. A longitudinal section across the point of entry of the electron beam will reveal a teardrop-shaped area where the electrons have spread. Two types of electron scattering result from the interaction of the primary beam electrons with the atoms of the specimen: elastic and inelastic scattering.

Elastic Scattering. An *elastically scattered electron* is one that has changed direction without losing velocity or energy. This type of scattering results when a beam electron collides with or passes close to the nucleus of an atom of the specimen. When such an elastically scattered electron exits the specimen back in the direction from which it came, it is termed a *backscattered electron*. Such high energy electrons may interact with atoms of the specimen prior to their exit to generate secondary electrons some distance from their initial point of entry in the specimen (Figure 7-8). If backscattered electrons strike parts of the microscope, this may also generate extraneous secondary electrons that are summed with the secondaries from the specimen to give rise to *noise* in the final image (see Signal versus Noise, in this chapter). It is possible to detect backscattered electrons using special detectors as described later.

Inelastic Scattering. During *inelastic scattering*, some beam electrons interact with the atoms of the specimen to produce low energy or *secondary electrons*. Secondary electrons have energy ranges of 0 to 50 eV and are used to generate the three-dimensional images when gathered by a secondary electron detector. Occasionally, tightly bound, inner shell electrons may interact with a beam electron to effect

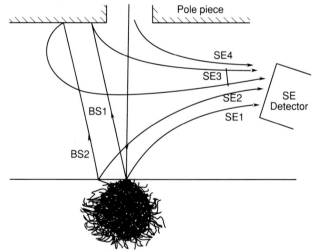

Figure 7-8 True signal is represented by secondary electrons that are generated from the spot of electrons focused on the specimen. True signal is represented by secondary electrons SE1 since they originate from the point where the beam strikes. SE2 are secondaries that originate some distance from the focused spot and represent spreading of the beam upon interaction with the specimen. Noise is generated when backscattered electrons BS1 and BS2 strike parts of the microscope chamber to generate SE3 electrons. SE4 secondaries originate if beam strikes parts of lens such as aperture or pole piece liner. (Courtesy of David C. Joy and The Royal Microscopical Society.)

the ejection of the inner shell electron. This event ionizes the atom until outer orbital electrons fill the inner void and energy is dissipated in the form of characteristic X rays and an Auger electron. These emanations can be detected by specialized X ray or electron spectrometers. Auger spectroscopy is rarely used in the biological sciences. It is an analytical method for determining the elemental composition of the uppermost atomic layers only. Other types of emanations resulting from the primary electron beam striking the specimen include heat, continuum X rays, and light. All of these emanations may be detected and displayed or recorded as described in Chapter 15.

Specimen Manipulation

The specimen is normally secured to a metal stub (usually aluminum) and is grounded to prevent the buildup of static high voltage charges when the beam electrons strike the specimen. It is necessary to orient the specimen precisely relative to the electron beam and electron detectors, so most SEMs have controls for rotating and traversing the specimen in x, y, and z (height) directions. Besides *rotational, lateral, and height adjustments*, it is also possible to *tilt* the specimen in order to enhance the collection of electrons by a particular detector. Judicious combinations of these movements not only permit accurate location of desired areas of the specimen, but they may have a large effect on magnification, contrast, resolution, and depth of field. Consequently, when poor images are encountered, some improvement may be gained by a simple reorientation of the specimen.

Electron Detector, Signal Processing, and Recording Systems

Signal Versus Noise

The emanations, or signals, generated as a result of the electron beam striking a specimen are used to convey different types of information about the specimen. True signal consists of information generated from the spot struck by the beam electrons. Unfortunately, extraneous emanations, termed *noise*, may be generated by a number of events.

The Origin of Noise in the SEM

In the case of secondary electrons (Figure 7-8), SE1 and SE2 represent signal coming from the beam spot. Noise is represented by SE3 and SE4 electrons that were gen-

erated as a result of backscattered electrons BS1 and BS2 striking metal parts of the SEM. Noise is usually evidenced as a "snowy" image devoid of clear details. Noise may also arise when the beam strikes parts of the microscope column (e.g., final apertures) to generate spurious secondary electrons. Faulty electronics may also generate noise during the processing of the signal. For more details, see the article by Joy (1984).

A way of expressing the relationship between true versus extraneous signal is the *signal to noise ratio*. Whenever noise rises to an unacceptable level (based on a poor quality image), the signal to noise ratio is said to be low. One must therefore either reduce the noise or raise the signal to achieve a satisfactory image. Since it is more difficult to reduce the noise level, most microscopists attempt to raise the amount of signal from the specimen. Ways of increasing signal from the specimen are described later in this chapter (Major Operational Modes of the SEM).

Secondary Electron Detector

After the beam electrons strike the specimen, low energy secondary electrons leave the specimen from many different angles. Because they are weakly negative, the secondary electrons will be attracted to any positive source. The secondary electron detector uses this phenomenon to gather electrons. The most common type of secondary electron detector is based on the original 1960 *scintillator-photomultiplier* design of Everhart and Thornley (Figures 7-9 and 7-10).

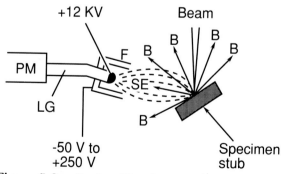

Figure 7-9   Everhart-Thornley secondary electron detector. Dashed lines show secondary electrons, SE, that are attracted to an aluminum-coated scintillator due to the strong positive pull exerted by the Faraday cage, F, and the +12,000v potential of the aluminum coating over the scintillator on the end of the light guide, LG. After striking a photocathode (not shown) the signal enters the photomultiplier, PM. Backscattered electrons, B, are not attracted by the detector. (Redrawn from Goldstein et al, 1981.)

Figure 7-10 Photograph of Everhart-Thornley secondary electron detector from a Kent-Cambridge SEM. The arrowheads show the paths secondary electrons might travel from the specimen (S) to the detector (not shown) housed inside the Faraday cage (F). After striking the scintillator of the detector, photons of light travel down the plastic light guide (L) to the photocathode of the photomultiplier (not shown).

In this system, the secondary electrons are attracted to a Faraday cage that is maintained at −100 to +300 V and surrounds the secondary electron detector. Upon reaching the cage, the electrons are more strongly attracted to the end of the detector (Figure 7-9) since its thin aluminum coating is placed at +12,000 V. The electrons strike the several nanometer thick aluminum coating with such impact that they pass through the aluminum layer, strike a phosphorescent *scintillator* material, and generate a brief burst of light, a scintilla, that travels down the lucite or quartz light guide. This burst of light then strikes a photocathode on the end of a photomultiplier. The *photocathode* is coated with a material that generates electrons on contact with light.

After being generated in the photocathode, the photoelectrons then enter the *photomultiplier* and travel down a series of dynodes or electrodes that proportionally increase the number of electrons at each stage (Figure 7-11). The yield of secondary electrons may be increased by increasing the voltage to the dynodes to effect a higher *gain* (on the order of 10^5 to 10^6). The SEM control that changes the gain of the photomultiplier is usually called the gain or, more commonly, the *contrast* control, since it affects the overall contrast of the image on the display monitor. *Increasing the photomultiplier gain will increase the contrast of the image* (the highlights are increased more than the shadow areas resulting in a greater brightening of the light areas compared to the dark areas).

Signal Processing

The small amount of voltage generated by the photomultiplier now enters a preamplifier-amplifier component of the SEM where the weak signal is amplified electronically. When one increases the output from the *preamplifier*, the *brightness* of the image increases overall (both the highlights and shadow areas are boosted).

Most SEMs have a control termed *gamma* that may be used to selectively extend the contrast range in either the highlights or the shadows without loss of information in the nonamplified component. For instance, at a gamma setting of 0.5, contrast in the highlights is expanded while the shadows are compressed in contrast. At gamma 1.0, contrast levels in both the highlights and shadows are expanded. At gamma 2.0, contrast in the shadows is expanded while the highlights are compressed. Gamma is most useful in situations where the specimen has excessive contrast (Figure 7-12).

Some Specialized Features of the SEM

A special control termed *dynamic focusing* may be used on specimens that are tilted relative to the beam. Without dynamic focusing, the beam spot would enlarge as it was scanned from the top to the bottom of the specimen (Figure 7-13). To maintain the spot size within the appropriate range, electronic modules are utilized to change the focal length of the final condenser lens as it is scanned over the specimen. In this manner, the spot size is the same at the top of the specimen as it is when it reaches the low points at the bottom of the tilted specimen.

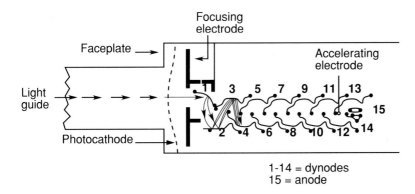

Figure 7-11 (top) Schematic of photomultiplier where photoelectrons (generated by photons from the light guide striking the photocathode) are multiplied by striking a series of high voltage dynodes to generate more secondary electrons. (bottom) Enlarged area of photomultiplier showing entry of electron and multiplication of signal along dynodes.

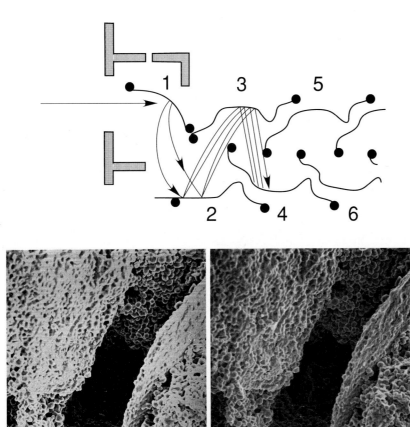

Figure 7-12 Demonstration of the use of gamma to diminish excessive contrast. The left-hand image is without gamma applied while the right-hand image shows the use of gamma. (Courtesy of S. Schmitt.)

Tilt correction is necessary on tilted specimens since the top of the specimen is at a higher magnification (shorter scan length) than the bottom of the specimen (longer scan length). This physical phenomenon may be corrected by reducing the scan lengths of the beam as it traverses from top to bottom. The factor by which the scan length is reduced depends on the degree of tilt and is adjustable by a control on the panel of the SEM. This correction must be used carefully since specimens with a

great deal of topography may be distorted by activating this control (Figure 7-14). Flat specimens work best with this mode.

Normally, the secondary electrons are displayed on the viewing monitor in the *brightness modulation* mode (i.e., the brightness of the pixel on the viewing monitor is proportional to the number of secondary electrons from the corresponding point on the specimen). It is also possible to display the secondary electrons in the *Y-modu-*

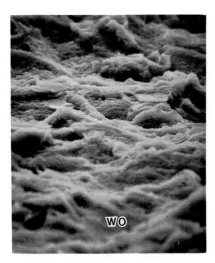
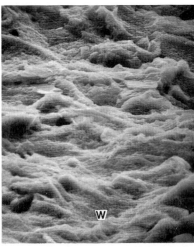

Figure 7-13 Tilted sample without (WO) and with (W) dynamic focusing applied. (Courtesy of S. Schmitt.)

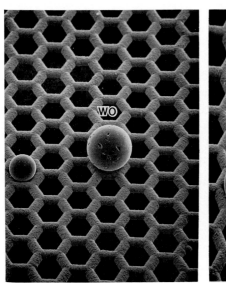
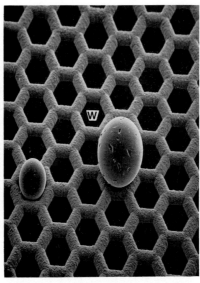

Figure 7-14 Tilted sample without (WO) and with (W) tilt correction applied. In these two micrographs, note how the hexagonal background pattern has been corrected with tilt correction applied but how the spherical objects have been distorted. Specimen consists of two glass beads deposited onto a TEM grid with hexagonal mesh pattern. (Courtesy of S. Schmitt.)

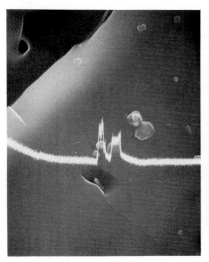
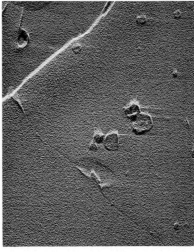

Figure 7-15 Y modulation mode of sample (left) in which only one scan line is y modulated and (right) where all scan lines are modulated on the specimen. (Courtesy of R. Tindall.)

lation mode. In this mode, the scan on the viewing monitor is proportionally displaced in the y direction on the viewing monitor depending on the number of secondaries detected at each location on the scan (Figure 7-15A). This image is similar to one displayed on an oscilloscope. If one accumulates a number of these scans on the viewing monitor, three dimensionality of the specimen is enhanced (Figure 7-15B). This mode is particularly useful for specimens that are flat and devoid of much topographic contrast. Interpretation of images generated using Y-modulation is difficult since bright areas of the specimen (which give rise to deflections in the y direction) may be due to other than topographic features (see Three Dimensionality of the SEM Image, later in this chapter).

Image Recording

Recording of the image displayed on the cathode ray tube or monitor differs from the methods followed in transmission electron microscopy. Unlike TEM, where the electrons interact directly with the photographic medium, SEM images are most often photographed directly from a monitor through the lens of either a 35 mm roll film camera or a larger format 4″ × 5″ sheet film camera. The shutter of the camera is not used. Instead, the camera shutter remains open as the electron beam is slowly scanned across the specimen. The much slower (90 seconds or so) photographic scan is similar to a focal plane shutter exposing the length of the film.

Films for Use in the SEM

Several types of conventional black and white films may be used to record the image (Kodak PLUS-X Panchromatic Professional sheet films 4147, 2147, Kodak Technical Pan Films 2415 in 35 mm rolls, 6415 in 120-size rolls, and 4415 in sheet films). Most often, researchers use a 4 × 5 instant type film, such as Polaroid Type 55, that gives both a 4 × 5 positive and a high quality negative. Such films are 4 to 5 times more expensive than standard films, but are favored since darkroom time is saved and the quality of the image may be evaluated in less than one minute. When numerous images are to be recorded, roll films may provide a significant savings but at the loss of immediate gratification. Additional information on films and negative handling is found in Chapter 8. The latest generation of SEMs digitally encode images making it possible to store images on disk or to print out the images on a variety of printers. Although these images are currently not of the same high quality as a photographic print in terms of resolution, the technology should improve in the next few years so that such printouts may eventually replace the photographic print.

Some Useful Features in Modern SEMs

A valuable addition to most SEMs is the *automatic data display* that permits the generation of informational data on the viewing and recording monitors. With this accessory, experiment numbers, dates, accelerating voltages, magnifications, and magnification scales may be displayed. Keyboards and alphanumeric generators permit the insertion of text and numbers anywhere on the image to be permanently recorded on the negatives.

Automatic brightness and contrast modules permit calibration of the exposure parameters for each type of film used in the SEM. Once calibrated, perfectly exposed negatives are generally the rule and very little recording medium is wasted. In addition, *autofocus* controls permit the precise focusing of the SEM even at very low magnifications.

Newer SEMs now routinely encode images digitally so that the information may later be processed and analyzed by computer. Typical applications might include processing to: enhance contrast, introduce color (pseudocoloration), effect three-dimensional renditions, and to apply statistical analyses.

Three Dimensionality of the SEM Image

The three-dimensional appearance of SEM images is due to differences in contrast between various structural features of the specimen when they are displayed on the viewing monitor. Contrast arises when different parts of the specimen generate differing amounts of secondary electrons when the electron beam strikes them. Areas that generate large numbers of secondary electrons will appear brighter than areas that generate fewer secondary electrons. The yield of secondary electrons by these various areas may be influenced by several conditions.

The *orientation of the specimen topography* relative to the electron beam and secondary electron detector greatly affects the yield of secondary electrons. As illustrated in Figure 7-16, certain areas of the specimen (designated "D" in the figure) will not be struck by the beam and will not yield any secondary electrons. These areas will appear dark on the display monitor. Areas such as "I" in Figure 7-16 will be struck by the beam, but since they face away from the detector, fewer secondary electrons will be collected and intermediate levels of brightness will be displayed. Optimal yields of secondary electrons would come from areas that are struck by

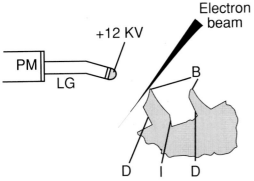

Figure 7-16 Three dimensionality and contrast are due to the yield of secondary electrons from various parts of the specimen. Areas marked B face the beam and are in line of sight with the detector so that they will appear bright, I (intermediate brightness) faces the beam but fewer secondaries reach the detector since it is not in line of sight, D is dark in appearance since the beam does not strike this area and no secondaries are generated.

the beam and face the detector ("B" in the figure). These areas would appear as highlights in the image.

A second condition that affects the yield of secondary electrons is the *angle that the beam enters the specimen surface*. If the beam enters a specimen at a 90 degree angle, the beam penetrates directly into the specimen and any secondaries generated below a certain depth will not be able to escape. On the other hand, if the beam strikes the specimen in a grazing manner, then the beam does not penetrate to a great depth and more secondaries will be able to escape since they are closer to the surface. Since rounded objects are more likely to be grazed by the electron beam than would flat objects, round areas usually appear to have a sharp bright line around them due to the enhanced yield of secondaries (see Figure 7-17).

In a third situation, thin, raised areas of the specimen usually appear much brighter than broad, flat areas. This phenomenon is termed the *edge effect* since it takes place along sharp edges or peaks in the specimen. These areas appear brighter because the secondary electrons are able to escape from all sides of the thin areas in the projection (Figure 7-18). An example of a specimen demonstrating pronounced edge effect is also shown in Figure 7-19.

Other conditions that may affect contrast in a specimen include:

• The distribution of elements with *different atomic numbers*. Higher atomic numbered elements have a greater yield of secondary and backscattered electrons than do elements with lower atomic

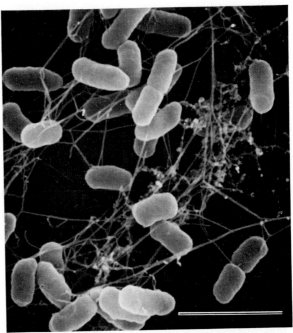

Figure 7-17 Rounded specimens demonstrate an enhanced emission of secondary electrons on their periphery since the electron beam grazes, rather than penetrates the surface coating on the specimen. Note bright periphery on all of the rounded bacterial cells. Marker bar, 2.3 μm.

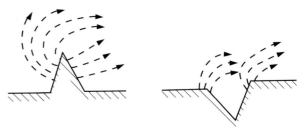

Figure 7-18 The edge effect, or enhanced electron emission, occurs along the edges of thin raised areas since secondary electrons may exit from both sides of the structure.

numbers. Higher atomic numbered elements therefore appear brighter in the SEM.

• Higher *accelerating voltages* result in lower contrast due to greater beam penetration and enhanced secondary yield from all parts of the topography. If more contrast is needed than can be obtained using the SEM contrast controls, then lower accelerating voltages should be used.

• *Charge accumulation* on incompletely coated or grounded areas of the specimen will result in an increase in contrast. For instance, large areas that are suspended by a thin stalk tend to build up a static charge from the electron beam and cannot dissipate the charge rapidly enough through the thinned portion. This may cause the deflection of the beam so that it strikes other areas to generate

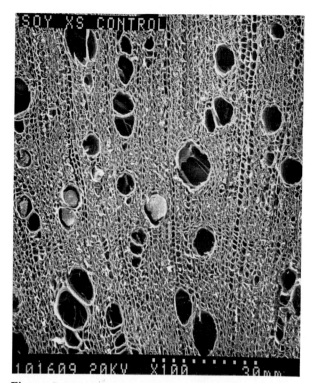

Figure 7-19 In this specimen of soybean stem, the cell walls are very bright due to enhanced emission of secondary electrons. The edge effect is diagrammed in Figure 7-18. The information bar on the bottom of the micrograph indicates that the image was recorded at 20 kV accelerating voltage and at a magnification of 100×. In addition, a magnification scale is recorded onto the negative to show that the total distance between the 11 white dots above the 100× mark is 0.30 mm.

an excessive amount of secondary electrons. Likewise, naturally *magnetic areas* in a specimen may either deflect or attract the beam to affect the yield of secondary electrons.

• When *crystals* are oriented along certain lattice planes relative to the beam, an enhanced yield of secondaries may result in an increase in brightness along these lattice planes so that certain crystals will appear much brighter than others.

Stereo Imaging with the SEM

The conditions described previously all contribute to the generation of an image that appears to have depth even though the image is recorded on a piece of photographic paper that has only two dimensions. It is possible to generate images with even more three dimensionality by a variety of other methods. The reader has undoubtedly experienced slide shows or even motion pictures in which special

glasses were used by the audience to enhance the three dimensionality of the images on screen. In the SEM, the enhancement of three dimensionality not only serves an esthetic function but it assists the viewer in discriminating between projections and depressions, and it elucidates spatial relationships to help determine distances between two objects (i.e., are two cells actually close enough to touch or is one simply in front of another one). In addition, some investigators feel that since one usually combines two separate micrographs in order to generate the single image (see next section) then the resolution and signal-to-noise ratio of the stereo image is enhanced.

The perception of depth is due to the *parallax phenomenon* wherein separate views from the right and left eyes are merged by the brain into one image. Since the left eye sees more of the left side of an object and less of the right side than does the right eye, the brain interprets this variation as depth. It is possible to achieve this same effect in the SEM by taking two separate micrographs of an object so that one of the micrographs shows more of the left side of the object while the other micrograph shows more of the right side. When the two micrographs are placed side by side so that the left eye views the micrograph showing more of the left side while the right eye views the other micrograph, and when the images are merged by the brain (usually with aid of an optical device), the perception of depth occurs in the micrograph. The key to the procedure is to (1) generate two different micrographs from separate viewpoints and (2) merge these separate views using the appropriate tools.

Generating Two Micrographs with Separate Views

Two micrographs (termed *stereo pairs*) with the appropriate separate views may be generated in several different ways. One would take one micrograph followed by a second that had a different view due to (a) shifting the specimen slightly to the left or right, (b) rotating the specimen, (c) deflecting the electron beam electronically, or (d) tilting the specimen before taking the second micrograph. The first two methods are less desirable since they are only effective at low magnifications and require a eucentric goniometer stage, respectively. Not all SEMs have the capability to tilt the beam between micrographs so that the third method may not always be possible. On the other hand, since it is possible to tilt the

specimen in all SEMs, the last method is probably the most prevalent one. A description of the general steps involved in generating the two micrographs by this method follows.

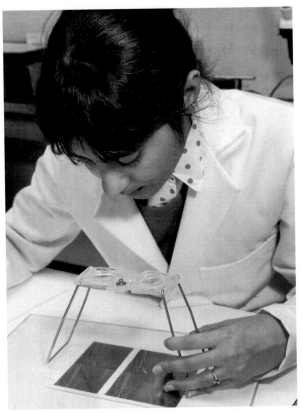

Figure 7-20   Setup for viewing stereo pairs. The two micrographs are placed alongside each other under the special stereo viewing glasses, which merge the two slightly different views into one image that has three dimensionality.

Using Tilt to Generate Stereo Views in the SEM

1. Take the first micrograph in the usual manner. The specimen should show reasonable contrast and be set up to show good depth of field (i.e., proper working distance and aperture sizes).
2. Without moving the specimen, take a wax marking pencil and trace over the outline of the specimen on the viewing CRT. Rather than marking directly on the CRT, carefully tape a piece of clear acetate plastic over the CRT and mark on this sheet.
3. Tilt the specimen slightly. The amount of tilt depends upon the magnification (more tilt at lower magnifications) and the topography of the specimen (more tilt for specimens with flat topographies). In general, most researchers use tilts ranging from 4 to 15 degrees.
4. Refocus the specimen using the Z control (specimen height) adjustment of the stage. One must not focus using the final condenser lens, as is the normal procedure, since this would affect the magnification of the subsequent micrograph.
5. Realign the image so that it coincides as closely as possible with the image traced on the acetate sheet using the wax pencil.
6. Adjust the brightness and contrast of the image in the SEM to match the first micrograph and take the second micrograph.

Merging Two Micrographs with Separate Views to Generate a Stereo Image

The most common method of presenting stereo images is to print the two separate images at the same contrast and density levels in the darkroom. The micrographs usually are small contact prints no larger than 4″ × 5″. The micrographs are then placed alongside each other and viewed with a special set of glasses (stereo viewers) placed directly over the two prints. By moving the two prints carefully, while looking through the viewers, it is possible to get the two images to converge to obtain a striking stereo effect. A very few individuals are able to place an index card between the two prints and by crossing their eyes force the two images to converge into the stereoscopic view. Figure 7-20 shows the setup and simple tools needed for viewing micrographs in stereo. This type of stereo viewer may be obtained from most EM supply houses at a relatively modest

cost. An example of two images that were prepared by tilting a specimen are shown in Figure 7-21. The stereo glasses shown in the preceding figure are needed to produce the desired image.

A very simple procedure requiring only a small mirror was recently described by Harrington and Welford (1990). The two contact prints are produced in the usual manner except that one of the prints (produced from the negative that was tilted closer to the detector) is printed with the emulsion side up so that a mirror image is generated. After aligning the two prints side by side with a gap of approximately ½″, a mirror approximately 6″ high is placed between the two images with the reflective side towards the print that was reversed (i.e., mirror facing the mirror image). After placing one's nose at the edge of the mirror and carefully orienting the mirror until the two images blend into one, it is possible to obtain a stereo image quite easily. The key to this procedure is the careful production of the mirror-imaged print. An example of these types of prints is given in Figure 7-22. A 6″ high mirror

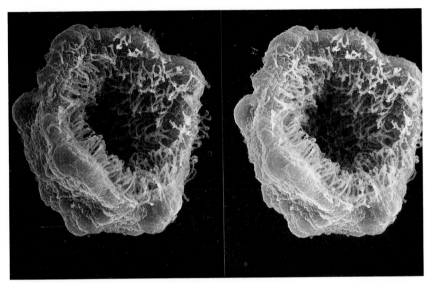

Figure 7-21 Stereo pairs ready for viewing using the special glasses shown in Figure 7-20. This illustration shows an inverting *Volvox* embryo magnified 1200×. The left view was tilted +7 degrees relative to the right hand view. (Courtesy of G. M. Veith.)

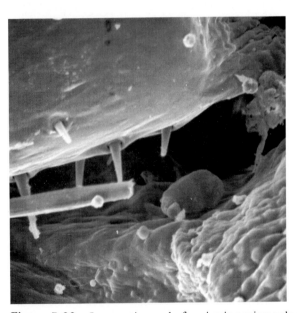 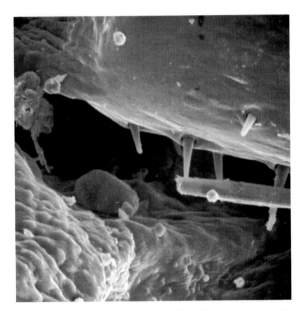

Figure 7-22 Stereo pairs ready for viewing using only a six inch high common mirror. The mirror is positioned between the micrographs and one views down the mirror as is recommended for viewing the two micrographs as indicated in Figure 7-23.

shown in Figure 7-23. (Courtesy of D. Harrington and A. Welford and Wiley-Liss Publishers.)

Preparing Stereo Images for Projection

If one wishes to project stereo images for an audience, several different (but equally demanding procedures) may be used. One method involves the use of glasses made up of lenses of complementary colors such that the left eye is covered by a red filter while the right eye is covered by a green filter. One then prepares two separate positive slides from the two stereo pairs and projects them using two separate projectors (the projector on the left with a red filter over its lens and the projector on the right with a green filter). The slide projected by the left-handed (red) projector can only be seen by the left (red lensed) eye, while the right eye (green lens) can see only the image from the right (green) projector. This technique can also be accomplished using polarizing, instead of colored filters, except that alignment of the two projectors is more critical and special screens are needed.

Another method of producing a slide for stereo imaging involves the production of a single colored (rather than black and white) slide that has merged the red (left) and green (right) images onto one slide. In this instance, one needs only the colored glasses and a single projector for viewing purposes. The disadvantage to this technique

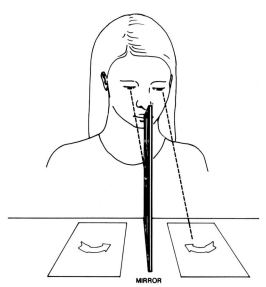

Figure 7-23   Method of viewing mirror image stereo pairs shown in Figure 7-22. (Courtesy of D. Harrington and A. Welford and Wiley-Liss Publishers.)

is that the production of the merged image on the colored slide is difficult, except in the latest generation of SEMs. Newer SEMs have color viewing screens and electronic beam tilting capabilities so that it is possible to automatically tilt and record appropriately colored images of the stereo pairs on a single slide. Such colored slides that contain the stereo image encoded in color are termed *anaglyphs*. For more information on stereo imaging in the SEM, see the references by Barber and Emerson (1980), Peachey (1978), and Wergin (1984).

Major Operational Modes of the SEM

The SEM may be operated in two major modes: high resolution or great depth of field. As will be seen, these modes require conflicting paramaters to be established so that it is impossible to optimize both conditions simultaneously.

High Resolution

High resolutions demand a small, coherent spot with minimal aberrations and adequate signal-to-noise ratios. A summary of some of the conditions necessary to achieve high resolutions follows.

Conditions for High Resolution

1. *Small spot sizes* are achieved by optimally energizing all condenser lenses. The first condenser lens is adjusted to give a small spot that will then be further

demagnified by the remaining lenses. As described previously, the small spot will have a diminished number of electrons (i.e., low current) due to the wide aperture angles needed to achieve diminished spot sizes, so take care not to over energize the first condenser lens (follow manufacturer's recommendations).

2. *Small apertures* should be used to help diminish the size of the spot as well as to minimize spherical aberration. Again, small apertures will exacerbate the loss of current in the beam spot and may lead to a fall off of signal.

3. *Adequate signal-to-noise ratios* are essential in order to reveal details present in the smaller spot. The operational principle here is: *Maximize the generation of secondaries from the specimen.* This may be accomplished by:

   a. Putting *more current in the beam* by getting higher emissions of electrons from the gun (alter bias settings, move filament closer to aperture in shield, move anode closer to filament, use lanthanum hexaboride filament or field emission gun).

   b. Using *slower scan rates* on the specimen. Longer dwell times of the beam on the specimen will generate more secondary electrons from the spot. Generally, when images are photographed, slow scan speeds are automatically utilized to maximize secondary electron emission, but it may be possible to slow the scan even more. One must remember, however, that this increase in current may damage sensitive specimens.

4. *Proper accelerating voltages* should be selected based on the nature of the specimen. The principle to follow here is: *Use the lowest possible kV that gives an acceptable signal-to-noise ratio.* Since beam penetration and enlargement of the spot size result from using increasingly higher accelerating voltages, one seeks a voltage high enough to minimize chromatic aberration (see Chapter 6), while not degrading resolution. If the specimen is dense, then lower kVs should be employed since the beam electrons tend to spread near the surface of the specimen, increasing the spot size. Most biological specimens are not overly dense so that accelerating voltages in the 10 to 15 kV range are normally employed. Some particularly fragile specimens may benefit from even lower kVs, but this puts severe demands on the cleanliness and vacuum requirements of the SEM in order to minimize chromatic aberration.

5. *Short working distance* is needed in order to achieve a small enough spot, as well as to minimize chromatic aberration. It may also be recalled from Chapter 6 that shorter focal length lenses have the wider aperture angles necessary to achieve higher resolutions.

6. *Proper stigmation* of the spot is necessary for the reasons described in the chapter section entitled Condenser Lenses. Any contamination, especially on the final aperture, will cause astigmatism and a loss of resolution. Lower accelerating voltages are more susceptible to astigmatism (as well as to chromatic aberration) so a thoroughly clean and well-evacuated SEM is nec-

essary to maintain proper stigmation at lower kVs. Since astigmatism builds up with time, one must continually check and correct for astigmatism during the microscopy session.

Great Depth of Field

Great depth of field may be achieved by using small apertures, long working distances, and lower magnifications. Since resolution is degraded by the latter two conditions, one may now increase the spot size to achieve a better signal-to-noise ratio. This operational mode is used to view specimens with extremes of topography. It should be realized that, from an esthetic point of view, it may be desirable to have only the area of interest in sharp focus while throwing a cluttered background slightly out of focus.

Imaging Other Types of Specimen Signals

So far, we have discussed only the more commonly used detector for imaging secondary electrons. Other types of emanations may also be detected and displayed on the monitor to reveal information in addition to topography. These detectors include those for backscattered electrons and visible light. The analysis of X ray emanations will be covered in Chapter 15 since the technique also applies to TEM and scanning transmission electron microscopy.

Backscattered Electrons

Backscattered electrons, as mentioned earlier, are elastically scattered electrons that exit from the specimen with energy levels similar to the primary or beam electrons. Being of high energy, they may escape from great depths in the specimen and exit some distance from the point of entry of the beam electrons. As they interact with the atoms of the specimen, they may generate secondary electrons that are quite some distance away from the secondaries generated by the beam spot (effectively increasing the size of the beam spot and degrading resolution). In addition, backscattered electrons may strike metallic parts of the SEM chamber to generate even more secondary electrons (Figure 7-8). All of these extraneous electrons may be detected by the secondary electron detector to raise the noise level in an image and ultimately degrade resolution.

Since backscattered electrons travel in straight lines at high velocities, most cannot be attracted over to the standard secondary electron detector. The few that do enter the secondary detector generate a high signal upon entering the photomultiplier. Consequently, some of the highlights seen in the standard secondary image may be due to backscattered electrons.

If one quantitates the yield of secondary and backscattered electrons in various atomic elements, it will be observed that as the atomic number increases, the yield of both secondary and backscattered electrons also increases. This is because larger atomic-numbered elements have more orbital electrons available to interact with beam electrons. Interestingly, a comparison of the ratio of secondary to backscattered electrons finds that there is a significantly higher proportion of backscattered electrons generated compared to secondaries as the atomic number increases. Both of these phenomena have practical implications to the operator: *Higher atomic numbered elements appear slightly brighter than lower atomic numbered elements in a secondary electron detector and significantly brighter in a backscattered electron detector.*

In practice, one should be aware of a few basic priciples when using backscatter imaging modes on biological specimens. Since the contrast in this mode is the result of differences in atomic number (so called *Z contrast* where Z refers to the atomic number), there should be an atomic number difference of 3 between the various elements in the sample, otherwise contrast differences may not be discernible. The specimen should be as flat as possible in order to reduce interfering topographic contrast caused by specimen terrain.

Backscattered Electron Detection

Secondary electrons are used to generate an image based on topographic contrast, whereas backscattered electrons are not normally used to study topographies. Instead, contrast is based on detecting areas of different atomic numbered elements (Z contrast). Unfortunately, in the biological sciences most of the elements present in specimens are of a relatively low atomic number and yield few backscattered electrons. However, if one applies selective stains of heavy metals such as silver or lead, the stained areas will appear very bright due to the high

Figure 7-24 (A) Secondary electron image of *Xenopus* (frog) optic nerve tract. (B) Same specimen as in 7-24(A), except viewed in the backscattered imaging mode. Since the nerves have been stained with silver, they appear much brighter than the background so that it is much easier to trace them throughout the tissue. (Courtesy of J. S. J. Taylor and The Williams and Wilkins Co.)

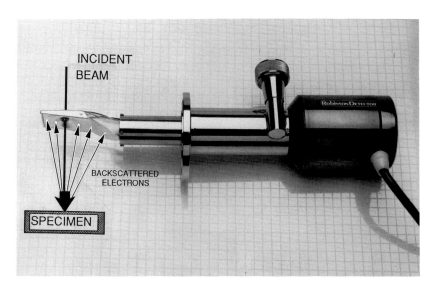

Figure 7-25 Photograph of a wide angled scintillator-photomultiplier backscattered electron detector. The electron beam passes through the hole in the center of the flattened portion (left) to strike the specimen and generate backscattered electrons. These electrons are detected when they strike the scintillator coating on the end of the wedge-shaped lucite light guide. (Courtesy of Electron Detectors, Inc.)

numbers of backscattered electrons (Figure 7-24).

Although relatively few backscattered electrons enter the secondary electron detector, it is possible to filter out secondary electrons by shutting off the positive voltages to the detector or by making the detector slightly negative to repel secondaries. Since only a relatively low number of backscattered electrons are collected using this method, the image will be rather noisy. The yield may be increased if the first condenser lens is adjusted to give a larger spot containing more primary beam electrons. Unfortunately, these highly energetic spots may damage

some specimens and still not effect a good yield of backscattered electrons.

Specially designed detectors may be installed into most SEMs to collect backscattered electrons. All of these detectors use one principle to enhance efficiency of collection: *Increase the surface area of the detector and place it high above the specimen where backscattered electrons most likely will be encountered.* Several different types of detectors are available, but the two main types used in biological sciences are the wide angled scintillator-photomultiplier and the solid state detector.

Types of Backscattered Electron Detectors

The *wide angled scintillator-photomultiplier* uses a technology similar to the conventional secondary detector, but with some important differences. The detector is located high and above the specimen rather than off to one side, and the scintillator surface is considerably larger than the conventional secondary detector (Figure 7-25). Since a photomultiplier is utilized, this type of detector has high overall performance with a high signal-to-noise ratio, and good resolution and discrimination capabilities for different atomic numbered elements. A variation of this basic design involves placing multiple scintillator-photomulti-

plier detectors high above the specimen to increase the detection area. Such multiple units are quite expensive due to the large amount of electronics involved.

Solid state detectors are probably the most common type of backscatter detector used since they are less expensive, reasonably sensitive to differences in atomic number, easily maintained, and take up less valuable space inside the specimen chamber. In one type of solid state detector, four quadrants of a circular silicon diode are mounted directly under the final condenser lens (Figure 7-26). Backscattered electrons that strike the diodes cause the ejection of electrons in the silicon and generate a flow of current proportional to the number of backscattered electrons striking it. This small amount of current is then amplified by electronics, and the signal is ultimately sent to the display monitor where large amounts of current are displayed as bright spots on the viewing screen. Some disadvantages include lower sensitivities (since photomultipliers are not involved) and a lower resolution than the scintillator types.

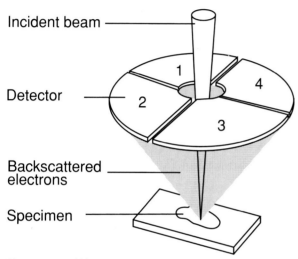

Figure 7-26(A) Diagram of solid state detector placed under final lens. Backscattered electrons strike the silicon diode and the current is detected and related to numbers of electrons striking it. (Courtesy of JEOL)

Figure 7-26(B) Photograph of backscatter detector showing the circular silicon diode that is divided into quadrants. (Courtesy of G.W. Electronics.)

Important Considerations when Using Backscattered Imaging with Biological Specimens

The beginning microscopist should be aware that backscattered electrons are most useful for discriminating areas of different atomic numbered elements in the specimen. For this reason, one should not coat specimens with heavy metals such as palladium or gold in order to render them conductive. Instead, if a conductive coating is needed, a layer of carbon is usually deposited over the specimen using a vacuum evaporator (see Chapter 5).

Backscattered electrons are not very plentiful, so one may wish to increase the yield using larger spot sizes with more current (i.e., having more electrons) or possibly higher accelerating voltages. Both of these conditions will improve the signal, but at the expense of resolving power. More backscatter electrons will be detected if the detector is placed high and above the specimen and the specimen is tilted toward the overhead detector.

Excessive topography in a specimen may interfere with the discrimination of areas of different atomic number. Consequently, specimens with low topographies are better suited for such Z-contrast investigations. This is not to exclude using the backscatter electron detector to improve topographic contrast. However, it is generally true that *secondary detectors are used primarily for conventional imaging using topographic contrast, while backscatter detectors are best used for differentiating different elements in a specimen based on Z-contrast.*

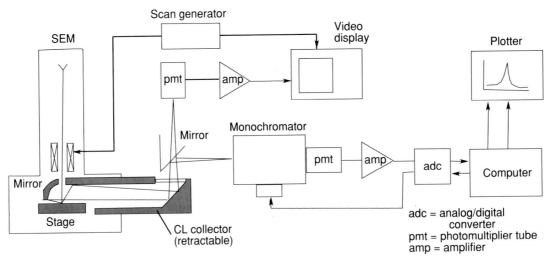

Figure 7-27 Diagram of one type of cathodoluminescence detector. The specimen is surrounded by a front-surfaced parabolic mirror. Light is reflected by the mirrors into the detector, a light pipe/photomultiplier system. (Courtesy of Oxford Instruments North America, Inc.)

Cathodoluminescence

Upon bombardment by beam electrons, certain materials such as phosphors, semiconductors, and insulators may emit photons in the ultraviolet and visible energy ranges. It is possible to detect this very low level of light by using appropriate equipment. A simple detector may be fashioned from some types of secondary electron detectors by removing the scintillator disc. In this manner, the light guide will pick up light emanating from the sample and photomultiply and process this signal in the same manner as it would the photons generated by secondaries striking the scintillator.

One Type of Cathodoluminescence Detector

Since a modified secondary electron detector only detects 4% of the light emitted from the specimen with a spatial resolution of approximately 0.5 μm (the light microscope is 0.2 μm!), this method is not satisfactory except for cursory investigation. Several specially designed cathodoluminescence detectors have been designed to enhance sensitivity. One such device is diagrammed in Figure 7-27. In this system, the collector consists of a parabolic mirror placed over the specimen. The parabolic mirror reflects light emitted from the specimen into a reflecting light guide and into a photomultiplier for signal amplification. In addition, an analysis system may be added to analyze the spectrum of light emitted by the specimen to aid in the identification of the chemical nature of the luminescing areas.

Only a few biological compounds cathodoluminesce, so there may be limited need for this type of detector. However, certain stains such as Calcofluor (a water soluble dye that reacts with newly deposited cell walls), Brilliant Yellow 6G, and fluorescein isothiocyanate exhibit cathodoluminescence and may be used to stain cells to tag certain structures. In addition, certain biological compounds naturally exhibit the phenomenon. These include: tryptophan, tyrosine, adenine, guanine, thymine, polymerized DNA, polyurethane, some herbicides, cotton, certain areas of adrenal and lung tissues and leukocytes treated with formaldehyde. Most SEM manufacturers are able to supply (or recommend a manufacturer of) such detectors. A bibliography of articles dealing with cathodoluminescence was published in 1976 by Brocker and Pfefferkorn.

Using Secondary and Backscattered Electrons in Biological Studies

As was discussed in the previous sections, secondary electrons are used to reveal the three-dimensional features of specimens, whereas backscattered electrons are generally used to differentiate areas that differ in atomic number. In some studies, it may be useful to combine both types of signal in order to study the distribution of a particular material within or on the surface of a specimen. In the following three studies, the investigators used both types of

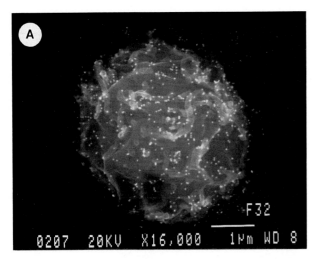

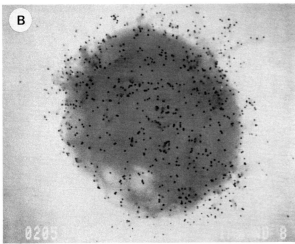

Figure 7-28   Normal human blood T cells that have been labeled with colloidal gold particles. The cells were first reacted with a purified mouse monoclonal antibody against a specific cell surface marker for interleukin-2 followed by gold particles coated with goat antimouse IgG. (A) This image is generated by combining the secondary signal to give the overall image of the cell with the backscattered image of the bright gold particles. (B) This image of the same cell is taken only in the backscattered electron imaging mode with reversed polarity so that the 40 nm gold particles appear dark rather than bright. Marker bar = 1 μm. (Courtesy of J. L. Pauly and Wiley-Liss, Inc.)

detectors in such capacities. For more detail, study the original articles.

Study 1: Localization of Cell Surface Receptors Using Gold Labels

In a 1990 study conducted by Helinski and coworkers, the goal was to localize cell surface receptors for interleukin-2 in human T leukocytes. To achieve this, the cells were initially fixed using an aldehyde fixative and rinsed extensively to remove reactive aldehyde groups. Mouse monoclonal antibody against the cell receptor was reacted with the fixed cells. After

rinsing to remove the unattached antibody, the cells were reacted with gold particles labeled with goat anti-mouse antibody. After an overnight fixation in glutaraldehyde to stabilize the label on the cell surfaces, the cells were rinsed in buffer, dehydrated, and placed into Peldri II. The Peldri II fluorocarbon was sublimated using the apparatus shown in Figure 3-8A, and the specimen was coated with carbon for conductivity and examined in the SEM using a standard secondary electron detector as well as a scintillator-photomultiplier backscatter detector. In the example shown in Figure 7-28, the secondary electron detector was used to give an overall image of the cells, whereas the backscatter detector was used to reveal the sites where the antibody gold is attached.

Study 2: Detection of Subsurface Structures

Since backscattered electrons may emerge from deep inside biological tissues, this study by Horiguchi, et al. (1984) took advantage of the penetrating power of the primary electron beam to examine nuclei inside of intact cells. Small pieces of tadpole tail muscle were fixed in an aldehyde fixative and rinsed extensively prior to submersion of the specimens in an ammoniacal silver solution. After rinsing in distilled water, the tissue was again fixed in glutaraldehyde, rinsed in acetic acid, dehydrated up to 70% alcohol, and frozen in liquid nitrogen. Tissues were fractured to expose blood vessels, further dehydrated, critical point dried, and coated with carbon for conductivity. A standard secondary electron detector was used for three-dimensional imaging purposes, whereas a solid state detector was used to image the silver-stained organelles. Cell nuclei were specifically stained with the silver and exhibited high contrast against an unstained background (Figure 7-29).

Study 3: Selective Staining of Neuronal Tissues

Some of the selective silver stains used in light microscopy may be employed for SEM. In a study by Taylor, Fawcett, and Hirst (1984), central nervous tissues of *Xenopus laevis* (newt) that had been fixed in a variety of light microscope fixatives were dehydrated, embedded in paraffin, sectioned at 7 μm, and mounted on coverslips. The sections were stained with a variety of silver stains for neuronal tissues, dehydrated, critical point dried, and carbon coated prior to examination in the SEM. The specimens were examined using both the standard secondary detector as well as a solid state high resolution backscattered electron detector using 20 to 25 kV accelerating voltage. In the secondary imaging mode, one was able to observe the surface features of such structures as the retina (Figure 7-30A). With the backscattered detector, one was able to observe the silver-stained nerve cells lying under the limiting membrane of the retina (Figure 7-30B). The procedure described in this

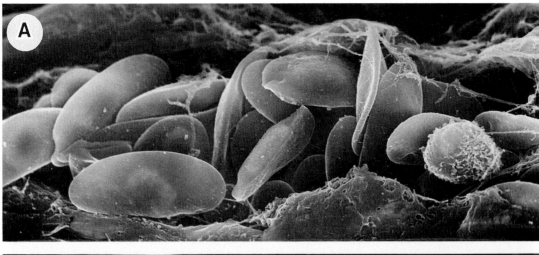

Figure 7-29 Red blood cells from tadpole as viewed in the (A) secondary electron mode and (B) backscattered electron mode. The nuclei were stained with a silver stain and appear very bright. Note that the backscattered image reveals structure beneath the surface of the cell membrane due to the penetrating power of the backscattered electrons. (Courtesy of Kyozo Watanabe and The Williams and Wilkins Co.)

paper makes it possible to more effectively distinguish and trace nerve fibers in a variety of tissues.

Specialized Instrumentation for Observing Unfixed Tissues

Observation of Frozen Specimens

The majority of biological specimens that are examined in the SEM undergo the standard procedures of fixation, dehydration, drying, mounting, and coating with a metal prior to observation. It is possible, however, to observe fresh, unfixed materials in a standard SEM under certain conditions. Cold stages have been developed to fit on most standard SEMs and permit one to observe a frozen specimen using low accelerating voltages for a short period of time. Such stages would be useful with certain types of specimens that are not adequately preserved by the standard methods or whenever speed of viewing is necessary (i.e., to rapidly freeze a dynamic cellular process and to view it with a minimum of disturbance). For example, the secretion of water-soluble polymers by an organism would be best observed using a rapidly frozen specimen. It may be obvious that certain types of specimens (snowflakes, ice cream, frozen biological specimens taken from Arctic areas) that occur in the frozen condition may be destroyed by thawing.

In a typical situation, a specimen is first cryoprotected using glycerol or DMSO and rapidly frozen to liquid nitrogen temperatures below $-170°$ C (see Chapter 14). After transferring the frozen specimen onto a prechilled specimen stage, the specimen

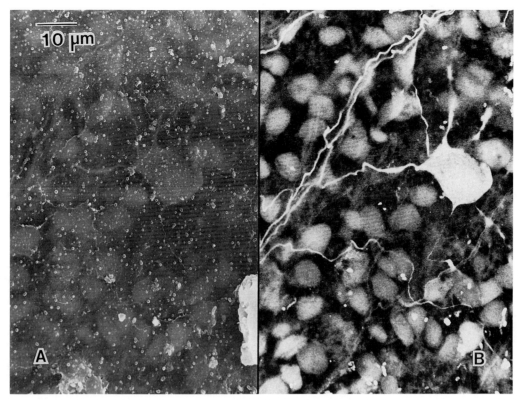

Figure 7-30 (A) Secondary electron image of optic nerve of frog. Since the axons were stained with a silver stain, they have an enhanced emission of backscattered electrons and are therefore brighter than surrounding cells. (B) Backscattered electron image of the same specimen shown in 7-30(A). In this imaging mode, it is possible to view through the limiting membrane of the retina and see the underlying nerve cells that have been stained with a silver compound. (Both micrographs courtesy of J. S. J. Taylor and The Williams and Wilkins Co.)

is evacuated and inserted into the SEM viewing chamber. A reservoir of liquid nitrogen is used to maintain the temperature of the cold stage throughout the viewing process. In many instances, after warming up the specimen slightly, any ice crystals that formed on the surface will be sublimed away by the vacuum system of the SEM. The specimen surfaces may then be observed at low to intermediate magnifications.

In order to minimize charging, accelerating voltages of 1 to 5 kV are used. Salts remaining in the frozen aqueous system of the specimen also impart some conductivity to the tissues. Metal coatings are still desirable; consequently, some cold stages contain a small set of electrodes or a sputter coater. Once metal coated, frozen specimens may be observed for much longer periods of time and with improved resolutions compared to uncoated specimens. Since such extremely cold specimens act as miniature cryopumps, they attract and condense contaminants onto the specimen surfaces. Clean vacuum systems (turbo or ion pumped) or good cry-

otraps are needed to prevent specimen contamination. Some cryostages provide the operators with manipulators for fracturing the frozen specimen to expose underlying structures. A soil nematode observed on a cryo stage is shown in Figure 7-31.

Observation of Fresh Specimens

In 1989 a unique type of SEM, termed an environmental SEM (ESEM), became available commercially. It allowed the observation of unfixed, nonfrozen, uncoated biological specimens at pressures of about one million times higher than the conventional SEM (e.g., 2.7×10^3 Pa, as described in Chapter 6). The instrument, manufactured by ElectroScan (Figure 7-32), operates in the standard accelerating voltage ranges of 0 to 30 kV with claimed resolution capabilities of 10 nm at a pressure of 6×10^2 Pa. The ESEM is differentially pumped so that the gun area has vacuums in the 10^{-4} Pa range common in most conventional SEMs. The pressure then rises (due to a differential pumping system) as one approaches the specimen chamber. As a result, the ESEM column is at conventional vacuum levels while only the specimen chamber is at the increased pressures needed to observe fresh, hydrated tissues. The detector used in this system is unlike the conventional one developed by Everhart and Thornley, but is based on the gaseous discharge detector described by Danilatos in 1982. It is anticipated that continued development and exploration of the capabilities of the ESEM by researchers will bring forth new discoveries of dynamic processes in living cells. Some examples of biological specimens viewed in the ESEM are shown in Figure 7-33.

Figure 7-31 A soil nematode viewed in the frozen hydrated state using an SEM equipped with a cryo stage that was chilled with liquid nitrogen. (Courtesy of Oxford Instruments.)

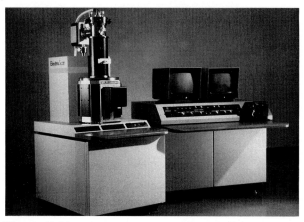

Figure 7-32 The environmental SEM (ESEM) for observing fresh, unfixed specimens. Since the instrument is able to operate at relatively low vacuum levels (in contrast to the conventional SEM), it is possible to observe live, hydrated specimens at claimed resolutions of 5–10 nm. A different type of detector than found in the conventional SEM is used in the ESEM. (Courtesy of ElectroScan Corporation.)

A   **B**   **C**

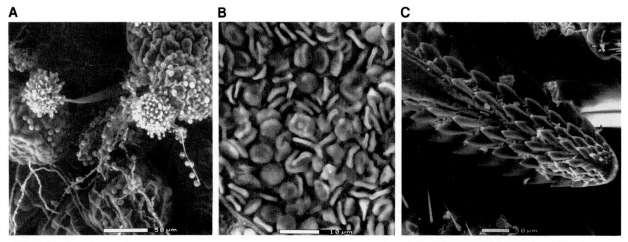

Figure 7-33 Examples of different types of unfixed, hydrated biological specimens observed in the Environmental SEM. (A) Bread mold viewed at 315×. Accelerating voltage, 12 kV; vacuum, 3.7 Torr; temperature, −15° C. (B) Fresh frozen red blood cells in lung tissue. 1,400×; 20kV; 2.8 Torr; −15° C. (C) Mouth parts of tick. 200×; 20kV; 3.7 Torr; 24° C.

References

Barber, V. C., and C. J. Emerson. 1980. Preparation of SEM anaglyph stereo material for use in teaching and research. *Scanning* 3:202–6.

Becker, R. P., and O. Johari, eds. 1979. *Cell surface labeling.* Scanning Electron Microscopy, Inc. (AMF O'Hare, IL) 344 pp.

Brocker, W., and G. Pfefferkorn. 1976. Bibliography on cathodoluminescence. *Scanning Electron Microsc* IV: 725–37.

Danilatos, G. D. 1982. Foundations of environmental scanning electron microscopy. *Advances in electronics and electron physics* 71:109–250.

Danilatos, G. D., and R. Postle. 1982. The environmental scanning electron microscope and its applications. *Scanning Electron Microsc* I:1–16.

DeHarven, E., R. Leung, and H. Christensen. 1984. A novel approach for scanning electron microscopy of colloidal gold-labeled cell surfaces. *J Cell Biol* 99:53–7.

Everhart, T. E., and T. L. Hayes. 1972. The scanning electron microscope. *Scientific American* 226: 54–68.

Everhart, T. E., and R. F. M. Thornley. 1960. Wide-band detector for microampere low-energy electron currents. *J Sci Inst* 37:246–8.

Goldstein, J. I., D. E. Newbury, P. Echlin, D. C. Joy, C. Fiori, and E. Lifshin. 1981. *Scanning electron microscopy and x-ray microanalysis: a text for biologists, materials scientists, and geologists.* New York: Plenum Publishing Corp. 673 pp.

Haggis, G. H., E. F. Bond, and R. G. Fulcher. 1976. Improved resolution in cathodoluminescent microscopy of biological material. *J Microsc* 108:177–84.

Harrington, D. A., and A. Welford. 1990. An alternate method of viewing stereo pairs from the SEM. *J Electron Microsc Tech* 15:101–2.

Helinski, E. H., G. H. Bootsma, R. J. McGroarty, G. M. Ovak, E. de Harven, and J. L. Pauly. 1990. Scanning electron microscopic study of immuno-gold-labeled human leukocytes. *J Electron Microsc Tech* 14:298–306.

Horiguchi, T., F. Sasaki, and H. Takahama. 1984. Identification of cells by backscattered electron imaging of silver stained bulk tissues in scanning electron microscopy. *Stain Technol* 59:143–8.

Joy, D. C. 1984. Beam interactions, contrast and resolution in the SEM. *J Microsc* 136:241–58.

Murphy, J. A. and G. M. Roomans, eds. 1984. *Preparation of biological specimens for scanning electron microscopy.* Scanning Electron Microscopy, Inc. (AMF O'Hare, IL) 344 pp.

Peachey, L. D. 1978. Stereoscopic electron microscopy: principles and methods. *Bull Electron Microsc Soc Am* 8(1):15–21.

Postek, M. T., K. S. Howard, A. Johnson, and K. L. McMichael. 1980. *Scanning electron microscopy: a student's handbook.* Ladd Research Industries, Inc. (Burlington, VT) 305 pp.

Taylor, J. S. H., J. W. Fawcett, and L. Hirst. 1984. The use of backscattered electrons to examine selectively stained nerve fibers in the scanning electron microscope. *Stain Technol* 59:335–41.

Wells, O. C. 1974. *Scanning electron microscopy.* New York: McGraw-Hill Book Company. 421 pp.

Wergin, W. P. 1984. Importance of incorporating stereopsis and stereometry into a scanning electron microscopy course. *Scanning Electron Microsc* III: 1225–35.

Photographic Principles

Negative Recording Medium

Exposure Process in the SEM

Exposure Process in the TEM

Speed and Resolution

Improving Resolution and Contrast
  in TEM Negatives

Commercial Films, Handling,
  Developing and Troubleshooting

Handling of Negative Materials and
  Processing

Darkroom Printing

Work Prints and Final Prints

The Enlarger and Accessories

Printing Papers

Enlarging

Print Processing

Burning-in and Dodging

Techniques to Enhance Contrast

Matte and Glossy Electron
  Micrographs

Preparing Micrographs for Publication

The Final Print

Trimming Prints

Mounting Prints

Labeling Prints

Reproducing Prints

Slide Presentations

Poster Presentations

Making Montages

Determining Print Magnification
  from a Negative

References

Production of the Electron Micrograph

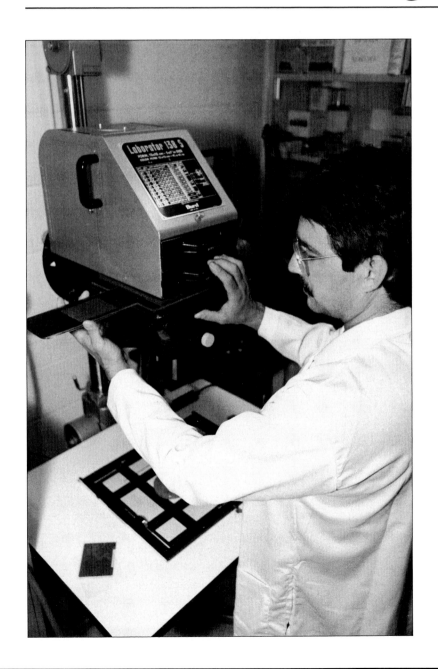

Photographic Principles

The final record of most electron microscope investigations is usually the positive photographic print that is produced from a black and white negative. This electron micrograph is the medium of presentation of one's work to the scientific community and serves as a basis for judging the quality of the research in electron microscopy. The production of a high quality negative and print is as important as any of the preceding preparatory steps. Perhaps because they may be regarded as technologically simple endeavors, the photographic aspects of electron microscopy are often slighted. This is unfortunate since excellent morphological data may be lost unless it is recorded properly. An unsuitably exposed or processed negative usually yields a poor quality print, whereas a properly managed negative nearly always produces a good print.

General Steps in Producing an Electron Micrograph

- Expose the sensitive photographic emulsion by means of electrons.
- Develop the exposed emulsion in the darkroom to generate a negative image.
- Treat the negative with fixer for stabilization.
- Wash and dry the negative.
- Project the negative onto a photographic paper using an enlarger.
- Develop the photographic paper.
- Stop development process and stabilize the image in photographic fixer.
- Wash and dry the photographic print.

A number of factors may result in the production of suboptimal negatives, necessitating rescue efforts in the darkroom or even the return to the electron microscope to re-record the images. It is very important, therefore, to be aware of the considerations necessary to produce both a high quality negative as well as a good photographic print.

Negative Recording Medium

The most prevalent method of recording images in electron microscopes involves the transfer of energy from either an electron or a photon into a sensitive crystal (or grain) of silver halide (usually silver bromide). The AgBr grains (Figure 8-1) used in negative materials for the electron microscope are typically around 0.5 μm in diameter and are suspended in a matrix of gelatin.

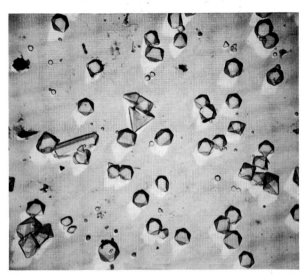

Figure 8-1   Transmission electron micrograph of a cubic silver bromide emulsion such as may be used on photographic recording media. The silver grains have been shadowed with platinum to generate the white "shadows". Magnification = 20,000×. (Courtesy of Eastman Kodak.)

During the manufacture of the recording material, the AgBr/gelatin suspension, or emulsion, is spread over the substrate (plastic film, glass, or paper) and allowed to dry as a 12 μm thick opalescent layer that is sensitive to both photons and electrons. In negative materials, the emulsion layer is usually overlayered with a 1 μm coating of plain gelatin to protect the underlying silver grains against abrasions that would physically render the grains developable (i.e., a black scratch on the negative). In some negative films, a second layer of plain gelatin is applied to the back side of the film to prevent curling of the developed film during the drying process. An antihalation dye may be added to the backing or substrate of certain negative materials to prevent reflection of light back through the sensitive emulsion layer. The dye makes the back side darker in appearance under a safelight and facilitates the recognition of the shinier emulsion-coated side. The dye is removed during the development process.

The silver halide crystals in an emulsion may be manufactured in various sizes. The size of the grains will affect the resolution of the negative-recording medium and may affect the speed of the film (see Resolution and Speed in this chapter). For instance, negative medium with large silver bromide grains will not have the resolving capabilities of a negative medium with very small silver halide grains. Electron microscope emulsions have small silver grains (e.g., fine grain) in order to enhance the resolution capabilities of the negative.

For the production of electron microscope negative recording media, the emulsion is coated over either a film base of plastic (usually polyester or cellulose acetate) or sometimes glass. Plastic, rather than glass, films are used most often since they are less expensive, easier to store and transport, safer to handle, and have better adherence of the emulsion to the plastic. On the other hand, polyester films take longer to evacuate when placed inside of the microscope (cellulose acetate films take considerably longer) and the plastic substrate is more easily scratched. Glass substrates are dimensionally more stable than plastic and are occasionally used when the most critical of measurements must be made from the negative, as in electron diffraction studies. It is safe to say that glass substrates are only rarely used these days. Figure 8-2 is a diagram of a typical recording medium used in electron microscopy.

Exposure Process in the SEM

In the SEM, images are usually recorded by photographing the image off a high resolution cathode ray tube (CRT). A camera, carrying 35 mm or sheet film of various sizes, is used in this process. For convenience, most researchers use an instant film such as Polaroid Type 55, which produces both a high quality negative and a positive print at the same time. Normally a 4″ × 5″ carrier is used for the individual packets that are slipped into the back of the camera. After development (by pulling the exposed film packet through two pressure rollers of the special film holder), the negative part of the packet is placed in a concentrated solution of sodium sulfite to remove the developer gel. After 30 seconds, the negative is washed in running tap water and dipped in a dilute wetting agent such as Photo Flo and dried in a dust-free environment. The positive print should be coated with an acidic polymer that neutralizes the developer and seals the print in a layer of plastic. Some types of instant print films do not have to be coated since they are covered with a thin layer of plastic during the manufacturing process.

Instant films are quite convenient and provide excellent negatives as well as acceptable prints. They are, however, expensive when one needs to take many micrographs. For economic reasons, it is possible to use other roll or sheet films to record the images from the CRT. Among the films that can be used are the slower speed panchromatic films (Kodak Plus-X Pan 4147 and 2147, Technical Pan Films 2415, 4415, and 6415) as well as Kodak Commercial Film 4127, specially designed for recording SEM images from a CRT. Other types of *panchromatic* films by other manufacturers are also acceptable and may be tested for suitability. The development process is normally conducted in total darkness for all except the 4127 film, which is *orthochromatic*. (Panchromatic films are sensitive to nearly all wavelengths of light and must be handled in total darkness, whereas orthochromatic films are not sensitive to certain wavelengths and can be handled under so-called "safelight" conditions.) The processing of these films is similar to the one described for TEM (next section), except that the type of developer, as well as the final working dilution, may differ. Since developers and processing procedures are continually being improved by the manufacturers, one should follow the current recommendations included with the particular recording medium.

Exposure Process in the TEM

Although photographic emulsions are sensitive to both photons and electrons, the emulsion will react differently when exposed to a photon compared to an electron. A photon from the light microscope typically has an energy of 2 to 3 electron volts. It takes approximately 10 of these low energy photons to expose a single silver halide grain. On the other hand, a single electron in a transmission electron microscope has energies several tens of thousands times greater (50 to 100 kV) than a photon, so that one electron will expose many different silver grains as it passes through the emulsion. For instance, a 10 kV electron will pass through about 1 μm of emulsion before its energy is dissipated, whereas a one

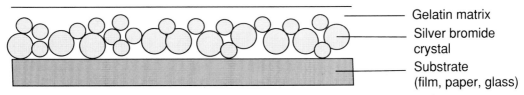

Figure 8-2 Diagram of recording medium (film or paper) used to capture electron or photon images.

million volt electron is capable of passing through approximately 2,000 μm of emulsion. The latter high energy electrons may pass through the emulsion, strike the negative substrate, and be backscattered through the emulsion to expose more grains. This results in a diffusion of the overall image so that it becomes less sharp. Fortunately, at the energy levels used in the standard TEM, the 50 to 75 kV electron is slowed down sufficiently by the emulsion so that only several silver grains are normally exposed by a single electron. Thus, photographic emulsions are exposed much more efficiently by electrons than by photons. In practice, the exposure of an emulsion by electrons may be considered to be 100% efficient.

When an electron strikes a silver halide grain, some of its energy is transferred to the crystal to generate aggregates of 3 to 4 metallic silver atoms, termed a latent-image speck. If the exposed silver halide crystal is placed in the proper developer, the latent-image specks serve as nucleation sites to rapidly convert the entire crystal from a silver halide (i.e., Ag^+) to a reduced grain of metallic silver (Figure 8-3). In contrast, silver halide crystals that were not struck by an electron will develop at a considerably slower rate . One must stop the development process after a certain period of time in order to prevent these unexposed crystals from developing and causing background "fog," or an increase in the overall density of the negative. It is in the photographic "fixer" that these undeveloped silver halide crystals are selectively made soluble and removed, leaving behind only the metallic silver grains still embedded in the gelatin.

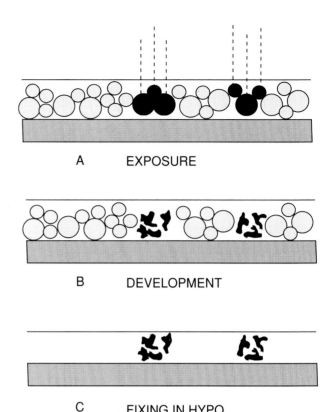

Figure 8-3   During the development process, AgBr emulsion grains that were exposed to electrons or photons are developed into metallic Ag grains. Unexposed and undeveloped AgBr grains are removed by the photographic fixer, leaving behind only the developed Ag grains suspended in the gelatin.

appear to be excessively large. In fact, one will not begin to see the actual silver grains until the negative is enlarged over 12 times.

Grain Versus Noise

A negative film may be thought of as a carrier of several layers of silver halide crystals that become developable if struck by an appropriate particle. In the ideal situation, the electrons that form the image upon interaction with the specimen will expose all of the proper silver halide crystals to generate a clean, noise-free image. However, unless enough electrons strike the emulsion (i.e., proper exposure), gaps or voids consisting of unexposed areas may be interspersed among exposed silver halide crystals. This will be evidenced as a "snowy" or noisy image when the negative is printed. The inexperienced microscopist may interpret this noise as a grainy picture (meaning that one is actually seeing overly large individual silver grains rather than gaps between groups of silver grains). One may mistakenly conclude that the resolution of the emulsion on the negative is inadequate, since the silver grains

Speed and Resolution

Photographic emulsions with large silver grains are said to be of *high speed*, since it takes fewer photons to expose a given area of emulsion. On the other hand, *high resolution* photographic emulsions have smaller silver bromide grains, requiring more photons to expose a given area, and are consequently said to be slower speed emulsions. Since the efficiency of exposure by electrons is essentially 100%, speed of emulsions per se is usually a minor consideration in transmission electron microscopy. *Resolution capability of the emulsion is the major concern.* For this reason, most electron microscope emulsions have small silver grains in order to be able to record fine details of the image.

Resolution is also affected by the accelerating voltage used. Higher energy electrons tend to scatter more throughout the emulsion so that the image is spread out and resolution is slightly degraded. Some negative materials for use at higher accelerating voltages have an increased thickness of plain gelatin overlying the emulsion in order to slow down electrons and diminish the image spread. A few emulsions may utilize grains that require more than one electron hit to render them developable.

Since the silver grains used in negative emulsions for the TEM are relatively small, it is essential that one *collect enough electrons on the TEM negative (proper exposure)* to fill in the image details in order to avoid image gaps and diminish the noise (i.e., improve the signal-to-noise ratio). *The benefits of collecting more electrons are an increase in the resolution capabilities and an increase in contrast of the TEM negative.* This is an important difference between emulsions exposed to photons versus electrons: Continued exposure to photons (as in the case of SEM images recorded from a CRT) will increase only the overall density and not the contrast. Since it is so important from a practical standpoint to collect more electrons to effect better resolution and contrast in TEM negatives, this is discussed in the next section.

Improving Resolution and Contrast in TEM Negatives

It is possible to collect more electrons on the negative by making several different adjustments to the TEM. The easiest way is to *increase the time of exposure* of the TEM negative to the electron beam. In other words, allow the beam to strike the specimen (and negative) for a longer period of time. In older microscopes, this might simply involve increasing the exposure time from 2 to 4 seconds, for example. In older instruments, this can be achieved by adjusting the electron beam intensity to the proper point for a 2 second exposure and then overriding the automatic exposure system by manually setting the exposure to 4 seconds.

Alternatively, one may reset the sensitivity setting of the exposure metering system so that the film speed is effectively lower. This will program the electronics to automatically allow more electrons to strike the negative. It will be necessary to calibrate the exposure meter by taking a number of trial exposures at different meter settings until the proper density/contrast levels are seen in the developed negative.

NOTE: To most photographers, electron microscope negatives that are properly exposed for contrast/resolution appear to be slightly overexposed (overly dense). If there is any doubt about exposure of TEM negatives, it is better to err on the side of slight overexposure since an underexposed negative lacks informational detail.

A recalibration of the sensitivity setting of the exposure meter may cause one to increase the number of electrons striking the specimen (the beam current) by: (a) adjusting the condenser lenses to focus more electrons per unit area on the specimen (i.e., adjust illumination or brightness controls on the TEM), or (b) changing the gun bias or filament height relative to the shield to increase the yield of electrons out of the gun (see Chapter 6).

In addition, specimen contrast may be increased by using lower accelerating voltages, smaller objective apertures, and pointed filaments (see Chapter 6).

Besides using various instrumental manipulations described above, contrast may be affected in other ways as well. It is possible to increase negative density and contrast by increasing development activity through: (a) increasing developer concentration, (b) increasing developer temperature, and (c) increasing development time. Unfortunately, since signal and noise both increase proportionally using these methods, the resolution of the negative will not be as good compared to the procedure of gathering more electrons.

The best method to obtain a high quality negative is to properly expose the negative in the electron microscope and to develop it according to the manufacturer's recommendations rather than resorting to rescue efforts in the darkroom.

Commercial Films, Handling, Developing, and Troubleshooting

Negative Recording Media for TEM

Several products exist for recording images in the TEM. Probably the most commonly used negative medium is Kodak 4489 Electron Microscope Film.

TEM Negative Media

Kodak 4489 film has a substrate of polyester of 0.18 mm in thickness and comes in four different sizes—with the the 3¼″ × 4″ being the most prevalent. It is composed of an extremely fine grained emulsion that is highly efficient in the detection of electrons in the 40 to 100 kV range. The film is not sensitive to several wavelengths (i.e., it is orthochromatic) and can be conveniently handled under OA (green/yellow), 1A (light red) or OC (light amber) safelights. It contains a dyed gel layer on the backside to facilitate recognition of the lighter emulsion side. This film is also available as a 35 or 70 mm roll film in 100 ft lengths, where it is designated as SO-281.

A second film, Kodak SO-163, has the same general characteristics 4489, except that it is twice as fast (slightly larger grains), requires longer to pump down, and has more of a tendency to curl when dry. The recommended safelights are: 1 (red), GBX-2 (red), or 6B (brown). SO-163 is not available in the roll format. Although different, the SO-163 emulsion closely resembles the one used on a third, rarely used product, the Kodak Electron Image Plates made of glass.

Like the Electron Image glass plates, SO-163 is more versatile than 4489 in terms of the development and exposure conditions. Whereas 4489 offers only intermediate speed and signal to noise when developed, SO-163 and the Electron Image Plates can additionally be processed for maximum speed and low signal to noise (with unstable specimens) or for minimum speed and maximum signal to noise (with stable specimens).

In general, TEM negatives are placed within the evacuated microscope just below the viewing screen. However, not all negative materials are placed inside the vacuum of the TEM.

In the Zeiss EM-109 electron microscope, the electrons strike a fiber optic faceplate to generate photons that travel outside the vacuum chamber to form a photonic image on a sensitive emulsion. The 120-size roll film normally used is the Kodak Technical Pan Film 6415 with a cellulose acetate base. The film is developed for 5 to 6 minutes in D-19 developer diluted 1:2. This film is sensitive to all visible wavelengths (panchromatic) and must, therefore, be handled in total darkness. Since this film is not inside the vacuum chamber, no evacuation of the film is necessary, and the film may be loaded directly from atmospheric conditions.

Handling of Negative Materials and Processing

Proper handling of film begins *prior* to opening the package. If the film was stored in a refrigerator or freezer (to extend its shelf life), it is necessary to allow several hours for the material to warm to room temperature prior to opening. Otherwise, moisture may condense on the chilled emulsion and ruin the recording medium. The film should be opened under appropriate safelight conditions as specified by the manufacturer. It is important that the proper size

bulb be used in the safelight and that the safelight be checked periodically for leaks or cracks in the filters. This is best done by placing a test sheet of film (emulsion side facing the safelight) in the work area. Several coins (penny, nickel, quarter, and half dollar) are placed in several locations over the film and removed after various times (perhaps 1, 5, 10, and 20 minutes). The film is developed normally and the negative examined over a backlighted viewbox. The outlines of the coins should not be visible if the appropriate safelight conditions were followed.

The work area over which the negatives are to be loaded/unloaded into the cassette carriers of the electron microscope must be dust- and static-free. This is easily accomplished by thoroughly wiping the working surface with a damp, lint-free towel. The damp towel not only removes lint, but helps to dissipate any static generated during the cleaning process.

Removal of the films or plates from the package must be done carefully so as not to abrade the surface of the emulsion. For instance, in an attempt to separate closely packed films, one may be tempted to slide one film over another. This may not only generate an abrasion, but enough static electricity may be produced to cause local fogging of the film in the shape of a "lightning bolt" (Figure 8-4). It must be realized that the abrasions need not be deep enough to gouge the emulsion, since light pressure may be sufficient to mechanically expose the silver halide grains. In fact, most "scratches" seen on negatives are probably caused by slight contacts with the emul-

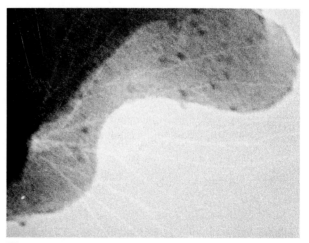

Figure 8-4 Static discharge artifact on a TEM negative. Due to low humidity in the TEM room static electricity generated by sliding one negative over another caused the buildup of static electricity which, when discharged, gives off a miniature "lightning bolt" that exposes the sensitive emulsion.

sion surfaces. Try not to contact the emulsion, but handle the film by the edges only.

Lintless nylon gloves should be worn when handling film to avoid transferring body oils and moisture to the medium. In fact, it is good practice to avoid skin contact with any object that is to go into the electron microscope. Lint and dust particles must be avoided, since they will be opaque to the electrons and recorded as transparent areas on the negative and black profiles on the final prints. One may remove the particles using either a camel's hair brush or special compressed air that is oil and moisture free. It is poor practice to blow on the films, since droplets of moisture (saliva) may be deposited onto the emulsions.

After removing the exposed negatives from the microscope and loading fresh films into cassettes, the unexposed film cassettes are placed into a chamber for pre-evacuation for several hours prior to insertion into the high vacuum of the TEM. Usually the films are evacuated with a rotary pump in a chamber near the TEM. Under the 10^{-1} to 10^{-2} Pa vacuums achieved, moisture and other contaminating gases are removed from the film. Prepumping will greatly speed up subsequent film exchanges in the microscope column itself, since it may take several hours of evacuation to achieve operational vacuums in the microscope. In this manner, one has a supply of pre-evacuated film ready to insert into the TEM. Most TEMs have a built-in prepumping chamber in the microscope chasis for storing loaded film casettes until needed.

The exposed negatives are usually transferred into plastic or metal racks or carriers prior to processing. Such carriers can accommodate an entire load of exposed films for passage through the development process (Figure 8-5).

> NOTE: Although it is possible to develop negatives in photographic trays, this is not a consistent, safe, or efficient way to manage more than 3 or 4 films or plates.

Although not necessary, nitrogen burst agitation systems are preferred whereby pulsed bursts of nitrogen gas are forced from the bottoms of the various solutions to agitate the processing chemicals and maintain a fresh supply of active chemicals over the emulsion surface.

Most microscopists manually move the racks of negatives through the various containers of solutions (Figure 8-6). It is important that a *consistent manner*

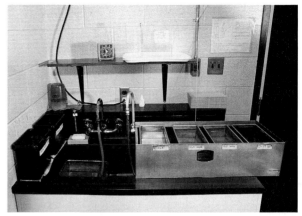

Figure 8-5 Typical setup used to develop sheet films (TEM or SEM). The containers of various solutions (developer, two water rinses, fixer) are placed into a large temperature controlled waterbath and kept covered when not in use in order to prevent oxidation. Negatives are placed into plastic racks or holders to be developed in groups of 15 or more.

Figure 8-6 Method of agitating sheet films for development of negatives. For uniform processing, follow this four-step agitation cycle. Each cycle should be completed in about 5 to 7 seconds and repeated every 30 seconds. Use this same agitation cycle for processing plates in a rack or holder. (Courtesy of Eastman Kodak)

of agitation be established in order to generate a uniformly and properly processed negative. The protocol illustrated in Figure 8-6 is highly recommended for beginners or for microscopists having problems developing negatives. Figure 8-7 shows a flow diagram for negative processing as used in a typical electron microscopy laboratory.

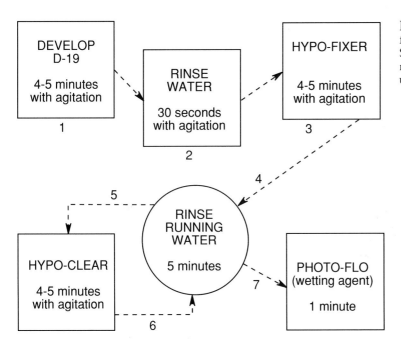

Figure 8-7 Flow diagram of typical process for developing sheet film negatives for TEM. SEM negatives are developed in a similar manner, except that a different developer would be used.

Standardization of Procedures

Once a satisfactory processing procedure has been established, it is important that this protocol be followed strictly. This becomes very important whenever several investigators are working on the same or related projects. Otherwise inconsistent negatives will be produced that will be difficult to match up during final darkroom printing for publication purposes. Several points to observe are

1. Use the same type of negative film whenever possible.
2. Standardize the type and dilution of developer used, as well as the temperature, time, and agitation scheme followed.
3. Evaluate the status of the chemicals used for processing. Developers have a stated capacity for a certain number of films, after which quality will significantly deteriorate. In addition, certain developers have a stated shelf life and should be discarded after that time—even though the full number of films may not have been run through the solution. Always record the date of preparation and numbers of negatives put through the process. Fixers are easily checked using a drop of hypo checker that turns turbid upon exhaustion of the fixer.
4. Clean vessels and tanks are essential to prevent contaminating the chemicals used in processing. Always rinse out measuring vessels before and after use to prevent cross-contamination. Better

yet, use separate vessels to measure out the various solutions.

Problems with Negatives

Occasionally, poor quality negatives may be produced due to either operator error or faulty chemicals. Table 8-1 should help to narrow down the causes of the problem.

Darkroom Printing

The last step in the production of an electron micrograph is the generation of a positive print on photographic paper. This is an exciting step because it is the first chance one has for a detailed analysis of the features revealed by the tissue section. Since electron microscope usage is expensive, generally only a preliminary survey is made of tissue features while one is operating the microscope. More detailed analysis can be accomplished at one's convenience using the printed electron micrograph. Electron micrographs represent the culmination of many hours of preparatory work and should be carefully prepared.

Micrograph enlargement or printing, as it is sometimes called, is carried out in a darkroom. Certain darkroom techniques can be used to enhance selected features of the negative and make the micrograph more pleasing to the eye. On the other hand, poor darkroom technique will be detrimental

Table 8-1   Common Problems with TEM Negatives

Appearance of Negative	Possible Causes	Remedies
Low contrast.	Incorrect sample preparation.	Use stains or shadowing to enhance contrast.
	Proper sized aperture not used.	Insert smaller objective aperture.
	Improper exposure.	Increase exposure in TEM.
	Improper development.	Check recommended procedures.
	Exhausted developer.	Check developer.
Hazy, unsharp image, lack of detail.	Sample, thermal drift.	Prepare samples correctly.
	Instrumental instability.	Check instruction manual.
	Contaminated column.	Clean EM column.
	Improper processing of negatives.	Use recommended procedure.
	Incorrect focus.	Refocus. Try auto-focus of TEM.
Bright fringe around specimen.	Incorrect focus.	Refocus. Try auto-focus of TEM.
Viewing screen not evenly illuminated.	Beam deflectors not centered.	Center condenser traverse/tilts.
	Condenser lens too near cross-over position.	Adjust condenser lens away from crossover.
Scratches and white spots on negatives.	Dirt or dust on negatives prior to exposure in EM.	Remove dust from unexposed negatives with brush or compressed air.
	Handling abrasion in loading.	Handle film more carefully.
Overall film density with low contrast.	Fogging of film or plate.	Use proper safelights, check for light leak.
	Overdevelopment of negative.	
Streaks.	Uneven development.	See recommended processing protocols in this chapter.
	Improper agitation.	
Higher density along edges of negatives.	Excessive agitation in developer.	Agitate properly (see Figure 8-6).
	Insufficient agitation in stop bath.	
Mottled or uneven appearance.	Use of acid stop bath before fixing.	Eliminate acid stop bath.
	Insufficient water rinse before fixing.	Rinse as recommended.
Negatives adhere to processing racks.	Too little agitation at beginning of processing.	Agitate properly in developer (see Figure 8-6).
Water spots.	Excess water not removed from negatives surfaces.	Use Photo Flo or other wetting agent before drying.

Source: Kodak Publication P-236

to the interpretation of the micrograph and will detract from the aesthetic appeal of the electron micrograph. In addition, poor darkroom techniques will effect the acceptability of the micrograph for publication.

Work Prints and Final Prints

Micrograph production is usually handled in two steps. *Work prints* of all negatives are made initially. These are used to analyze specimen features and to determine which micrographs should be given more attention in the printing process and which should be enlarged. From work prints, a decision is made about which of the prints should be printed again for publication. Usually, great care is not exercised in making work prints. Work prints may be made in a small size to save photographic paper. Some investigators will make *contact prints* of their negatives by placing the negative directly on photographic paper and shining light through the negative onto the photographic paper. Contact prints are small (the same size as the negative), but may be used for analysis in a rough way. Selected micrographs may be printed at a larger size at a later time or may be printed in rapid succession to save time. The quality of work prints is usually slightly suboptimal, but of sufficient quality that their detail may be analyzed. Usually, the standard 3¼″ × 4″ negative is magnified about 2.6 to 2.7 times to obtain an 8″ × 10″ print. Prints of this size are generally considered adequate for survey purposes.

Final prints or *publication prints* are high quality photographic prints made from selected negatives for the purposes of making a scientific presentation such as a printed publication, poster, or a slide talk. Usually several prints that vary only slightly in contrast and/or exposure are made from one negative in order to have the capability of selecting one that is considered optimal. Great care is exercised in making these prints. Well-focused prints may be magnified up to ten times negative size to obtain the maximum resolution from the negative, although such an enlargement is rare and will require special equipment and large size photographic paper. Usually only a portion of the negative is enlarged.

The Enlarger and Accessories

The photographic enlarger is used to project a magnified image of the negative onto photographic paper coated with a photosensitive emulsion. Upon exposure, the emulsion, containing a *latent image*, will produce a positive image during development. What was a shadow (darkened) on the fluorescent screen of the transmission electron micrograph became a reversed image (light area) on the negative is returned to its original shadowlike appearance on the print. The darkened images one sees on a transmission electron micrograph represent the areas where electrons were deflected to various degrees by tissue components that had been reacted with osmium or heavy metal stains. In the long run, the density on the photographic paper is related to the number of electrons striking a particular area of the negative.

There are four types of enlargers used (Figure 8-8). The *diffuse* light source enlarger employs either a frosted light bulb or a fluorescent bulb as the illumination source. The light is diffused by a ground glass placed between the light source and the negative. The diffuse nature of the light source produces light rays that are scattered. Such enlargers are not used for printing electron micrographs. A *condenser* enlarger is similar to a diffuse light source enlarger

in utilizing a light bulb, but in place of the ground glass are condenser lenses that focus the light onto the negative. Condenser enlargers are commonly used in the printing of scanning electron micrographs since they soften the image. The *point light source* enlarger employs an intense tungsten or halogen light source housed in a clear bulb. The source acts if it were a single, concentrated, light source, allowing the condenser lenses of the enlarger to generate light that is more coherent than a condenser enlarger and not diffused. The point source, which is also more complex to use, is considered optimal for transmission electron microscopy, producing micrographs more representative of the negative image. The overall resolution and contrast of the print are increased using a point light source enlarger. Point source enlargers are the most commonly used type for printing TEM micrographs. A *cold light source* enlarger is similar to a diffuse light source enlarger, only the light source is usually a fluorescent tube. It is suitable for printing scanning electron micrographs, the result being an overall softened image.

Negatives used in most TEM investigations are of an uncommon size (3¼″ × 4″). The negative is situated in an appropriate negative holder, usually without glass supporting windows in order to avoid dust and scratches, even though the negative sandwiched between glass will insure a flatter image (all areas will be evenly focused).

The enlarging lens must be compatible with the size of the negative to be enlarged to achieve the appropriate field enlargement. For a 3¼″ × 4″ electron microscope negative, the enlarging lens should have a focal length of around 135 mm. Lenses that will not enlarge the negative more than a few times or that cut out the corners of the negative are inappropriate. The enlarging lens on a condenser enlarger has a feature, the aperture, which will determine the amount of light that is projected. F/stop numbers indicated on the side of the enlarger lens regulate the amount of light passing through the lens. Each higher number indicated allows half the amount of light (as the previous smaller number) to pass through the lens. For example, an f8 aperture allows twice the amount of light as an f11 aperture. One focuses the projected negative image usually at the lowest numbered f/stop to obtain the maximum light possible and to minimize the depth of field (see Chapter 6). On a point light source enlarger, a variable transformer, instead of the aperture, is used to vary the illumination. Closing down the aperture

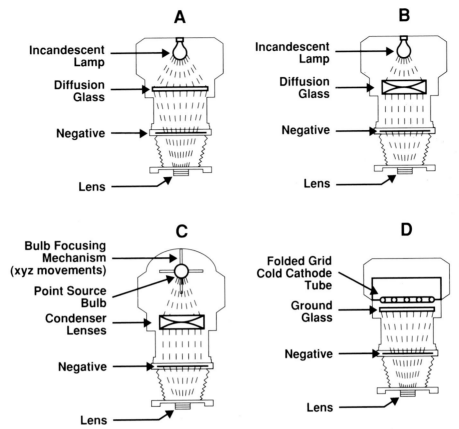

Figure 8-8 Drawings of (a) a diffuse light source enlarger, (b) a condenser enlarger, (c) a point light source enlarger, and (d) a cold light source enlarger. (Diagram adapted from Postek et al., 1980, *SEM: A Students Handbook*, reprinted courtesy of M. T. Postek.)

will cause the point source filament to become visible on the enlarger stand.

The position of the enlarger on the support stand is the major practical factor that determines the enlargement. For each enlargement, the focus knob must be manipulated to bring the image into sharp focus. Critical focus is achieved using a small magnifying device to image the silver grain in the negative on the projected image until the grains of the negative are sharply focused by the enlarger controls. An automatic timer attached to the enlarger determines the length of time the light is exposed to the photographic paper.

An easel holds the photographic paper under the enlarger. Some easels may allow one to make the print of any size or dimension up to the size of the photographic paper used. Other easels fit only one size of photographic paper. Special orange or red safelights, which do not expose the photographic emulsion, are used in the darkroom to allow one to see the surroundings while working with the enlarger and while viewing the projected image.

Printing Papers

When choosing a photographic paper for electron microscopy, one has several options:

Fiber base or resin coated Fiber-based papers have a paper base, whereas resin coated (also known as RC) are fiber-based papers coated with a surface layer of clear plastic. Although fiber-based papers have been the traditional paper of choice, the newer resin-coated papers are more convenient to process. Both are acceptable and commonly used for electron microscopy.

Paper weight The thickness of the base determines the weight. Less wrinkling of the paper is seen with heavier weight papers, although single weight paper is more convenient to process and less costly.

Tone The warmness or coldness of papers is referred to as the tone of the paper. Warm papers are more brown in tone; cold papers are more blue-black in tone. A cold paper is often used for transmission electron microscopy, whereas a warm paper is often used for scanning electron microscopy. Common brands of fiber-based paper for electron microscopy are Kodabromide, Kodak Polycontrast, Ilfobrom, and Agfa Brovira. Comparable resin-coated papers are produced by Kodak, Ilford, and Agfa.

Surface Paper surfaces extend from glossy to matte (flat). Glossy is customarily used for publication purposes; however, poster presentations are easier to view if nonreflecting, matte surfaces are used.

Contrast This refers to the range of blacks and whites in the micrograph. Sharp blacks along with sharp whites and fewer intermediate grey tones constitute *high contrast*. The presence of greys in the micrograph produces *low contrast* and gives the micrograph a "muddy" appearance. There are many possible contrast ranges between high and low contrast.

In general, stained tissue sections are of a low contrast. Thus, for transmission electron microscopy, high-contrast-producing photographic papers are usually required to obtain a pleasing micrograph. Within the overall constraint for the need to increase contrast, some electron microscopists prefer more grey tones and others opt for fewer grey tones, preferring instead sharp blacks and whites. This is simply a matter of taste. Within certain limits, both high contrast and low contrast micrographs are seen in publications. It is not aesthetically pleasing to publish a high contrast micrograph next to one of much lower contrast. Micrographs should match each other in contrast.

Contrast may be regulated by either the grade of paper employed or with filters and variable grade paper. *Graded papers* have numbers from #1 to #5 (or sometimes #6) with #1 producing the lowest contrast and #6 the highest contrast. Most transmission electron microscopists find that grades #3 to #6 give the desired results with the usual range of negatives produced by the electron microscope. A contrast series shows the differences in image contrast one may expect when graded papers are utilized (Figure 8-9). When using *variable contrast papers*, only one paper type is used. The contrast is regulated with filters inserted into the enlarger during the printing process. Both methods for varying contrast are in common usage for electron microscopy. Expect that a higher contrast filter or graded paper will require more exposure than one of a lower contrast.

A typical problem that surfaces, especially with beginners, is determining how to manipulate contrast and exposure (see below) simultaneously to obtain a suitable micrograph. To some degree, exposure and contrast are a matter of taste, but this is only true within limits. This problem is solved if one first learns, through examination of micrographs, the difference between contrast and exposure, and one corrects micrograph problem(s) in a systematic manner. With experience the problem will be minimized.

Enlarging

Basic photography texts should be consulted for the detailed steps in enlarging the negative image to make a print (see references). Fortunately, many individuals are familiar with the enlarging process and help is usually available.

The intensity of light on the photographic paper is one factor that determines the *exposure* or the overall darkening of the photographic paper. The more total light, the darker the image. All other variables being constant, the amount of light coming from the enlarger at any instant may be regulated by the enlarging lens in a condenser enlarger and a variable rheostat in the point light source enlarger. The amount of light striking the photographic emulsion is also dependent on the time the photographic paper is exposed to the light. Figure 8-10 is a series of prints made on the same grade paper where only the exposure time is varied while the contrast is generally similar for all exposures. To determine if a problem is a contrast or exposure problem or both, one should analyze a micrograph systematically for differences in blacks and whites (contrast) and for the degree of general darkening of the paper (exposure). Problems may exist with both contrast and exposure.

Both contrast and exposure can be manipulated to produce an acceptable background density of a print. As a general rule, where there are no tissue elements (for example, intercellular space or clear vacuoles), these areas of the print should match the unexposed border of the micrograph or be darkened (grayed) slightly as compared with the border. When this is evident, the exposure is correct. If the contrast is too great or too little, use a photographic paper that will compensate.

A good way to determine how a print will appear is to expose a small piece of printing paper or *test strip* (Figure 8-11) before the final print is made. The test strip can be covered partially by opaque cardboard and uncovered systematically in timed

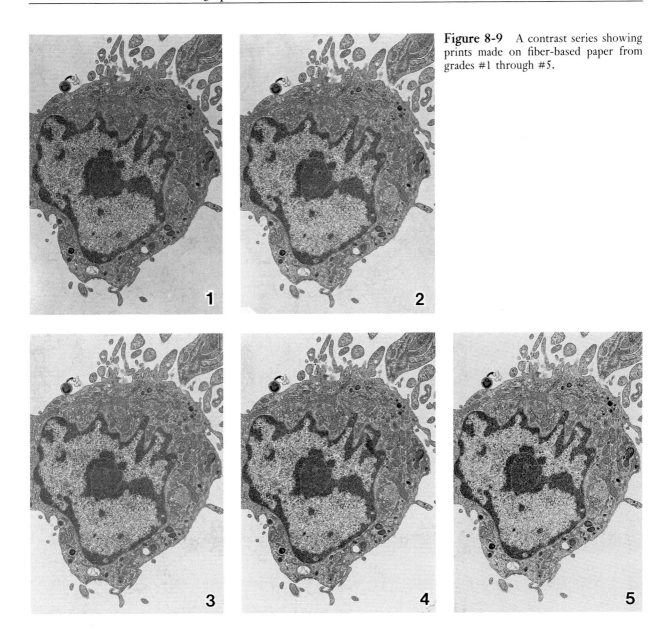

Figure 8-9 A contrast series showing prints made on fiber-based paper from grades #1 through #5.

steps from one end to the other (e.g., at 3 second intervals) during the exposure to determine optimal exposure time. When the print is developed, it will show a series of exposures along the test strip. Examination of the strip will allow one to determine the appropriate exposure. Examining the area of the print with the proper exposure will then allow one to determine the suitability of the contrast.

Since considerable time and effort may go into producing a print, one should keep a record of the exposure time, light setting, and intensity of the light as determined by a rheostat on the enlarger. A soft pencil may be used to record this and other information on the back of the photographic paper used to make the print.

Print Processing

The standard print processing setup is shown in Figure 8-12. Recommended times and brief instructions follow.

Developer: Use the manufacturer's recommended developer. Dektol made by Kodak is a common developer. Develop 1 to 2.5 minutes for fiber-based papers and 1 minute for resin-coated papers. Agitate gently but constantly.

Stop bath: The stop bath will neutralize the developer and allow the fixer to last longer. A commercial acetic acid (4%) bath is recommended as a stop bath for fiber-based papers. A water bath is used

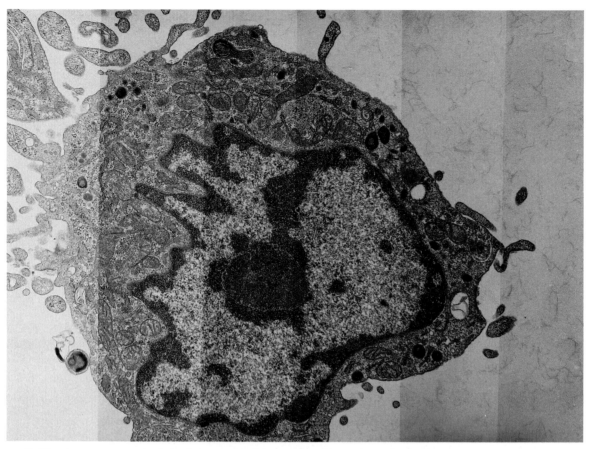

Figure 8-10   An exposure series is made on paper of the same grade, varying only the time the light was projected onto the photographic paper.

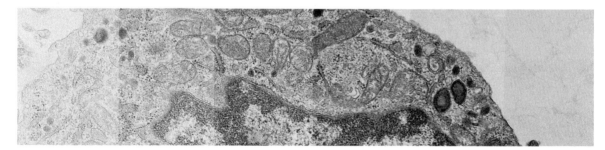

Figure 8-11   A test strip showing several exposures that are made in order to determine an approximate range for a final exposure.

for resin-coated papers. About 15 seconds is needed in the stop bath.

Fixer: One or two baths of a product such as Kodak Rapid Fix are used. Fiber-based papers require about 5 minutes in fixer, while resin-coated papers require about 2 minutes. If two fixer baths are used, then as the first fixer bath becomes old, it is replaced with the second bath and a new bath is made to replace the second bath. An exhausted fixer solution is easily determined with a drop of test solution (Edwal or

Kodak) that will turn cloudy if the fixer is in need of replenishment.

Water rinse: The micrographs are washed for 2 to 5 minutes in water.

Fixer remover: For fiber-based papers only, use one of several commercial solutions (for example, Perma Wash by Kodak) for 2 to 4 minutes.

Wash: Fiber-based papers should be washed from 8 to 15 minutes, whereas resin-coated papers can

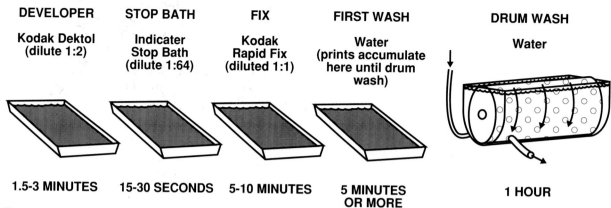

Figure 8-12   Print processing set-up.

be washed for 4 to 5 minutes. It is preferable to use a system in which running water is used to agitate the prints. Special rotating washers or swirl washing sinks have been designed for this purpose. However, care must be exercised with the resin coated-papers or the emulsion will be scratched or abraded (i.e., do not wash prints in a tumbling drum system).

Burning-in and Dodging

Negatives are not always evenly exposed. There are many causes of uneven exposure, some of which relate to variations in the thickness of the tissue section, others to unevenness of the electron illumination, and yet others to errors made in the developing and enlarging process. Whatever the cause, the final print can be made to show more of an even exposure by the rescue processes of dodging and burning-in. When a print shows one underexposed, lighter area, an opaque material (for example, cardboard) with a hole cut in it (1″ to 3″) is held between the light source and the printing paper during a fraction of the entire exposure (for example, 5 seconds of 20 seconds). The area that was underexposed is allowed to expose beyond the time that is optimal for the remainder of the print. The hole is moved slowly to cover all underexposed areas of the print and "feather" the burned-in edges. Thus, the underexposed area is "burned-in" to compensate for negative imperfections.

Often a portion of an electron microscope print will be overexposed (darker) as compared with the remainder of the print. This artifact may have the same causes as those mentioned previously for underexposure. The circular piece of cardboard, cut out from the opaque cardboard described above, can be mounted onto a narrow rod (for example, a long

pencil) of some type and used to "dodge" the overexposed area. The cardboard is placed between the enlarging lens and the printing paper and moved slowly to cover the overexposed area. Usually dodging is effective if under 30–40% of the exposure time is blocked to a selected area. Several attempts may be necessary before the desired result is produced. It is possible that the same print may require both burning-in and dodging.

Techniques to Enhance Contrast

There are several procedures that allow the small increases in contrast one often seeks in a transmission electron micrograph. Use of a point source enlarger will increase contrast. Exposure times that are long (30 seconds to 1 minute) at low light levels will produce a slightly more contrasted micrograph. Development in undiluted developer will increase contrast. All of the above used in concert will provide the equivalent of one paper grade of higher contrast (e.g., grade #5 to #6).

Matte and Glossy Electron Micrographs

Even though a glossy paper is used to print micrographs, fiber-based prints can be made either glossy or matte (nonshiny) during the drying process. If heated drum dryers are used and the emulsion surface is placed in contact with the drum, the print will be glossy. Most work prints and publication prints are dried in this manner. A matte finish is obtained by placing the emulsion side toward the drying cloth. Prints displayed in a poster session of a scientific meeting are dried for a matte finish. They reflect light poorly and thus are easier to view. Resin-coated papers are purchased with specified

matte or gloss finishes and never dried in a conventional drum dryer.

Preparing Micrographs for Publication

Electron micrographs are made for viewing. If one's goal is publishing, it is hoped the micrographs will contain new information that will add to the existing body of scientific literature. The goal of every biological electron microscopist is to publish data in the best form possible. The data for the microscopist *is* the electron micrograph. Therefore, it is imperative to produce micrographs of the highest quality for presentation at scientific meetings and for publication.

The Final Print

Choosing a representative print. A criticism commonly heard is that electron microscopists publish micrographs that are not representative of tissue features. This criticism should be taken into account when selecting prints to portray the findings of a study. The final print should not be misleading as regards the typical findings of the study. If one can easily avoid misrepresentation, then why is the criticism often leveled at microscopists? The answer lies in the microscopist's desire to publish the most aesthetically pleasing print possible. In doing so, there may be a tendency to select a beautiful print that may, at the same time, show an atypical or infrequently encountered structural feature. Clearly, this tendency should be avoided.

Matching several prints placed together. Final or publication prints are usually based on scrutiny of a work print and after a determination of how to improve the work print. This involves meticulous attention to detail in the darkroom, and the selection of one print from several that are almost identical (slight variations in contrast and exposure). In choosing the final prints from several available prints, one should attempt to match prints according to density and contrast as closely as possible since high contrast prints placed adjacent to low contrast prints (or underexposed next to properly exposed prints) are not pleasing to the eye and pose difficulties when they are reproduced for publication.

Selecting the dimensions of a final print. Usually, there are limits or specified sizes for publication prints. Work print micrographs of the 8″ × 10″ size usually do not fit the format specified by the journal. One relatively simple way to obtain a print of the appropriate size is to draw a box on white paper of the same size as the micrograph that will be submitted to the journal. The paper is placed on the enlarging easel and the negative enlarged to occupy the space within the box. The easel may be rotated to fit the desired information from the micrograph into the box.

Often micrographs are grouped into a plate on a journal page and, to save space and reduce journal page charges, micrographs must be strategically arranged. (The publication charges for micrographs can be a major concern.) It is not unusual to see five to ten electron micrographs published on a single journal page. Considerable planning is involved in determining the dimensions of the final publication micrograph.

Trimming Prints

Most final prints are usually prepared in a rectangular or square shape with the corners at 90 degree angles. To trim a print in one of these shapes, use a high quality levered (guillotine) paper cutter or a rotary (roller) paper cutter. The surface of the paper cutter must be engraved with lines that allow one to measure distances from a cut surface. Decide the position of one border of the print and trim this border with the cutter. This border becomes the straight edge from which to trim the three other edges. Being conscious of the dimensions of the micrograph, use the straight edge and align it precisely with one of the horizontal lines on the cutter. The next cut will be a perfect 90 degree angle from the first. Paying continued attention to dimensions, use the original straight edge aligned with the ruled surface of the cutter to make the remaining cuts. The final micrograph should be perfectly squared (i.e., 90° angles) and of an appropriate size.

Mounting Prints

Some journals require that electron micrographs be mounted. The most common method of mounting uses a mounting medium called "dry mount tissue" (Figure 8-13). It has a texture and appearance like wax paper, and becomes adhesive when heat is applied. Dry mount tissue is usually applied to the back of the paper *before* trimming. The trimming process assures that the size of the adhesive paper will match that of the final print. To apply dry mount tissue,

A

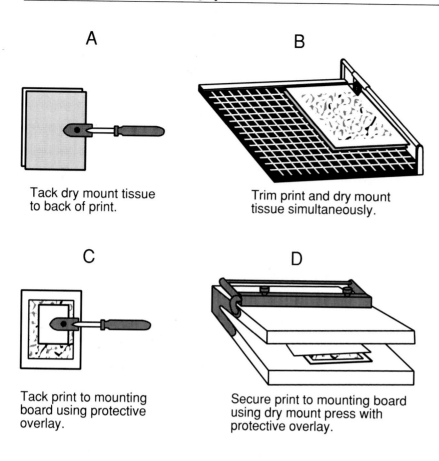

Tack dry mount tissue
to back of print.

B

Trim print and dry mount
tissue simultaneously.

C

Tack print to mounting
board using protective
overlay.

D

Secure print to mounting board
using dry mount press with
protective overlay.

Figure 8-13 The dry mounting process shown allows electron micrographs to be mounted on a stiff mounting board.

use a heated iron termed a "tacking iron." The tacking iron is heated and touched to a small area of the dry mount tissue lying on the back of the print.

CAUTION: The tacking iron gathers adhesive from the dry mount tissue and thus should never be touched directly to the face of the print!

To mount the micrograph on stiff mounting board, the micrograph should be positioned on the board and adhered to the board with the tacking iron. Use another piece of paper on top of the micrograph to keep the tacking iron from soiling the micrograph. Do not press the tacking iron too hard against the micrograph or it will indent the micrograph.

With the micrograph spot-fastened to the mounting board, a dry mount press should be used to seal all areas of the print to the mounting board. At this point, the mounted print should have a flat, smooth surface. Print mounting should be performed before labeling since some labels will come off under the heat of the dry mount press.

Labeling Prints

Virtually all journals require that prints be labeled with symbols or letters to indicate the structures of interest that are described in the legends to the fig-

ures. Most art or stationary stores sell cut-out or press-on symbols or lettering that can be pressed onto the micrograph. Labels should be placed carefully to indicate or point to the structure of interest and not interfere with the display of the structure itself. They should contrast sharply with the background of the micrograph so as to stand out to the viewer's eye. One may obtain black labels or white labels or in some instances labels that are black, but highlighted in white. Figure numbers also come in a variety of sizes and types.

Reproducing Prints

If one is sending a series of plates to an editor for review for a scientific journal, it is inconvenient to construct three or four sets of plates the number of copies often required by the journal. Original plates are of the highest quality, but the time and expense of making them must be weighed against the benefit of having original plates. In addition, plates with many figures may take many days to prepare. Most investigators send an original set of plates and a photographic copy of the original. The copy is made using a continuous tone negative, which is a film that allows all the tones of grey to be reproduced.

Although not as high quality as the original, these "reviewer's copies" are usually adequate for evaluation by the 2 or 3 peer reviewers.

Slide Presentations

The lengths of the short axis to the long axis of a 35 mm slide is $1'' \times 1\frac{1}{2}''$, respectively. Therefore, the information to be shown should be in proportion to the slide dimensions ($1 \times 1\frac{1}{2}$) to fill the entire viewing area of the slide. Slides of micrographs should be made using a film that cannot only pick up the blacks and whites of the micrograph, but the intermediate grey tones as well. There are few black and white slide films that have a continuous tone capability.

To prepare continuous tone slides for presentation, one must begin with a high quality print with the ratio of dimensions mentioned above. The print must be properly exposed and labeled as necessary, since this is not readily accomplished after the slide has been prepared. A convenient, continuous tone film is Kodak 2468 Direct Reversal Film, which is available in 100 foot rolls from some professional graphic arts suppliers, as well as 36-exposure rolls from some EM supply houses. A high resolution, flat-field macro lens and a stable copy stand with four 150 W flood lamps is also recommended.

The electron micrograph is positioned on the copy stand, the camera critically focused, and an exposure is made (try f/8 for 38–40 sec). It is important to fill as much of the viewing field as possible to avoid wasting space on the slide. If the proportions of the $8'' \times 10''$ print do not match the 35 mm film format, some in-camera cropping is necessary. The film, which is safe under a red 1A filter, is developed in a 1:1 dilution of Kodak Dektol for 5 minutes at 75° F, rinsed in a water or acetic acid stop bath for 1 minute and fixed for 5 minutes in Kodak Rapid Fixer. The exposed film should be thoroughly washed, rinsed in a wetting agent such as Kodak Photo Flo, and dried in a dust-free environment. For highest quality work, one should place the transparencies between glass mounts to assure a flat field, since unsupported transparencies may warp enough so that portions of the image may appear to be out of focus during projection.

Poster Presentations

Scientific meeting organizers tend more and more toward scheduling poster presentations. Therefore, micrographs must be enlarged considerably to be viewed from a minimum distance of 4 feet. Usually, they are dried matte to minimize light reflections from the room. The mounting procedure for poster micrographs is the same as previously described for publication micrographs. Posters are organized as a logical "story" of the research beginning with an introduction, then materials and methods, then results (micrographs and charts), followed by interpretations and a conclusion statement. To be effective, the poster must be simple, pleasing to the eye, and be easily followed by an individual inexperienced in the field.

Making Montages

A montage is made from several micrographs in which the edges of each photograph overlap (Figure 8-14). Montages are effective ways of portraying large areas of the specimen at a relatively high magnification. The overlapping regions of electron micrographs blend together when they are appropriately trimmed. Although montages should be simple to make, they rarely turn out to be perfect. Sometimes the negatives do not match perfectly due to a variety of technical reasons relating to the microscope (for example, improper centering or defocusing of the condenser lens) or due to optical distortions in the lenses (Chapter 6) and to tissue changes under the electron beam (for example, changes in the density of the section due to differential exposure under the electron beam). In addition, prints may not be enlarged exactly the same or may dry to slightly differing sizes. Exposures for adjacent micrographs are often uneven, contributing to unsightly regions of the montage where two micrographs meet. The juncture of two micrographs may be made less noticeable by overlapping adjacent micrographs and cutting both micrographs simultaneously with pinking shears and then matching and taping each surface from the nonemulsion side of the photograph. In spite of all of the problems in making a montage, they prove useful when large areas of tissue must be analyzed. Crang (1987) has described a good method of making montages.

Determining Print Magnification from a Negative

Determining the enlargement of a print and its final magnification are straightforward procedures. Photographic enlargement is a simple ratio of a distance between two objects on the negative and the distance

Figure 8-14 An example of a montage in which several overlapping electron micrographs were spliced to display an extended view of the tissue at high magnification.

between the same two objects on the print. (To be as accurate as possible, measure objects that are relatively far apart on the negative.) As an example, if objects that are 15.0 mm on the negative are printed 45.0 mm apart on the micrograph, the print is magnified three times the negative image. The magnification of the micrograph is calculated by multiplying the magnification of the negative times the print enlargement. The following formula expresses how to determine magnification given a knowledge of the negative magnification.

$$\text{Print magnification} = \text{Negative magnification}$$
$$\times \frac{\text{distance between two points on the negative}}{\text{distance between two points on the enlarged micrograph}}$$

References

Crang, R. E. 1987. Montaging electron micrographs. *J Electron Microsc Tech* 7:53–60.

Electron microscopy and photography. 1973. Kodak Data Book No. P-236. (Rochester, N.Y.)

Farnell, G. C., and R. B. Flint. 1975. Photographic aspects of electron microscopy. In *Principles and techniques of electron microscopy. Biological applications*, Vol. 5. M. A. Hayat, ed. New York: Van Nostrand Reinhold Co. pp. 19–61.

Horenstein, H. 1983. *Black & white photography. A basic manual.* 2d ed. Boston: Little Brown and Co.

Postek, M.T., K. S. Howard, A. Johnson, and K. L. McMichael. 1980. *Scanning electron microscopy: a student's handbook.* Ladd Research Industries, Inc. (Burlington, VT) 305 pp.

Shipman, C. 1974. *Understanding photography.* Tucson: H. P. Books.

Immunocytochemistry

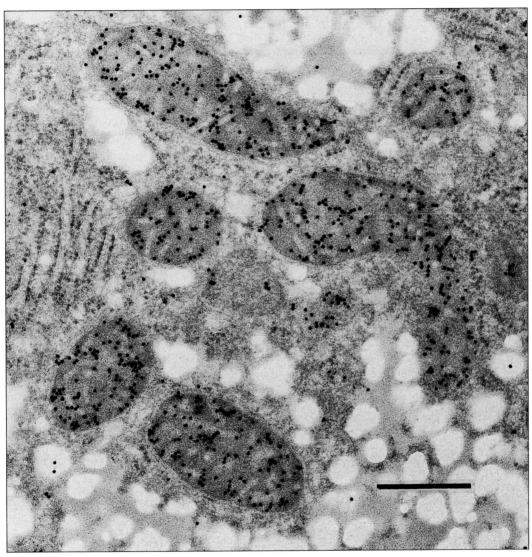

Courtesy of M. Bendayan.

The process for detecting antigens using antibodies directed specifically against them is called *immunochemistry*. When the electron microscope is used to detect the localization, the technique is usually referred to as *immunocytochemistry*, although sometimes immunocytochemistry is said to also be performed at the light microscope level. The molecular recognition properties of the antibody for the antigen in the *antigen-antibody reaction* give specificity to the localization process. An antigen is localized by one or more antibodies that are applied to the cells. A *tag* recognizable under the light or electron microscope is attached directly or indirectly to the last of the added antibodies. In theory, a single molecule of a particular antigen can be localized by a tagged antibody or series of antibodies.

Immunocytochemistry has made valuable contributions to our knowledge of the localization of a variety of substances. It is the most powerful and potentially specific of all of the localization techniques described in this text. Immunocytochemistry has permitted the localization of substances where the only major requirement for localization is that a specific antibody be developed against the antigen. Of the various localization techniques currently available for electron microscopic investigations (see Chapters 10 through 12), immunocytochemistry is not only the most widely used, but holds the most promise for the future.

Immunocytochemistry was a natural outgrowth of *immunohistochemistry*, a term used to describe immune localization obtained primarily at the light microscope level. Most of the dyes and fluorescent compounds used to visualize the immune complexes at the light microscope level are not appropriate tags for use with the electron microscope since they are not electron-dense nor do they produce an electron dense product.

In 1959, Singer used an antibody tagged with the electron-dense protein ferritin to obtain ultrastructural localization. The tag, which was electron dense due to the presence of iron in the protein, was applied prior to embedding and came to be known as a *preembedding stain*. The first *postembedding stain* applied to a tissue section was uranium (Sternberger et al., 1965). The development, by Graham and Karnovsky (1966), of cytochemical methods to localize enzyme (peroxidase) activity allowed Nakane and Pierce (1967) and others to label an antibody with an enzyme and to use the *enzyme-labeled tag* for electron microscopy. Subsequently, the peroxidase-antiperoxidase (PAP) technique was developed by Sternberger et al. (1970) to improve the sensitivity and resolution of the enzyme method. Ultrastructural tags of more recent vintage include the *avidin-biotin system* (Heitzmann and Richards, 1974) and the currently very popular *colloidal gold technique* (Faulk and Taylor, 1971).

The Antigen-Antibody Reaction

When higher organisms (most vertebrates) encounter a foreign molecule or *antigen* repeatedly, they will develop *antibodies*, a process termed immunogenesis. For instance, antibodies may be formed against antigens, such as might be on red blood cells from a different organism. Complex molecules under a molecular weight of about 5,000 are usually not immunogenic, but may be made so if conjugated, or chemically coupled, to a larger molecule. Proteins and complex branched chains of sugars are very immunogenic, whereas lipids, nucleic acids, and less complex molecules are poorly immunogenic. A *determinant* is the unique portion of the antigen, either molecular or conformational or both, which is recognized by the antibody.

Antibodies may be harvested from animals that have become immunogenic in the natural course of events to some environmental stimulus or disease. Humans with certain autoimmune diseases have a high level of antibodies to some of their own tissue components. For example, actin (a filamentous protein) antibodies are often present in the sera of patients with certain forms of hepatitis.

An *adjuvant* is a substance or treatment that may be used to augment the immune response. Antibodies may also be produced artificially by repeated injection of antigens with, or without, accompanying adjuvant. Repeated injection is, by far, the most common source of antibodies for use in immunocytochemistry. Animals that are commonly used to produce antibodies for immunocytochemistry are the rabbit, guinea pig, mouse, and goat. The response to antigens, with or without adjuvant, in these species is the production of *polyclonal antisera*, which is always a mixture of antibodies against different sites (determinants) on the same antigen. Since polyclonal antibodies bind to an antigen at multiple determinants, several antibodies may bind a single antigen molecule. *Monoclonal antibodies* are produced first by animal immunization and secondly by *in vitro* fusion of an individual antibody secreting cell from the immunized animal with specific tumor cells. The cell that is formed is called a *hybridoma cell*. It not only secretes clones of antibodies binding to a single determinant of the antigen, but divides in culture and may be frozen and thawed so that a continuous source of monoclonal antibody is available. The advantage of using monoclonal antibodies is that they are highly specific. The disadvantages of using monoclonal antibodies for immunocytochemistry are that they are often of low affinity and they bind at one site only, a site that may be altered during tissue preparation or may be hidden and not accessible to the antibody.

Antibodies, while generally thought of as being specific for the molecule that elicited their formation, may *cross-react* with other molecules having the same or similar determinant. Cross reactions may be with related antigens in another species or molecules that have little functional or structural similarity to the antigen except at the binding antibody site.

The antibodies present in a subfraction of the serum (noncellular portion of the blood) of mammals are collectively called *immunoglobulins*. Structural subclassifications of immunoglobulins, abbreviated IgG, IgA, IgM, IgD, and IgE, have a variety of immune and nonimmune functions. We will focus on the structure of the most common of these, the *IgG* molecule, since other classes of immunoglobulins have many structural similarities to IgG. IgG is the major immunoglobulin in serum of most mammals and the most common immunological probe for immunocytochemistry.

Figure 9-1 depicts the general structure of an IgG molecule, having a molecular weight of about 150,000. Of its four polypeptide chains, two are *heavy chains* (MW 50,000) and two are *light chains* (MW 25,000). The identical heavy chains are linked to each other by disulfide bridges, and the two light chains are also attached to each heavy chain by disulfide bridges. The molecule has the general appearance of a *Y*, which is usually depicted diagrammatically as an inverted *Y*. The two short limbs of the Y, composed of both light and heavy chains, may bend somewhat at the hinge region of the heavy chain (shown in Figure 9-1 as a coil). Two identical *antigen binding sites* are found at the ends of two short limbs of the Y on the light and heavy chains, whereas the stem of the Y serves, among other functions, to notify the immune system that it has recognized a foreign substance that needs to be eliminated. Enzymatic cleavage with papain severs the

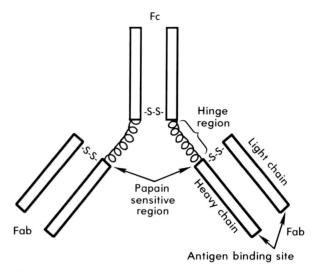

Figure 9-1 Structure of the IgG molecule (Fab, antigen binding fragment; Fc, crystallizable fragment; S-S, disulfide bonds).

heavy chains, yielding three fragments (Figure 9-2): two *Fab* (antibody binding) fragments and a single *Fc* (crystallizable) fragment. Fab fragments may be used in place of IgG in immunocytochemical procedures. Subsequent figures will depict the IgG molecule in the simple Y configuration.

It is important for our purposes to realize that *antibodies themselves, being complex protein molecules, are immunogenic*. For example, a goat repeatedly injected with mouse IgG will form antibodies to this immunoglobulin fraction. The antibody produced is called goat anti-mouse IgG (Figure 9-3). This terminology for naming antibodies, by first specifying the species in which the antibody was produced and, second, specifying the species against which it was directed, will be employed throughout this chapter. Although technically incomplete, in some instances the name of the species in which the antibody was produced is omitted.

Approaches to Labeling

The general approach to immunocytochemistry is to attach electron or light microscopically detectable *tags* to antibody molecules. The tagged antibodies are then exposed to the tissue antigen of interest to which they specifically attach. There are several ways in which this may be accomplished.

The *direct method* for antibody labeling is straightforward, in principle. An antibody is first prepared against the antigen. This antibody is conjugated to a tag, visible (or later made visible) in the

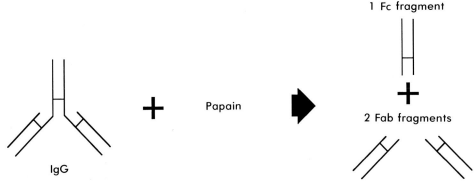

Figure 9-2   Cleavage products of IgG after treatments with papain.

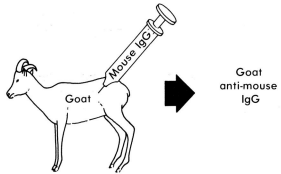

Figure 9-3   Goat anti-mouse IgG is produced after injection of a goat with mouse IgG. Antibodies produced in this manner are specifically directed against other antibodies (proteins), and thus are commonly used as secondary antibodies in immunocytochemical reactions.

electron microscope. The *primary antibody*, as it is called, is then reacted with the tissue antigen (Figure 9-4).

By far, the most common strategy for antigen localization is the *indirect method* (Figure 9-5). The tissue antigen is first exposed to a primary antibody that will attach to it. After washing away unbound antibody, a *secondary antibody*, prepared in a different species and directed against the immunoglobulins of the class of the primary antibody (e.g., goat anti-mouse IgG), is allowed to bind. Usually the second antibody contains the tag that may be visualized by electron microscopy, although tertiary, and so forth, antibodies that contain the tag may be added.

The indirect method is the most commonly used procedure for several reasons. The primary antibody is usually difficult to obtain, since frequently only small amounts of purified antigen may be available to elicit its formation. Conjugation of any primary antibody to a tag is often difficult and inefficient; consequently it is often a "hit or miss" procedure that may exhaust the precious little primary antibody available. Addition of tags may also severely interfere with the binding of the primary antibody to the antigen. Secondary antibodies, on the other hand, are usually plentiful since they are easily obtained in bulk from commercial sources. The ultrastructural tagging is, therefore, not as risky if plenty of secondary antibody is available. Furthermore, localization is amplified by the number of secondary antibodies containing tags, which can bind to a single antigen (the primary antibody) due to the polyclonal nature of most antisera (Figure 9-6). Thus several secondary antibodies may bind to a single primary antibody.

Protein A, produced by certain strains of *Staphylococcus aureus* bacteria, or related proteins such as

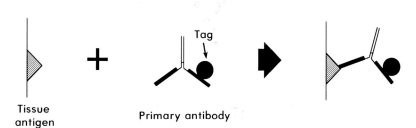

Tissue antigen   **Primary antibody**

Figure 9-4   The *direct method* of antibody labeling. A tissue antigen is exposed to a primary antibody that has been previously conjugated to a tag.

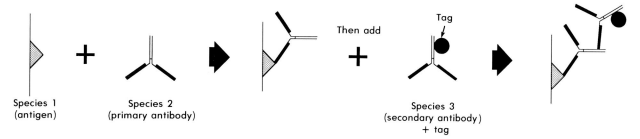

Species 1
(antigen)

Species 2
(primary antibody)

Then add

Tag

Species 3
(secondary antibody)
+ tag

Figure 9-5 The *indirect method* for antibody labeling. A tissue antigen (from species 1) is exposed to a primary antibody (generally made in a second species) that has been made to bind the antigen. After binding of the primary antibody, a tagged secondary antibody (made in a third species)

is exposed to the bound antigen-antibody complex. The secondary antibody was produced to react against all IgGs of the second species (Figure 9-4) and contains a tag. The result is a two-layered antibody sequence with an attached tag.

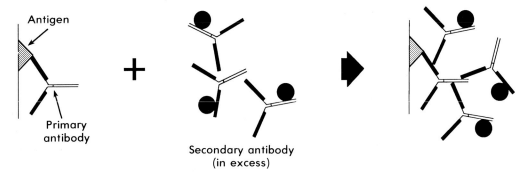

Antigen

Primary
antibody

Secondary antibody
(in excess)

Figure 9-6 Amplification of localization using polyclonal antisera. Since the tagged secondary antibodies can react with multiple determinants on the primary antibody, the response

is greatly amplified by the increased binding of secondary antibody to primary antibody and the presence of multiple tags.

protein G from a streptococcal organism, have the interesting ability to bind to the Fc portion of some immunoglobulins (notably IgG) from several species. A variety of ultrastructural tags may be conjugated to protein A. Sequential addition of primary antibody and protein A-tag represents a major ultrastructural form of localization, albeit based only partially on immunologic principles (Figure 9-7). The affinities of protein A and G for various classes of immunoglobulins are given in Table 9-1.

The *hybrid antibody method* employs a primary antibody that has been artificially constituted from Fab fragments of two separate antibodies: one directed against the antigen and one directed against

a tag. First the hybrid antibody is exposed to the tissue followed by the tag (Figure 9-8). Due to technical difficulties and loss of reactivity during production of the hybrid antibody, this technique is used only occasionally.

Ultrastructural Tags

Under the proper conditions, immunoglobulins and immunoglobulin fragments are directly visible under the electron microscope (Figure 9-9), but their use for most localization purposes requires an attached electron-dense substance. Labels, stains, or tags, as

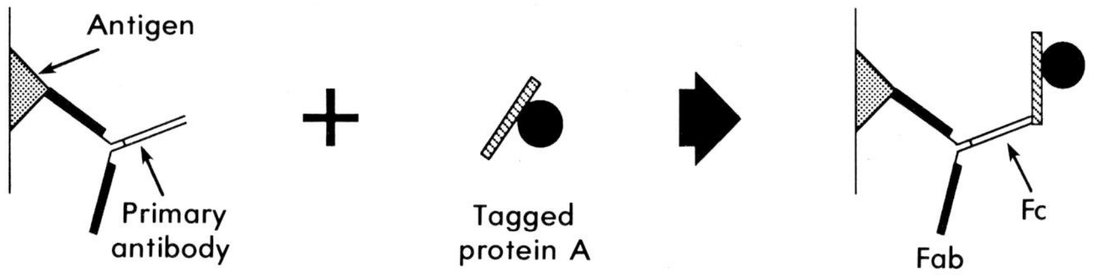

Antigen

Primary
antibody

Tagged
protein A

Fc

Fab

Figure 9-7 Localization using *protein A*. Tagged protein A is added to the antigen-antibody complex. Protein A binds to the Fc portion of the IgG molecule.

Table 9-1   The Relative Affinities of Protein A and G for Various Antibody Subclasses in Various Species

Antibody	Affinity for protein A	Affinity for protein G
Human IgG$_1$	+ + + +	+ + + +
Human IgG$_2$	+ + + +	+ + + +
Human IgG$_3$	−	+ + + +
Human IgG$_4$	+ + + +	+ + + +
Mouse IgG$_1$	+	+ + + +
Mouse IgG$_{2a}$	+ + + +	+ + + +
Mouse IgG$_{2b}$	+ + +	+ + +
Mouse IgG$_3$	+ +	+ + +
Rat IgG$_1$	−	+
Rat IgG$_{2a}$	−	+ + + +
Rat IgG$_{2b}$	−	+ +
Rat IgG$_{2c}$	+	+ +

Summarized data taken from Harlow and Lane (1988).

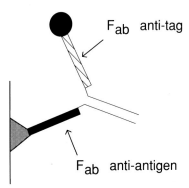

F$_{ab}$ anti-tag

F$_{ab}$ anti-antigen

Figure 9-8   Hybrid antibody binding to antigen. The IgG molecule shown here bound to the antigen was first reconstituted from portions of two different antibodies resulting in one Fab region directed against the antigen and one Fab region directed against the tag.

they are frequently called, allow the site of antigen localization to be readily visualized. The types of tags fall into three major categories, based on the nature of the tag. Some tags are organic molecules that possess *structured electron opacity*, others are *enzymes* whose reaction product can be detected after the addition of the substrate, and yet others are *heavy metals* that can be visualized directly.

Some of the first tags discovered were those that possessed structured electron opacity. *Ferritin*, the principal storage protein for iron in mammals, is obtained primarily from horse spleen. The ferritin molecule is about 10 to 12 nm in diameter and has a molecular weight of about 450,000. The core of

the molecule is rich in iron, which is responsible for its electron opacity (Figure 9-10). Several methods are available for conjugation of ferritin to immunoglobulins (Williams, 1977), although most ferritin-secondary antibody conjugates are usually procured commercially. Ferritin was used extensively in the early phase of immunocytochemistry; however, the development of better tags have made its recent use less popular. An example of ferritin labeling on a cell surface is shown in Figure 9-11. Other tags in the structured electron opacity category with ferritin are *hemocyanin* and *viruses*. The latter two tags are more often used in surface localization studies using the scanning electron microscope since they are quite large.

An enzyme tag is commonly attached to an immunoglobulin with the intent of using the catalytic properties of the enzyme to produce an insoluble reaction product visible in the electron microscope. *Horseradish peroxidase* (MW 40,000) is used almost exclusively in this category since its insoluble reaction product can be made electron dense with the addition of osmium. In the presence of the hydrogen peroxidase substrate (H_2O_2), added 3-3′diaminobenzidine molecules serve as an electron donor to the reaction. The oxidation products of diaminobenzidine form an insoluble precipitate that appears colored red/brown at the light microscope level. For visualization of the precipitate at the electron microscope level, it is necessary to render it electron dense by chelation with osmium tetroxide where it appears as an extemely dense precipitate (Figure 9-12). Thus, the reaction can be visualized at both the light and electron microscope levels. The immunocytochemical reaction employing horseradish peroxidase is summarized in Figure 9-13.

An advantage of the enzyme method is that its reaction product is amplified by being continually deposited at the reaction site. For the same reason, the localization may be less precise than desired in relation to the antigen, since there is an indeterminate amount of reaction product present. A modification of the enzyme method described above achieved popularity in the 1970s. The *PAP* (peroxidase-anti-peroxidase; MW = 420,000) *technique*, as it is called, uses a soluble complex of peroxidase and anti-peroxidase (Figure 9-14), which is available commercially. The PAP complex is essentially three enzyme molecules linked by two antibody molecules directed against peroxidase.

The PAP protocol, as shown in Figure 9-14, first calls for the binding of a primary antibody that

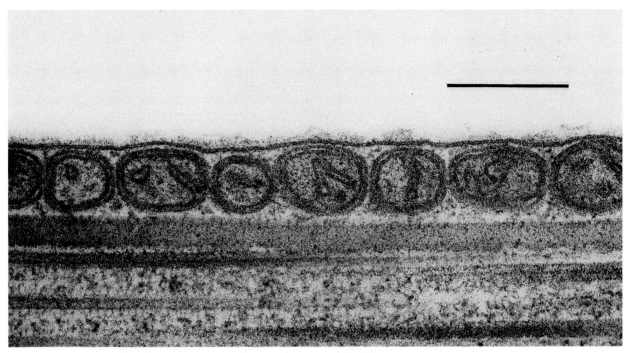

Figure 9-9 Micrographs showing IgG directly labeling a plasma membrane of a sperm tail. IgG appears as a fuzzy material seen on the plasma membrane. Bar = 0.25 μm.

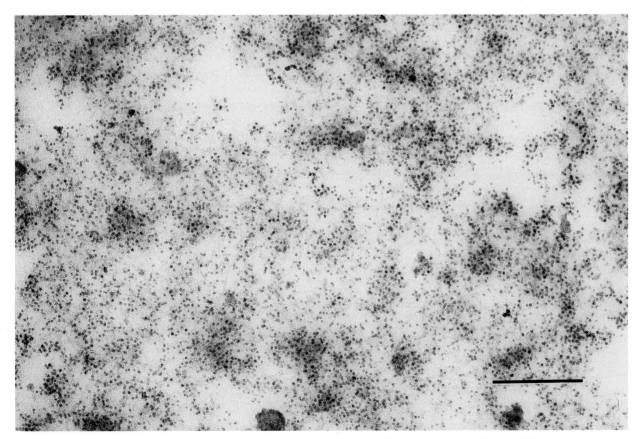

Figure 9-10 *Ferritin* molecules. Each ferritin molecule measures about 10 nm in diameter. Bar = 0.25 μm.

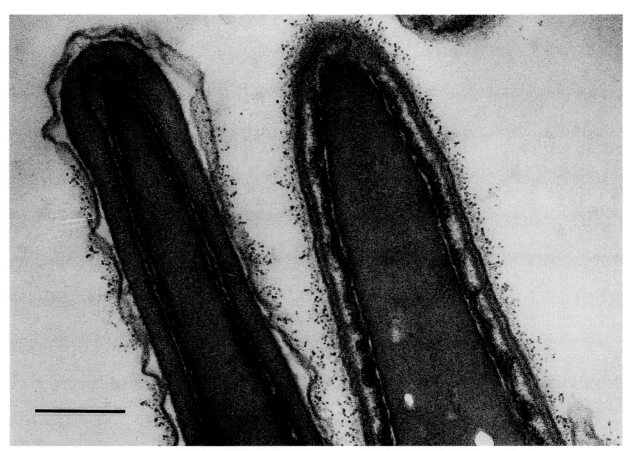

Figure 9-11   Labeling of the head region of two sperm with ferritin. An indirect technique was used to label cell surface antigens. Bar = 0.25 μm.

has been made to react with the antigen. Next, one adds a secondary antibody, from another species, which is directed against immunoglobulins of the class of the primary antibody. The PAP complex is then added. The PAP complex binds to the secondary antibody because the anti-peroxidase molecules were prepared in the same species as the primary antibody, but they exhibit different determinants than the primary antibody does. Thus, the secondary antibody, being directed against all immunoglobulins of the first species, binds both the primary antibody and the PAP complex. With the addition of diaminobenzidine and later osmium, a characteristic electron-dense reaction product is visible under the electron microscope (Figure 9-15; for examples see Sternberger, 1979, and Moriarity and Garner, 1977). Gold or nickel grids must be used in this procedure to avoid reaction of copper grid bars with osmium. The PAP technique and the other aforementioned tags are now rarely used at the ultrastructural level and have been replaced by the more popular colloidal gold tags.

Although several heavy metals have been used occasionally as immune tags, the most common tag currently used is *colloidal gold*. Colloidal suspensions of gold may be prepared easily and readily tagged to immunoglobulins or protein-A (Williams, 1977). Of particular note is that these discrete, highly electron-dense particles can be made in sizes from approximately 3 nm in diameter and upwards. Gold particles are easily detected on tissue sections because of their high electron scattering properties. Furthermore, they can be made small enough to allow clear visualization of surrounding tissue structure (Figure 9-16 and 9-17). Antigen binding sites are more easily quantifiable using colloidal gold tags than with other techniques because of the discrete properties of this tag. Many colloidal gold-immunoglobulin conjugates are available commercially, as is protein A-gold. Relatively simple methods also are available for the adsorption of immunoglobulins onto the surface of colloidal gold. Caution is in order since colloidal gold is not irreversibly bound to an antibody and may be released from its antibody by a change in pH.

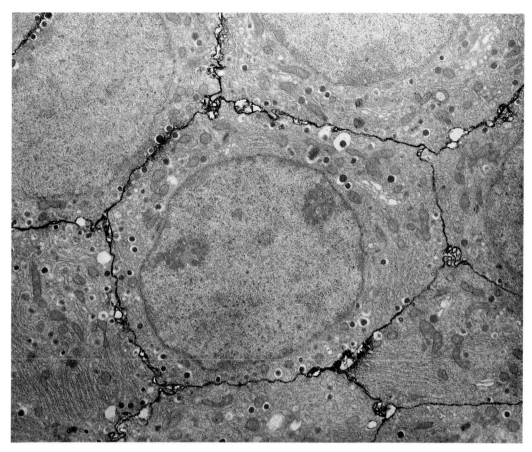

Figure 9-12 *Horseradish peroxidase* used in an indirect localization protocol for a plasma membrane protein in islet cells of the pancreas. The tissue was not stained with lead or uranium, thus rendering the localization more prominent. (Micrograph courtesy of O. K. Langley.)

The ease with which colloidal gold tags may be made or obtained commercially and their stability, when stored according to recommendations, also contribute to their being the most popular immune localization tag. Several current references employing colloidal gold are provided at the end of this chapter.

General Considerations in Performing an Immunocytochemical Experiment

Rarely are any two experiments employing immunocytochemistry alike. There are many factors that govern the course and specific details of an experiment. Unfortunately, trial and error play a large role in the eventual outcome of an experiment, although the methodology section of published papers may not indicate all the "trials and tribulations" that the investigator has experienced to produce the final re-

sult. Some of the factors to consider in an experiment follow.

Immunohistochemistry/Immunofluorescence

A sound knowledge of light microscope immunohistochemistry serves as an important introduction to immunocytochemistry. As with most localization procedures, it is often best to try experiments at the *light microscope level* initially. In some cases, the tag may be the same as used at the electron microscope level, but most of the time FITC (fluorescene isothiocyanate) is employed for immunofluorescence. Nevertheless, a rough idea of antigen localization can be achieved readily (in about 4 hours) using light microscopy. Controls may be employed at the light microscope level to establish the validity of the localization with generally similar methods as proposed herein for electron microscopy. Many of the conditions (fixation, etc.) that optimize localization may be determined with light microscopy. Several

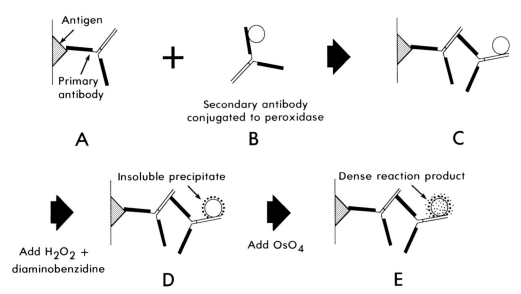

Figure 9-13 Steps in indirect labeling with peroxidase: (A) The antigen is attached to the primary antibody; (B) a secondary antibody from another species, directed against the IgG class of the primary antibody, is labeled with horse-radish peroxidase and (C) allowed to bind. The addition of hydrogen peroxide and 3-3′diaminobenzidine (D) causes an insoluble precipitate to form (E), which appears dense under the electron microscope after the addition of osmium.

tags that are used for electron microscopy can also be visualized by light microscopy (e.g., colloidal gold).

Obtaining and Applying the Primary Antibody

It is important to obtain a primary antibody directed against the specific antigen to be localized. Although the antigen may be named similarly in several species, that does not imply that it is identical in these species. An antibody may or may not cross-react with antigens from one species to the next when there are slight molecular or conformational dissimilarities in the antigen.

Many primary antibodies are available commercially. They are usually expensive and provided in small quantities. Although commercial sources may indicate that the antibody meets certain specifications, it is important to verify this independently. Many investigators are generous in supplying high quality antibodies to other investigators at no cost. These are often of better quality than those commercially obtained.

Preparing one's own antibody from scratch may be a necessity, especially if one has more than a casual interest in localizing the antigen. This is time consuming and often costly. Polyclonal antibodies are made by immunizing a species, other than the one in which localization will be attempted, with either pure antigen or a mixture of antigens. This immunogen is the antigen that is to be localized. An adjuvant is frequently used to augment the immune response. The mouse, rabbit, and goat are commonly used species for producing antibodies. The purer the antigen injected, the more specific the response. Several booster shots may be required to increase the *titer* (concentration) of the antibody in the antisera. Because some animals may not respond at all, several animals must be injected to find one that is a good responder or immunogenic. It is important to use standard immunological procedures to determine the presence and specificity of the antibody (Harlow and Lane, 1988).

After obtaining blood from the animal, the serum fraction should be isolated. Serum may be used for localization purposes, or it may be further purified into an immunoglobulin fraction or even to IgG or IgG directed only against the antigen, for example. Purification of immunoglobulins is by standard immunological techniques (Harlow and Lane, 1988). The primary antibody is then diluted serially (e.g., 1:100; 1:200; 1:400; 1:800) for use in immunocytochemistry. If the postembedding technique (see below) is used, the primary antibody is often applied over a long period, from a few hours up to a day. This increases the possibility that all available antigenic sites will be bound by the primary antibody.

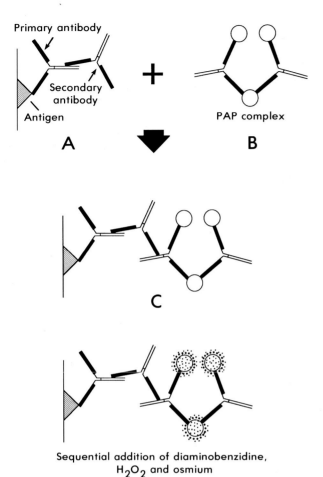

Primary antibody

Secondary antibody

Antigen

A

PAP complex

B

C

Sequential addition of diaminobenzidine, H_2O_2 and osmium

D

Figure 9-14 Steps in the peroxidase-antiperoxidase (PAP) technique: (A) The antigen has been sequentially exposed to the primary and secondary antibodies; (B) the PAP complex is added to the complex after excess primary and secondary antibodies have been washed away (C). The PAP complex reacts with the secondary antibody (since the PAP complex was produced in the same species as the primary antibody); (D) a dense reaction product is formed after osmium impregnation of the insoluble diaminobenzidine precipitate (see Figure 9-13).

Obtaining the Secondary Antibody

Unlike the primary antibody, the secondary antibody is usually available commercially in large amounts and at reasonable cost. It is relatively simple to produce large amounts of a secondary antibody against IgG from one species (mouse, for example) in the goat, an animal from which large amounts of serum can be obtained. The secondary antibody must be chosen to react against both the immunoglobulins of the species (mouse) and the class (e.g., IgG or IgM, etc.) of the primary antibody employed. The secondary antibody is usually only briefly exposed to the tissue (1–2 hours).

Tissue Fixation

The overall goal of tissue fixation for immunocytochemistry is to preserve the tissue as close to its natural state as possible, while at the same time maintaining antigenicity. This is often much easier said than done, since the antigenic determinant is easily altered (denatured) or masked during fixation. As it turns out, for most antigens, glutaraldehyde must often be used in concentrations less than 1% since it is a strong denaturing agent. Osmium is rarely used prior to carrying out the labeling reaction. It is clear that antigens do not respond uniformly to fixation processes, necessitating experimentation with a variety of fixatives and fixative combinations. Formaldehyde, in concentrations as high as 4%, has been used in combination with low percentages (e.g., 0.5%) of glutaraldehyde to give good results for localization of many tissue antigens. Tissue contrast in postembedding procedures where osmium is not used is, not surprisingly, very low.

Preembedding or Postembedding Labeling

A major decision to be made is whether to perform the immunocytochemical applications prior to or after the embedding procedure (see Table 9-2). There are distinct advantages and disadvantages to both types of procedures.

Preembedding labeling or "staining," as it is sometimes erroneously called, is advantageous for *surface labeling* of live cells or cell fractions, although cell interior labeling is also conducted. Caution should be exercised because the labeling procedure itself may cause redistribution of the antigen on the cell surface. Mild fixation with 0.2% glutaraldehyde, for example, may stabilize antigens somewhat prior to labeling. Preembedding labeling is useful when the label has easy access to the antigen.

Since antibodies do not penetrate cell membranes, live cells may be broken open with *detergents* to reveal antigenic determinants within the interior of a live cell. Of course, ultrastructural preservation is severely compromised under such conditions. Sections, about 50 μm in thickness, made from solid tissue with a special vibrating or chopping microtome knife can be used to expose some cells to immunolabels. It is also possible to use *frozen ultrathin sections* (see Chapter 4) for immunocytochemistry. Fresh frozen sections are placed on grids and immunolabeled. Sections can then be fixed and stained.

The main advantage of preembedding labeling is that antigenicity is maintained during the labeling

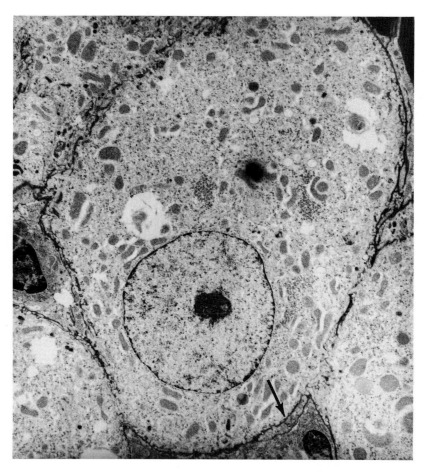

Figure 9-15 Use of the PAP technique to localize a surface antigen. Dense staining is seen along the cell surface (arrow). (Micrograph courtesy of A. Mayerhofer.)

procedure. Tissue contrast is enhanced since the tissue may be osmicated, having already been labeled. Labeling with ferritin, colloidal gold, and peroxidase have been commonly used with this technique. An example of preembedding surface labeling is provided (Russell et al. 1982).

In the postembedding technique, the labeling procedure is carried out directly on tissue sections from embedded specimens. When using the postembedding technique, it is always presumed that antigenicity has been partially compromised by the fixation protocol. Tissue sections may be *etched* (partially eaten away) with materials such as hydrogen peroxide to expose hidden antigenic sites. The colloidal gold and PAP technique are especially suited for postembedding labeling. Tissue section contrast may be enhanced somewhat if osmication is performed on the thin section after labeling.

Embedding

In postembedding labeling procedures, the embedding process and the embedding medium itself may influence the antibody binding. Dehydration, infil-

tration, and embedding steps must be carried out under conditions that optimally preserve antigenicity. In postembedding procedures, embedding media (LR White, methylacrylate, JB-4, Lowcryl) with special characteristics (e.g., low temperature embedding) that will facilitate the preservation of antigenicity can be purchased.

Blocking

At several steps during a typical protocol, the tissue is exposed to normal, or nonimmune serum (i.e., serum from animals that have not been exposed to the antigen), or albumin for the purpose of *blocking* sites that may react nonspecifically with any antibody. Normal or nonimmune serum and albumin proteins are used to react with, for example, any residual nonreacted aldehyde sites. If this step had not been performed, these sites may have reacted with one of the antibodies and given false results. (Since aldehydes such as glutaraldehyde are bifunctional agents [see Chapter 2], commonly reacting at

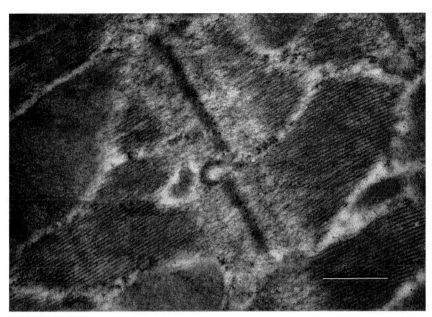

Figure 9-16 Localization using colloidal gold. Localization of alpha-actinin was achieved using an indirect technique. Gold particles are seen primarily on the Z-band of skeletal muscle. A few gold particles are scattered throughout the tissue and represent nonspecific localization. The tissue lacks contrast because osmium was not utilized in the fixation process. Bar = 0.25 μm.

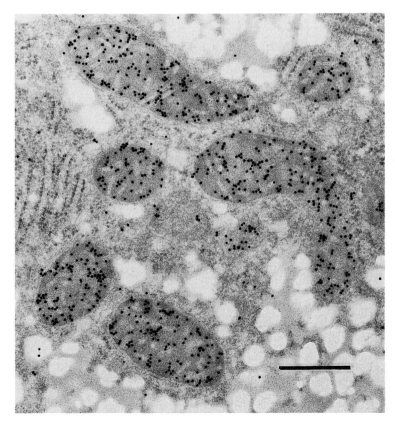

Figure 9-17 Localization using colloidal gold. Localization of carbamyl phosphate synthetase in rat liver tissue fixed with 1% glutaraldehyde and embedded in Lowicryl K4M. The protein A gold tag was employed. Gold particles are seen within mitochondria. Bar = 0.5 μm. (Micrograph courtesy of M. Bendayan.)

Table 9-2 Two Major Types of Labeling Protocols

Steps in a Postembedding Protocol	Steps in a Preembedding Protocol
1. Primary fixation	1. Mild primary fixation, blocking and application of primary antibody
2. Dehydration, infiltration, sectioning, and embedding	*or*
3. Etching of embedding media (optional)	1. Application of primary antibody to live cells
4. Blocking	2. Washing
5. Application of primary antibody	3. Blocking
6. Washing	4. Application of secondary antibody with tag
7. Blocking	5. Washing
8. Application of secondary antibody conjugated to a tag	6. Fixation (including osmium)
9. Washing	7. Dehydration, infiltration, embedding, sectioning
10. Postfixation with osmium	8. Staining with lead and uranyl acetate
11. Washing	
12. Staining with lead and uranyl acetate	

one end, but remaining unreacted at the other end, they may bind IgG molecules nonspecifically.)

Staining

Lead and uranium staining are usually employed at the end of the protocol to enhance tissue contrast. In some instances, staining is omitted if the tissue contrast obscures the electron-dense tag employed.

Controls

Immunocytochemical experiments are virtually worthless without a battery of controls that should be conducted simultaneously with the localization attempt. There is a tendency for the novice to believe, initially, that any aggregation of label represents specific localization. Unfortunately, the controls will often reveal considerable localization artifact that might lead one to an erroneous conclusion if only experimentals (showing the same localization artifact) were examined. Localization "veterans" tend to be skeptical about their results until all the controls have verified that the locali-

zation seen is real. Several types of controls follow. Not all of the controls must be performed in every experiment.

Adsorption
The primary antibody is reacted with an excess of antigen, a process termed adsorption. After elimination of the antigen-antibody complex, the remaining solution, free of antibody (it is hoped), is exposed to the tissue. This control, although not always feasible due to sparse antigen supply, will show that the antigen and not some other substance is responsible for the localization seen. It is also possible to adsorb the primary antibody with homogenized tissue that contains the antigen. Essentially, the tissue antigen, which is in excess of the antibody, precipitates the antibody leaving the solution free of antibody. The adsorbed solution should not react with the antigen in localization protocols.

Use of Tag or Unlabeled Antibody
The tag and unlabeled antibody are used separately in place of the conjugated antibody. This establishes that the specific properties of the labeled antibody are responsible for the localization and the antigen or the tag alone are not responsible for the localization.

Omission of Primary or Secondary Antibodies
In theory, the labeling should *not* be successful if either the primary or subsequent antibodies are deleted. If labeling is seen under these conditions, then it is not the labeling intended as the result of the sequence of antibodies exposed to the antigen.

Use of Pre-Immune Sera
Collection of sera from animals prior to production of the primary antisera (so-called pre-immune sera) and its use in place of the primary antiserum or antibody will show if some component in the immune sera, other than the specific IgG, is responsible for antibody binding and the localization pattern recorded. If pre-immune sera cannot be obtained, then it is possible to employ normal serum from non-immunized animals of the same species.

Dealing with Soluble Antigens

Soluble molecules are those freely capable of movement within the cell. For example, steroid hormones or soluble proteins are free to move about the cell within its aqueous environment. The processes of

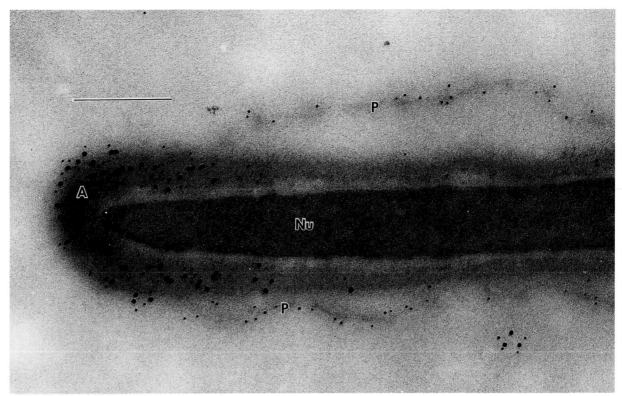

Figure 9-18 Localization of two proteins using gold tags of 5 nm and 10 nm diameter. Note the two antigens are present in two different sites. The larger particles localize an enzyme (proacrosin) in a boar sperm cell acrosome (A) while the smaller particles localize a protein (AP_Z) found in the cytoplasmic membrane (P) as well as the acrosome. Bar = 0.5 μm. (Courtesy of R. N. Peterson and K. Polakoski.)

fixation, dehydration, etc., allow the further movement and/or diffusion of molecules away from their original site; thus, these processes are detrimental to localization of soluble molecules. *Ultrathin frozen sections* (see Chapter 4), *freeze-substitution*, and special embedding media have been used occasionally to localize soluble antigens.

Multiple Labeling Option

Colloidal gold markers can be purchased in uniform sizes. It is easy to distinguish 5 nm gold particles from 10 nm gold, or the latter from 20 nm gold. Thus one may localize two (or more) different antigens on a tissue section, one with a small gold tag and one with a larger gold tag (Doerr-Schott and Lichte, 1986). Figure 9-18 shows a double labeling micrograph where two different proteins are localized on a thin section of boar sperm.

Interpretation of Micrographs

If control experiments are adequately conducted, there should be little difficulty in interpreting micrographs. It is important to remember that, although localization using immunocytochemistry is very specific, the tag may lie a slight distance away from the antigen. Each IgG molecule is about 12 nm long. If the PAP method is employed, then the tag may be several times this distance in any direction from the antigen source. It is possible to quantify some labeling, especially colloidal gold.

References

Classic Literature

Faulk, W. P., and G. M. Taylor. 1971. An immunocolloid method for the electron microscope. *Immunochemistry* 8:1081–3.

Graham, R. C., and M. J. Karnovsky. 1966. The early stages of adsorption of injected horseradish peroxidase in the proximal tubule of the mouse kidney: ultrastructural cytochemistry by a new technique. *J Histochem Cytochem* 14:291–302.

Heitzman, H., and F. M. Richards. 1974. Use of the avidin-biotin complex for specific staining of biological membranes in electron microscopy. *Proc Natl Acad Sci USA* 71:3537–9.

Nakane, P. K., and G. B. Pierce. 1967. Enzyme-labelled antibodies for light and electron microscopic localization of tissue antigens. *J Cell Biol* 33:308–18.

Singer, S. J. 1959. Preparation of an electron-dense antibody conjugate. *Nature* 183:1523–5.

Sternberger, L. A., et al. 1965. Indirect immunouranium technique for staining embedded antigens in electron microscopy. *Exp Mol Pathol* 4:112–25.

Sternberger, L. A., et al. 1970. The unlabeled antibody enzyme method of immunohistochemistry. Preparation and properties of soluble antibody-antigen complex (horseradish peroxidase-antihorseradish peroxidase) and its use in identification of spirochetes. *J Histochem Cytochem* 18:315–33.

Methods References

Bullock, G. R., and P. Petrusz. 1982. *Techniques in immunocytochemistry*. London: Academic Press.

Griffiths, G., and H. Hoppeler. 1987. Quantitation in immunocytochemistry: Correlation of immunogold labeling to absolute number of membrane antigens. *J Histochem Cytochem* 34:1389–98.

Harlow, E., and D. Lane. 1988. *Antibodies: a laboratory manual*. New York: Cold Spring Harbor Laboratory.

Mollenhauer, H. H., and D. J. Morré. 1991. Golgi apparatus form and function. *J Electron Microsc Tech* 17:2–14.

Moriarty, G. C., and L. L. Garner. 1977. Immunocytochemical studies of cells in the rat adenohypophysis containing both ACTH and FSH. *Nature* 265:356–8.

Polak, J. M., and I. M. Varndell. 1984. *Immunolabeling for electron microscopy*. Amsterdam: The Netherlands: Elsevier Science Publishers B.V.

Russell, L. D., R. N. Peterson, and T. A. Russell. 1982. Visualization of anti-sperm plasma membrane IgG and Fab as a method for localization of boar sperm membrane antigens. *J Histochem and Cytochem* 30:1217–27.

Silver, M. M., and S. A. Hearn. 1987. Post embedding immunoelectron microscopy using protein A-gold. *Ultrastruct Pathol* 11:693–703.

Sternberger, L. A. 1979. *Immunocytochemistry*. New York: John Wiley and Sons.

Verkleij, A. J., and J. L. M. Leunissen. 1989. *Immuno-gold labeling in cell biology*. Boca Raton, Fl: CRC Press Inc. 368 pp.

Williams, M. A. 1977. *Autoradiography and immunocytochemistry*, A. M. Glauert, ed. New York: North Holland Pub. Co. pp 1–76.

Recent Localization Reports

Bendayan, M. et al. 1987. Effect of tissue processing on colloidal gold cytochemistry. *J Histochem Cytochem* 35:983–96.

Doerr-Schott, J., and C. M. Lichte. 1986. A triple ultrastructural immunogold staining method. Application to the simultaneous demonstration of three hypophyseal hormones. *J Histochem Cytochem* 34:1101–4.

Jorgensen, A. O., and L. J. McGuffee. 1987. Immunoelectron microscopic localization of sarcoplasmic reticulum proteins in cryofixed, freeze-dried, and low temperature-embedded tissue. *J Histochem Cytochem* 35:723–32.

Mar, H. et al. 1987. Correlative light and electron microscopic immunocytochemistry on the same section with colloidal gold. *J Histochem Cytochem* 35: 419–25.

Vila-Porcile, E. et al. 1987. Cellular and subcellular distribution of laminin in adult rat anterior pituitary. *J Histochem Cytochem* 35:287–99.

Enzyme Cytochemistry

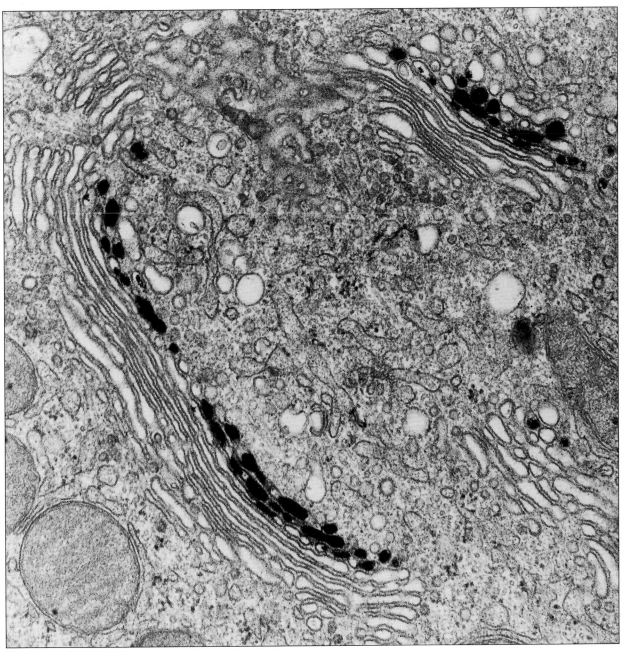

Courtesy of D. Friend.

The modification of organic molecules in cells takes place via enzyme-catalyzed reactions. These reactions play a major role in regulating the metabolic pathways within cells. *Enzyme cytochemistry* utilizes the functional properties of enzymes to localize the site(s) of their activity. It is one of several localization modalities used in both light and electron microscopy. *Enzyme histochemistry* is a term applied to the localization of enzymes in tissues and cells and is generally undertaken at the light microscope level. Since the electron microscope affords increased resolution, the study of the localization of enzymes in subcellular compartments is usually undertaken with this tool. Most studies at the ultrastructural level are termed *enzyme cytochemistry.*

Enzyme cytochemistry has played a major role in helping us understand the structure and functions of cells. It has been an effective bridge between morphology and biochemistry and has helped elucidate pathways for uptake, processing, and discharge of various cell products. Since cytochemical reactions are associated with specific organelles, cytochemistry may be used with subcellular fractions to identify the organelle one wishes to study and to determine the purity of a subcellular fraction. Enzyme cytochemistry facilitates the determination of the purity of cultured cells as they grow and differentiate in tissue culture. One cell type may contain a marker enzyme that differentiates it from contaminating cell types. This technique is also important in pathology, where it can be used to distinguish cells by the specific enzymes they contain. Otherwise, cell types may be difficult to distinguish, leading to confusion in diagnosis and treatment.

A large proportion of cellular activity is mediated by enzymes. The cell may be viewed as a chemical machine containing enzymes that determine the types and rates of chemical reactions that occur. Metabolic reactions are numerous and their locations are varied. Generally, enzymatic reactions are compartmentalized within the cell. Enzymes may be bound to a cellular constituent, such as the mitochondrion or the plasma membrane, or may catalyze reactions within the soluble fraction of the cell. To understand cellular activity, it is important to know the site of cellular reactions.

Basis of Enzyme Cytochemistry

Enzymes speed up a reaction, but are not consumed by the reaction. The substance acted upon by the enzyme is the *substrate*. Enzyme cytochemistry targets enzymes by adding a large amount of substrate from an exogenous source to generate large amounts of the specific reaction product of interest. An exogenously added *trapping agent* is used to bind to and contain or trap the reaction product and make it directly visible under the electron microscope. Sometimes further reactions with the trapping agent are necessary to make it visible. The general scheme of enzyme cytochemical reactions may be summarized as follows:

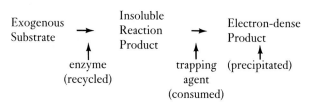

Enzyme cytochemistry is a natural outgrowth of enzyme histochemistry. To improve the resolution of localization, investigators turned to the tool of electron microscopy. The enzyme methods for light microscopy of the 1930s and 1940s (reviewed by Barka and Anderson, 1963; Davenport, 1960; Pearse, 1985) developed into techniques for electron microscopy in the 1950s, 1960s, and 1970s. The addition of new methodologies in recent years has made available a large reservoir of methods for localization of an even larger number of enzymes. Selected publications that had a major impact on the field are cited at the end of this chapter (see Classic References). Although enzyme cytochemistry reached its peak in the 1970s, it remains a powerful tool for the electron microscopist. Papers employing enzyme localization at the electron microscope level may be found in *The Journal of Histochemistry and Cytochemistry* and various other journals cited in Chapter 1.

Requirements for Performing Enzyme Cytochemistry

As described above, the principle behind enzyme cytochemistry is inherently simple—a substrate and trapping agent are added to the biological material. The nonreacted components are later washed out and the tissue prepared for electron microscopy using more or less standard methods. Difficulties with cytochemical procedures arise when the conditions for the experiment are not optimal. In fact, it is rare that the conditions are optimal, sometimes making the technique technically demanding. There are at least five basic points to be considered if one is to perform enzyme cytochemistry successfully:

Preservation of Tissue Structure and Enzymatic Activity

Enzymes, being proteins, are vulnerable to fixatives that will cross-link them and cause denaturation. For most purposes, tissues must be fixed prior to performing cytochemistry. Thus, there is a delicate balance between preservation of tissue structure and preservation of enzymatic activity. Generally, percentages of glutaraldehyde from 0.2% to 2% are employed. Occasionally, formaldehyde is used in conjunction with glutaraldehyde to facilitate rapid fixation. The fixation protocol must be adjusted for each enzyme since not all enzymes are equally vulnerable. An advantage of fixation is that it breaks down membrane permeability to allow entrance of substrate that would not normally penetrate the cell.

Maximization of Reaction Conditions

Enzymatically catalyzed reactions proceed *in vivo* under carefully controlled conditions. They occur at *temperatures, pH,* and in the presence of *substrate concentrations* that are optimal for the production of the reaction product. Therefore, one normally would reproduce these conditions during the cytochemical procedure. Additionally, the *concentration of the trapping agent* should be optimized. Buffers used to maintain pH should be compatible with the reaction, and none of the chemicals used in tissue processing should interfere with or extract the localization products. The trapping agent should not diffuse from the initial reaction site. Considering all the factors that are important for the reaction to take place, it is sometimes surprising when cytochemical procedures work as planned.

Facilitation of Substrate Penetration

The ability of the substrate and/or trapping agent to diffuse through the tissue is frequently limited to less than 100 μm. Therefore, solid tissues must be thinly sliced prior to incubation in the reaction medium. This presents a problem, since most fresh or weakly fixed tissue is often soft and therefore difficult to slice. Special instruments with sharp vibrating razor blades on simple microtomes, termed *vibrating microtomes,* have been developed to slice unfixed or weakly fixed tissue. Usually, tissue to be sliced is loosely embedded in a warmed agar mixture prior to sectioning. Agar, when cooled, provides a moderate amount of support for the vibrating knife to cut the tissue.

Use of Appropriate Controls

As in all scientific experiments, it is imperative that controls be performed. Usually controls are carried out concomitantly with the experimental localization. There have been many instances where controls have revealed that the localization was all or, in part, artifactual. Whenever possible, localization experiments are first carried out at the light microscope level for a preliminary determination of the adequacy of controls and to provide, as well, a tentative hint as to localization. If either the substrate or trapping agent are deleted, the reaction should not occur. Inhibitors or potentiators of the enzyme should modify the amount of reaction product present.

If it is possible to analyze biochemically cell fractions for the enzymatic reaction products, then the electron microscopic localization in tissues at this same site is usually real.

Visualization of Reaction Product

Tissue sections should always be viewed first with no staining reagent other than that provided by the enzymatic reaction product. In stained tissue, the reaction product may not be sharply contrasted against the tissue. On the other hand, if the reaction product is sufficiently electron-dense it may appear prominent even after tissues have been stained. Obviously, stain contamination may be mistaken for reaction product.

Trapping Agents

Several types of trapping agents are used in enzyme cyotochemical reactions. By far, the most common is the *insoluble heavy metal salt,* which is inherently electron dense. Lead especially, but also manganese, barium, cadmium cobalt, and copper are favorites for this purpose. For example, the phosphate group released by a phosphatase is commonly trapped by lead to form insoluble lead phosphate.

Trapping agents may be *enzymes* themselves, which are capable of reacting with other substrates

Table 10-1 Compartmentalization of Marker Enzymes in Mammalian Cells

Enzyme	General Location
acid phosphatase	Golgi, endoplasmic reticulum and lysosomes (GERL)
acetyl cholinesterase	plasma membrane
adenosine triphosphatase	plasma membrane
adenylate cyclase	plasma membrane
alkaline phosphatase	plasma membrane
catalase	peroxisome (microbody)
cytochrome oxidase	mitochondrion
5′ nucleotidase	plasma membrane
nucleoside diphosphatase	Golgi
peroxidase	peroxisome (microbody; Figure 10-1)
succinic dehydrogenase	mitochondrion
thiamine pyrophosphatase	Golgi (non-GERL; Figure 10-2)

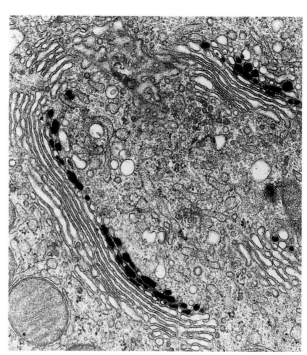

Figure 10-2 Thyamine pyrophosphate activity (dense deposit) is located in specific cisternae of this Golgi apparatus. (Micrograph courtesy of D. Friend.)

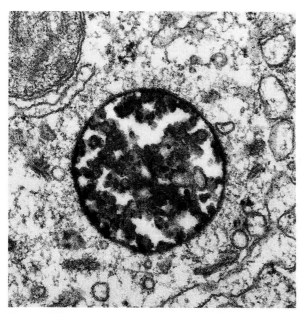

Figure 10-1 Peroxidase activity is represented by the dense material within this multivesicular body. (Micrograph courtesy of D. Friend.)

to produce an electron-dense precipitate. The most common of these is *peroxidase*, which may be attached to the reaction product. In the presence of hydrogen peroxide, diaminobenzidine, and then os-

mium, a dense reaction product is formed (see Chapter 9 for reaction of peroxidase).

Marker Enzymes

Cytochemists have found that enzymes tend to be *compartmentalized* such that each organelle may perform a specific task. Thus, an enzyme may serve as a *marker enzyme* for an organelle or subcellular compartment. When a subcellular fraction is isolated for biochemical studies, it is often useful to perform cytochemistry to determine if a specific cell fraction is contaminated with other cell fractions. For example, if biochemical studies on plasma membranes are contemplated, it is first important to know if the cell fraction is pure or contaminated with another organelle such as endoplasmic reticulum. Enzyme cytochemistry will also allow one to visualize the cytochemical reaction in a subcellular fraction and to compare it with the reaction in the intact cell.

It is not a hard and fast rule that an enzyme associated with, for example, the plasma membrane is indeed so in every cell type. However, cell biologists have examined many cell types and have found regular associations of cellular enzymes with specific compartments. The location of the marker

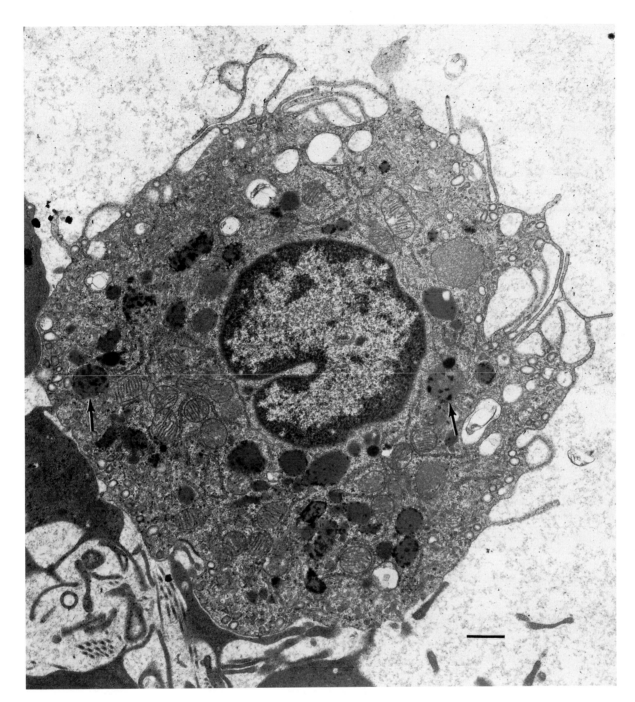

Figure 10-3 Acid phosphatase activity (arrows) within lysosomes of a macrophage. Numerous lysosomes are expected to be present in macrophages since these cells are active scavengers. Lead and uranium stained preparation. Bar = 1 μm.

enzymes listed in Table 10-1 is accurate for most mammalian cell types and have been confirmed by biochemical analyses of cell fractions. (See also Figures 10-1 and 10-2).

A Typical Protocol

In this section, a commonly used cytochemical method is summarized to give an example of how an enzyme cytochemical procedure is performed.

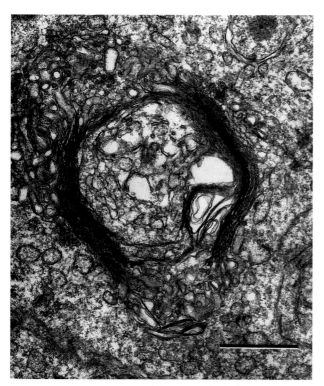

Figure 10-4   Acid phosphatase activity within certain of the Golgi cisternae of a germ cell. Lead and uranium stained preparation. Bar = 1.0 μm.

For more details, consult the original reference sources.

The original electron microscopic method for localization of phophatases, which are active at acid pH, was developed by Barka and Anderson (1962) for light microscopy. It was modified by Frank and Christensen (1968) for use at the ultrastructural level. The authors used this method to localize acid phosphatase in guinea pig Leydig cells. This localization report is reviewed briefly, emphasizing the steps of the protocol.

Fixation
The testis was perfused with cacodylate-buffered glutaraldehyde for a period up to 2.5 hours. The concentration of glutaraldehyde (1.4%) was purposely kept very low to not destroy enzyme activity. After perfusion, the tissue was placed in fixative for up to 1 hour. After buffer washes, a vibrating microtome was used to make 50 to 100 μm sections to permeabilize the cells and effect substrate penetration.

Incubation
Sections were incubated for 20 minutes in an incubation medium modified by Barka and Anderson (1962) from an earlier enzyme histochemistry procedure. The medium consisted of the *substrate*, sodium β-glycerol phosphate in tris-maleate buffer, and a *trapping agent*, lead nitrate. The rationale of the procedure is that phosphate ions will be liberated by the enzyme-catalyzed reaction and trapped by lead to form an insoluble lead phosphate precipitate.

Controls
Controls consisted of deletion of the substrate from the incubation medium and use of sodium fluoride in the substrate, an inhibitor of enzyme activity. After incubation, the tissue was washed in acetate-veronol buffer containing sucrose.

Postfixation and Tissue Processing
Osmium tetroxide (1%) was used as a postfixative. Some tissue was stained *en bloc* with uranyl acetate after postfixation, and other tissue was stained after dehydration, embedding, and sectioning. Note that osmium tetroxide would have destroyed enzyme activity had it been exposed to the tissue before the incubation.

Results
Acid phosphatase activity of Leydig cells was seen in the inner cisternae of the Golgi apparatus, within lipofuscin pigment granules, and autophagic vacuoles. The authors proposed a mechanism by which vacuoles are formed.

Examples of Cytochemistry for Selected Enzymes

The method just described was employed by one of the authors of this text (L.D.R.) to localize acid phosphatase activity in seminiferous tubules of the testis. Photographic examples of acid phophatase localization are illustrated (Figures 10-3 through 10-5).

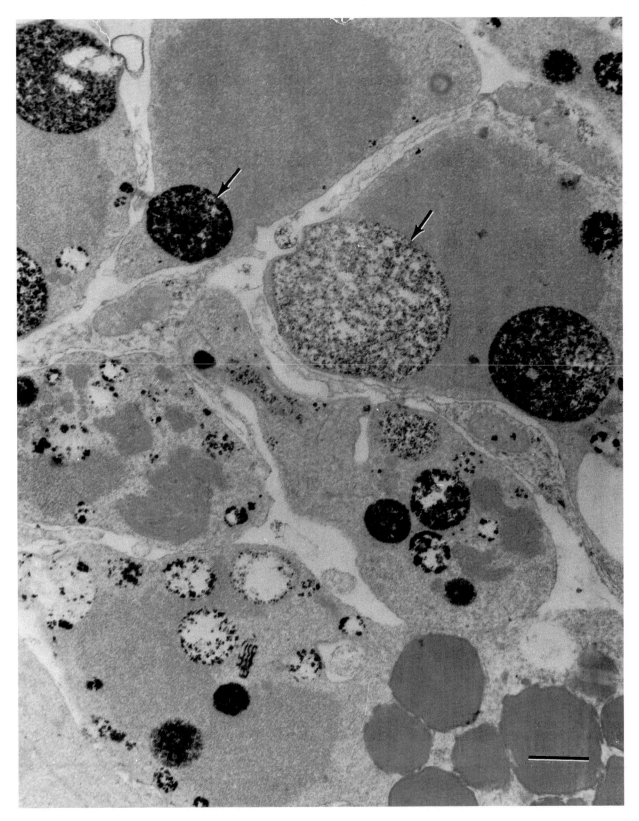

Figure 10-5 Acid phosphatase activity (arrows) is present within that portion of the germ cell cytoplasm that is discarded (termed the residual body) when sperm are formed. The presence of enzyme activity suggests that degradation of the excess sperm's cytoplasm is, at least, partially accomplished by the lysosomes within the cell or by a process known as autophagy. (Unstained preparation) Bar = 1.0 μm.

References

Classics

Barnett, R. J., and G. E. Palade. 1958. Application of histochemistry to electron microscopy. *J Histochem Cytochem* 6:1–12.

Beard, M. E., and A. B. Novikoff. 1969. Distribution of peroxisomes (microbodies) in the nephron of the rat. *J Cell Biol* 42:501–18.

Brandes, D., et al. 1956. Histochemical techniques for electron microscopy: Alkaline phosphatase. *Nature* 177:382–3.

Essner, E., and A. B. Novikoff. 1962. Cytological studies on two functional hepatomas. Interrelations of endoplasmic reticulum, Golgi apparatus and lysosomes. *J Cell Biol* 15:289–312.

Seligman, A. M., et al. 1967. Ultrastructural demonstration of cytochrome oxidase activity by the Nadi reaction with osmiophilic reagent. *J Cell Biol* 34:787–800.

Sheldon, H., et al. 1955. Histochemical reactions for electron microscopy: Acid phosphatase. *Exptl Cell Res* 9:592–6.

Tice, L. W., and R. Barnett. 1962. Fine structural localization of adenosine triphosphatase in heart muscle myofibrils. *J Cell Biol* 15:401–16.

Reviews

Borgers, M., and A. Verheyen. 1985. Enzyme cytochemistry. In *Int Rev Cytol*, G. H. Bourne and J. F. Danielli, eds. 95:163–227.

Enzyme Histochemistry

Barka, T., and P. J. Anderson. 1963. *Histochemistry: theory practice and bibliography*. New York: Harper and Row.

Davenport, H. A. 1960. *Histological and histochemical techniques*. Philadelphia: W. B. Saunders Co.

Pearse, A. G. E. 1985. *Histochemistry: theoretical and applied*. Vol 1–3. Edinburgh: Churchill Livingston.

Acetylcholinesterase

Friedenberg, R. M., and A. M. Seligman. 1972. Acetylcholinesterase at the myoneural junction: cytochemical ultrastructure and some biochemical considerations. *J Histochem Cytochem* 20:771–92.

Lewis, P. R., and C. C. D. Shute. 1966. The distribution of cholinesterase in cholinergic neurons demonstrated with the electron microscope. *J Cell Sci* 1:381–97.

Acid Phosphatase

Barka, T., and P. J. Anderson. 1962. Histochemical methods for acid phosphatase using hexazonium pararosonilin as a coupler. *J Histochem Cytochem* 10:741–53.

Barka, T. 1964. Electron histochemical localization of acid phosphatase activity in the small intestine of mouse. *J Histochem Cytochem* 12:229–39.

Frank, A. L., and A. K. Christensen. 1968. Localization of acid phosphatase in lipofuscin granules and possible autophagic vacuoles in interstitial cells of the guinea pig testis. *J Cell Biol* 36:1–13.

Adenylyl Cyclase

Cutler, et al. 1978. Cytochemical localization of adenylyl cyclase and of calcium ion activated magnesium ATPase in the dense tubular system of human blood platelets. *Biochem Biophys Acta* 542:357–71.

Alkaline Phosphatase

Hugon, J., and M. Borgers. 1966. A direct lead method for the electron microscopic visualization of alkaline phosphatase activity. *J Histochem Cytochem* 44:429–32.

Arylsulfatase

Kawano, J., and E. Akiwa. 1987. Ultrastructural localization of arylsulfatase C activity in rat kidney. *J Histochem Cytochem* 35:523–30.

ATP-ASES

Ando, T., et al. 1981. A new, one-step method for histochemistry and cytochemistry of Ca^{2+}-ATPase activity. *Acta Histochem Cytochem* 14:705–26.

Ernst, S. A., and S. R. Hootman. 1981. Microscopical methods for the localization of Na+, K+-ATPase. *J Histochem Cytochem* 20:23–38.

Catalase

Novikoff, A. B., and R. J. Goldfischer. 1969. Visualization of microbodies (peroxisomes) and mitochondria with diaminobenzidine. *J Histochem Cytochem* 17:675–80.

Cytochrome Oxidase

Seligman, A. M., et al. 1968. Nondroplet ultrastructural demonstration of cytochrome oxidase activity with a polymerizing osmiophilic agent, diaminobenzidine (DAB). *J Cell Biol* 38:1–14.

Cytosine Monophosphatase

Novikoff, A. B. 1963. Lysosomes in the physiology and pathology of cells. Contributions of staining methods. In *Ciba foundation symposium on lysosomes*. A.V. S. de Reuck and M. P. Cameron, eds. Boston: Little Brown and Co. pp 36–77.

Esterases, Nonspecific

Monahan, R. A., et al. 1981. Ultrastructural localization of non-specific esterase activity in guinea pig and human monocytes, macrophages and lymphocytes. *Blood* 58:1089–99.

Glucose-6-Phosphatase

Tice, L. W., and R. J. Barnett. 1961. Localization of glucose-6-phosphatase activity in hepatic cells with the electron microscope. *J Histochem Cytochem* 9:635–6.

Nicotine Adenine Dinucleotide Phosphatase

Smith, C. E. 1980. Ultrastructural localization of nicotinamide dinucleotide phosphatase (NADPase) activity in the intermediate saccules of the Golgi apparatus in rat incisor ameloblasts. *J Histochem Cytochem* 28:10–26.

5'-Nucleotidase

Kreutzberg, G. W., and S. T. Hussain. 1982. Cytochemical heterogeneity of the glial plasma membranes: 5' nucleotidase activity in retinal photoreceptor cells. *Neuroscience* 11:857–66.

Peroxidase (Figure 10-4)

Graham, R. C., and M. J. Karnovsky. 1966. The early stages of absorption on injected horseradish peroxidase in the proximal tubules of mouse kidney. Ultrastructural cytochemisty by a new technique. *J Histochem Cytochem* 14:291–302.

Thiamine Pyrophosphatase (Figure 10-5)

Novikoff, A. B., and S. Goldfischer. 1961. Nucleoside diphosphatase activity in the Golgi apparatus and its usefulness for cytological studies. *Proc Natl Acad Sci* USA 47:802–10.

C H A P T E R   11

Autoradiography/ Radioautography

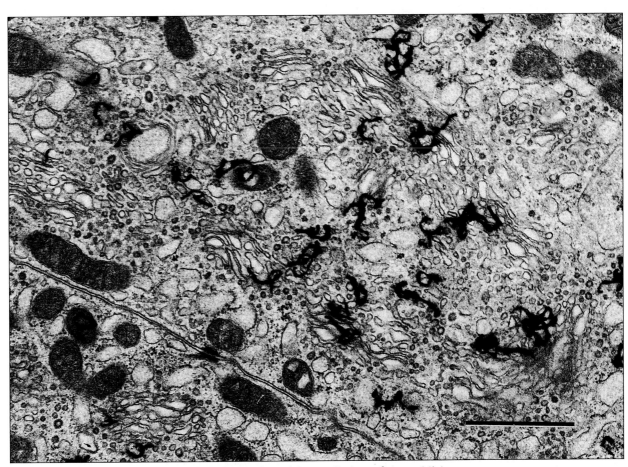

Courtesy of C. Flickinger; taken from *Anat Rec* with permission of the publisher.

Autoradiography and radioautography are interchangeable terms that refer to the use of radioactive isotopes and a radio-sensitive emulsion to trace cell processes or to localize substances within biological specimens. In brief, radioactive compounds are first administered to live organisms (Figure 11-1). Then the tissue is fixed, embedded, and sectioned, and the tissue section is brought into direct contact with a photographic emulsion. Over a period of weeks or months, as ionizing radiation is released from the radioactive substances, the radiation energy causes changes in the emulsion that are visualized only after a photographic development process. In development, the photographic emulsion (usually silver bromide) is reduced to metallic silver in a way very similar to the development of emulsions used in the ordinary photographic process. The developed emulsion and tissue are viewed simultaneously. The result is an *autoradiograph*, a picture of both the tissue and the darkened spots on the photographic emulsion. The darkened spots or *grains* that appear on the tissue

localize the general site of the radioactive emission. Autoradiography may be used either at the light or electron microscope levels; however, the resolution of the latter is naturally much greater (Figure 11-2).

Autoradiography has contributed to our knowledge of the functions of organelles. It has made a significant impact on our understanding of *dynamic* aspects of cells and tissues. There are numerous examples of how the intercellular pathways of compounds or drugs have been traced. In many instances, autoradiography has been key in our understanding of the site(s) of synthetic and metabolic pathways and has detailed secretory pathways in a very precise manner. Cell divisions have been mapped and quantified, and the fate of labeled cells has been traced throughout the organism. It has been an important localization technique, often used in circumstances where no other techniques were available. Receptors, vitamins, and a variety of other substances have been localized and their fate determined through autoradiography.

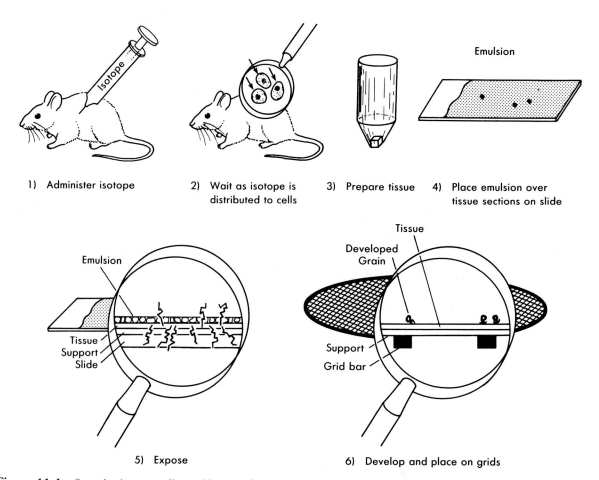

1) Administer isotope

2) Wait as isotope is distributed to cells

3) Prepare tissue

4) Place emulsion over tissue sections on slide

5) Expose

6) Develop and place on grids

Figure 11-1 Steps in the autoradiographic procedure.

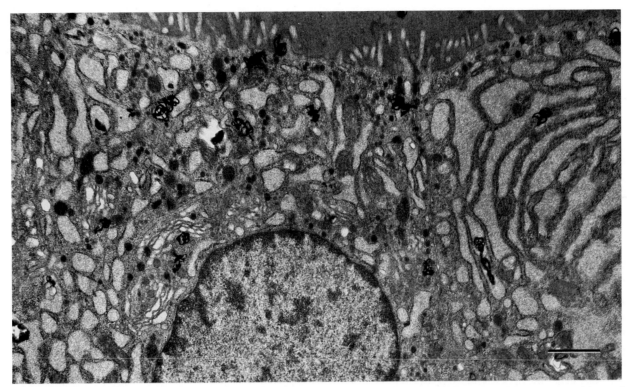

Figure 11-2 An electron microscope autoradiograph showing labeling of a thyroid follicular cell 90 min. after injection of ³H fucose. Grains (black squiggles) are located both over the Golgi apparatus and secretory vesicles and colloid. Bar = 1.0 μm. (Micrograph provided by Gary Bennett.)

Autoradiography has been an indispensable technique to the biologist. It is exquisitely sensitive, allowing the detection and thus localization of single molecules. Electron microscope autoradiography is a difficult, time-consuming, and sometimes very costly technique. As such, it is not in widespread use. To appreciate this technique fully, it is first necessary to understand the nature of radioactivity and how it interacts with the photographic emulsion.

Radioactivity

Some atomic nuclei spontaneously emit rays or subatomic particles with the release of energy. Atomic nuclei that are chemically similar, but possess different masses are termed *isotopes*. Isotopes are named by including their mass number as a superscript (e.g., ¹³¹I), generally written to the left of the atomic symbol. Some isotopes are stable (e.g., ¹²C and ¹³C), whereas others are not (¹⁴C) as evidenced by radioactive emissions.

Radioactive emissions of unstable isotopes are of two kinds. Some emit *X rays* or *gamma rays* and others emit charged particles. The emission of par-

ticles is often referred to as *disintegrations*. When radiation has the property of releasing electrons as it passes through a substance, it is termed *ionizing radiation*. The charged particles are most commonly employed in autoradiography. *Alpha* emitters such as uranium and thorium discharge a helium nucleus (two protons and two electrons), which because of their large mass and energy travel in a straight line. *Beta* emitters discharge the equivalent of an electron, a negatively charged particle, which takes a nonlinear path. Since alpha emitters are usually heavy metals and therefore toxic to most biological systems, beta emitters are most often used in autoradiography (Table 11-1); these include tritium (or ³H), carbon 14, phosphorus 32, sulphur 35, and iodine 131. The path of electrons from beta emitter disintegrations is easily detected by photographic emulsion.

Radioactivity is quantified in *curies* (Ci), such that one curie is equal to 3.7×10^{10} disintegrations per second.[1] The strength of a radioactive substance is related to its *specific activity*, which is defined as

---

[1] The International Commission of Radiation Units and Measurements (ICRU) has recommended that the *becquerel* be substituted for the curie as the unit for measurement of radioactivity. One curie equals 2.7×11^{-11} becquerels.

Table 11-1 Characteristics of Some Beta Emitters

Radioisotope	Mass	Half-life	Maximum Energy (MeV)	Frequency of Use in Autoradiography
Calcium	45	153 d	0.25	low
Carbon	14	5760 y	0.15	low
Chlorine	36	31,000 y	0.714	low
Iodine	131	8 d	0.61	moderate
Iron	59	45 d	0.46	moderate
Mercury	203	47 d	0.21	low
Phosphorus	32	14.2 d	1.71	moderate
Sulphur	35	87.2 d	0.167	moderate
Tritium	^{3}H	12.26 y	0.018	extremely heavy

the amount of radioactivity in curies in one millimole of the substance. Knowledge of the specific activity of a radioactive mixture is important in planning autoradiographic experiments: Too high a level of radioactivity will cause tissue damage, insufficient quantities will greatly lengthen the time needed to expose the emulsion.

The *half-life* of a radioactive substance is the period in which disintegrations reduce the radioactivity of a compound to half its original value. Half-lives of substances vary considerably, extending from small fractions of a second to quadrillions of years. Because of their short half-life, some isotopes are impractical for use in autoradiography. Some isotopes with half-lives that make them practical for use in autoradiography are carbon 14 (^{14}C = 5,760 years), tritium (^{3}H = 12.26 years), calcium 45 (^{45}Ca = 153 days), Iron 59 (^{59}Fe = 45 days), and iodine 131 (^{131}I = 8.04 days). Compounds may be made specifically with one of their stable isotopes replaced by a radioactive one. Compounds tagged with the isotope of choice are available commercially or, less commonly, may be custom-synthesized by a commercial source at a premium cost.

Emulsion Used in Autoradiography

Photographic film has a light-sensitive layer called the *emulsion* that will darken when developed in photographic chemicals (Chapter 8). Emulsion is sensitive not only to light, but also to various rays and particles such as electrons emitted from a radioactive source. The emulsion contains mainly silver bromide crystals with occasional silver sulfide imperfections or aggregations in the crystalline structure. The crystals are suspended in a gelatin matrix to

form the emulsion. When light or electrons strike a silver bromide crystal, a release of electrons occurs within the crystal. Some electrons strike the silver sulfide aggregation and become trapped. The added negative charge of the silver sulfide crystal produces a *latent image*, which promotes the reduction of silver bromide to metallic silver during the development of the emulsion. Only silver bromide crystals that have been struck by a particle are reduced to silver, whereas, unexposed crystals do not develop into silver grains. The silver appears as a *grain* in the autoradiograph (Figure 11-3).

Unlike photons, which must strike a crystal numerous times to expose a grain, ionizing radiation will produce a latent image much more readily. Thus, there is a direct relationship of the number of emissions to the grains produced. The emission may pass randomly through the emulsion, and different development methods will either record the latent image as a track (squiggle; Figure 11-4A) or as a rounded density (Figures 11-4B and C), both of which are metallic silver.

In 1861, Niepce de St. Victor found that uranium salts darkened an undeveloped photographic plate. In 1924, Lacassagne was the first to place photographic plates in contact with tissues containing radioactivity. When he removed and developed the plate, he saw that it had darkened in selective areas. Photographic emulsions at room temperature are gels, but when heated slightly they become liquids, similar to dessert gelatins that liquify as the temperature increases. Belanger and Leblond, in 1946, melted emulsion from a photographic plate and deposited it onto a tissue section. They were able to visualize both the section and the developed silver grains simultaneously. Thus, emulsions may be placed in direct contact with radioactive tissues by

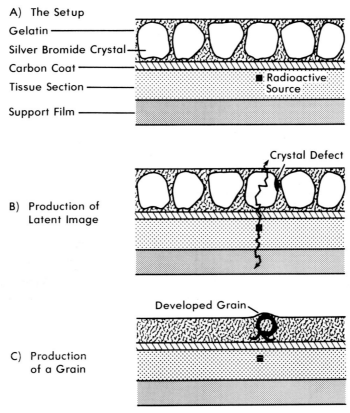

A) The Setup
Gelatin
Silver Bromide Crystal
Carbon Coat
Tissue Section
Radioactive Source
Support Film

Crystal Defect
B) Production of Latent Image

Developed Grain
C) Production of a Grain

Figure 11-3 Production of latent images by a radioactive source and the developed grain. Silver sulfide aggregations within the silver bromide crystals are depicted as a "crystal defect".

How to Perform Autoradiography

coating the tissue with the liquid emulsion. In 1956, Liquier-Milward was the first to demonstrate that this principle could be applied to the electron microscope. Today, special emulsions are commercially produced for exclusive use in autoradiography.

How to Perform Autoradiography

Before beginning autoradiography, it is necessary to determine if electron microscopic autoradiography is the appropriate technique to solve the particular problem. Would more readily performed and less expensive localization techniques, such as immunocytochemical methods (Chapter 9), work as well?

It is important to know if the radioactive compound being administered is reasonably *specific* to allow meaningful conclusions to be made about localization or determination of sites of synthetic or metabolic activity. As it turns out, relatively few compounds have unique specificity. ^{3}H leucine and ^{35}S methionine are labeled compounds for the demonstration of protein synthesis; ^{3}H uridine and ^{3}H thymidine are incorporated into RNA and DNA,

respectively; ^{3}H fucose is incorporated primarily into glycoproteins, especially cell membrane glycoproteins, as well as secretory material and lysosomal enzymes. The isotope ^{131}I is concentrated in the thyroid gland where it is incorporated into thyroglobulin. There are other examples of isotopes that are specific for biological processes.

It is also important to determine if the compound is accessible to intracellular sites. The experiment at the end of this chapter, which illustrates how autoradiography is performed, shows how, in some circumstances, such a problem may be overcome. Also, labeled compounds may be given as a *pulse* (single exposure), or as a pulse followed by administration of the same, but nonradioactive, compound (*pulse-chase*), or as a *continuous infusion* or multiple infusions. The time that the labeled compound is available, in any significant quantity, to the tissue differs in each of the previous situations (pulse chase, pulse, and continuous infusion from short to long availability), and the interpretation of the experiment in each of these circumstances may differ.

A little thought during the planning phase of an experiment may mean the difference between a successful and an unsuccessful experiment.

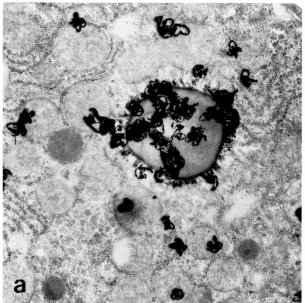

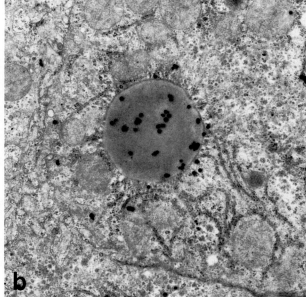

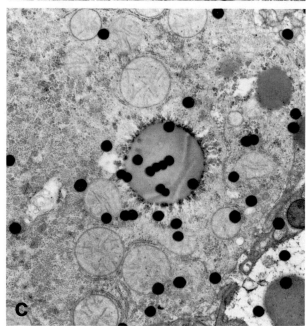

Figure 11-4 Appearance of grains using different developing solutions: (a) chemical development Ilford L4 emulsion in D-19 developer; (b) and (c) physical development of Ilford L4 emulsion.

Planning the Experiment

Once the autoradiography technique has been chosen, it is important to plan the experiment well in advance. Radioactive compounds are costly and some have short half-lives, so it is desirable to make the first attempt a successful one.

CAUTION: Radioactive materials are extremely hazardous substances. At most institutions, licensing and/or certification procedures are required before an individual is allowed to handle them.

There are several considerations in planning the experiment. Specific activity of the radiolabel will be one factor in determining how much of the labeled compound is to be ordered. Too little radioactivity will not allow for practical exposure periods, and too much may cause damage to tissues. What is the half-life of the labeled compound? How much labeled compound should be used per administration? How many administrations of the compound? How many animals/plants and how many sampling intervals should be employed? All of these questions should be answered prior to beginning the experiment and especially prior to ordering the radiolabeled compound.

To determine sampling interval, it is important to estimate how long it will take the compound to reach the site(s) of interest. What route of administration will allow the labeled compound to reach the general site to be studied? Will the labeled compound be stabilized during fixation, or does its solubility pose problems with localization? Soluble compounds may be removed from the specimen during ordinary tissue preparation and may require specialized tissue preparative techniques such as freeze-drying (see Chapter 3).

The Experimental Protocol

It is always important to remember that radioactive compounds are *hazardous substances* and that their use is governed by strict federal and state rules and regulations designed to minimize health risks. Given that adequate precautions are taken with radiolabeled compounds, the following steps are usually followed in a typical experiment.

Administration of the Radioactive Substance and Tissue Preparation

The current literature is often the best guide to the method and route of administration of radiolabeled substances. After administration, there is a variable period in which the labeled substance is taken up or incorporated into cells. If the experimental design is appropriate, the substance is specifically bound to the receptor or incorporated into the metabolic pathway of interest. At the sampling time, the tissue is fixed rapidly by methods that provide the best ultrastructural results. It is then dehydrated and eventually embedded in the medium of choice.

Light Microscope Autoradiography

Light microscope autoradiography, using thick (0.5 to 1.0 μm) sections, is usually undertaken initially. Thick sections contain more radioactivity and will expose more quickly. Since the exposure period for light microscopy is considerably less than for electron microscopy, valuable time will be saved if the thin (electron microscopic) sections are exposed simultaneously. The exposure time for electron microscopy may be estimated from the relative abundance of grains seen in light microscopy. About one

week of light microscope exposure is equivalent to one month of exposure for electron microscope autoradiography. Developed grains at the light microscope level appear as spots. A preliminary determination of localization of grains may be made at the light microscope level. The pattern of grain distribution may be compared at various sacrifice intervals to determine which sacrifice intervals will be most interesting at the electron microscope level.

Application of Emulsion

The most technically demanding step is the application of the emulsion to the thin sections. Part of the difficulty is that this operation must take place in a darkened room. There are several methods for applying emulsion to tissue; none are easy for the novice. The thin section is placed on a grid that has has been coated with a collodion support film. A partially gelled emulsion film is layered onto the grid using a wire loop that has been dipped in liquid emulsion so that a thin film of it is stretched across the loop. Alternatively, the grid with supporting film may be dipped in a liquid emulsion. These methods sometimes provide highly variable results often causing a nonuniform coat of emulsion to be layered onto the tissue.

More commonly used is a *stripping film* method in which a microscope slide is first coated with collodion, thin sections placed on the coating, and then a thin layer of carbon evaporated onto the tissue. (The carbon is deposited to provide a better substrate for emulsion deposition and adhesion.) An emulsion layer is coated onto the slide by dipping the slide into warmed (liquified) emulsion. Special equipment may be used (Figure 11-5) to assure that the emulsion is applied in a thin, uniform layer. Regardless of the technique used to apply emulsion, it is desirable to have a single molecular layer of closely packed silver halide crystals uniformly covering the tissue section (Figure 11-6).

Exposure of the Autoradiograph

After the emulsion is applied, the tissue is placed in a light-tight box where radioactive disintegrations occur creating latent images. The direction of radioactive emissions is random; only about 50% are thought to strike the emulsion, and even fewer strike silver bromide crystals to produce latent images. The half-life amount of compound administered, the amount and rate of uptake and turnover, as well as

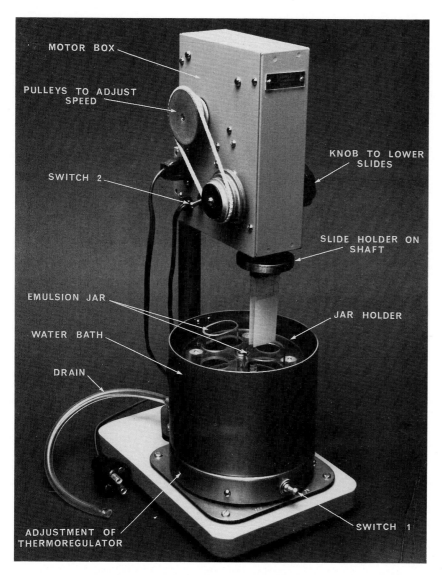

Figure 11-5 Semiautomatic coating device for autoradiography used to insure application of a monolayer of silver bromide emulsion. (Courtesy of B. Kopriwa; taken from *J Histochem Cytochem* 14:923 with permission of the publisher.)

the specific activity all influence the exposure time. Exposure periods from one to ten months are common, although exposures may be shorter or longer. It is important that the autoradiograph show sufficient grains to provide information, yet not too many grains as to obscure tissue structure. Labeled compounds with half-lives under two weeks are difficult to work with given the constraint of rapid loss in radioactivity. ^{131}I is an exception if one is studying the thyroid, since the majority of injected iodine becomes rapidly concentrated in the thyroid.

Development of the Autoradiograph

Grids, or slides containing tissue, are successively transferred through developer, stop bath, fixer, and distilled water washes to produce grains from latent images and to remove unexposed grains. The fixation process removes the unexposed and undeveloped silver halide crystals but leaves the gelatin. After the fixer step, the lights may be turned on.

Staining of the Tissue

Tissue may be "prestained "or "poststained" or both. In prestaining, tissues are usually stained with uranyl acetate prior to the exposure phase. Poststaining with lead citrate (usually) takes place after development. It is important to obtain good staining that results in a contrasty image, since specimens supported with collodion and carbon coats will decrease the contrast.

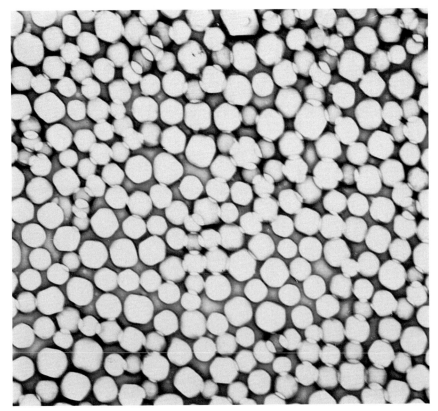

Figure 11-6 Appearance of a monolayer of grains under the electron microscope. Shown are the holes replacing silver bromide crystals in the gelatin matrix of the undeveloped emulsion after the crystals have been destroyed by the electron beam. ×50,000. (Micrograph courtesy of B. Kopriwa; taken from *J Histochem Cytochem* 14:923 with permission of the publisher.)

Placement of the Exposed Tissue on Grids

In the stripping film method, the collodion coat containing the tissue section is floated free of the slide (Figure 11-7). Grids are placed on top of the tissue, and the collodion holding the tissue and grids is lifted from the water from above. After drying, the excess collodion is trimmed from the grid.

Interpretation of Autoradiographs

In transmission electron micrographs, developed grains appear as irregular tracks or single spots depending on the specific developing procedure utilized (see Figure 11-4). All autoradiographs show some nonspecific grains or background, but excessive background is an indication of aging of the emulsion, excessive heating of the emulsion, exposure of the emulsion to a light leak, or spurious radioactive disintegrations.

The major obstacle to accurate interpretation of autoradiographs is the precise determination of the source of the disintegration. If all disintegrations were directed straight above, then the grains would directly overlay the source of emission. However, disintegrations are randomly directed and may travel some distance before striking a silver halide crystal. A circle with the source as the center, or *resolution circle*, represents the possible span of radioactive emissions (Figure 11-8).

Numerous methods have been developed to assess the probability that specific structures are labeled. The most commonly used isotope, tritium (^{3}H), has a range of less than 1 μm from the source. Some investigators draw a resolution circle around a grain that encompasses all of the possible sources of emission. It is then possible to calculate the probability that a given structure is the source of the emission. Others simply score the organelle that is under the grain. Yet others determine the grain density by counting the number of grains over a structure and dividing this by the area occupied by the structure. The larger the cell structure acting as the source of radioactive emissions, the easier it is to identify the source (Figure 11-9). In some experiments, it is obvious to the eye that grains are exclusively associated with one organelle. One must resort to sophisticated methods of quantitation when the source of radioactivity is a small structural element or when there are several sources of emissions.

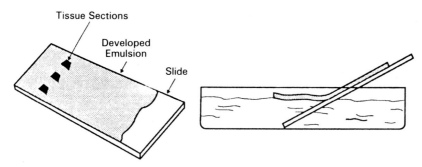

1) Developed emulsion, tissue and collodion on slide

2) Floatation of collodion containing tissue and developed emulsion

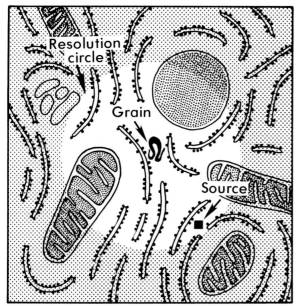

3) Placement of grids on tissue and pick up with filter paper

4) Dry in oven and separate tissue bearing grids

Figure 11-8 A resolution circle may be drawn with a grain as its center to indicate the probable boundaries of the location of the source. Any of the organelles within the resolution circle may be the source of radioactivity.

A Typical Experiment Employing Autoradiography

As an example of the application of autoradiography, the article by Bennett and O'Shaughnessy (1981) is used to show how information is gained through the technique of autoradiography.

The authors point out that *sialic acid* residues are common peripheral constituents of carbohydrate side chains of secretory and membrane proteins. Furthermore, sialic acid residues have been implicated in important cell functions such as determining the life span of circulating cells. The purpose of the experiment was to determine the subcellular site at which sialic acid residues are added to carbohydrate moieties within hepatocytes.

Sialic acid does not, itself, enter cells. Therefore, a chemically related radiolabeled precursor product, [³H] N-acetylmannosamine, which readily enters cells and is converted by cells to sialic acid, was utilized. It was important first to determine that this precursor was specifically incorporated into protein as sialic acid. Biochemical studies verified this, allowing the interpretation that, after a period of time, the precursor substance that would appear as grains in autoradiographs would be converted to sialic acid.

Given this specificity, the authors proceeded with autoradiography at both the light and electron microscope

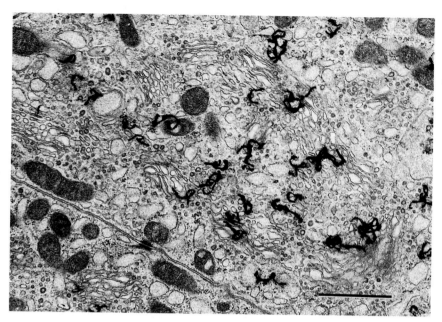

Figure 11-9 For large organelles such as this Golgi apparatus the source of grains is apparent. (Micrograph courtesy of C. Flickinger; taken from *Anat Rec* 210:435 with permission of the publisher.) Bar = 1.0 μm.

levels. Animals were killed at varying intervals after injection of the radioactive sialic acid precursor. Light microscopic autoradiography (Figure 11-10) revealed abundant radioactive incorporation into hepatocytes and specific intracellular localization of grains within the cytoplasm of hepatocytes. With time, grain localization was seen at the cell surface.

To perform electron microscopic autoradiography, thin sections were placed on collodion-coated glass slides; the sections were carbon coated by thermal evaporation (see Chapter 5) and stained with uranyl acetate. Slides were coated with emulsion (dipping technique) in a darkroom and allowed to expose in the dark for a period up to eight months. Slides were then developed and portions

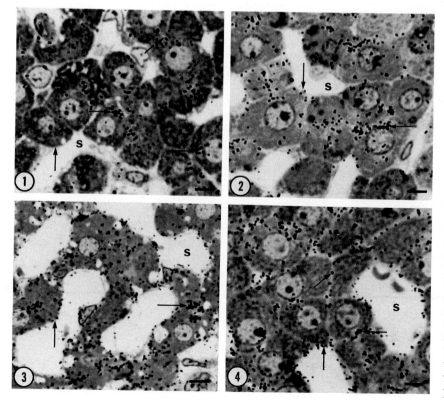

Figure 11-10 Light microscope autoradiographs of sialic acid ([³H] N-acetylmannosamine) incorporation into rat liver. (1) At 10 minutes after injection, grains are seen over what appears to be the Golgi region (horizontal arrows) and sinusoidal surface (vertical arrows). (2) At 30 minutes after injection, some grains are over the Golgi region and some are over the sinusoidal region. (3) and (4) At 4 hours and 24 hours, respectively, after injection most grains are over the sinusoidal surface of hepatocytes (vertical arrows) and lateral surfaces (oblique arrow). Few remain over the bile canalicular region (horizontal arrow). (This and subsequent micrographs in this chapter courtesy of G. Bennett, taken from *J Cell Biol* 88:1, 1981 with permission of the publisher.)

of the collodion support containing sections were circumscribed, placed on 300 mesh grids, and poststained with lead citrate.

At 10 minutes after injection, grains were located primarily over the Golgi region (Figure 11-11), specifically the trans face of the Golgi apparatus, but some were over secretory granules and other organelles. At longer intervals (1 to 24 hours) after the death of the animal, the proportion of grains over the Golgi apparatus decreased whereas the proportion over the cell surface at the hepatic sinusoidal region and within lysosomes increased (Figure 11-12). At three and nine days after injection of precursor, the location of grains was similar to the one-day interval, although few total grains were recorded.

The authors concluded that, because grains were concentrated in the Golgi apparatus, sialic acid is primarily incorporated into glycoproteins at this site. More specifically, the incorporation of glycoproteins was in the trans face of the Golgi apparatus indicating that only a portion of the Golgi was actively incorporating sialic acid. Once formed, the glycoproteins migrated to secretory products and lysosomes and to the plasma membrane. That label was diminished at longer time intervals suggesting that there was continuous renewal of sialic-acid-bearing glycoproteins.

This journal article is typical of many good autoradiographic experiments found in the literature. Although, in this particular experiment, a dynamic synthetic process is traced using a labeled precursor, it should not be inferred that all autoradiographic experiments are designed similarly. Some are designed for localization purposes and do not involve multiple sacrifice intervals. The highlights of the article are only briefly summarized here; consequently, full treatment of the subject may be appreciated only by reading the article in its entirety.

Source: Bennett, G., and D. O'Shaughnessy. 1981. The site of incorporation of sialic acid residues into glycoproteins and the subsequent fate of these molecules in various rat and mouse cell types as shown by radioautography after injection of [³H]N-acetylmannosamine I. Observations in hepatocytes. *J Cell Biol* 88:1-15.

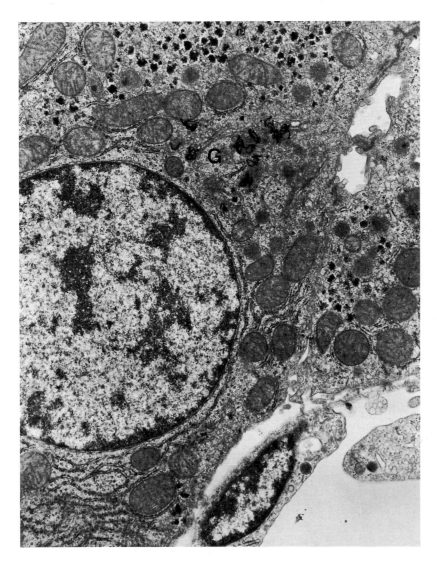

Figure 11-11 Ten minute autoradiograph of sialic acid ([³H] N-acetylmannosamine precursor) incorporation into a rat hepatocyte showing grains exclusively over the Golgi region (G).

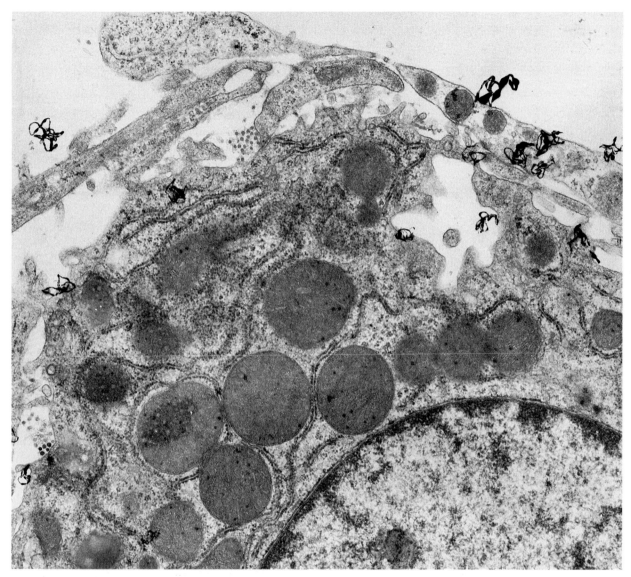

Figure 11-12 A twenty-four hour autoradiograph of sialic acid ([³H] N-acetylmannosamine precursor) incorporation into a rat hepatocyte showing that most of the grains are over the sinusoidal surface of the cell. From the forgoing series of micrographs it was shown that the radioactive tracer injected has dynamic properties within the cell.

References

Classic and Historical References

Belanger, L. F., and C. P. Leblond. 1946. A method for locating radioactive elements in tissues by covering histological sections with photographic emulsions. *Endocrinology* 9:386–400.

Jamieson, J. D., and G. E. Palade. 1967. Intracellular transport of secretory proteins in pancreatic exocrine cell. *J Cell Biol* 34:597–615.

Leblond, C. P. 1987. Radioautography: Role played by anatomists in the development and application of the technique. In *The American association of anatomists: essays on the history of anatomy in America.* J. E. Pauly, ed., Baltimore: Williams and Wilkins, pp 89–103.

Liquier-Milward, J. 1956. Electron microscopy and radioautography as coupled techniques in tracer experiments. *Nature* (London) 177:619.

Revel, J. P., and E. D. Hay. 1961. Autoradiographic localization of DNA synthesis in a specific ultrastructural component of the interphase nucleus. *Exptl Cell Res* 25:474–97.

Technique References

Becker, W., and A. Bruce. 1985. Autoradiographic studies with fatty acids and some other lipids: a review. *Prog Lipid Res* 24:325–46.

Kopriwa, B. M. 1973. A reliable, standardized method for ultrastructural electron microscopic radioautography. *Histochemie* 37:1–7.

Roth, J. L., and W. E. Stumpf. 1969. *Autoradiography of diffusible substances*. New York: Academic Press.

Salpeter, M. M., and L. Bachmann. 1972. Autoradiography. In *Principles and techniques of electron microscopy*, M.A. Hayat, ed. New York: Van Nostrand Reinhold Co., pp 219–78.

Salpeter, M. M., and F. A. McHenry. 1973. Electron microscope autoradiography. In *Advanced techniques in biological electron microscopy*, J. K. Koehler, ed. New York: Springer-Verlag, pp 113–52.

Williams, M. A. 1977. *Autoradiography and immunocytochemistry*, A. M. Glauert, ed. North Holland, Amsterdam: Elsevier, pp 77–197.

Recent Reports

Bennett, G., and D. O'Shaughnessy. 1981. The site of incorporation of sialic acid residues into glycoproteins and the subsequent fate of these molecules in various rat and mouse cell types as shown by radioautography after injection of [^{3}H] N-acetyl-mannosamine I. Observations in hepatocytes. *J Cell Biol* 88:1–15.

Bissionnette, R., et al. 1987. Radioautographic comparison of RNA synthesis patterns in epithelial cells of mouse pyloric antrum following ^{3}H-uridine and ^{3}H-orotic acid injections. *Am J Anat* 180:209–25.

Haddad, A., et al. 1987. Localization of glycoproteins in insulin secretory granules by ultrastructural autoradiography. *J Histochem Cytochem* 35:1059–62.

Mazariegos, M. R., et al. 1987. Radioautographic tracing of ^{3}H proline in the endodermal cells of the parietal yolk sac as an indicator of the biogenesis of basement membrane components. *Am J Anat* 179:79–93.

Miscellaneous
Localization Techniques

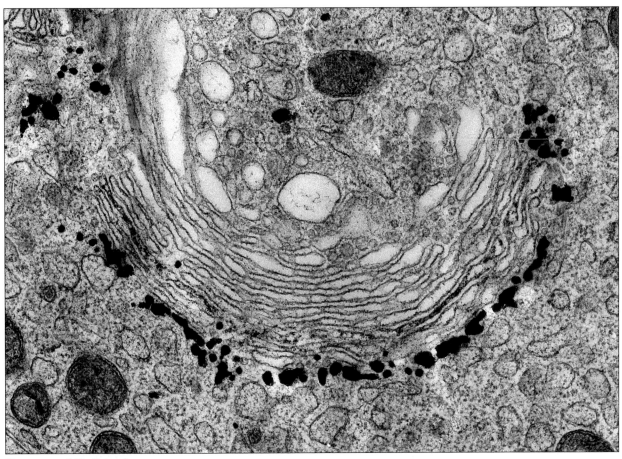

Courtesy of D. Friend.

Localization techniques based on enzymatic, antigenic, and radioactive properties molecules have been covered previously (Chapters 9, 10, and 11). However, there are several other methods for localization of specific substances or cell structures that have proven useful for electron microscopy. These, in general, are based on some property of the particular molecule/structure that is to be localized.

Some localization techniques may be based on preferential staining of particular structures by what are called *cytochemical stains*. These stains are generally less specific than has been demonstrated for most other forms of localization. Even the routine stains used for electron microscopy, lead and uranium, are, to some degree, a means for localization of cellular constituents based on the chemical properties of these constituents (see Chapter 5). They are less specific than the cytochemical stains that will be considered in this chapter, although, as will be seen, there is a wide range of specificities of cytochemical stains and other forms of localization.

What follows is a list of selected localization and staining techniques to illustrate how methods have been developed to localize specific molecules or cell structures.

Actin

The natural affinity of the heads of the myosin molecule for actin causes myosin to bind to actin and provides the basis for a localization scheme for actin. Fragments (the heads) of myosin molecules are first isolated and purified (Ishikawa et al., 1969). They are then exposed to actin in cells that have been broken open, usually by detergents such as Triton X-100 or Nonidet NP 40. In the electron microscope, filaments containing actin appear "decorated" with an arrowhead pattern formed by the head region of myosin molecules (Figures 12-1 and 12-2).

Carbohydrates/Oligosaccharides

Using Lectins

Lectins are plant agglutinins from various species that have affinity for specific sugar sequences in oligosaccharide residues. They may be conjugated to a variety of ultrastructural tags and used for localization purposes. Over fifty lectins are available, and these possess a wide range of specificities. Selected lectins and their specificities are shown in Table 12-1.

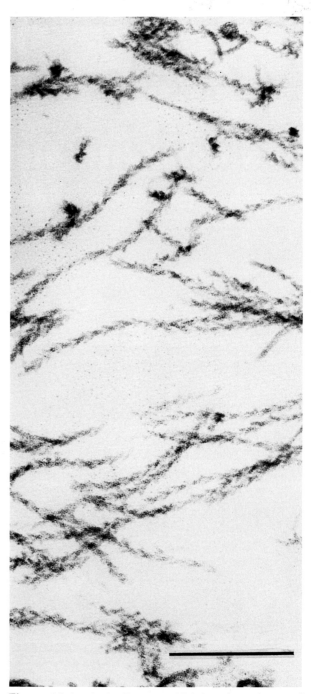

Figure 12-1 Actin filaments which have been decorated with the heads (S-1 fragment) of the myosin molecule. The filaments are fuzzy, appearing to display uniformly oriented arrowheads at periodic intervals along the filament. Bar = 0.25 μm.

Using Tannic Acid and Metals

The mechanism of the reaction of tannic acid and metals with carbohydrates is uncertain, but it may involve hydrogen bonding as described in a paper by Sannes et al., 1978.

Figure 12-2 Actin filaments (indicated by arrows) of cell 1 appear decorated with the S-1 fragment of myosin as shown in Figure 12-1. Filaments in an adjoining cell are not dec-orated indicating that they have not been exposed to myosin due to the impermeability of the cell. Bar = 0.25 μm.

Table 1 Specificities of Selected Plant Lectins for Oligosaccharide Residues

Lectin	Oligosaccharide Bound
Concanavalin A	a-D-glucose
Soy bean lectin	a-galactose and N-acetylglucosamine
Wheat germ lectin	N-acetylglucosamine
Lotus seed lectin	fucose

Using a Modified PAS-Schiff Reaction

Aldehyde groups are produced when periodic acid is used to oxidize carbohydrates. These are then reacted with alkaline silver solutions in which silver is deposited at the reaction site (Figures 12-3 and 12-4).

Alkaline bismuth stains certain polysaccharides incorporating the PAS-Schiff reaction.

Using Colloidal Iron and Colloidal Thorium

Colloidal solutions of some metals will allow a clear demonstration of some carbohydrates. Acid muco-polysaccharide components will react with colloidal solutions of iron and thorium. A dense amorphous precipitate is seen at the site of carbohydrates.

Golgi Complex/Multivesicular Body

It has long been recognized from light microscope studies that osmium will stain the Golgi apparatus. Using the method described by Friend (1969), os-mium is reduced and deposited at specific saccules of the Golgi by a mechanism that remains unknown (Figure 12-5).

Glycogen/Membranes

This method replaces postfixation with osmium te-troxide by postfixation in a mixture of osmium te-troxide and potassium ferrocyanide (Karnovsky, 1971; Russell and Burguet, 1977). When the two solutions are mixed, the mixture appears dark brown as the osmium tetroxide is reduced by the potassium ferrocyanide. The mechanism of enhancement of glycogen and membranes is unknown, but is thought to be due to reduction of osmium by potassium fer-rocyanide (See Chapter 19, Figure 19-98). Mem-branes appear trilaminar (see Chapter 18, Figure 19-5), and glycogen stands out in sharp contrast to the cytoplasmic matrix. The matrix is leached out during the postfixation procedure, which further contrasts membrane preservation. In addition, ribosomes are poorly visualized. The technique often imparts a pleasing appearance to micrographs.

Ions

Anions and cations are frequently employed as probes to localize other ions. The techniques are based on charge affinity of tissue ions (electrostatic forces) for the oppositely charged heavy metal.

One method employs pyroantimonate, which in the presence of cations selectively precipitates as the cation-antimonate complex (Figure 12-6).

Nucleic Acids (DNA/RNA)

Techniques for staining nucleic acids, which are ref-erenced at the end of this chapter, have several dif-ferent bases for their ability to localize nucleic acids.

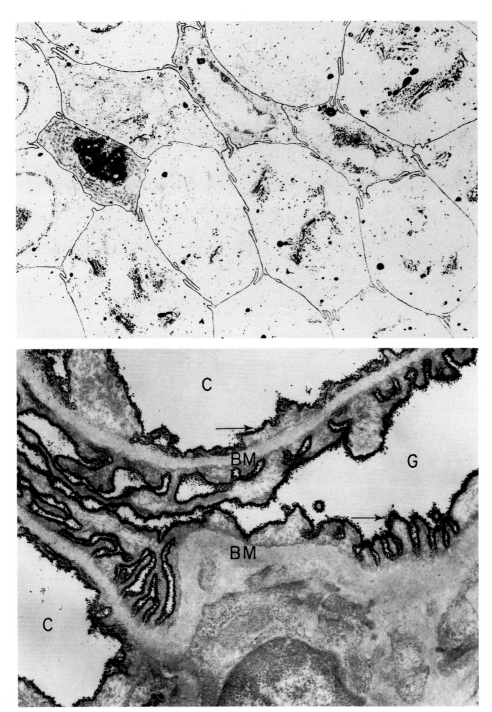

Figures 12-3 and 12-4 The PAS-chromic acid-silver technique is used to localize complex carbohydrates on the intestinal epithelium (Figure 12-3, top) and in the kidney (Figure 12-4, bottom) in unstained sections. In the top figure, which is at low magnification, the glycocalyx is stained as are the Golgi apparatus and lysosomes and rough endoplasmic reticulum. A heavily stained Golgi of a goblet cell is seen at the right of the figure. In the bottom figure, which is at a moderate magnification, the glycocalyx of the capillary (C) and of the glomerulus (G) of the kidney is heavily stained (arrow). (Micrograph courtesy of C. P. Leblond taken from *J Cell Biol*, Vol. 40:395-414 and Vol. 32:27-53 respectively, and used with permission of the publisher.)

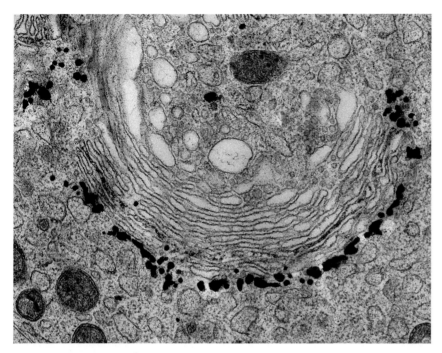

Figure 12-5 This Golgi apparatus impregnated with osmium indicates that the Golgi is not a chemically homogeneous organelle. Osmium is only impregnated on the cis-face of the Golgi. (Micrograph courtesy of D. Friend.)

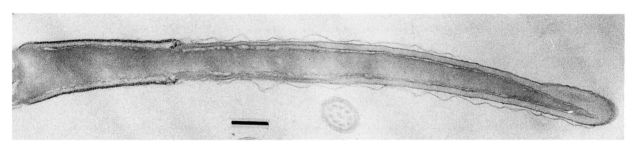

Figure 12-6 Localization of cations on sperm with pyroantimonate. This unstained preparation shows that cations are not evenly distributed over the cell surface. Bar = 1.0 μm.

Each paper should be consulted for details of the localization protocols as well as for the theoretical basis of staining.

photunstic acid imparts the electron density to the protein molecules being localized.

Protein

The phosphotungstic technique is generally regarded as a rather nonspecific technique for protein localization. In the referenced technique (Silverman and Glick, 1969), phosphotungstic acid most likely acts as an anionic stain for positively charged groups of proteins. The heavy tungsten molecule of phos-

Sterols

Two compounds, filipin (polyene antibiotic) and saponin (plant glycoside) are known to form a complex with certain sterols such as cholesterol. In doing so, they cause a perturbation of the membrane, which appears distinctive in thin sections and in freeze fracture preparations (Figures 12-7 and 12-8).

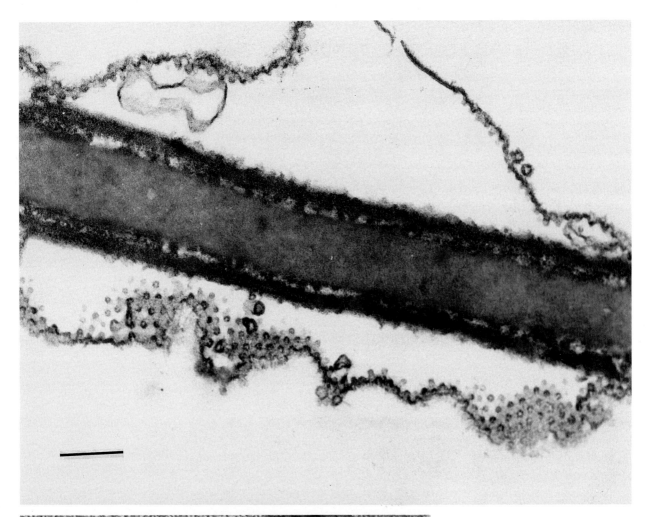

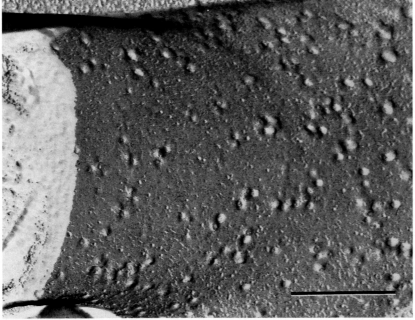

Figure 12-7 and 12-8 Filipin sterol membrane perturbations shown in thin section and freeze fracture. In thin section, filipin sterol complexes within the membrane appear as small corrugations of the membrane (arrowhead) and in freeze fracture they appear as bumps and indentations (arrows) that are much larger than intramembranous particles. Bar = 0.25 μm.

References

Actin

Ishikawa, H., R. Bischoff, and H., Holtzer, 1969. Formation of arrowhead complexes with heavy meromyosin in a variety of cell types. *J Cell Biol* 43:312–28.

Russell, L. D. 1986. Characterization of filaments within the subacrosomal space of rat spermatids during spermiogenesis. *Tissue Cell* 18:887–98.

Carbohydrates/Oligosaccharides

Bernhard, W., and S. Avrameas. 1971. Ultrastructural visualization of cellular carbohydrate components by means of Concanavalin A. *Exptl Cell Res* 64:232–36.

Nicolson, G. L. 1978. Ultrastructural localization of lectin receptors. In *Advanced techniques in biological electron microscopy. Specific ultrastructural probes*, J. K. Koehler, ed. Heidelberg: Springer-Verlag, pp 1–38.

Roth, J. 1983. Application of lectin-gold complexes for electron microscopic localization of glycoconjugates on thin sections. *J Histochem Cytochem* 31:987–99.

Using Tannic Acid and Metals

Sannes, P. L., et al. 1978. Tannic acid-metal salt sequences for light and electron microscopic localization of complex carbohydrates. *J Histochem Cytochem* 26:55–61.

Using a Modified PAS-Schiff Reaction

Ainsworth, S. K., et al. 1972. Alkaline bismuth reagent for high resolution demonstration of periodate-reactive sites. *J Histochem Cytochem* 20:995–1005.

Rambourg, A., et al. 1969. Detection of complex carbohydrates in the Golgi apparatus of rat cells. *J Cell Biol* 40:395–414.

Using Colloidal Iron and Colloidal Thorium

Albersheim, P., et al. 1960. Stained pectin as seen in the electron microscope. *J Biophys Biochem Cytol* 8:501–6.

Rambourg, A., and C. P. Leblond. 1967. Electron microscopic observations on the carbohydrate-rich cell coat present on the surface of cells in the rat. *J Cell Biol* 32:27–53.

Golgi Complex/Multivesicular Body

Friend, D. S. 1969. Cytochemical staining of multivesicular body and Golgi vesicles. *J Cell Biol* 41:269–79.

Glycogen Membranes

Karnovsky, M. J. 1971. Use of ferrocyanide-reduced osmium tetroxide in electron microscopy. *J Cell Biol, Abstract #284.*

Russell, L. D., and S. Burguet. 1977. Ultrastructure of Leydig cells as revealed by secondary tissue treatment with a ferrocyanide-osmium mixture. *Tissue Cell* 9:751–66.

Ions

Gasic, G. J., et al. 1968. Positive and negative colloidal iron as cell surface stains. *Lab Invest* 18:63–71.

Happel, R. D., and J. A. V. Simpson. 1982. Distribution of mitochondrial calcium: Pyroantimonate precipitation and atomic absorption spectroscopy. *J Histochem Cytochem* 30:305–11.

Nucleic Acids (DNA/RNA)

Bendayan, M. 1981. Electron microscopical localization of nucleic acids by the use of enzyme-gold complexes. *J Histochem Cytochem* 29:531–41.

Bendayan, M., and E. Puvion. 1984. Ultrastructural localization of nucleic acids through several cytochemical techniques on osmium-fixed tissues: Comparative evaluation of the different labelings. *J Histochem Cytochem* 32:1185–91.

Protein

Silverman, L., and D. Glick. 1969. The reactivity and staining of tissue protein with phosphotungstic acid. *J Cell Biol* 40:761–7.

Sterols

Elias, P. M., et al. 1979. Membrane sterol heterogeneity: Freeze-fracture detection with saponin and filipin. *J Histochem Cytochem* 27:1247–60.

Quantitative Electron Microscopy

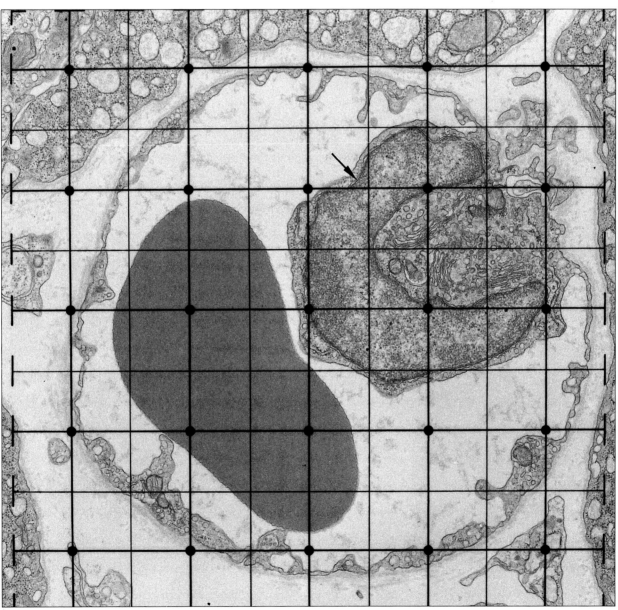

Courtesy of R. P. Bolender with permission of the Anatomical Record.

The purely descriptive studies that predominated at the beginning of biological electron microscopy have gradually evolved to include more experimental approaches and specialized techniques. One such area of specialization is the analysis of thin sections to obtain *quantitative information*. During the period of descriptive morphology, vague terms such as "few," "sparse," "numerous," "abundant," "plentiful," and "many" were frequently used as the only expression of quantity. Imprecise expressions of this type are no longer considered appropriate in situations where it is important to make a definitive statement about cell or tissue features. It is well known that micrographs may be selected arbitrarily by the investigator to portray a finding as being representative when, in fact, the micrograph shows atypical features. Unfortunately, this has a negative impact on the reputation of all microscopists. Additionally, Chapter 18 shows how tissue sections may give misleading information about the size and shape of structures. Thus, for a finding to be a credible one, it is important to use quantitative techniques, whenever possible, to assure the reader that what is being shown is representative of the entire sample.

In some experimental situations, it is important to express findings quantitatively, but in a *relative* way, e.g., "the volume of a particular structure was increased three-fold after treatment." This statement is a quantitative measure and, as such, is much preferred to estimates that are subjective and based on a simple examination of tissue features. It provides a relative measure of volume and is useful when made in comparison to the volume of another structure. It also is necessary to express quantity in an *absolute* sense to correlate structural findings with physiological responses or biochemical measurements, e.g., "the volume of a particular structure was 210 μm^3. After a particular treatment of the animal, its volume increased three-fold to 630 μm^3." Absolute volumes may either stand alone as a descriptive finding or be used to make comparisons in an experiment. Expression of structural features in quantitative terms, using either relative or absolute measures, has greatly enhanced the credibility of many morphological studies.

Quantitative measurements may be obtained directly from the microscope (see Chapter 15), from negatives using a densitometer or a measuring device, from micrographs using a ruler, or from serial thin section reconstruction. The obvious challenge is to obtain quantitative information about three-dimensional biological structure from thin sections. Because of its extreme thinness in relation to most of the elements contained within it, the thin section is considered essentially a *two-dimensional structure* having length and width, but no thickness. Figure 13-1 shows that a structure with three dimensions (e.g., expressed as μm^3) in life will reveal only two dimensions (expressed as μm^2) in section. In a similar way, a two-dimensional object (μm^2) will reveal only one dimension (μm) in all thin sections except in grazing sections. One-dimensional objects appear as points and are essentially dimensionless.

The major aspect of quantitative electron microscopy that deals with quantitative interpretations of thin sections is called *morphometry* or *stereology*. Technically, stereology is the geometric interpretation of sections, which are for most purposes two-dimensional images. Morphometry is the quantitative analysis of structures using, for example, stereological principles. In a practical sense, both terms apply to studies in which geometric and/or quantitative information may be derived from thin sections. Some investigators use the terms interchangeably. Stereology will be employed in the following discussion.

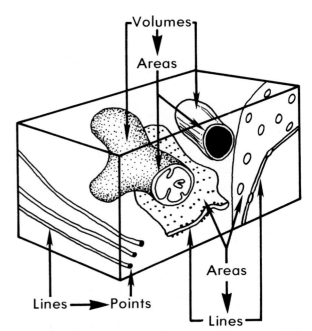

Figure 13-1 Images in 3-dimension versus images in section as diagramed on the cut face of a portion of a cell. On the section face, volumes are viewed as areas, areas are viewed as lines, and lines are viewed as points. The surface of a section shows N minus 1 dimensions as compared with the three-dimensional object.

Basic stereological principles, as they apply to biological specimens, are largely borrowed from the field of geology. In geology, it is necessary to determine the composition of rocks based on a view of a cut surface of the rock. The principle formulated by Delesse (1847), which directly relates fraction area, as viewed on the surface of a geologic specimen, to volume density of the internal aspects of a geological specimen, is used heavily in biology. Biological stereologists have expanded on the early geological principles to provide new and innovative ways to apply stereological principles. The *International Society of Stereology* has standardized the language of stereology and promotes this field of quantitative biology. The *Society for Quantitative Morphology* has recently been formed in the United States to communicate advances in the field.

When to Use Stereology

The number of quantitative determinations that may be made on biological specimens is virtually unlimited. Stereological determinations are usually very time consuming, so it is important first to decide if stereology is the proper procedure, and what should be the scope of the stereological determinations.

There are two major types of stereological analyses, descriptive and experimental. Descriptive stereology involves determinations for many tissue features. A *descriptive* study is used to quantitate many features of a structural component. This kind of study serves as a reference base for future studies that have an experimental or comparative basis. It also allows a comparison between structure and function (see the example of stereology at the end of the chapter).

Experimental stereology may be used to prove or disprove a particular point (also called *targeted* stereology) or may be used to determine any difference between control and treatment groups. The difference between these two types of studies is the breadth of the stereological determination. In the former, a very limited number of stereological determinations are indicated. In this instance, one sets out with a hypothesis about some quantitative tissue feature. There may be a debate, for example, of whether or not smooth endoplasmic reticulum increased after a particular treatment. Although a study of this kind targets only smooth endoplasmic reticulum, one must usually use both light and electron microscopy to obtain the final answer. In a study to determine differences between control and treatment samples, there are many parameters potentially examined. Essentially, the entire gamut of tissue features must be quantitated to answer this question. The important point is to determine which parameters are likely to differ. In this case, before electron microscope studies are undertaken, it is usually beneficial to perform a subjective study at both the light and electron microscope levels. Sometimes preliminary (rough) quantitative studies are useful. Based on the findings, one is usually able to limit the stereological study to keep it from becoming unwieldy.

As mentioned previously, the need to use stereology is usually indicated from subjective observations in a descriptive study. Once the decision is made to proceed with stereology, then one usually conducts a light microscopic study followed by an electron microscopic study. An example of sequential stereology undertaken in this manner has been provided by Sinha Hikim et al. (descriptive study, 1988a; light microscope study, 1988b; electron microscope studies, 1989a and b).

General Scheme of Stereology

There are many ways to obtain quantitative information about biological specimens. Some are simple and some are exceedingly complex. There is no way that an introductory text can cover them all satisfactorily. Therefore, only basic principles will be discussed for the kinds of stereology of primary interest to biologists. If more information is required or if one is planning a stereological experiment, consult the references at the end of this chapter for more detailed information.

There is a standardized schematic or plan that applies to most stereological determinations (Figure 13-2). The *specimen* is prepared for electron microscopy and sections are made. *Micrographs* are taken and *test systems* (points, intersections, lines, etc.) applied to them in a systematic manner. From the *counts* obtained, formulas are applied to obtain three-dimensional information or *results*. *Statistical tests* are performed to analyze the quality of the data or to make comparisons between groups or formulate *conclusions*. Although, at a glance, this plan appears simple, there are numerous considerations at every step of the way.

In planning a stereological experiment, it is generally best to work backwards. Begin by determining

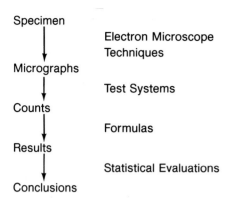

Specimen

↓   Electron Microscope
         Techniques
Micrographs

↓   Test Systems
Counts

↓   Formulas
Results

↓   Statistical Evaluations
Conclusions

Figure 13-2   Scheme of Stereology.

what kind of structural features will be evaluated statistically. Determine what parameter (volume, surface area, etc.) is desired. Determine what formulas will lead to a given parameter and, finally, what test systems will give the desired counts, and so forth.

Parameters Measured and Symbols Used in Stereology

What are some of the parameters obtainable through stereological techniques? Basically, these are *area, volume, surface area, length,* and *number* although, less commonly, other parameters are determined. Each of these is abbreviated according to convention (Table 13-1). The kinds of counts used (points, intersections, etc.) in test systems are also abbreviated for convenient use in formulas. Subscripts identify how a particular value is expressed. For example, V, by itself, stands for volume (e. g., μm^3) and V_V stands for a volume of a structure per unit test volume or volume fraction ($\mu m^3/\mu m^3$ or expressed $\times$ 100 as a volume percentage; V_v %). For example, a nucleus might be said to have a volume (V) of 600 μm^3 or to occupy 21% (V_v %) of the entire cell. Sometimes additional subscripts are added to refer to a particular cell or organelle (e.g., V_{Vcell} means volume fraction of a cell).

Tissue Compartments or Spaces

Although it may seem obvious, it is always important to remember that a living organism occupies space. An organ system occupies part of the space or a *compartment* within the organism. Each organ can

Table 13-1   Convention for Abbreviation of Stereological Symbols

Symbol	Meaning	Expressed As
V	Volume of a structure	μm^3
V_V	Volume fraction or density	$\mu m^3/\mu m^3$
v	Mean volume of an individual element	μm^3
A	Area	μm^2
A_A	Area fraction or density	$\mu m^2/\mu m^2$
$\overline{A}$	Mean profile area (A_A/N_A)	μm^2
L	Length of a test line	μm
L_L	Linear fraction, i.e., length of a line on a feature per unit test length	$\mu m/\mu m$
L_A	Length per unit test area	$\mu m/\mu m^2$
L_V	Length per unit test volume	$\mu m/\mu m^3$
S	Surface area	μm^2
S_V	Surface area per unit test volume	$\mu m^2/\mu m^3$
$\overline{S}$	Mean surface area	μm^2
S_v	Surface area to volume ratio	μm^{-1}
P	Number of points	self-explanatory
P_P	Number of points on structure/total points(P_T)	self-explanatory
N	Number of features	self-explanatory
N_A	Number of profiles of a feature/unit test area	μm^{-2}
N_V	Number of profiles of a feature/unit test volume	μm^{-3}
Q	Number of transection points	self-explanatory
I_L	Number of line intersections/line length	cm^{-1}
d	Profile diameter	μm
$\overline{d}$	Mean diameter of profiles	μm
$\overline{D}$	Mean diameter	μm
T	Section thickness	μm

then be subdivided into a number of smaller compartments (e.g., tissues, tissue spaces, cells, organelles, etc.) based on structural properties seen with the naked eye or with the light and electron microscopes. The smaller compartments may be divided again into even smaller compartments based on finer structural subdivisions, until the smallest compartments that can be resolved at the electron microscope level are identified (for an example of compartments in one organ, see Figure 13-3). Generally, stereological analyses focus on only one aspect of a

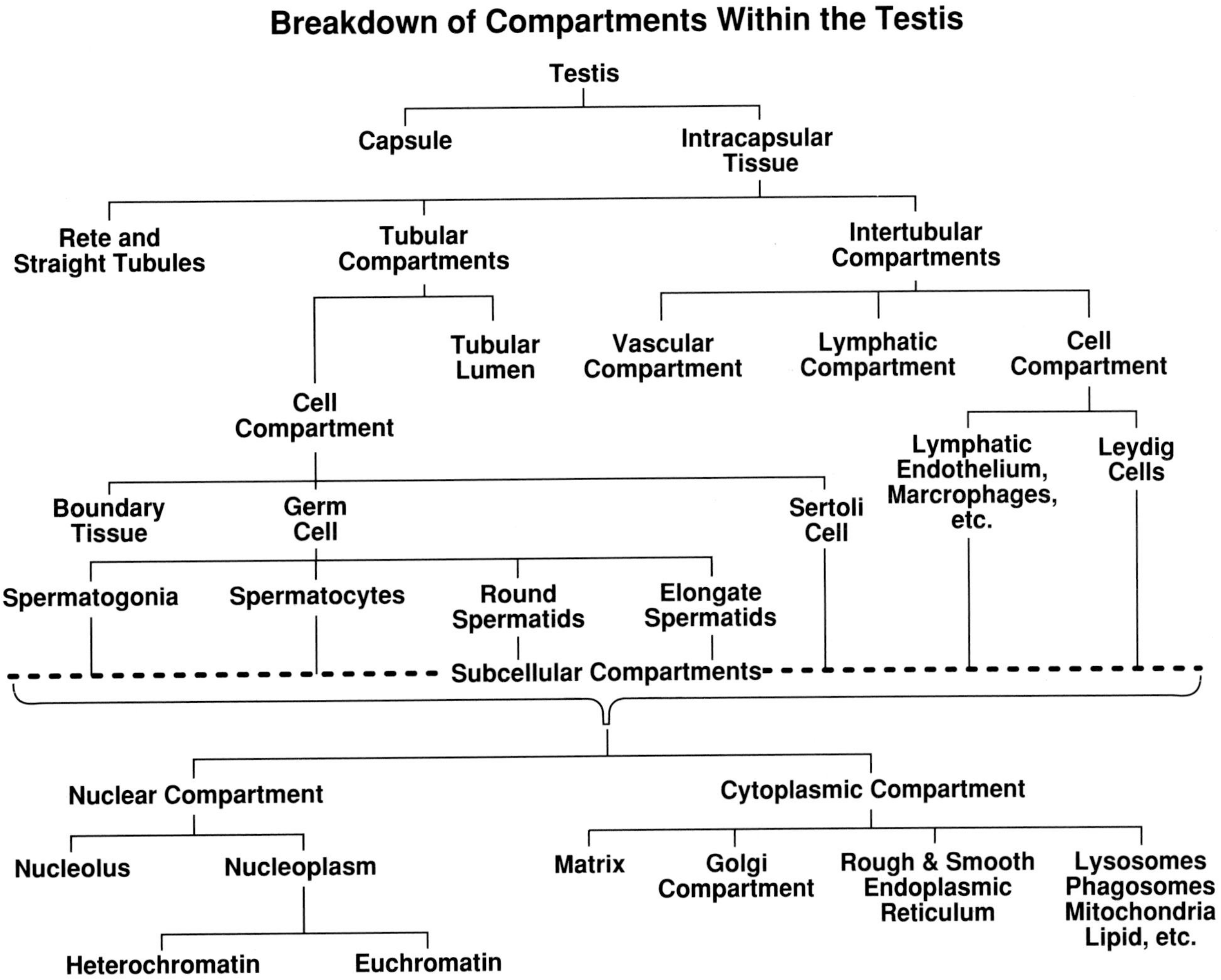

Breakdown of Compartments Within the Testis

Testis

- **Capsule**
- **Intracapsular Tissue**
  - **Rete and Straight Tubules**
  - **Tubular Compartments**
    - **Cell Compartment**
      - **Boundary Tissue**
      - **Germ Cell**
        - Spermatogonia
        - Spermatocytes
        - Round Spermatids
        - Elongate Spermatids
      - **Sertoli Cell**
    - **Tubular Lumen**
  - **Intertubular Compartments**
    - **Vascular Compartment**
    - **Lymphatic Compartment**
    - **Cell Compartment**
      - **Lymphatic Endothelium, Marcrophages, etc.**
      - **Leydig Cells**

- - - Subcellular Compartments - - -

- **Nuclear Compartment**
  - **Nucleolus**
  - **Nucleoplasm**
    - **Heterochromatin**
    - **Euchromatin**
- **Cytoplasmic Compartment**
  - **Matrix**
  - **Golgi Compartment**
  - **Rough & Smooth Endoplasmic Reticulum**
  - **Lysosomes Phagosomes Mitochondria Lipid, etc.**

Figure 13-3 Compartments within an organ using the testis as an example. For each cell type listed immediately above the dotted line, the subcellular compartments are shown below the dotted line.

particular organ or tissue, due to the *labor-intensive* nature of conducting even a limited stereological study. For every tissue compartment in Figure 13-3, one may determine volume, surface area, number, etc., making the possible data accumulation unwieldy. It is best to limit the information obtained to tissue spaces that are likely to prove most interesting and to expend one's efforts in performing a limited, but quality, stereological study.

Since many compartments can be visualized in their entirety at the light microscope level, their evaluation is best conducted at that level. Large tissue spaces may be evaluated using low magnification light microscopy, whereas smaller ones require higher magnification light microscopy. The same principle applies to the electron microscope evaluation for measurements of cellular and subcellular compartments where stereologic evaluations may take place at two or more different magnification ranges. It is important to note that even limited stereological evaluations commonly use both light and electron microscopy in concert.

Test Systems

A variety of test systems may be overlaid on micrographs to obtain counts that will be substituted into formulas. Each of these test systems is to be used only under specified conditions. For now, the test systems will be described briefly and illustrated. Their full significance will not be appreciated until their use in obtaining counts is discussed later.

The most commonly used test system is the *point lattice* (ordered system of points), which is either directly overlaid on the micrograph as a transparency, or fitted directly into the eyepiece of a microscope. In practice, the same effect as a point lattice is usually obtained with a *square lattice*, where points are considered to be the intersection of any two lines. Square lattices may be designed with widely spaced bold lines indicating broad subdivisions and much finer lines for closely spaced subdivisions (*double lattice*; Figure 13-4). These systems are primarily used in volume density determinations. In performing volume density determinations, one might use only the intersections of broad lines for objects that, because they are either numerous or large, occupy a large percentage of a micrograph. All intersections (bold and fine) should be used for objects that occupy a small portion of the micrograph. A double lattice system thus allows simultaneous and efficient determinations of volume density of bigger objects with the bold grid and determinations of smaller or sparse structures with the finer grid.

The *parallel line test system* or a specially designed system of short lines known as a *multipurpose test system* (Figure 13-5) can be overlaid on micrographs to determine intersections of a test line with structures that appear to be single dimensional on a micrograph (e.g., a membrane). Both the length of the test line and the number of intersections that the test line makes with the structure of interest will be needed to obtain surface density and surface area. The multipurpose test system was designed to allow both volume density (the intersections at the line ends are used to count point hits) and surface area determinations (intersections of the line with pertinent surface contours on the micrograph structure) on a single test system. The *Mertz curvilinear test system*, based on similar principles, but employing curved lines, is used in certain circumstances (anisotropic tissues are tissues that vary in the orientation of their components depending on the plane of section, e.g, epithelia, muscle) where surface density and surface area determinations are important (Figure 13-6).

Basic Types of Determinations and Associated Formulas

Under the appropriate conditions (see Assumptions and Conditions, next section), counts obtained from test systems may be entered into mathematically derived formulas that are used to obtain parameter estimates. The following parameter estimates described are volume, area, surface area, and number.

Volume Determinations

The point or square lattice test systems are placed over and affixed firmly to the micrograph. Points or intersections coincident with (overlaying) the particular structure (e.g., mitochondria, P_{mit}) of interest are counted. These are commonly termed "hits." Figure 13-7 illustrates a micrograph with a superimposed square lattice. The total number of hits or points overlying the cell cytoplasm containing the structure (P_{tot}) are also ascertained. To determine the relative volume of the mitochondria or volume fraction (V_V), the following formula is applied:

Equation 13.1

$$V_V = \frac{P_{mit}}{P_{tot}}$$

The result obtained may be multiplied by 100 to express the data as a volume percentage (V_V %).

Having obtained the volume fraction, it is next often desirable to know the absolute volume (V) of the structure in question. Before the absolute volume can be obtained, it is necessary to know the absolute volume (V) of the larger tissue space in which the smaller is contained. To continue with the example used above, mitochondria are contained within the cell cytoplasm so that the absolute mitochondrial volume per cell ($V_{mit/cell\ cyto}$) can be expressed as the product of the volume of the cell cytoplasm ($V_{cell\ cyto}$) and the volume fraction of mitochondria (V_{Vmit}) as follows:

Equation 13.2

$$V_{mit} = V_{Vmit} \times V_{cell\ cyto}$$

One can continue to extend the reasoning pattern set forth in finding the volume of the mitochondria in the particular cell cytoplasm to finding the volume of those mitochondria in the cell (nucleus and cytoplasm) and even the volume of mitochondria for that particular cell type in the organism. Equations 13.1 and 13.2 are, therefore, basic for most stereologic determinations.

At some point in the sequence of making determinations, it becomes important to obtain the *volume of a cell*, often one of the more difficult determinations in stereology. The approaches for accomplishing this are numerous and sometimes difficult. All possess some degree of inherent error. A few approaches are described below.

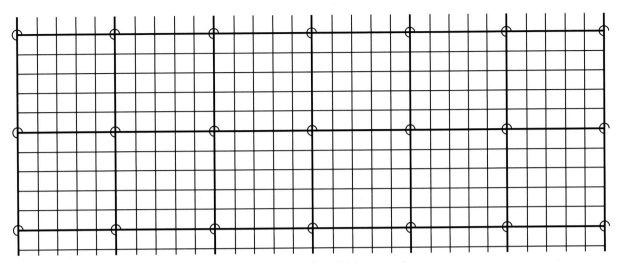

Figure 13-4 A double square lattice test system. Bold and fine divisions are shown.

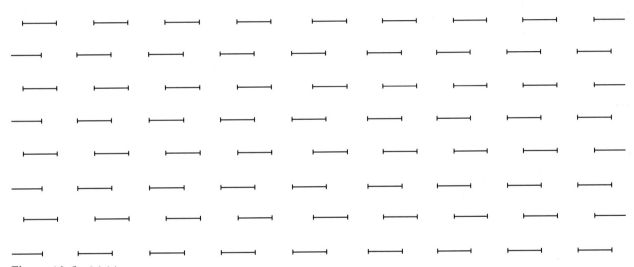

Figure 13-5 Multipurpose test system.

Figure 13-6 Mertz curvilinear test system.

Volume Determination Methods for Spherical Objects

If the cell or object (e.g., nucleus) in question is spherical and each cell of one type is of uniform size, then one simply need measure the diameter (or radius) of the cell and apply the appropriate formula:

Equations 13.3 and 13.4

$$V_{cell} = 4/3 \; \Pi \; r^3 \;\; or \;\; V_{cell} = 1/6\Pi \; d^3$$

(Cells, especially those in compact tissues, are usually not isolated such that their cell boundaries are clearly distinguishable by light microscopy, nor are they usually spherical. However, their nuclei are usually clearly distinguishable and frequently spherical. This is why most stereologic determinations are directed first at determining the volume of nuclei as a prelude to determining cell volume.)

Equations 13.3 and 13.4 imply that one can determine the diameter of cells or nuclei from their appearance in sections. The real diameter (Figure 13-8), as measured in sections, or *profile diameter* (d) is obtained by selecting and measuring the largest profiles in thick sections (0.5–2 μm). This is only true if the largest profiles are selected from a much larger population since only a certain percentage of cells or nuclei will have been sectioned through their middle or through their real diameter. By systematically searching through serial sections, the largest profile diameter or real diameter may be selected. Also, by cutting extremely thick sections and staining them heavily, one is able to find the focal plane, which reveals the largest profile diameter. In the last technique, focusing up and down on the section should cause the spherical cell (or nucleus) to appear small, then appear progressively larger, reach a maximum diameter, and then progressively diminish in size. The maximum profile diameter is to be measured and used as the real diameter.

Most light microscopes can be equipped with a special measuring device in the eyepiece called an *ocular micrometer*, which first must be calibrated with another micrometer that fits on the stage of the microscope, the *stage micrometer*. The stage micrometer is an absolute scale that is divided into microns (or fractions of millimeters), whereas the ocular micrometer is a ruler with an arbitrary scale that must be calibrated for each magnification. The tissue elements are measured with the aid of the ocular micrometer and then converted into absolute measurements using the stage micrometer.

When using spherical cells or nuclei, the option is also available to calculate a mean diameter ($\overline{D}$) from a population where the mean profile diameter ($\overline{d}$) is known. The formula is as follows:

Equation 13.5

$$\overline{D} = 4/\Pi \times \overline{d}$$

For this method to be accurate, it is important that all profile diameters in a randomly selected area be measured. In a strict sense, this formula applies only to profiles from a population of uniform size.

Mean cell volume (V_{cell}) in systems with spherical nuclei can be determined by using both the numerical density ($N_{V\,nuc}$) measurements of the nucleus (assuming cells are mononucleate) obtained, for example, by the Floderus equation (Equation 13.12), and the volume density (V_{Vcell}) measurements of the cell obtained by point counting. The relationship used to obtain cell volume is as follows:

Equation 13.6

$$V_{cell} = \frac{V_{Vcell}}{N_{Vnuc}}$$

The cell volume can also be obtained from the nuclear volume, if the volume fraction of the nucleus (N_{Vnuc}) within the cell is known. Any of the previously mentioned methods to calculate nuclear volume (V_{nuc}) can be employed. To determine the cell volume (V_{cell}), the following expression is used:

Equation 13.7

$$V_{cell} = \frac{V_{nuc}}{N_{Vnuc}}$$

The volume fraction of the nucleus can also be obtained by point counting, usually at the light microscope level. V_{Vnuc} is expressed as the the points over the nucleus divided by the total points over the nucleus and cytoplasm. Such a determination is one of the most common in stereology.

Volume Determinations for Nonspherical Objects

Unfortunately for the stereologist, most cells are not spherical and/or they vary in size. Many nuclear types are also nonspherical. Determining the volume of nonspherical structures necessitates using different approaches than for spherical objects to obtain mean cell or nuclear volume determinations. If the cells are of regular geometric shapes and their dimensions known, specific geometrical formulas may be applied to calculate volume after measuring some parameter at the light microscope level. Individual size differences are handled by giving a mean size accompanied by some measure of variability such as standard error or standard deviation.

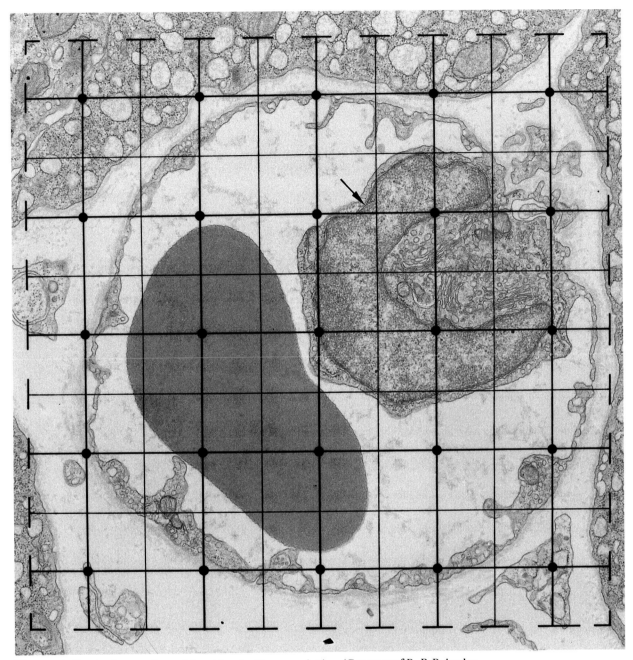

Figure 13-7   A micrograph overlain with a double square lattice. (Courtesy of R. P. Bolender with permission of the Anatomical Record.)

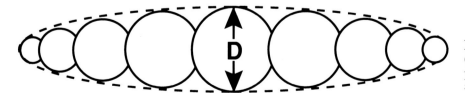

Figure 13-8   The real diameter (D) is measured from the largest profile diameters in a set of serial sections.

A time-consuming, but relatively accurate method for obtaining object size is to *reconstruct an object from serial sections* of known thickness. The

areas of individual profiles are summed to obtain the total area. The volume (μm^3) is then the product of the summed areas (μm^2) of the profiles of interest

Figure 13-9 Tracings of the profiles of a serially sectioned nucleus that is of an irregular shape. The summed areas (ΣA) of all profiles times the section thickness (t) equals the nuclear volume (V$_n$).

$$V_N = \Sigma \ \text{Areas (A)} \ \times \ \text{section thickness (t)}$$

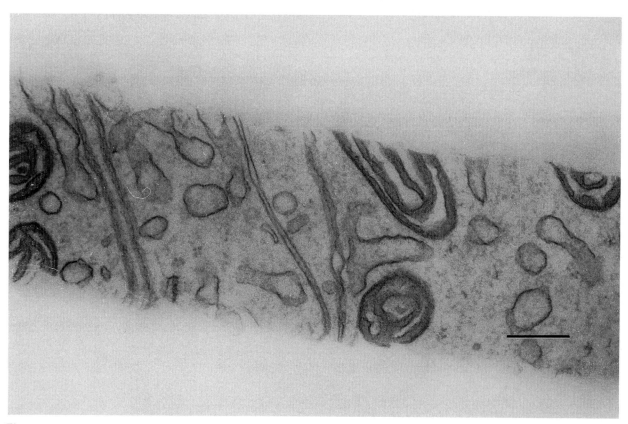

Figure 13-10   Re-embedded thick section measuring 0.09 μm in width (thickness). Bar = 0.25 μm.

and the section thicknesses (μm). Figure 13-9 is a summary diagram of this method. Thick sections are often used to make serial reconstructions and areas determined at the light microscope level by one of the methods discussed below.

Although microtomes have various gauges indicating the thickness of the sections produced at a particular setting, the scales on the microtome may not be an accurate measure of section thickness. Many investigators will reembed thick (0.5 to 2.0 μm) sections in epoxy and make thin sections perpendicular to them to measure section thickness in

the electron microscope. Figure 13-10 shows such a section, which is approximately 0.9 μm in thickness.

The thickness of paraffin sections may be obtained readily by high power light microscopy. If the bottom and top portions of the sections are brought into focus separately, the distance traveled on the fine focus knob in going from one to another is directly proportional to the section thickness. The measurement equivalent to the gauge reading may be determined from the microscope manual. The accuracy of this method is sometimes questioned.

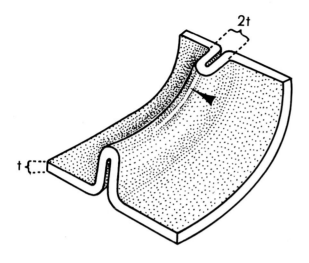

Fold method for measuring section thickness

Figure 13-11 Diagram of a small fold in a tissue section of the kind utilized to make section thickness determinations. The thickness of the fold is approximately twice as thick as the section thickness. (See also, Figure 13-12)

Various methods for the determination of the thickness of ultrathin sections have been reviewed by De Groot (1988). The thickness of ultrathin sections, as determined by their interference color (see Chapter 4), may only be approximated. A simple technique has been developed to determine the thickness of ultrathin sections, but only in situations where an elevated fold appears in the tissue section (Figures 13-11 and 13-12). At high magnification, this fold has an electron dense line that is the middle space of the fold that is occupied by stain. The section thickness is roughly 1/2 the measured width of the fold. If one is making serial sections of spherical objects, the mean section thickness may be determined by finding out how many sections are necessary to traverse the sphere of a given diameter. Determine the diameter of the sphere from the largest profile section of the sphere and divide it by the number of sections that traversed the sphere to obtain section thickness.

Recent developments in light microscopy have opened the door to making certain volume determinations possibly a routine and relatively easy task. With the new *con-focal microscopes* (several commercial versions are now available, see Figure 13-15), it is possible to examine fresh or fixed cells and to "section them optically" with lasers. A computer interfaced with the con-focal microscope will determine the areas of the sections and compute the

volumes based on the sum of the thicknesses. Although relatively expensive at present, it is expected that con-focal microscopes will soon facilitate the difficult problem of volume determinations, especially when cells and nuclei are irregularly shaped.

Area Determinations

Rarely do investigators wish to determine *area* parameters (A or A_A) as an end point in their measurements. They do, however, use area density (A_A) measurements from micrographs and equate them with volume density or fraction ($A_A = V_v$) if certain conditions are met, such as random sectioning and sampling from compartments of the same size (see Assumptions and Conditions later in this chapter). Volume density is then used in a relative way to make comparisons between samples. For example, area density (A_A) measurements of lipid droplets can be compared to determine if there is a difference in the amount of lipid in one sample as compared to another. There are many ways to determine area density and, in general, they are some of the least difficult stereological procedures. Micrographs are two-dimensional images from which area density measurements may be made directly.

Area density (A_A) is converted to area (A) using the following formula:

Equation 13.8

$$A = A_A \times A_{ref\,vol}$$

where $A_{ref\,vol}$ is a known area, such as the area of a cell profile. To calculate area (A) from area density measurements, one must take into account the magnification of the micrograph (equation 13.9).

Equation 13.9

$$\text{Area (A)} = \frac{\text{Area (A) as measured on the micrograph}}{\text{magnification}^2}$$

Area at the magnification of the micrograph may be determined with a simple measuring device called a *planimeter* or a slightly more sophisticated instrument called a *digitizer*. For a relative measure, some investigators have actually cut out images from micrographs and weighed them on a sensitive balance (*cut-and-weigh method*). The additive weights for one sample may be used to make comparisons with those from another sample if the total test area of the two are similar. To obtain area (at the micrograph magnification) from A_A using the cut-and-weigh method, a known area of photographic paper

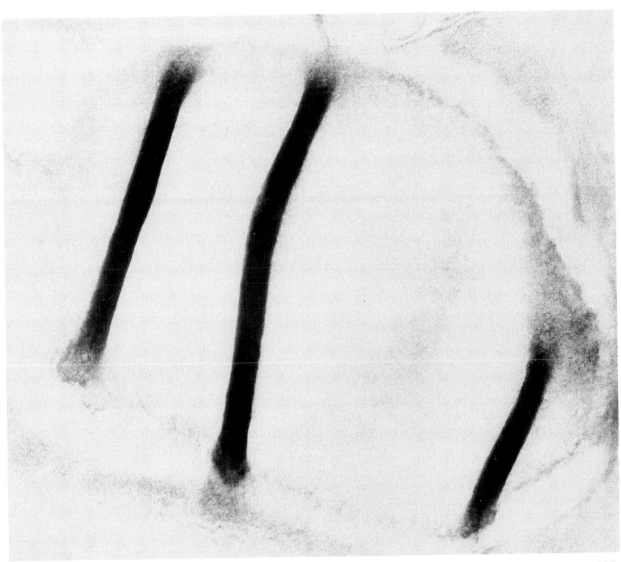

Figure 13-12 A small fold in a tissue section. The micrograph is purposely underexposed to show the internal detail of the fold. There is a slight space through the middle of the fold occupied by stain.

(1 cm²) is weighed and used as a reference [A = sample wt × reference wt (1 cm²)]

Areas may be determined from microscope projections (i.e., *camera lucida*) drawings, although the area obtained must be divided by the magnification² (Equation 13.9) to determine unmagnified area. A_A measurements can be made utilizing a *point counting test system*. As mentioned above, A_A and V_V are equivalent.

Surface Density and Surface-to-Volume Determinations

The *parallel line test system* is used for surface density determinations. The general principle employed is that there is a relationship between the number of times that test lines intersect a one-dimensional structure (line) in a micrograph and the surface area of a two-dimensional structure. Of course, in the micrograph, the two-dimensional structure was sectioned to appear in the micrograph as a line (see Figure 13-1). The formula used is:

Equation 13.10

$$S_V = 2 \times I_L$$

where S_V is the surface area per unit volume ($\mu m^2/\mu m^3$) and I_L is the density of intersections on a test line length, i.e., the number of intersections of the linear probes (test line) with the surface contour of a given profile divided by total length of the test lines.

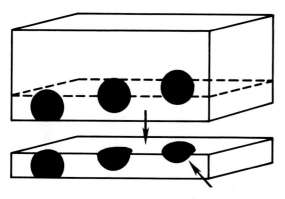

Polar Region or Cap

Figure 13-13 The polar region or *cap* of the nucleus of a spherical object in a section removed from the block face.

A *multipurpose test system* (Figure 13-5) is used to determine surface-to-volume ratio. Generally, this test system is used to measure organelle surface area per cell volume. This system has lines that can be used to determine intersections with the structure of interest as well as end of line intersections that may be used as points for volume density determinations. The surface-to-volume ratio (S/V) is expressed as cm²/cm³ and is obtained as follows:

Equation 13.11

$$S/V = \frac{4 \times I}{P_t \times Z}$$

where I is the number of intersections. P_t is the number of points at the ends of the lines and Z is the length of the line expressed in relationship to the magnification of the micrograph (i.e., the distance in micrometer on the micrograph equivalent to the line length measured).

Numerical Density Determinations

The number of structures may be expressed per organ (N), per area (N_A in cm²) or per volume (N_V in cm³). Once the *numerical density* (N_V) is obtained, it is easy to determine the number of structures per organ by multiplying the numerical density (which is the number of a given structure in a unit volume of an organ) by the organ volume. Let us first examine how numerical density can be obtained.

The *Floderus equation* is the most commonly used means to determine numerical density of spherical cells. This equation is frequently employed at the light microscope level to provide data essential for the logical extension of the study to the ultrastructural level (see A Typical Published Report

later in this chapter). The formula is based on counts of nuclei as seen in section (profile counts).

Equation 13.12

$$N_V = \frac{N_A}{(T + d - 2h)}$$

A calibrated device fitted into the ocular of a microscope is used to determine the number of nuclei per unit area (N_A). The average nuclear profile diameter ($\bar{d}$) and section thickness (T) are determined by one of the methods described above. The height (h) of the *cap* (which is the height of the smallest observed nuclear profile in the section) is placed in the formula. Cap size serves as a correction for the small polar regions of a nucleus, which become lost to the eye since they are grazed during sectioning and therefore are only present in part of one section. A cap is shown in Figure 13-13. There is no practical way to determine cap size. Operationally, many investigators simply assume cap size to be about 1/3 of the section thickness or 1/10 of the diameter of the spherical object.

Sometimes the investigator knows both the mean nuclear volume (v_N) and the total volume of a population of cells (V_{Vnuc}). It is then relatively simple to calculate the numerical density:

Equation 13.13

$$N_V = \frac{V_{Vnuc}}{v_N}$$

Given the test systems and work necessary to obtain the counts to enter into formulas, it is easy to see why stereology is often considered tedious. Only after considerable effort and calculations are the final results obtained; however, these may validate the time and effort spent in the project. Only after working with the various equations set forth above will one really understand stereologic schemes.

Assumptions and Conditions

Stereological methods are statistical in nature, so it follows that the investigator can inadvertently prejudice the data through selection and bias. If one wants to determine, for example, the volume density of a structure, it is important that the particular structure neither be selected for nor avoided. Stereological formulas were derived with the idea that structures will be encountered randomly. *Randomness* in selecting animals, tissue blocks, sectioning

planes, and areas for morphometry will assure that the likelihood of encountering an object or a surface is proportional to its occurrence in the tissue section. When performing stereological determinations, the added work to assure randomness is well worth the effort. Being very careful to insure randomness will instill confidence in the final outcome.

Many tissues exhibit an ordered pattern of their elements. The choice of a sampling method becomes particularly important in these systems so that a particular element is neither selected nor avoided. Since it is difficult to generalize broadly about procedures for all possible conditions, it is best to examine the literature and determine how sampling has been conducted for a particular tissue.

Multiple Stage Sampling

As mentioned above, stereology at the electron microscope level is often performed initially at the light microscope level. This is usually a requirement when absolute, rather than relative, measurements are required at the electron microscope level. For example, the volume fraction or relative area (A_A) of mitochondria may be obtained rather readily by point counting in both normal and treated animals at the ultrastructural level. Suppose an investigator recorded a 50% decrease in V_V of mitochondria. If an investigator based a conclusion solely on the initial volume density determinations, he or she might have erroneously suggested that there are fewer mitochondria. This data indicates only the area density of mitochondria, but not necessarily whether the cells containing the mitochondria have grown, shrunk, multiplied, etc., to produce this effect. It is very possible that this same treatment resulted in a 50% decrease in the volume density (V_V) of mitochondria, but that the total volume (V) of a population of mitochondria has remained unchanged (Figure 13-14). This could occur if the cell grew in size. There are other possibilities as well, one of which would be a change in size of mitochondria. This example illustrates that volume density data could not, by itself, provide this type of information. For most tissues, it is only possible to determine both relative and absolute parameters such as V_V and V if light and electron microscopy are employed together in a *multiple stage sampling protocol*.

To obtain an absolute volume, the volume of the largest tissue compartment must be determined at the onset of the morphometric study. This volume

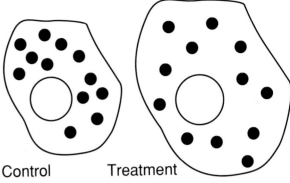

Figure 13-14   In the two cells shown, V_V measurements would have recorded a volume density for the solid structure in the control as being twice that seen after treatment. Measurement of the cell size at the light microscope level using multiple stage sampling indicates that the volume (V) of the structure has not changed on a per cell basis; the cell has simply grown.

serves as a reference for all of the volume determinations made for smaller compartments (Figure 13-3). The largest tissue space (usually an organ) is weighed and/or its volume is determined by submersion. The volume of the fluid displaced by submersion is equal to the volume of the submerged object. Often it is inadvisable to use water to submerse biologicals, since it is nonphysiologic and may have deleterious effects on tissue structure. Isotonic solutions are preferable.

A more accurate way to determine volume is to measure fluid displacement by a weighing method. The weight difference (W) of an isotonic solution before and during submersion of the tissue can be used to determine the tissue compartment volume (V) if the specific gravity (sp gr) of the fluid is known. The relationship is as follows:

Equation 13.14

$$V = \frac{W}{sp\ gr}$$

Processing and sectioning will usually cause some change (usually a decrease) in a tissue's original volume, such that another determination of volume may be necessary and a correction factor employed in the calculations.

After tissue processing and sectioning, the volume fraction (V_V) of the compartment of interest is determined by point counting (see above). The volume (V) of that smaller compartment is the product of the volume fraction of the smaller compartment and the volume of the larger tissue compartment (or organ) in which it is contained.

Equation 13.15

$$V_{smaller\ comp} = V_{V\ smaller\ comp} \times V_{larger\ comp}$$

Once the volume of the larger compartment is known, it is evident that one may tackle progressively smaller and smaller compartments in a systematic manner to obtain absolute volume data. Equation 13.15 is similar to Equation 13.2 and is one of the most commonly utilized formulas in stereology.

Light microscopy is useful to obtain point counting information from the larger tissue compartments. As far as analysis of smaller compartments, one cannot usually advance beyond nuclear and cytoplasmic compartments (using $\times 1,000$ magnification) to obtain useful point counting information simply due to the limited resolution of the light microscope. Electron microscopy is then used beyond this point. Two or three steps (magnification levels) of electron microscopy may be necessary to determine volume fractions and surface areas for both large and small organelles.

How Much Data from Test Systems is Needed?

When performing point counting or using the line intersection method, one must consider how many points need to be counted or intersections measured to obtain adequate results. As a general rule, as data is added, the standard error reaches a value that changes little with the addition of more data. Thus, the standard error first will lower rapidly, and then begin to level off as more data are added.

Most stereologists accept that a standard error (e_A^2) of less than 10% of the mean is adequate. In such circumstances, the number of points (P) to be counted, where x equals the investigator's preliminary estimate of the percentage of the sample occupied by the structure of interest, is given in Equation 13-16. To achieve a standard error of 10% of the mean with the tissue component estimated to occupy about 5% of the sample, it is necessary to record about 1,900 points. The equation for calculation of points is as follows:

Equation 13.16

$$P = \frac{1 - x}{E_A^2 \cdot x} = \frac{1 - 0.05}{(0.1)^2 \times 0.05} = 1,900$$

Of course, the actual standard error should be calculated after the actual data are obtained.

Computer-Assisted Stereology

An advantage of employing stereological techniques is that they are not particularly expensive. They may be performed by hand and calculations made on a hand calculator. Computers and software, if purchased specifically for stereology, are a major expense. They will speed up certain aspects of the project such as data acquisition, storage, tabulation, and computations using stereological formulas and also statistical evaluation of the data. *Automatic image analyzers* take information from two-dimensional images such as video tapes and then record the counts more accurately than simple point counting by eye. Stereological formulas may then be applied by directing the computer to do so. Automated image analyzers have limited abilities to recognize micrograph images, so it is first important to determine if this equipment will aid in the specific project under consideration. Both computer-assisted methods and automatic image analyzers are recommended when stereology will be used on a routine basis (Figure 13-15).

A Typical Published Report

As an example of how stereology is performed in an experimental setting, the published article by Mori and Christensen (1980) follows, in brief. To be fully appreciated, the article must be read in its entirety. The methods section is especially helpful in describing how the stereology was performed.

Rationale for the Study

This study is a good example of descriptive stereology. Besides providing baseline data on the Leydig cell population in the rat testis, the authors wished to correlate the cellular and subcellular structure of the *Leydig cell* with its major function of *steroid production* (testosterone). Since testosterone production rates are known and the particular cell type and the organelles involved in this process are well known, the authors targeted the Leydig cell and particularly its mitochondria and smooth endoplasmic reticulum for morphometric determinations.

Methods

Figure 13-3 shows the space or compartment occupied by the Leydig cell and its organelles. The *organelles* are within the *Leydig cell* compartment and the Leydig cell is within the *intertubular compartment*. The intertubular compartment is considered to be within a *parenchymal compartment* within the *testis*. With this information, it was possible to plan a stereological experiment. The basic plan of the methods employed were as follows:

Tissue Preparation

Specific gravity of the tissue was determined by floating the testis in sucrose solutions of known concentration

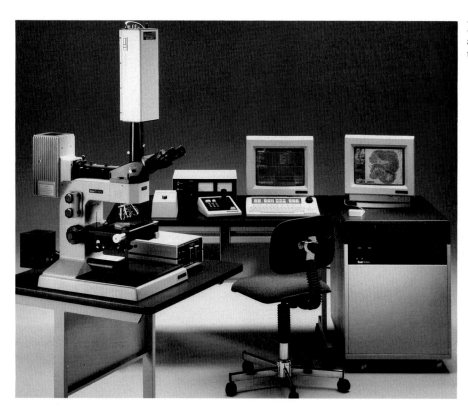

Figure 13-15 A modern-image analysis system. This system utilizes a con-focal microscope.

and then dividing the weight difference (weight after addition of the testis − weight before addition of the testis) by the specific gravity of the fluid. Tissues were fixed in glutaraldehyde and the osmolarity controlled to prevent excessive tissue shrinkage. After fixation, the tissue was sliced and alternating slices used for light microscopy (methacrylate embedding) and electron microscopy (epoxy embedding). Measurements were made during tissue processing to determine extent of shrinkage. For electron microscopy, ferrocyanide was added to the osmium fixative to enhance membrane contrast since many membrane compartments would be measured. Methacrylate sections were cut at 2 μm for light microscopy and epoxy-embedded thin sections used showed silver to pale gold interference colors.

Morphometric Procedures at the Light Microscope Level

The volume of the parenchyma was determined by subtracting the volume of the capsule (connective tissue covering of the testis) from the total volume (parenchyma and capsule). The capsule volume was estimated as the product of the thickness of the capsule and the surface area of the testis. A point counting system, employing a square lattice, was used for low magnification microscopy to obtain the volume density (V_v) of the interstitial cell compartment. At higher magnification, the volume density of the Leydig cell compartment was also determined by point counting. Leydig cell nuclear volume density was measured in the same manner. The numerical density of Leydig cells was determined by the Floderus equation.

(Strictly speaking, this formula is for spherical nuclei, although most Leydig cell nuclei are not spherical.) The mean volume of the average Leydig cell was obtained by dividing the volume density (V_v) by the numerical density (N_v). The number of Leydig cells in the testis was found as the product of the numerical density and the testis weight.

Stereologic Determination at the Electron Microscopic Level

Since Leydig cells occupied only a small portion (2.7%) of the tissue, sampling throughout the tissue could not be performed with random electron micrographs. Instead, Leydig cells were photographed, taking care not to duplicate micrographs of the same cell. (This sampling procedure was not a good example of random sampling, but it is difficult to sample randomly when such a small percentage of volume of a tissue has the cell type of interest.) A two-stage sampling system was utilized to perform analyses on Leydig cells at the ultrastructural level. Micrographs printed at a final magnification of ×10,800 were used to determine the volume density of organelles and to determine the surface density of the plasma membrane. The surface density of organelles was determined at high magnification (×72,000). Point counting with a double lattice grid was used for volume density measurements, and line intersections on a multipurpose grid were used to determine surface density. The appropriate formula for each was employed (for V_v see Equation 13.1 above; for S_v see Equation 13.11). The authors used a complicated scheme to approximate a determination of numerical density of mitochondria (see the actual publication). There is an excellent discussion of actual and possible sources of error in the procedures used, which appears in the methods section. In some cases, correction factors were derived to be applied to raw data.

Results and Significance

The number of published parameters obtained for Leydig cells in this descriptive study were numerous and too extensive to reiterate; thus, only a few will be listed here. The number (N) of Leydig cells in one testis was 22 million although this population of cells comprised only 2.7% of the testis volume ($V_v\%$). The mean Leydig cell volume ($\bar{v}$) was 1,210 μm^3 and the mean surface area ($\bar{s}$) was 1,520μm^2. The two sites of steroidogenic enzymes, the mitochondria and smooth endoplasmic reticulum,

were examined in detail. The surface area (S_{cell}) of smooth endoplasmic reticulum, was about 10,500 μm^2/cell or expressed on a per organ basis about 2,300 μm^2/cm^3 of testicular tissue. The mitochondrial inner membrane demonstrated a surface area of 2,920 μm^2 and there was 644 cm^2 of inner mitochondrial membrane/cm^3 (S/V) of testis. When correlated with published endocrine data, the average Leydig cell secretes about 0.44 pg of testosterone per day. Each square centimeter of smooth endoplasmic reticulum and inner mitochondrial membrane produces 4.2 ng and 15 ng testosterone per day, respectively. This study was the first to correlate cell and organelle structural parameters with functional parameters. As a descriptive morphometric study, it established baseline values for future studies on Leydig cell structure and function.

References

Technical Articles and References

Bolender, R. P. 1981. Stereology: applications to pharmacology. *Ann Rev Pharmacol Toxicol* 21:549–73.

De Groot, D. G. 1988. Comparison of methods for the estimation of the thickness of ultrathin tissue sections. *J Microsc* 151:23–42.

Elias, H., et al. 1971. Stereology: applications to biological research. *Physiol Rev* 51:158–200.

Elias, H., and D. M. Hyde. 1980. An elementary introduction to stereology (quantitative microscopy). *Am J Anat* 159:412–46.

Loud, A.V. 1968. A quantitative stereological description of the ultrastructure of normal rat liver parenchymal cells. *J Cell Biol* 37:27–46.

Mayhew, T. M., and F. H. White. 1980. Ultrastructural morphometry of isolated cells: methods, models and applications. *Pathol Res Pract* 166:239–59.

Toth, R. 1982. An introduction to morphometric cytology and its application to botanical research. *Am J Bot* 69:1694–1706.

Weibel, E. R. 1969. Stereological principles for morphometry in electron microscopic cytology. *Int Rev Cytol* 26:235–302.

Weibel, E. R. 1973. Stereological techniques for electron microscopic morphometry. In *Principles and techniques of electron microscopy*, M. A Hayat, ed. New York: Van Nostrand Reinhold Co.

Weibel, E. R. 1979. Stereological methods. In *Practical methods for biological morphometry*, New York: Academic Press.

Williams, M. A. 1977. Quantitative methods in biology. In *Practical methods in electron microscopy*, A. M. Glauert, ed., Amsterdam: North Holland Pub. pp. 1–84.

Classic Articles

Bolender, R. P. 1978. Correlation of morphometry and stereology with biochemical analysis of cell fractions. *Int Rev Cytol* 55:247–89.

Bolender, R. P., et al. 1978. Integrated stereological and biochemical studies on hepatocyte membranes. Membrane recovery in subcellular fractions. *J Cell Biol* 77:565–83.

Delesse, A. 1847. Procede mechanique pour determines la composition de roches (extrait). *CR Acad Sci* (Paris) 25:544–60.

Eisenberg, B. R., et al. 1974. Stereologic analysis of mammalian skeletal muscle. *J Cell Biol* 60:732–54.

Floderus, S. 1944. Untersuchungen uber den bau der menschlicken hypophyse mit besonderer Berucksichtigung der quantitativen mikromorphologischen Verhailtnisse. *Acta Pathol Microbiol Scand Suppl* 53:1–276.

Loud, A. V., et al. 1965. Quantitative evaluation of cytoplasmic structures in electron micrographs. *Lab Invest* 14:996–1008.

Weibel, E. R., et al. 1969. Correlated morphometric and biochemical studies on the liver cell. *J Cell Biol* 42:68–91.

Recent Reports

Bowers, B., et al. 1981. Morphometric analysis of volumes and surface areas in membrane compartments during exocytosis in *Acanthomeba*. *J Cell Biol* 88:509–15.

Buschmann, R. J. 1983. Morphology of the small intestinal enterocytes of the fasted rat and the effects of colchicine. *Cell Tissue Res* 231:289–99.

Mori, H., and A. K. Christensen. 1980. Morphometric analysis of Leydig cells. *J Cell Biol* 80:340–54.

Ryoo, J. W., and R. J. Buschmann. 1983. A morphometric analysis of the hypertrophy of experimental liver cirrhosis. *Virchow's Arch Pathol Anat* 400:173–86.

Sinha Hikim, A. P., A. Bartke, and L. D. Russell. 1988a. The seasonal breeding hamster as a model to study structure-function relationships in the testis. *Tissue Cell* 20:63–78.

———. 1988b. Morphometric studies on hamster testes in gonadally active and inactive states: light microscope findings. *Biol Reprod* 39:1225–37.

———. 1989a. Correlative morphology and endocrinology of Sertoli cells in hamster testes in active and inactive states of spermatogenesis. *Enodocrinology* 125:1829–43.

———. 1989b. Structure/function relationships in active and inactive hamster Leydig cells: a correlative morphometric and endocrine study. *Endocrinology* 125:1844–56

Weber, J., et al. 1983. Three-dimensional reconstruction of a rat stage V Sertoli cell. II Morphometry of Sertoli-Sertoli and Sertoli-germ cell relationships. *Am J Anat* 167:163–79.

Wong, V., and L. D. Russell. 1983. Three-dimensional reconstruction of a rat stage V Sertoli cell. I Methods, basic configuration and dimensions. *Am J Anat* 167:133–61.

Freeze Fracture Replication

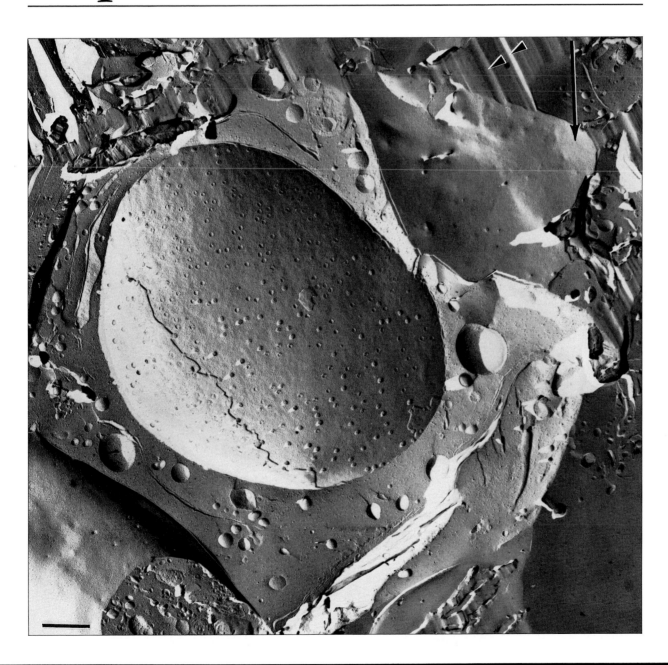

Freeze fracture is a technique for the replication of fractured surfaces of frozen specimens for their examination in the transmission electron microscope. It is especially suited for studies of *membrane structure* and particularly the internal aspect of the lipid bilayer of the membrane. Freeze fracture provides three-dimensional information encompassing the general organization of tissues and cells to the macromolecular organization of the membrane. The technique has contributed much of what we know about the structure of cell *junctions* and has confirmed that many cell types, such as the epithelial cell or sperm cell, have organized regions of their plasma membrane that are functionally diverse.

To understand how freeze fracture derives its usefulness, it is important to recognize how a typical *membrane* is organized. The current view of membrane organization was proposed by Singer and Nicolson (1972). In this model, proteins exist in a sea of *phospholipid* molecules. The phospholipid of membranes is organized into a bilayer with the polar heads of the molecules facing and interacting with the polar aqueous environment of both the cell in-

terior and cell exterior. The hydrophobic nonpolar carbon chains of each phospholipid layer are directed inward to face each other (Figure 14-1). Thus, the physicochemical properties of the internal and external aspects of the lipid bilayer differ.

If a very sharp knife is used to cut a frozen specimen, the specimen acts like a brittle object and consequently *fractures* more often than it produces a clean cut along the path of the knife (Figure 14-2). In the frozen state, the hydrophobic membrane interior is most subject to fracture when external forces are exerted on the specimen. Since the energy required to split the lipid bilayer is less than is necessary to fracture ice or cytoplasm, the lipid bilayer is frequently split to reveal the *membrane interior* or *membrane faces*.

An example of how a simple cell, the red blood cell, might fracture after it had been frozen is illustrated in Figure 14-2. The ice surrounding the cell is fractured erratically; however, when the fracture reaches the plasma membrane, the lipid bilayer splits yielding two complementary fragments, each attached to the ice. After the fragments are separated from each other, it is possible to look down on each

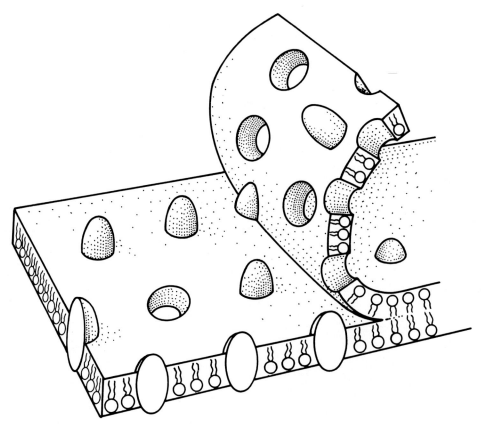

Figure 14-1 A lipid bilayer is split open to reveal its internal faces. After splitting, large proteins embedded in a phospholipid matrix remain with one of the two membrane halves.

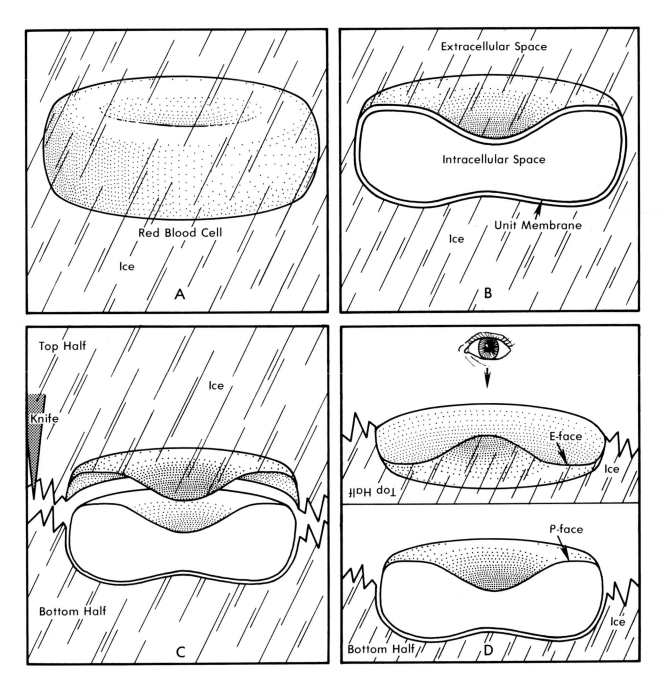

Figure 14-2 Fracture of a frozen red blood cell. (A) A cell is seen frozen in ice. (B) For the reader's convenience, it is depicted split in half from top to bottom allowing its lipid bilayer to be visualized during the subsequent fracturing process. The bilayer separates the intracellular space, containing frozen cytoplasm, and the extracellular space containing ice. Upon fracturing (C) with a knife the lipid bilayer, in the upper portion of the cell, is split down the middle. When both halves of the fractured surface are viewed from above after the top half is flipped over (D), the bottom fragment is seen to contain the face associated with the cytoplasm (protoplasm), or *P-face*, whereas the top fragment (now turned over) contains the face associated with the extracellular space, or the *E-face*.

of them (if one is turned over) and view the split surfaces or fracture *faces*. Specific names have been given to each of the fracture faces. The portion of the bilayer associated with the exterior of the cell is termed the *E-face* or *extracellular face*. The portion

associated with the interior of the cell is termed the *P-face* or *protoplasmic face*[1]. Another way of thinking

---

[1] In the early freeze fracture literature, the P-face was designated the A-face and the E-face the B-face. The present designation allows a better association of the face with its nearest neighbor.

of this is to imagine poking a pin through a particular face. Since the micrograph shows three dimesionality, one's eye is the judge of what appears closer to the viewer (convex) and is the structure first encountered by the pin. If the point of the pin exits the cell, then the face being examined is an E-face. On the other hand, a pin poked through the P-face or protoplasmic face would enter the cell interior or protoplasm.

Assume the fracture traversed a red cell plasma membrane in the manner shown in Figure 14-3. Follow the fracture faces in each surface view to see how complementary regions of the same membrane are viewed and named.

To this point, only examples of fractured plasma membranes have been provided. Of course, other organelles are bounded with unit membranes, which,

like the plasma membrane, may be given P or E designations. In membrane-bounded structures such as lysosomes, secretory granules, Golgi, and endoplasmic reticulum, the interior of the structure is considered extracellular space and the exterior is considered protoplasm, thus allowing the associated faces to be named E-face and P-face, respectively. The nucleus and mitochondrion are bounded by a double lipid bilayer. The internal aspect of the mitochondrium and nucleus is considered protoplasm and the space between each bounding membrane is considered extracellular space. Figure 14-4 shows how internal membrane faces are designated.

To provide an accurate interpretation of freeze fracture micrographs is a challenge. Even experienced investigators have mistakenly misnamed fracture faces in the scientific literature. Since the trans-

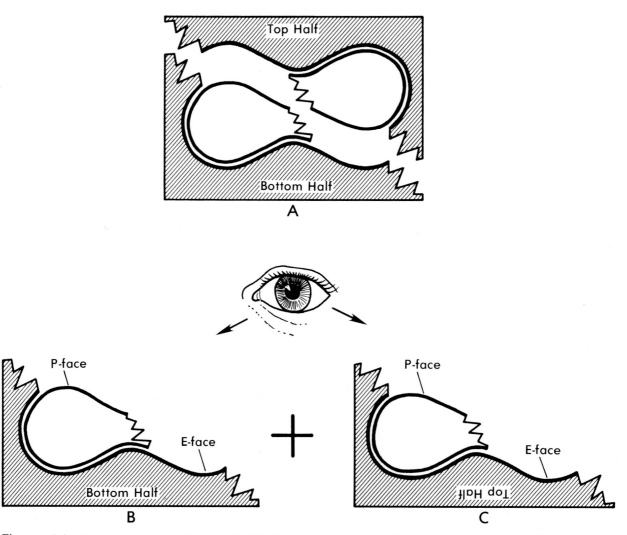

Figure 14-3 Fracture across a red blood cell. (A) The fracture describes a path through both the top and bottom surfaces of the cell. The bottom (B) and top (C) fracture fragments are displayed separately to show how membrane faces are designated. In C, the fracture fragment is turned over so that the fracture surface can be viewed from above.

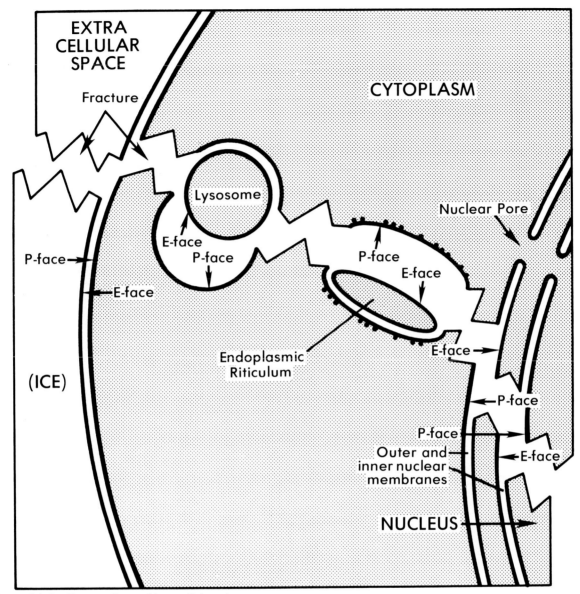

Figure 14-4 Fracture faces of intercellular components.

mission electron microscope does not view the surface of fractured tissues, but rather examines a *replica* of the surface, some guidelines to interpretation of the final image are necessary. Interpretation of freeze fracture is aided by a knowledge of how replicas are produced from biological materials.

How to Produce a Replica

A replica is a thin *platinum-carbon cast* of the fractured tissue surface that is viewed in the transmission electron microscope. Areas of high metal density (those facing the platinum source) are dark in

appearance since the electrons are deflected by heavy metals. Areas of low metal density allow electrons to pass through the replica. The topological features of the biological specimen are conveyed by the replica. An image is formed on the fluorescent screen after penetration of the replica by electrons.

The tissue (epithelia, cell suspensions, cell fractions, cell cultures, or solid tissue) is prepared for fracturing after mild fixation in 0.5% to 3% glutaraldehyde. Small pieces of tissue are infiltrated with 20% to 30% glycerol, a cryoprotectant that inhibits ice crystal formation by hydrogen bonding to water, interfering with its ability to crystallize. Due to the low thermal conductivity of most tissues, it is necessary to freeze tissues extremely rapidly in order to prevent ice crystal formation (Figure 14-5). Rapid freezing takes place in Freon that has been chilled to a slushy consistency at −150° to −155° C using liquid nitrogen. Although it is not as cold, freon is more effective in rapidly freezing the tissue than liquid nitrogen (−196° C) since its contact (wetting property) with the specimen optimizes heat transfer. The specimen is rapidly transferred into a liquid nitrogen storage vessel until needed for fracturing.

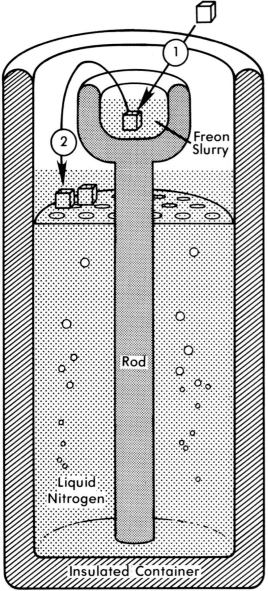

Figure 14-5 Freezing of a biological specimen for freeze fracture. The specimen is first (1) placed in a slurry of liquid Freon which has been cooled by a rod extending down into liquid nitrogen. Next (2), the specimen is rapidly transferred to liquid nitrogen where it remains frozen on a perforated tray until utilized in the freeze fracture apparatus.

Replicas are produced in a freeze fracture apparatus of the kind diagrammed in Figure 14-6. A modern commercial instrument with the same basic features is shown in Figure 14-7. All of the activity related to the specimen takes place in a chamber at the top of the apparatus. The chamber contains a stage on which the specimen is mounted, a moveable knife (razor blade), and electrodes for vaporizing platinum/carbon and carbon onto the specimen. Vacuum lines feed into the chamber from below to allow the carbon and platinum evaporation processes to take place under high vacuum (10^{-6} or 10^{-7} torr) and also to prevent excessive water condensation on the specimen. Liquid nitrogen lines are used to cool both the stage and the knife. In the transfer of the specimen, fracturing, and replication processes, tissues must remain solidly frozen. If they were to thaw in the transfer process, they would refreeze on the specimen stage, but this time more slowly with unwanted ice crystal formation.

Replicas are produced from frozen tissue in the following manner (Figure 14-8):

1. The frozen specimen is rapidly transferred to a precooled (−150° C) stage.
2. The chamber is closed and pumped to high vacuum.
3. The stage is warmed to −100° C and the tissue fractured with a razor blade attached to a rotating arm.
4. Platinum/carbon are evaporated at an angle of about 45° relative to the surface of the specimen.
5. Carbon is evaporated onto the specimen from directly overhead to add strength to the replica.
6. The chamber is brought to room temperature and pressure.
7. The thawed specimen is gently placed in 5% sodium hypochlorite[2] to digest all organic material from the replica. The tissue will bubble as it is digested from the replica. In some tissues, the replica may float free of the specimen.
8. The replica is moved to an additional chlorine bleach change and then through 2 or 3 distilled water washes.
9. The replica is picked up on a mesh grid and allowed to dry.
10. The replica is examined using a transmission electron microscope.

_____

[2] Grocery store chlorine bleach will suffice, although in some instances strong acids may be necessary to digest away organic matter.

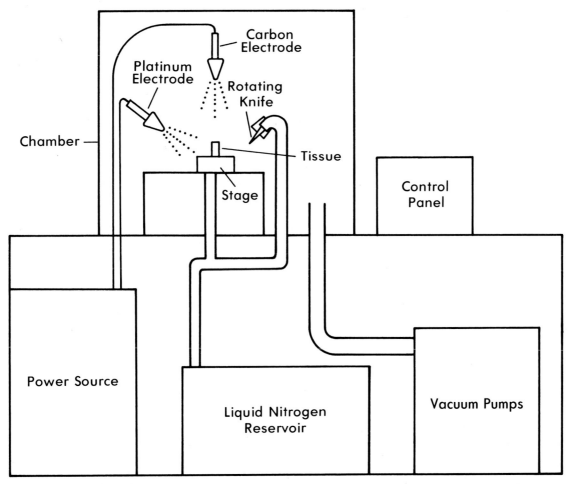

Figure 14-6 Diagrammatic representation of the essential features of a freeze-fracture apparatus. Fracturing of tissue takes place in a *chamber*. The chamber is maintained at high vacuum by vacuum pumps. Both the stage and the knife are cooled by liquid nitrogen. Platinum and carbon electrodes have a power source below. All of these features are operated from the control panel.

Interpretation of Freeze Fracture Replicas

A replica is a delicate, tissue-free, platinum carbon foil that displays the surface contours of the fractured tissue. Figure 14-9 is an electron micrograph that shows that platinum has been deposited differentially on the tissue surfaces that are perpendicular to, or closely aligned with, the direction of platinum deposition. A thicker deposition of platinum retards the penetration of the electron beam more so than does an area with sparse platinum. Some light areas may have no platinum and other darker areas may have platinum up to 5 nm in thickness. The variable and graded nature of platinum deposition imparts a three-dimensional appearance to freeze fracture micrographs, which mimics the actual contour of the fractured surface. At first glance, images appear like the lunar surface and display what appear to be cra-

ters, mountains, and valleys. The platinum particles that have been evaporated onto the specimen are about 2 nm in diameter. Resolution is limited by their large size.

In order to relate the replica image to actual tissue structure, certain guidelines have been developed, which, in most instances, will allow the correct membrane and membrane face to be identified. With experience, the investigator gains added confidence and the reasons for specific rules become clear. The guidelines for interpretation of freeze fracture replicas follow.

Finding a Membrane Face

Membrane faces are smoother areas than where ice alone or cytoplasm is fractured. Membrane faces contain *intramembranous particles (IMPs)*. IMPs are thought to be proteins embedded in the lipid bilayer,

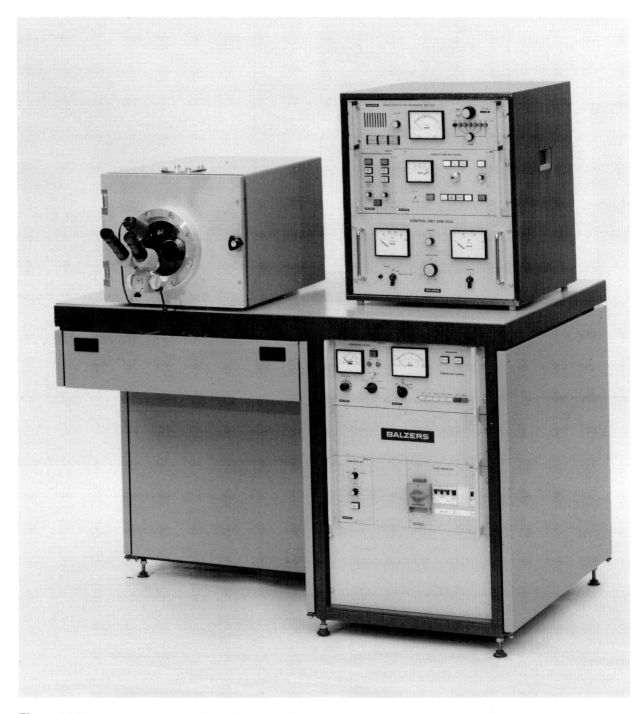

Figure 14-7   Modern commercial freeze-fracture apparatus. (Photograph provided courtesy of the Balzers AG.)

the three-dimensional structure of which is revealed upon fracture. These generally are the smallest distinct units (7 to 15 nm across or most frequently 8.5 nm) visualized in the bilayer (Figure 14-10). At low magnification, intramembranous particles may be difficult to resolve.

Orienting the Micrograph so that the Shading of IMPs is from Below

First, a determination of the direction of platinum shadowing is made by observing which side of the IMP appears darkest. The micrograph is then ori-

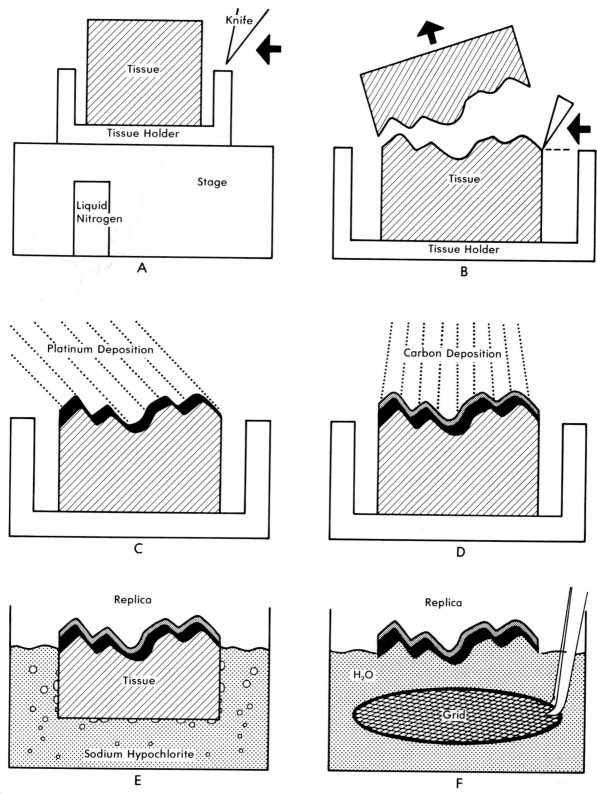

Figure 14-8 Diagram showing the steps in the fracturing and replication process. The frozen, immobilized tissue (A) is fractured (B) by a moving knife. Platinum (C) is deposited at a 45° angle and carbon is deposited (D) directly from above. The replica has a non-uniform layer of platinum and a relatively uniform carbon support layer. (E) The tissue is digested from the replica in sodium hypochlorite. (F) After water washes, the replica is picked up by a grid from below.

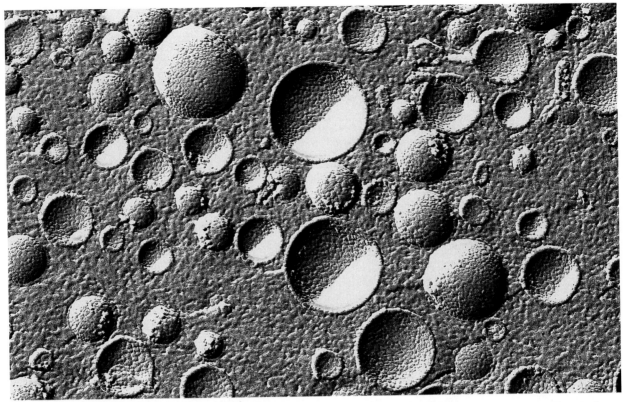

Figure 14-9 Electron micrograph showing differential deposition of platinum in membrane vesicles. The light areas have little platinum whereas the dark areas have a heavy deposition of platinum.

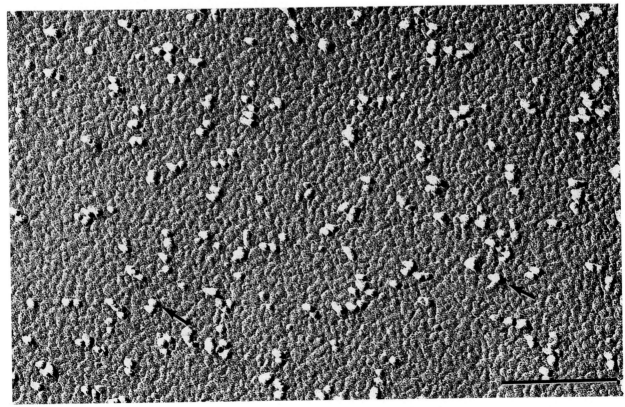

Figure 14-10 Intramembranous particles at high magnification (arrows). Bar = 0.25 μm.

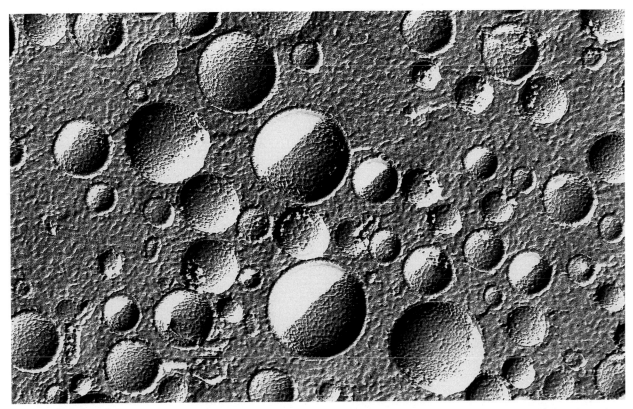

Figure 14-11 Upside down view of the freeze-fracture micrograph shown in Figure 14-9. The topography of one micrograph appears exactly the opposite of the other micrograph, however only Figure 14-9 is oriented correctly.

ented with the dark portion of the IMP facing downward. When the micrograph is oriented in this manner, what appears concave to the eye was indeed a concave feature of the replicated frozen tissue. The same is true of convex features. Figure 14-11 is an upside down view of Figure 14-9. Note how the concavities and convexities of the two micrographs are reversed when the micrographs are reversed in position.

Determining Which Membrane is being Viewed

The interpreter's basic knowledge of cells will help to determine what is being viewed. Cells, of course, are generally rounded or convex overall (Figure 14-12). Protoplasm (cytoplasm) intervenes between the cell membrane and the nucleus and contains numerous organelles and inclusions. This is also the topography represented by the replica. The plasma membrane has the features characteristic of the cell, such as microvilli, surface ruffles, junctions, etc., and can be so recognized. The nuclear membrane is identified by nuclear pores. Simple logic dictates that if the interior of the nucleus is being viewed, then to find the cell exterior it is first necessary to visualize sequentially the nuclear membrane, the cytoplasm, and the plasma membrane. Imagine that you are a minute creature able to walk from cell to cell, and imagine what logical steps must be taken to do so.

Determining Whether the Membrane Face being Viewed is a P-Face or an E-Face

When the surfaces of two cells are adjacent to each other and the plasma membrane faces of both cells are visualized, the E-face is always topographically the P-face (Figure 14-13). This is logical since, to exit one cell, it is necessary to first go through an E-face to enter the extracellular space. Then go through the P-face to enter the cytoplasm of the adjacent cell. This rule holds for the double lipid bilayer forming the nuclear and mitochondrial membranes. For example, if the fracture enters the nucleus from the cytoplasm, the E-face of the outer nuclear membrane will lie on top of the P-face of the inner nuclear membrane.

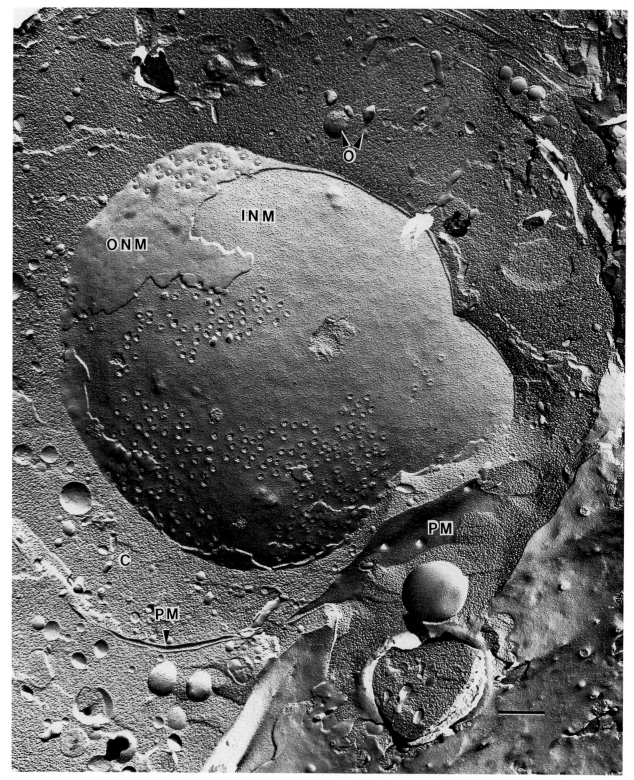

Figure 14-12 General view of a cell with freeze fracture. The cell has a rounded nucleus that is central in the figure and which is bounded by inner (INM) and outer nuclear membranes (ONM). The plasma membrane (PM) and nu-clear envelope sandwich the cytoplasm. Profiles of various organelles (O) are seen sparely scattered within the cytoplasm (C). Other cells adjoin this cell. Bar = 1 μm.

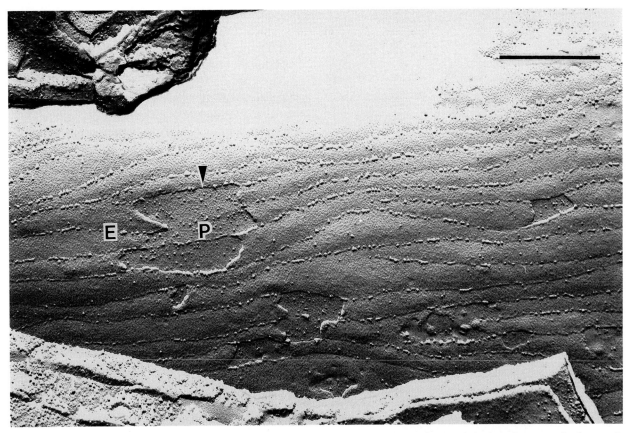

Figure 14-13 Membrane faces of adjacent cells. Where the fracture plane leaves one plasma membrane and jumps down to another, there is a distinct "step down" at this site (arrowhead). The membrane on top is the E-face of the cell that is more superficial and the membrane on the bottom is the P-face of the deeper cell. The rows of particles seen are sites of tight junctions. Bar = 1 μm.

Generally, the P-face contains more IMPs than does the E-face (Figure 14-14).

P-face and E-face may be ascertained by determining the next deepest structure. If the next deepest structure is the extracellular space or its equivalent, then an E-face is being examined. If, on the other hand, a poke through a membrane face leads to protoplasm, a P-face is the face in question.

Figures 14-15 and 14-16 are provided as examples of how to interpret freeze fracture images. The legends to each of these figures allow the reader to follow a pattern of reasoning that might be used to interpret each micrograph. Further examples of freeze fractured tissues are provided in Chapter 19.

Complementary Replicas

A technically difficult, but fascinating, procedure allows one to retrieve two replicas from the same tissue that will show both faces of the same membrane. Instead of fracturing the tissue by scraping the surface, as described above, the frozen tissue may be fractured by breaking it in half. The two surfaces that are replicated form *complementary replicas* (Figures 14-17 and 14-18). The technique for producing complementary replicas is added proof that the lipid bilayer is split down the middle during tissue fracturing. It has also been shown using complementary replicas that, in some cases, there are complementary pits to match IMP particles on the opposing face. Finding areas where replica faces match is a supreme test of an individual's patience.

Freeze Etching

This variation of freeze fracture is employed to examine actual membrane surfaces and not membrane faces as described above. The term freeze etching is often used erroneously to describe the entire freeze fracture process, whereas it is only a modi-

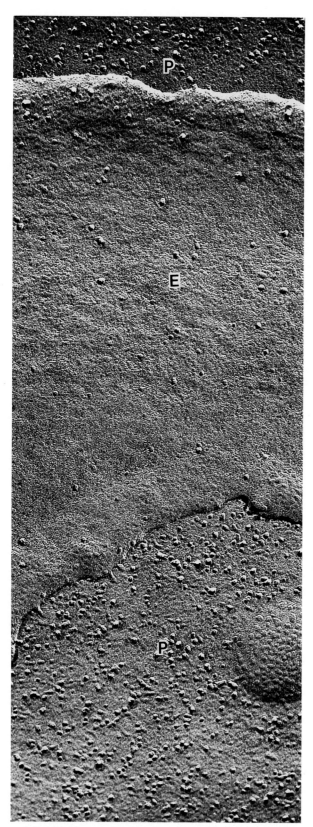

Figure 14-14 Relative abundance of intramembranous particles on the P- and E-faces.

fication of the freeze fracture process. To etch a specimen, the tissue is first cryoprotected, frozen, and fractured in the way described above. The knife is placed directly over the specimen, and the slightly cooler knife facilitates sublimation of the ice from the surface of the specimen onto the knife. Etching of ice from the surface occurs at 2 to 3 nm/sec at a temperature of −100° C. Removal of ice by sublimation allows the *true surface* of the membrane to be exposed (Figures 14-19 and 14-20). Herein lies the value of the technique, since the true surface inevitably provides different information about membrane organization than either one of its faces.

Quick Freeze, Deep Etch, Rotary Shadow Techniques

This fascinating technique, exploited relatively recently, allows an examination of certain insoluble surface and intracellular structures. The three-dimensional detail of membranous structures and cytoskeletal elements is quite striking (Figure 14-21).

Tissue is frozen fresh or after mild glutaraldehyde fixation by a technique known as *quick freezing*. To do this, the tissue is slammed against a polished copper block, which itself is cooled with liquid helium. Amazingly, the tissue does not crush in this process, although the surface 10 to 15 μm is frozen extremely rapidly without ice crystal damage. Subsequently, the frozen tissue is fractured in a freeze fracture apparatus and *etched* (see above) for a period not to exceed 10 minutes. Platinum and carbon are deposited on the specimen as the specimen holder rotates, a process known as *rotary shadowing*. The result is a three-dimensional view of internal cell structures such as that shown in Figure 14-21.

Freeze Fracture Cytochemistry

More recently, techniques have been developed to localize substances using variations in the freeze fracture technique coupled with cytochemical techniques. These techniques are relatively simple, but their full potential has not been exploited. They include *label-fracture* (Pinto da Silva and Kan, 1984) and *fracture-label* (Pinto da Silva, et al., 1981). It is not within the scope of this text to describe these methods, and the reader is referred to the cited methods. In addition, specific localization of lipid

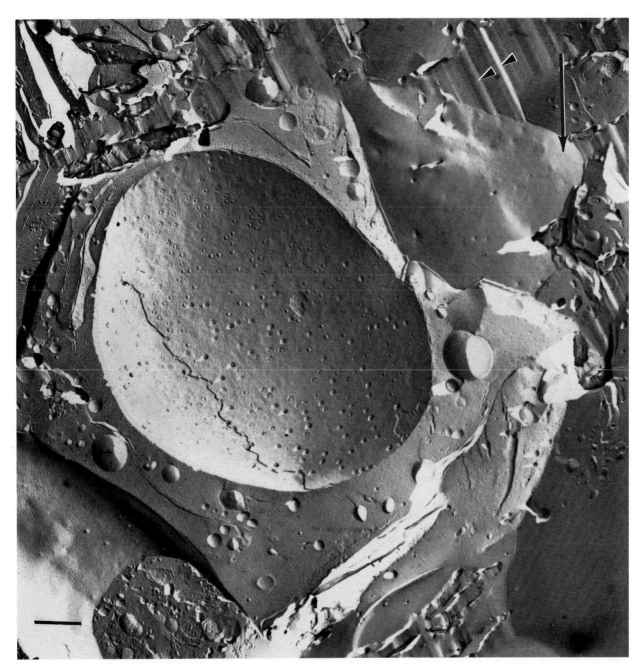

Figure 14-15   In this figure, the magnification is too low to observe intramembranous particles, so the direction of shading has been indicated by placement of the large arrow. Orient the book so that you are looking at the micrograph to make the arrow vertical with its apex (pointed end) upward. (The micrograph has been purposely placed in an atypical orientation so that the viewer must rotate the book. Moreover, most of the features in the figure have not been labeled in order for one to determine for oneself the structures in the micrograph.) The nucleus of a cell is readily identified as a convex structure with numerous pores. Since the nucleus is convex, what is being viewed is the outside of the nucleus. Outer and inner membranes forming the nuclear envelope can be identified using the logic that the outer nuclear membrane would appear to lie on top of the inner nuclear membrane. The E-face of the outer membrane of the nuclear envelope lies on top of the the P-face of the inner membrane. Only small portions of the plasma membrane (the first continuous membrane out from the nucleus) of the cell are fractured in the lipid bilayer. Along one region an adjoining cell has its fractured membrane (E-face) apposing the plasma membrane (P-face) of the cell just described. Knife marks (arrows), which have been replicated, appear as parallel lines on the micrograph. Bar = 1 μm.

Figure 14-16 Orient the figure using the shading of intramembranous particles as a guide. A few prominent particles have been indicated (arrowheads) to assist you. When this is accomplished, the large membrane faces appear *concave*, suggesting that one is within the cell looking outward. The nuclear membrane is identified by its pores and double lipid bilayer. The E-face of the outer membrane of the nuclear envelope is on top of the P-face of the inner membrane of the nuclear envelope. Cytoplasm intervenes between the nuclear envelope and the cell's plasma membrane, which here shows a large fratured area. In exiting the cell, one must go through the E-face of the plasma membrane. The P-face of an adjoining cell (arrow) shows only a small fractured area and lies deep to the E-face of the plasma membrane just described.

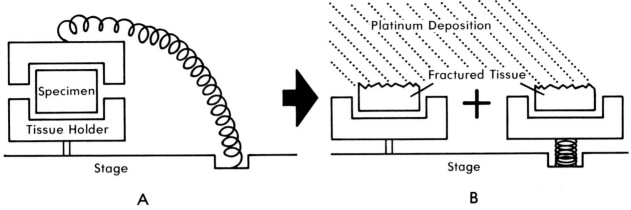

Figure 14-17 Production of complementary replicas. The tissue is held on the stage by a device, similar to that shown in A. Fracturing occurs as the tissue is broken apart by placing tension on the specimen. Subsequently (B), the two matching surfaces are shaded with platinum and carbon to form complementary replicas.

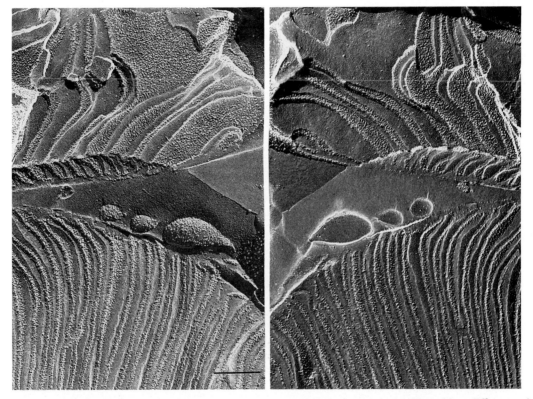

Figure 14-18 Complementary replicas. The two micrographs shown are membrane faces from guinea pig retina outer segments. As the replica was split, both portions of the split tissue were saved and replicated. Match each membrane component on the two replicas. Note differences in the relative abundance of intramembranous particles on complementary faces. Bar = 0.25 μm. (Micrograph courtesy of R. L. Steere, Advanced Biotechnologies, Inc.)

moieties has been achieved using freeze fracture techniques (see Menco, 1986).

Problems and Artifacts

It is possible to produce many replicas and take pictures of them, all in one day—that is, if everything goes as planned. This rarely happens. Of the many problems that may be encountered, machine problems (especially the carbon and platinum guns), tissue problems, freezing problems, and replica problems contribute to making the technique technically demanding.

Poor freezing and accidental warming of tissue may lead to ice crystal formation in tissues, which causes deformation of structure and obscures tissue

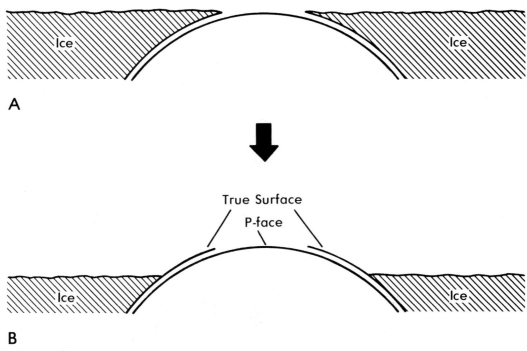

A

B

Figure 14-19   The process of etching. A typical fractured specimen (A) reveals the P-face of a membrane. After etch-ing, (B) the ice level has receded revealing the *true surface* of the membrane.

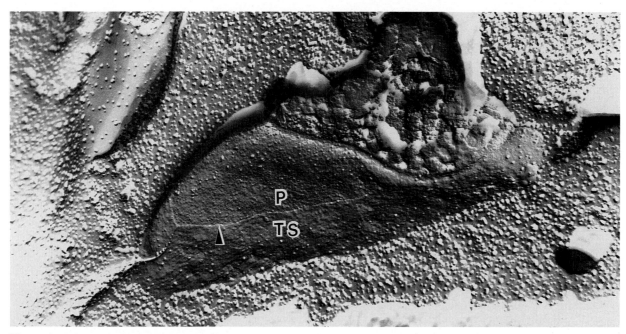

Figure 14-20   A micrograph revealing the true surface of a cell after etching. A thin line or "step down" (arrowhead) separates the P-face of a cell from the true surface.

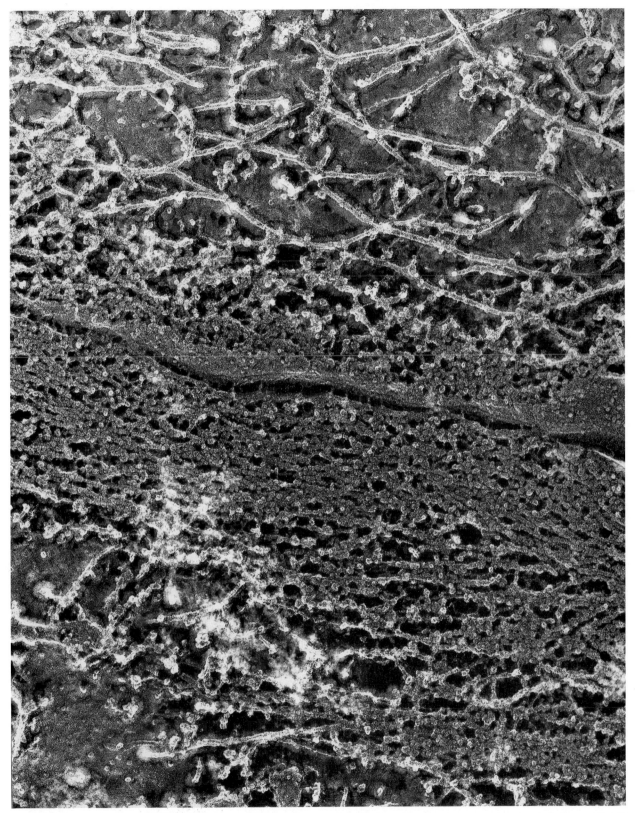

Figure 14-21   A micrograph showing an example of quick freezing, deep etching, and rotary shadowing. Two adjacent cells depict filaments. On the right, intermediate filaments (10 nm) are seen whereas at the center left the smaller (6 nm) actin filaments are visualized. (Micrograph courtesy of R. Kelly.)

Figure 14-22   An example of numerous ice crystals that have formed on improperly frozen tissue.

components (Figure 14-22). If tissues are not infiltrated adequately with glycerol, ice crystals will develop in all but the hardiest of cells (bacteria, yeasts, and spores). During the fracturing process, tissues may be torn from their tissue holder. Fortunately, most modern freeze fracture equipment allows several specimens to be fractured at one time. The platinum and carbon guns frequently malfunction, leading to either too little or excessive deposition of metal on the specimen. More recently produced guns are more reliable.

By far, the most annoying problem is breaking and shattering of the replica during the final steps necessary to free the replica from the tissue and clean its surface. The replica is fragile and brittle and may fragment as the underlying tissue is dissolved. Transfer of the replica from solution to solution during the cleaning process frequently causes it to fragment, a problem that is especially dismaying given the effort already expended to produce the replica.

Micrographs may show evidence of knife marks produced as the knife shaved, rather than fractured, the specimen (Figure 14-15). Extremely dense areas on the micrograph may indicate that tissue has been incompletely removed from the replica during the hypochlorite washes (Figure 14-23).

Membrane faces should occupy about 20% to 60% of the image in replicas of solid tissues. If there are large numbers of *cross-fractures* of membranous elements, it is an indication that too high a concentration of glutaraldehyde was utilized or the fixation process was too long (Figure 14-24).

Replicated membrane faces may show a fine cobblestone effect, giving the appearance of closely packed, but extremely small, intramembranous particles (Figure 14-25). The exact cause of this defect is not known, but is thought to be due to vapor condensation on the specimen. It is important to begin the replication process as soon as possible after fracturing to prevent this artifact.

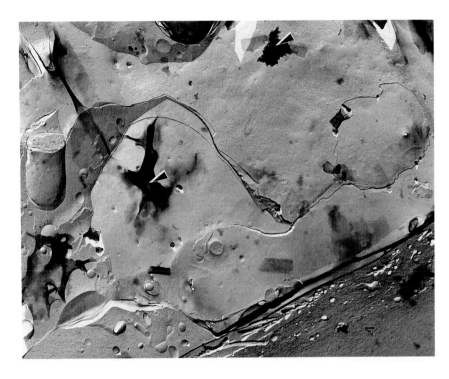

Figure 14-23 Undigested tissue and contamination appear dense on this freeze-fracture replica.

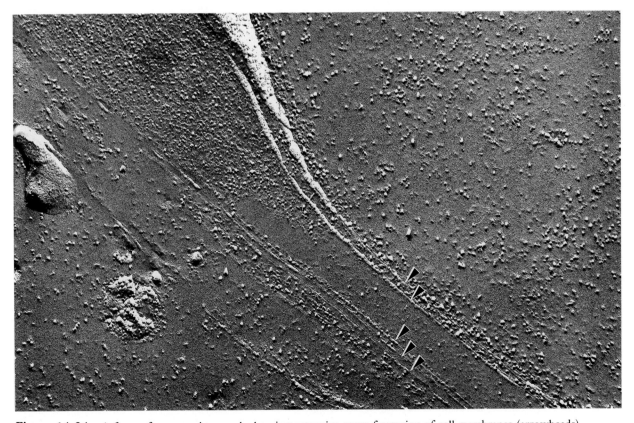

Figure 14-24 A freeze fracture micrograph showing extensive cross fracturing of cell membranes (arrowheads).

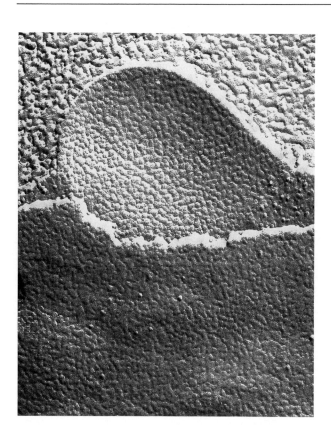

Figure 14-25 Background artifact on a membrane face. The background condensation gives a cobblestone appearance to the membrane face.

References

General References

Heuser, J. 1981. Quick-freeze, deep-etch preparation of samples for 3-D electron microscopy. *Trends Biochem Sci* 6:64–68.

Hui, S. W. 1989. *Freeze-fracture. Studies of membranes.* Boca Raton, Fl: CRC Press, Inc., 208 pp.

Orci, L., and A. Perrelet. 1975. *Freeze- etch histology. A comparison between thin sections and freeze-etch replicas.* New York: Springer-Verlag.

Technique and Methodological References

Branton, D., et al. 1975. Freeze-etching nomenclature. *Science* 190:54–6.

Branton, D., and S. Kirchanski. 1977. Interpreting the results of freeze-etching. *J Microsc* 111:117–24.

Bullivant, S. 1973. Freeze etching and freeze-fracturing. In *Advanced techniques in biological electron microscopy,* J. K. Koehler, ed. New York: Springer-Verlag, pp 67–112.

Hui, S. W. 1989. *Freeze-fracture. Studies of membrane.* Boca Raton, FL: CRC Press, Inc.

Menco, B. M. 1986. A survey of ultra-rapid cryo-fixation methods with particular emphasis on applications to freeze-fracturing, freeze etching, and freeze substitution. *J Electron Microsc Tech* 4:177–240.

Pinto da Silva, P., C. Parkison, and N. Dwyer. 1981. Fracture-label: cytochemistry of freeze-fractured faces in the erythrocyte membrane. *Proc Natl Acad Sci (USA)* 78:343–7.

Pinto da Silva, P., and F. W. K. Kan. 1984. Label-fracture: A method for high resolution labeling of cell surfaces. *J Cell Biol* 99:1156–61.

Rash, J. E., and C. S. Hudson. 1979. *Freeze fracture: Methods, artifacts and interpretations.* New York: Raven Press.

Rebhun, L. I. 1972. Freeze-substitution and freeze drying. In *Principles and techniques of electron micro-*

scopy, M. A. Hayat, ed., New York: Van Nostrand Reinhold Co. Vol. 2. pp 3–49.

Classic References

Branton, D. 1966. Fracture faces of frozen membranes. *Proc Natl Acad Sci* (USA) 55:1048–56.

Bullivant, S., and A. Ames. 1966. A simple freeze-fracture replication method for electron microscopy. *J Cell Biol* 29:435–47.

Friend, D. S., and D. W. Fawcett. 1974. Membrane differentiations in freeze-fractured mammalian sperm. *J Cell Biol* 63:641–64.

Heuser, J. E., et al. 1976. Preservation of synaptic structure by rapid freezing. *Cold Spring Harbor Symp Quant Biol* 40:17–24.

Moor, H., and K. Muhlethaler. 1963. Fine structure of frozen etched yeast cells. *J Cell Biol* 17:609–28.

Moor, H., et al. 1961. A new freezing-ultramicrotome. *J Biophys Biochem Cytol* 10:1–13.

Singer, J., and G. L. Nicolson. 1972. The fluid mosaic model of the structure of cell membranes. *Science* 175:720–31.

Steere, R. L. 1957. Electron microscopy of structural detail in frozen biological specimens. *J Biophys Biochem Cytol* 3:45–60.

Recent Publications

Bearer, E. L., and D. Friend. 1990. Morphology of mammalian sperm membranes during differentia-tion, maturation and capacitation. *J Electron Microsc* 16:281–97.

Friend, D. S. 1982. Plasma-membrane diversity in a highly polarized cell. *J Cell Biol* 93:243–9.

Heuser, J. E., and J. Keen. 1988. Deep-etch visualization of proteins involved in clathrin assembly. *J Cell Biol* 107:877–86.

Ru-Long, S., and P. Pinta da Silva. 1990. Simulcast: contiguous views of fracture faces and membrane surfaces in a single cell. *European J Cell Biol* 53: 122–30.

Shivers, R. R., et al. 1986. Integument of the tapeworm Scolex. Freeze-fracture of the syncytial layer, microvilli and discoid bodies. *Tissue Cell* 18:869–85.

Toyama, Y., and T. Nagano. 1988. Maturation changes of the plasma membrane of rat spermatozoa observed by surface replica, rapid freeze and deep-etch, and freeze-fracture methods. *Anat Rec* 220: 43–50.

Weber, J. E., et al. 1988. Effect of cytochalasin D on the integrity of the Sertoli cell barrier. *Am J Anat* 182:130–47.

Note: *The Journal of Electron Microscopic Technique* has recently published two excellent journal issues (Vol. 13:137–158 and Vol. 13:277–371, 1989) on freeze fracture history, technique, and results of freeze fracture studies.

The Analytical Electron Microscope

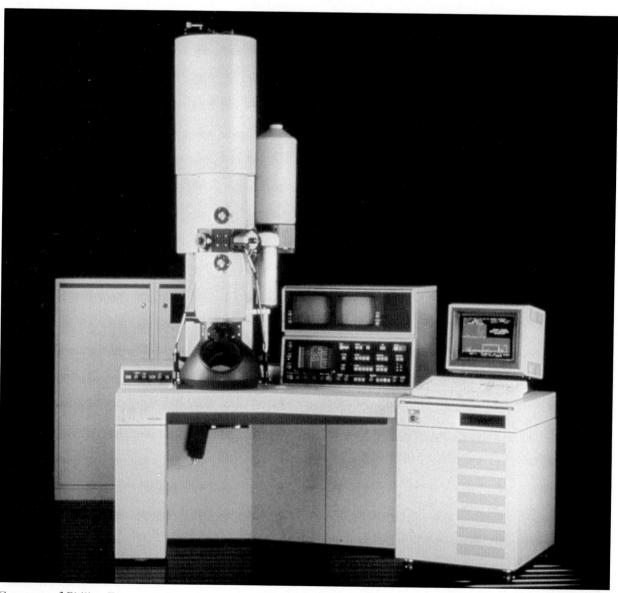

Courtesy of Philips Electronic Instruments.

The conventional transmission and scanning electron microscopes can generate images that reveal primarily morphological information in biological specimens. As has been illustrated in other chapters of this book, the use of these images goes beyond the investigation of various cytological features (descriptive cytology). Powerful quantitative and localization techniques (autoradiography, cytochemistry, immunocytochemistry) may be used in combination with electron microscopes to map the distribution of various morphological and macromolecular entities in the tissues under investigation. These techniques have revealed the architecture of the eukaryotic cell, and they have helped clarify the mechanisms of certain normal physiological processes (protein synthesis, packaging of DNA into chromosomes, recognition of antigenic sites by antibodies, structure of macromolecules such as enzymes—to name a few).

Electron microscopes can provide other data besides important morphological information. When a high energy beam of electrons interacts with a specimen, the atoms of the specimen may cause the electrons to decelerate. If this occurs, the kinetic energy associated with the electron is then converted into other forms of energy, and the electron may emerge with diminished kinetic energy and follow a different trajectory than it would have followed had it not interacted with the atoms of the specimen. The nature and the spectrum of the energy liberated, as well as the images formed by the new trajectories of the electron, may be captured using special detectors fitted to the electron microscope. Such detectors may reveal the atomic composition of the specimen struck by the beam of electrons. Under some circumstances, the quantity of the elements present may be ascertained. (As mentioned in Chapter 6, special detectors are not always needed since the transmission electron microscope is always able to utilize the diffraction mode to determine the identity of a crystalline inclusion in biological specimens.)

When electron microscopes are used to identify or characterize the chemical nature of components found in biological tissues, they are said to be *analytical electron microscopes.* A few applications of analytical electron microscopy might include: the localization of ions and electrolytes in various parts of the cell; a study of the changes in the distribution of ions during various physiological processes (growth, secretion, cell division, death); the identification of an unknown crystalline inclusion in a cell (asbestos fiber in a lung cell); investigating the pathways followed in the incorporation of toxic ions into the cell and in the detoxification process; and the confirmation of an enzymatic reaction product as a lead or osmium precipitate.

Interaction of an Electron Beam with a Specimen

Several different emanations or signals may be generated as a result of the electron beam striking a specimen. As illustrated in Figure 15-1, some electrons may pass through a suitably thin specimen with the loss of some energy. These are termed *inelastically scattered, transmitted electrons* and may be used in the conventional transmission electron microscope to reveal morphological information about a thin specimen. They also may be separated into various energy levels in an electron energy loss spectrometer for determination of elemental composition.

Other electrons may lose little or no energy upon interaction with the specimen. Such *elastically scattered* electrons may pass through the specimen as transmitted electrons or may be deflected back in the direction of the beam as *backscattered electrons.* Backscattered electrons may be detected using special detectors in scanning and transmission electron microscopes. As described in Chapter 7, such detectors may be used to discriminate areas of different atomic numbered elements: higher atomic numbered elements give off more backscattered electrons and appear brighter than lower numbered elements.

Secondary electrons, with energies typically under 50 eV, are a type of inelastically scattered electron that may be collected and imaged using secondary electron detectors in scanning or transmission electron microscopes. They are used primarily to reveal topographical features of a specimen in a scanning electron microscope, as described in Chapter 7.

Auger electrons are special types of low energy electrons that carry information about the chemical nature (atomic composition) of the specimen. Auger spectroscopy involves the collection and analysis of the spectrum of energy levels of these reflected electrons to give elemental composition from the upper atomic layers of the specimen. It is a powerful tool in the materials sciences for studying the distribution of the lighter numbered atomic elements on the surface of specimens. Since it has limited application in the biological sciences, and since a specialized

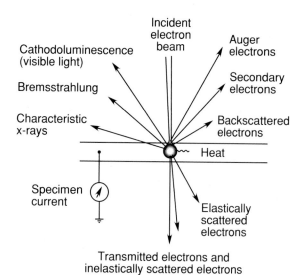

Figure 15-1 When an electron beam strikes a specimen, some of the kinetic energy is converted into various types of x-rays, visible light, and heat. Some electrons may be transmitted through the specimen with the loss of some energy (inelastically scattered) or no loss of energy (elastically scattered). Other electrons may be given off from the top of the specimen as high energy (backscattered) electrons or lower energy (auger, secondary) electrons. (Courtesy of Kevex Instruments, Inc.)

instrument (scanning auger electron spectrometer) is needed in this analysis, it will not be discussed further in this textbook.

Cathodoluminescence results when the energy of the impinging electrons is converted into visible light. Certain types of compounds are capable of cathodoluminescence and may be detected using special detectors, as described in Chapter 7. Since relatively few naturally occurring biological macromolecules are cathodoluminescent, and since the resolution currently obtained is similar to the light microscope, the detection of such signals is only occasionally done.

Two types of signals that currently are being exploited in analytical studies are *characteristic X rays* and certain of the inelastically scattered transmitted electrons. These signals are detected using either energy or wavelength dispersive *X-ray analyzers* or *electron energy loss spectrometers*, respectively. Transmitted electrons may also be used to give compositional information when the electron microscope is used in the *diffraction* mode. Since these are the three most widely used analytical techniques, they will be discussed separately in this chapter.

Microscopes Used for Detecting Analytical Signals

It is possible to attach a number of different detectors to electron microscopes. For instance, the scanning electron microscope may be fitted with secondary, backscattered, and transmitted electron detectors as well as detectors for X rays and cath-

odoluminescence. Besides these signal detectors, the transmission electron microscope may be equipped with electron energy loss spectrometers of various designs. Some accessories require little modification to the standard microscope, while others may require additional lenses and beam control electronics for optimal performance.

Obviously, the most versatile analytical instrument would be one that combined the features of a scanning and a transmission electron microscope. Such an instrument became available in the mid 1960s. Termed the *scanning transmission electron microscope,* or STEM, the instrument is able to generate and precisely position a very fine probe of high energy electrons and to raster the fine probe (as in the SEM) over a thin specimen while still being able to obtain diffraction patterns (as in the TEM). The miniaturization of the various detectors, as well as the small probe sizes generated, made possible the detection of the various signals from quite small areas of the specimen and thereby increased the resolution of the analytical techniques. Figure 15-2 is a photograph of an analytical electron microscope. A schematic diagram of the column of an analytical STEM is shown in Figure 15-3.

A number of design features of the STEM will be readily recognized. An illuminating system consisting of an electron gun and several condenser lenses is used to initiate and demagnify the fine electron probe. Double deflection coils are employed to generate the scanning raster (or to position the spot over the proper location on the specimen). The next lens of the STEM is unique in that it is a combination condenser/objective lens in which the upper

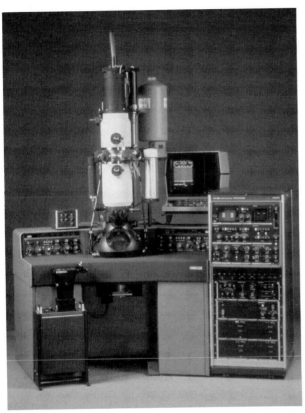

Figure 15-2 An analytical scanning transmission electron microscope equipped with energy dispersive x-ray detector, secondary electron detector, and electron energy loss spectrometer (under viewing screen). The large console on the right houses electronics associated with the STEM and SEM beam controls. (Courtesy of Philips Electronics Instruments).

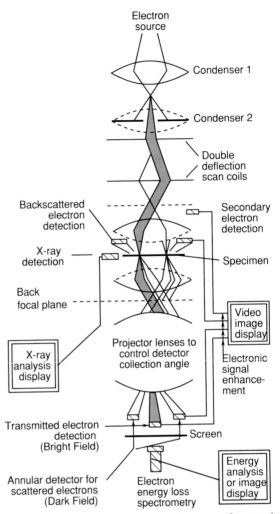

Figure 15-3 Schematic diagram of column of a scanning transmission electron microscope showing lenses and various detectors. (Courtesy of D. B. Williams and Philips Electronics Instruments.)

polepiece acts as a strong third condenser lens to focus the electron probe onto the specimen while the lower polepiece is used as a standard objective lens to form an image of the thin specimen. A number of detectors (secondary, backscattered, X-ray) are positioned near this lens in close proximity to the specimen to enhance the sensitivity of detection. Several projector lenses (intermediate, P1 and P2) follow the condenser/objective lens, and an electron energy loss spectrometer may be positioned underneath the viewing screen and camera.

Because of the expense involved, it is unusual for a STEM to be equipped with all of the analytical detectors. Instead, the most common configuration is a STEM equipped with energy dispersive X-ray analyzer. Of course, all STEMs have a variety of built-in diffraction capabilities since this is a lens function. Table 15-1 summarizes the instrumentation necessary and results obtainable using various analytical techniques. Most STEM instruments are

TEMs that have had the appropriate minilens installed so that they may function in either the TEM, SEM, or STEM modes. There are also specialized instruments, such as the Vacuum Generators HB-series STEM units, that are capable of extremely high resolutions in the STEM mode. However, they are less versatile for biologists since they are not readily operated in any of the TEM modes and are very expensive.

X-Ray Microanalysis

Two important types of X rays may be generated when the beam electron encounters the atoms of the specimen, continuum or bremsstrahlung X rays, and characteristic X rays.

Table 15-1   Summary of Main Features of Various Analytical Procedures

Procedure	Energy Dispersive X-ray Analysis (EDX)	Wavelength Dispersive Electron X-ray Analysis (WDX)	Electron Energy Loss Diffraction Spectroscopy (EELS)	Electron Diffraction
Microscope System Needed	TEM or STEM or SEM	TEM or STEM or SEM	TEM or STEM	TEM or STEM
What Identified	Elements with atomic numbers greater than 6 (11, normally)	Elements with atomic numbers greater than 3	Elements with atomic numbers greater than 3	Chemical identity of crystal
Quantitative	YES	YES	YES	NO
Smallest Area Analyzed (nm)	10 nm	10–100 nm	0.3–0.4 nm	1–10 nm
Detection Limit	10^{-16} gm	10^{-16} gm	10^3–10^4 atoms	Not applicable
Specimen Type	Thin, Ultrathin, Bulk	Thin, Ultrathin, Bulk	Ultrathin	Ultrathin

Continuum (Bremsstrahlung) X Rays

Continuum or bremsstrahlung X rays (Figure 15-4) are generated when an incoming, beam electron passing close to the atomic nucleus is slowed by the coulomb field of the nucleus (i.e., scattered inelastically) with the release of X-ray energy. *The intensity of X-ray energy released depends on how close the electron comes to the nucleus—closer passes decelerate the electron more and yield higher energy X rays.* Since this event is random and various electrons will lose varying amounts of energy, depending on their proximity to the nucleus, a plot of the theoretical intensity versus energy levels yields a graph as shown in Figure 15-5, dashed line.

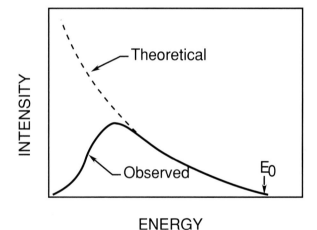

Figure 15-5   Plot of various energy levels of decelerated (bremsstrahlung) electrons versus intensity or amount of each energy level. The theoretical plot is shown as a dashed line while the energy that is measurable with current instruments is shown in the solid line labeled "observed". (Courtesy of Kevex Instruments, Inc.)

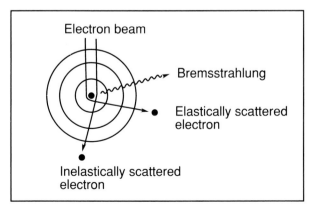

Figure 15-4   Generation of bremsstrahlung radiation upon deceleration of beam electron by atom of specimen. (Courtesy of Kevex Instruments, Inc.)

Continuum (Bremsstrahlung) X Rays

Theoretically, it is more likely that a higher number of electrons will miss the nucleus by a wide margin (to yield low energy X rays) than it is that an electron will pass close to or hit the nucleus. While this is true, in practical terms these electrons are of such low energy that they are not detected. Therefore, the observed energy distribution is as shown in the solid line in Figure 15-5. Since the energy distribution is variable, these X rays constitute what is called the *X-ray continuum*. Because these X rays result from the deceleration of the electron, they are sometimes termed bremsstrahlung (German for "braking radiation").

The bremsstrahlung are always part of the X rays generated from a specimen and sometimes may mask the discrete X rays that are used for elemental analysis. The bremsstrahlung may be used to measure specimen mass thickness when quantitative analysis is performed on thin sections. Other terms used in the literature to refer to the continuum X rays are *background* or *white radiation*.

Characteristic X Rays

The more useful types of X rays are generated when the high energy beam electrons interact with the shell electrons of the specimen atoms so that an inner shell electron is ejected. The removal of this electron temporarily ionizes the atom until an outer shell electron drops into the vacancy to stabilize the atom. Since this electron comes from a higher energy level, a certain amount of energy must be given off before it will be accommodated in the inner shell. The energy is released as an X ray, the energy of which equals the difference in energy between the two shells (Figure 15-6). Since this X ray is of a

discrete energy level, rather than a continuum, this event may be plotted as discrete peaks (Figure 15-7). Different elements will fill the vacancies in shells in unique ways. This means that since each element will generate a unique series of peaks, the spectrum may be used to identify the element. Such discrete X rays are termed *characteristic X rays*.

The Filling of Inner Orbital Electrons is an Orderly Process

The shells may be filled in a number of ways, depending on the element that interacts with the beam electrons. In the very simplified view shown in Figure 15-8, it may be seen that a K shell electron void may be filled by electrons from the L, M, or N shells. The filling of the K shell by the outer shell electrons is considerably more complicated

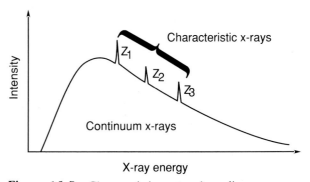

Figure 15-7 Characteristic x-rays show discrete energy levels (peaks) whereas continuum x-rays show the typical broad distribution of energies.

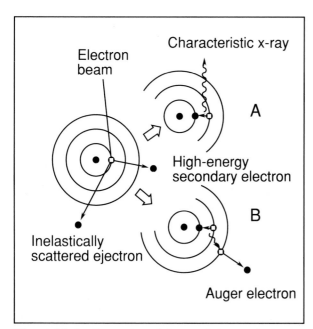

Figure 15-6 Generation of characteristic x-ray by impact of beam electron with orbital electron of specimen atom. The specimen electron may be ejected with either: (A) the replacement of this electron by an outer orbital electron and the generation of characteristic x-ray, (B) absorption of the x-ray by an outer orbital electron and the ejection of the electron as an Auger electron. (Courtesy of Kevex Instruments, Inc.)

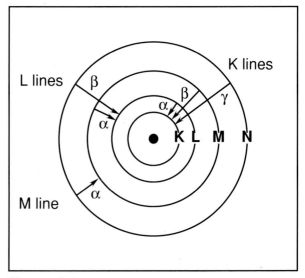

Figure 15-8 Various electron vacancies may be replaced by more peripheral electrons in the simple scheme shown. A K shell electron may be filled by an electron from the L, M, or N shell, depending on the atomic number of the element. (Courtesy of Kevex Instruments, Inc.)

than shown in this figure. For instance, depending on the atomic number of the element, a K shell electron void may be filled by 1 of the 12 electrons shown in Figure 15-9. A shorthand way of designating the specific X-ray line followed to fill the void is written as follows:

$$K_{ab}$$

where **K** indicates the shell filled and **ab** refers to the specific electron that fills the void (possibly $\beta 3$, which originates from a particular orbital in the M shell). Among the 11 other possible ways of filling the K shell electron (see Figure 15-9), one might encounter K_{a1}, K_{a2}, $K_{\beta 1}$, $K_{\beta 4}$, $K_{\gamma 1}$, etc. To reiterate, the designation $\beta 3$ refers to a specific electron, while the **K** refers to the shell that is filled by the specified electron.

It is important to note that a β electron may originate from shells M, N, and O, and that the specific electron that fills a vacancy depends upon the atomic number of the element. Elements with lower atomic numbers should have simple spectra, whereas higher atomic numbered elements would be expected to have more complex spectra since more electrons would be available to fill the void. Fortunately, computer software is available that will analyze the energy spectrum of X rays to give a plot of the X rays along with an identification of the elements present in the area sampled (Figure 15-10).

X-ray Microanalysis may be Conducted to Achieve Several Goals

The simpler and more commonly used procedure is a *qualitative analysis* to determine which elements are present in a particular location. Although this technique is challenging, it is less demanding than *quantitative analysis* where one seeks to find out either relative amounts of the elements (i.e., which parts of the cell have more Ca than others) or how much of an element is actually present in the organelle (i.e., 10^{-10} gm/vol sampled). The stringent specimen preparation and instrumental finesse needed for the latter undertaking are formidable, and relatively few studies have been carried out using true quantitative microanalysis.

Equipment for Detecting X Rays
Energy Dispersive X-Ray (EDX) Detectors
These detectors are the predominant types used in biological studies. The sensor consists of a disc-shaped semiconductor manufactured from a single crystal of silicon into which some lithium atoms are diffused (to correct for impurities and imperfections in the silicon crystal structure). When an X ray strikes the semiconductor

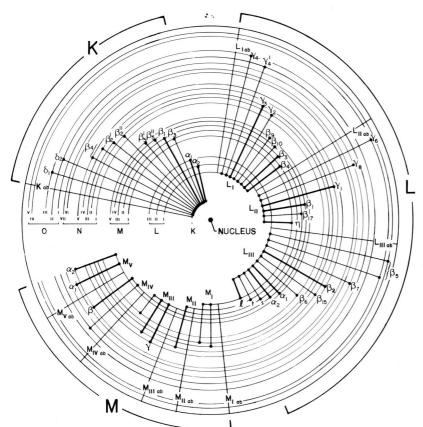

Figure 15-9 As the atomic number of the element increases, the increased number of outer orbitals permits a more complex filling of vacated orbitals. (Courtesy of R. Woldseth and Kevex Instruments, Inc.)

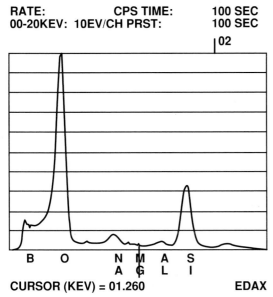

RATE: **CPS TIME:**   **100 SEC**
00-20KEV: 10EV/CH PRST: **100 SEC**

CURSOR (KEV) = 01.260   **EDAX**

Figure 15-10(A) Characteristic x-ray analysis of glass containing 17% B_2O_3. Note the presence of boron, oxygen, and silicon peaks on the spectrum. Traces of sodium are also indicated. (Courtesy of EDAX International.)

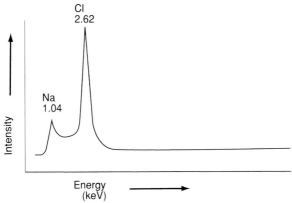

Figure 15-10(B) Absorption of the weaker Na x-rays by the window of the energy dispersive detector makes quantitation difficult. Other factors that make quantitation difficult result from the fact that elements with high atomic numbers, Z, give off more x-rays than lighter elements and due to the absorption, A, of the x-rays by atoms of the sample and to a phenomenon termed secondary fluorescence, F, which results when an absorbed x-ray gives rise to secondary x-rays. For accurate quantitation, the ZAF correction factors must be entered into the quantitative equation using computers.

crystal, the absorbed energy alters the ability of the crystal to conduct a charge. Since the crystal is maintained at a bias voltage of 100 to 1,000 volts, an increase in conductivity of the crystal can be readily detected and quantitated. *Since the energy of the X ray is directly proportional to the increase in conductivity in the silicon crystal, it is possible to collect and measure the conducted current over a period of time and determine the intensity of the X ray emission.*

The detector crystal must be cooled with liquid nitrogen (for maximum resolution and minimum noise) and maintained in an ultraclean high vacuum condition. Normally it is sealed apart from the vacuum of the electron microscope since contaminants may be present even in the electron microscope column. A 5 to 8 µm thick window of beryllium is normally used to seal off the crystal. Such windows allow X rays with energies greater than 2 eV to pass. However, lower energy X rays cannot pass through the window. This means that X rays from lighter elements (with atomic numbers lower than sodium) will be absorbed by the window.

The *absorption* of X-ray energy by the window of the detector poses a problem when quantitation is desired. For example, if one were analyzing NaCl, where both elements were present in equal amounts, the sodium peak would be considerably smaller (due to absorption) than the chlorine peak since the energies of the X rays would be 1.04 and 2.62 KeV, respectively (see Figure 15-10B).

In an attempt to permit detection of lower energy X rays, some manufacturers have fabricated lower density windows of plastics and even diamond, or have installed turrets on the end of the detector housing to permit the use of a windowless detector for a short period of time. These detectors are capable of detecting lighter elements down to lithium (atomic number 3) and should prove useful in biological research. Windowless units are prone to rapid contamination of the liquid nitrogen cooled crystal unless very clean microscope operating conditions are maintained. Venting the microscope column to atmospheric pressures with the window to the detector open will damage the detector. Figure 15-11A diagrams an EDX detector while Figure 15-11B shows the liquid nitrogen reservoir used to cool the detector that is sealed in the stainless steel tube.

Advantages and Disadvantages. EDX detectors have several advantages: (1) simple, robust design that does not take up too much space; (2) high sensitivity with nearly 100% efficiency; (3) ability to detect and display the entire elemental spectrum at once; and (4) ability to use smaller probe sizes with less damaging beam currents. Such detectors also have a number of disadvantages: (1) quantitative accuracy falls off at low concentrations; (2) resolution limited to 100 to 120 eV, so that closely placed X-ray energy peaks will not be discriminated; (3) decreased sensitivity for light elements; and (4) must be operated at liquid nitrogen temperatures.

Wavelength Dispersive X-Ray (WDX) Detectors
In this special type of detector, a small fraction of the X rays leaving the specimen strike a crystal and are reflected into the detector. The crystal will reflect only a narrow wavelength of X ray as determined by the Bragg equation (see Equation 15-1 in Electron Diffraction section of this chapter). The better types of crystal are curved to more effectively focus the X rays into the detector (Figure 15-12A).

Since a particular crystal will diffract only a very narrow wavelength of X rays, one crystal may be useful for detecting only a few atomic numbers. For example, a

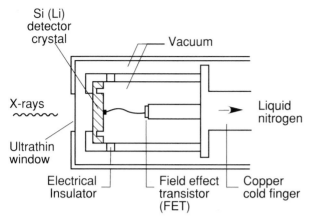

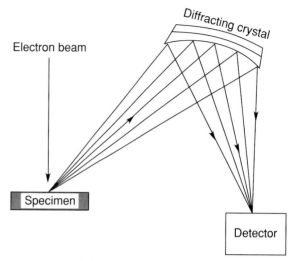

Figure 15-11(A)   An energy dispersive X-ray detector. Diagram showing the basic components making up the detector.

Figure 15-12(A)   Principle of operation of wavelength dispersive x-ray (WDX) spectrometer. X-rays generated by the specimen are "reflected" by the crystalline lattice into the detector. Very specific wavelengths of x-ray are reflected by different lattice spacings so different crystals are needed in the spectrometer to cover various energies of x-rays.

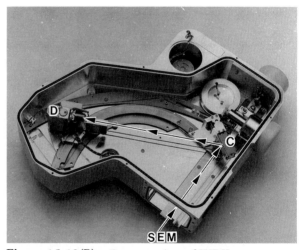

Figure 15-11(B)   The detector is housed inside of the stainless steel rod and chilled from the large vessel containing liquid nitrogen. (Both illustrations courtesy of Kevex Instruments, Inc.)

Figure 15-12(B)   Cut-away view of WDX spectrometer showing location of crystal (C) and detector (D). Arrow shows direction x-rays travel from specimen chamber to detector. (Courtesy of Microspec Corporation.)

gypsum crystal will diffract wavelengths of 0.26 to 1.5 nm to cover atomic numbered elements 11 to 14, while sodium chloride crystals diffract wavelengths of 0.09 to 0.53 nm to detect elements with atomic numbers ranging from 16 to 37. Different types of crystals are therefore needed in the spectrometer in order to cover the ranges of wavelengths of a particular study. By studying the Bragg equation, it will be understood why larger crystal

lattice spacings are needed to diffract the longer wavelength X rays generated by light elements with low atomic numbers.

The focused X rays are detected when they pass through a thin plastic window and enter a gas-filled cylinder containing a collector wire kept at a positive high voltage. The X rays cause an ionization of the argon/methane gas and generate a flow of electrons to the wire. This current is measured and quantitated over time. Since the amount of current is directly related to the energy of the original X ray, it is possible to determine the energy of the X ray. An internal view of a WDX detector is shown in Figure 15-12B.

Advantages and Disadvantages. The WDX detector is used less often in biological studies than the EDX is because the WDX detector is: (1) very large, more complex mechanically, and prone to error unless properly adjusted and calibrated; (2) able to detect only one element at a time; (3) less efficient than the energy dispersive detectors; (4) somewhat more expensive; and (5) tedious to operate. On the positive side, the WDX detectors: (1) offer ten times better capability than EDX to discriminate closely spaced X-ray energy peaks; (2) are better suited to light element analysis; (3) are better suited for trace element detection; and (4) do not require liquid nitrogen cooling of the detector.

Information Obtainable Using X-Ray Analysis

The electronics of the energy dispersive (EDX) and wavelength dispersive (WDX) detectors are involved in the acquisition, sorting, and display of data as a spectrum of energies. Computers greatly expedite this process and assist in the interpretation of the spectrum. Figure 15-13 compares the output of a sample that was analyzed by EDX and WDX. The sharp separation of closely placed X-ray energies is evident in the largest peak displayed in the EDX spectrum. Here the two elements barium and titanium have not been resolved since their characteristic X rays are so close in energy. On the other hand, WDX clearly resolved the two elements. In spite of this major advantage, WDX is seldom used in biological studies due to the excessive beam currents needed, which result in damage to most biological specimens.

Data may be presented in several different ways from both types of detectors. In the *spot analysis* mode, a fine probe of electrons is focused on the area of interest and a spectrum is generated, as shown in Figure 15-14. When this information is presented for publication, one first takes a micrograph of the area that was analyzed and places a pointer on the actual spot that was probed. The spectrum may then be displayed in a separate photograph, or it may be superimposed over the electron micrograph of the specimen.

In the *line scan analysis* mode, the electron probe is moved in a straight line across the specimen (pausing for 100 seconds or so at each point to generate enough X rays), and the amount of a specified element is superimposed as a line graph over the micrograph (Figure 15-15).

In a *dot map*, the beam is moved across a large area of the specimen, pausing for a fixed amount of time at each point to generate X rays. This may take many hours, depending on the area scanned, so computer control of the beam is very important to facilitate this process. Figure 15-16 is a dot map showing the distribution of iron in the same liver cell shown in Figure 15-14. Whenever a particular element (iron, in this case) is found in the specimen, this is indicated by a bright spot. Such data becomes quite informative when the dots are superimposed over an actual electron micrograph. With modern energy dispersive X-ray analysis systems, it is possible to simultaneously map many different elements by assigning various colors to the elements (i.e., sodium may be displayed as red areas, while phosphorus may be shown in green).

Specimen Preparation for X-Ray Microanalysis

Two of the goals of specimen preparation for X-ray analysis are to retain the elements of interest in their normal locations in the cell and to preserve the ultrastructure to the extent that it will be recognizable. These two goals are conflicting, since the fixatives, dehydrants, and embedments used in conventional specimen preparation procedures displace or completely remove diffusible ions. Obviously, the use of unfixed tissue presents an equally difficult problem of interpreting where in the cell the element is actually being localized.

For optimal results, the specimen should be thin, smooth, electrically and thermally conductive, and with discrete inclusions or compartments in the cell containing high concentrations of ions. Ideally the compartments should be surrounded by areas of lower atomic numbered elements or water so that interfering background X rays would not be present. These conditions are far removed from the actual situation in cytological material, so that X-ray microanalysis in biological systems has yet to fulfill its promise. Nonetheless, some useful information may be obtained under certain circumstances and with appropriate preparatory techniques. Several categories of specimens may be analyzed.

Bulk Samples

It is possible to examine bulk samples (thick slices, intact tissues, pieces of fractured specimens) in the SEM or STEM instrument operating in the SEM

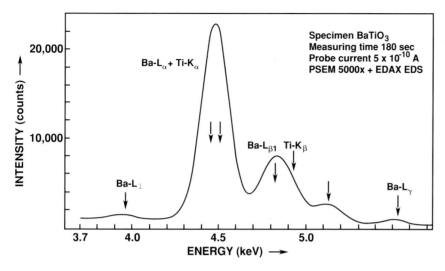

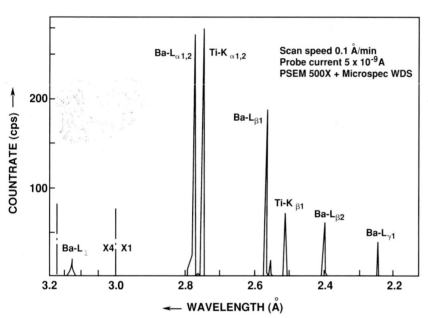

Figure 15-13 Spectral output from same sample analyzed by energy dispersive x-ray procedure (top) versus wavelength dispersive procedure (bottom). Note that WDX is able to resolve the two closely placed peaks that were summed together by EDX. (Courtesy of D. B. Williams.)

mode. Such specimens are mounted onto aluminum or carbon stubs and coated with a conducting layer of carbon. With the advent of new types of cold stages that maintain the specimen in the frozen state while under observation, it is possible to examine quick-frozen, uncoated specimens directly in the SEM or STEM without any drying procedures. Unfortunately, the irregular surfaces of the specimen restrict quantitative microanalysis, while the deep penetration of the probe into the specimen limits the resolution to 4 to 8 μm (Figure 15-17). In addition, the probe may melt locally certain areas of the specimen, leading to redistribution of diffusible ions.

The Challenge: Quantitative Analysis of Bulk Specimens

Quantitative microanalysis in bulk samples is a laborious, exacting undertaking that requires the use of a standard specimen (containing known amounts of the elements to be analyzed) that closely approximates the properties of the sample to be analyzed. In addition, one must apply correction factors that take into account differences in mean atomic number between specimen and the standard, the absorption of some of the X rays by the specimen, and a correction for extraneous X rays generated in the specimen by other X rays in the specimen (*X-ray fluorescence* phenomenon) (Figure 15-18).

One method of correction for the various variables in bulk biological specimens is the *ZAF correction method*

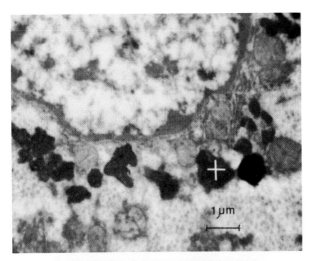

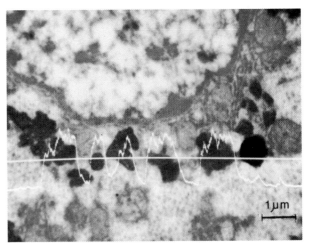

Figure 15-15 X-ray analysis was accomplished on the same liver cell shown in Figure 15-14 by scanning a straight line across the specimen as shown. As the specimen was analyzed for iron along the line scanned, the quantity of iron was indicated by the level of deflection of the peaks. (Courtesy of Dr. G. Schwalbach and Carl Zeiss, Inc.)

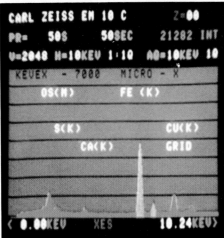

Figure 15-14 (Top panel) Energy dispersive x-ray analysis was conducted by focusing the probe on the spot indicated in the electron micrograph. This liver cell demonstrates hemochromatosis. (Bottom panel) The dense bodies are rich in iron as indicated in the x-ray spectrum taken from the marked spot. (Courtesy of Dr. G. Schwalbach and Carl Zeiss, Inc.)

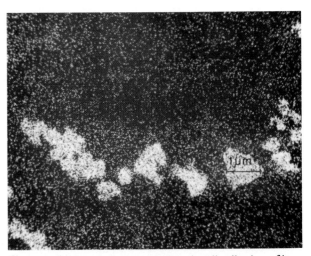

Figure 15-16 X-ray dot map showing distribution of iron in same liver cell shown in Figure 15-14. (Courtesy of Dr. G. Schwalbach and Carl Zeiss, Inc.)

(Philibert and Tixier, 1968), as summarized in the following equation:

$$C_i = (ZAF)_i \, I_i/I_{(i)}$$

where C_i is the amount of the element i present, $I_{(i)}$ is the intensity of the X rays from a standard composed only of element i; I_i is the X-ray intensity measured in the specimen under analysis; and ZAF refers to the three corrections applied for *atomic number*, *self-absorption*, and *fluorescence*, respectively.

Single Cells, Isolated Organelles, Liquid Secretions or Extracts

These types of specimen may be obtained from individual cells or cellular inclusions (particulate or fibrous) and deposited onto a carbon-coated grid and examined in a suitably equipped TEM or in a STEM operating in the transmitted mode. If the cells and constituents are thin enough, it is possible to obtain qualitative and semiquantitative data (Figure 15-19).

Sectioned Materials

The sections may range in thickness from 0.5 μm to less than 100 nm. They are particularly useful specimens for X-ray microanalysis since complicated correction equations (ZAF correction, for example) may not be necessary. Although thinner specimens are more desirable from a morphological and

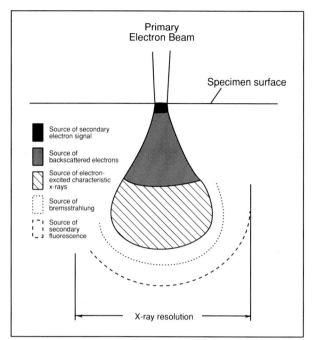

Figure 15-17 Diagram showing the depth and relative size of areas from which various signals may emanate. The enlargement of the signal source effectively diminishes resolution. Note that x-ray signals have the poorest resolution. Thin sections of specimens do not suffer from this problem since the probe does not spread out to this extent. (Courtesy of Kevex Instruments, Inc.)

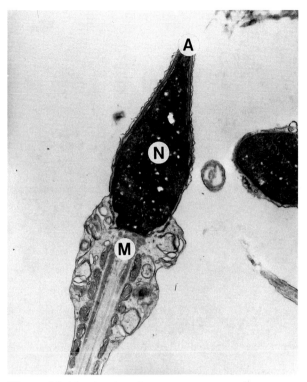

Figure 15-18 Section of human sperm cell showing areas where elemental x-ray analysis was conducted and quantitated in Figure 15-19.

a quantitative standpoint, the levels of elements present may be too small to detect.

Some Precautions with Sectioned Materials for X-Ray Microanalysis

If it is necessary to fix tissues and embed them in plastic prior to sectioning, one should use glutaraldehyde alone (since osmium may mask some elements) and avoid buffers containing ions that are to be localized. The specimens should be embedded in an acrylic resin such as LR White since it has a lower background reading of certain elements (especially sulfur and chlorine) than do the epoxy resins.

Sections on the order of 100 nm may be cut and mounted on Formvar-coated grids. If copper is one of the elements to be analyzed, grids composed of noninterfering metals (carbon, nylon, aluminum, beryllium, titanium, gold, nickel) should be used. It is desirable to carbon-coat the sections to prevent the buildup of static charges and to stabilize the sections during the analytical procedure.

When it is necessary to localize diffusible ions, then the standard fixation, dehydration, and embedding process must be avoided and alternative techniques should be used. If the specimen is inherently very thin—such as some individual cells or cell fractions—it may be possible to deposit the material or to grow the cells directly on plastic and carbon-coated grids. Specimens may then be rapidly

frozen and freeze-dried prior to examination, or one may use cold stages to maintain the specimen at 100° K. Obviously, ultrastructural details will be sacrificed, but it may be possible to recognize gross features as mitochondria, membranous systems, vesicles, etc., in the dried or frozen specimens.

Cryoultramicrotomy (see Chapter 4) may be beneficial with thicker specimens when diffusible ions are to be localized. The best approach is to quickly freeze the unfixed specimen, cut thin sections, and mount them onto a coated grid. This must be done without using any liquids and while maintaining the frozen state of the sections.

If one is able to accomplish this task, then the frozen sections may be freeze-dried (usually while still in the cryochamber of the ultramicrotome) or a special transfer stage may be used to move the frozen-hydrated specimen onto a cold stage in the TEM (Figure 15-20). Freeze-drying of hydrated specimens must be undertaken with trepidation, since the removal of water will leave unsupported ions in such an unattached condition that displacement of the ions is likely. These procedures are quite difficult and require expensive and sophisticated accessories for the ultramicrotome as well as for the electron microscope.

Very few laboratories are able to accomplish microanalytical procedures on frozen-hydrated specimens. Nonetheless, this combination of procedures is probably the only way that one may quantitate low concentrations of diffusible ions in biological specimens.

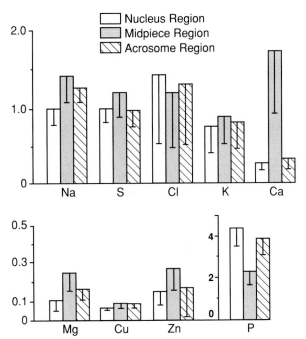

Figure 15-19 Quantitative data obtained from acrosome, midpiece and nucleus of sperm cell similar to the one shown in Figure 15-18. (Adapted from Chandler, J. A. 1977. X-ray Microanalysis in the Electron Microscope. Practical Methods in Electron Microscopy Vol. 5, Pt. II, A. M. Glauert, ed., North-Holland/American Elsevier Publishing Co.)

Electron Energy Loss Spectroscopy (EELS)

This technique is used to detect and differentiate various energy levels of electrons that have been transmitted through a thin specimen. As in EDX and WDX, the spectrum of electron energies is displayed and may be used to determine the elemental composition of the atoms in the specimen that caused the loss in energy of the beam electrons.

Differentiation of the various energy levels of electrons is carried out using an *electromagnetic spectrometer* placed either after the specimen or under the viewing screen of the TEM or STEM. As the beam electrons enter the electromagnetic field of the spectrometer, they are bent to various degrees and brought to focus some distance from the spectrometer. Lower energy electrons are deflected to a greater degree than are high energy electrons, so that the focal points of the various energy groups of electrons are physically separated. A movable plate with a narrow slit is positioned to permit electrons of a specific energy range to pass into a detector. The detector is similar to the scintillator/photomultiplier type used to detect secondary elec-

trons in the SEM. Figure 15-21 is a schematic representation of a system commonly used in EELS.

A typical range of energy loss for 100 kV beam electrons is from 100 to 1000 eV. Modern spectrometers can detect energy losses in the 0 to 2000 eV range with resolutions of better than 5 eV, so that the spectrometer is able to accommodate and adequately resolve the spectrum of anticipated energies in most biological specimens. Since EELS detectors detect a primary event (loss of energy in transmitted electrons) rather than a secondary event (X-ray emission), EELS is 10 to 100 times more efficient than EDX. Spatial resolution in EELS (10 nm) is comparable to that obtainable in thin specimens analyzed by EDX methods (10 to 50 nm). Figure 15-22 shows some electron micrographs in which EELS is used to map the distribution of phosphorus in a mitochondrion and endosplasmic reticulum, as well as to reveal the location of iron, calcium, and oxygen in a section of a lung biopsy.

Quantitation in EELS is more straightforward than with X-ray techniques: ZAF corrections are not needed and comparable standards need not be run each time. In addition, despite some current limitations, it eventually should be possible to obtain true quantitation of the number of atoms/cm^2 present in a thin specimen.

Theoretically, EELS should be ideally suited for the detection of low atomic numbered elements that make up biological tissues. In fact, EELS is less frequently used than EDX and WDX for detection and quantitation. The principal obstacle has been the inability to produce thin enough specimens—several times thinner than the 60 to 80 nm slices obtained by ultramicrotomy—since thicker specimens increase the background levels due to multiple scattering events. This is the same situation as with chromatic aberration (Chapter 6) in thicker sections.

Besides giving elemental composition and quantitation, the electron spectrometer may be used as an energy filter to enhance contrast and resolution in thicker specimens. In practice, an EEL spectrometer is placed between the objective and intermediate lenses where it is used to filter out electrons of particular energies with the remainder being used to form the image. This would permit one to examine thicker specimens, since selected wavelength electrons (that give rise to chromatic aberration) may be removed (Figures 15-23 and 15-24). One commercial electron microscope manufactured by Zeiss has such capabilities (Figure 15-25).

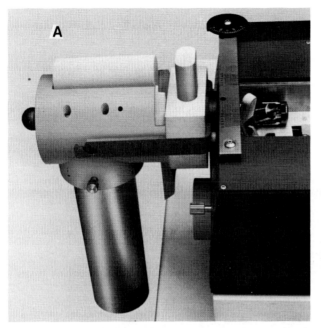
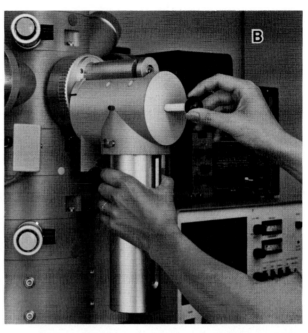

Figure 15-20 (A) Cryotransfer device attached to back of cryochamber of ultramicrotome. Frozen sections are transferred into the chamber and maintained at liquid nitrogen temperatures. (B) Frozen hydrated sections are moved into the electron microscope column for viewing. (Courtesy of Gatan.)

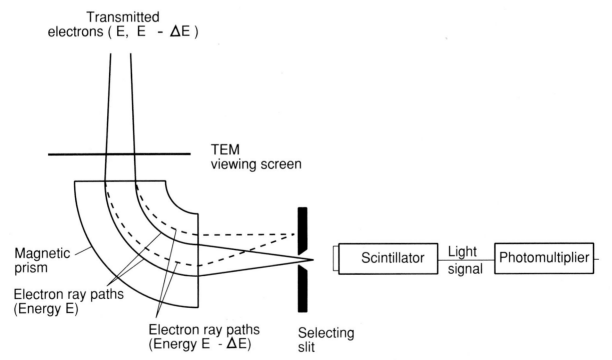

Figure 15-21 Schematic diagram of an electromagnetic spectrometer for electron energy loss investigations. Only two different energy levels of electrons are shown being focused in the plane of the selecting slit. Lower energy electrons (dashed line) are deflected to a greater extent than higher energy ones and thereby may be separated by the spectrometer. The separated electrons are then sampled by positioning the selecting slit in the proper location to allow the electrons to pass through the slit. (Courtesy of D. B. Williams and Philips Electronics Instruments.)

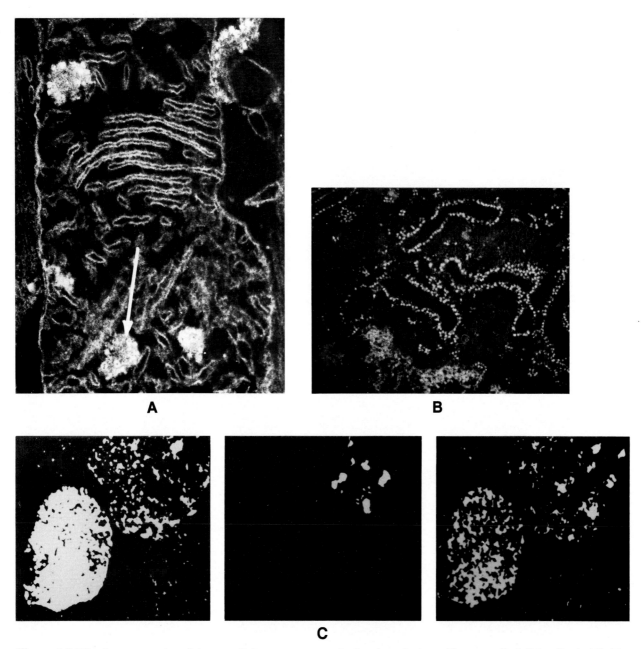

Figure 15-22 Some examples of the use of electron energy loss spectroscopy to detect particular elements in sectioned cells. (A) Portion of mitochondrion with areas rich in phosphorous appearing very bright. (Courtesy of P.A. Schnabel). (B) Phosphorous localized in ribosomes along endoplasmic reticulum. (Courtesy Carl Zeiss, Inc.). (C) Elemental distribution of Fe, Ca, and O in lung biopsy. (Courtesy of C. H. W. Horne). (All micrographs were provided by Carl Zeiss, Inc.)

A potentially exciting use of EELS lies in the imaging of single, heavy atoms on low atomic numbered substrates. For example, if one could prepare specific DNA or RNA probes that have been labeled with uranium atoms, one may be able to directly image a particular labeled gene along a strand of DNA in the chromosome.

In summary, EELS may be used to detect and quantitate the lower numbered atomic elements as well as to improve the imaging capabilities of a TEM or STEM. It is not as popular as X-ray analytical procedures in spite of its powerful capabilities primarily because most biological specimens are too thick and readily damaged by the beam during analysis.

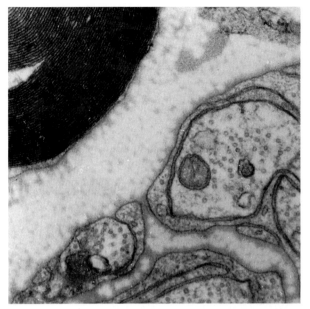

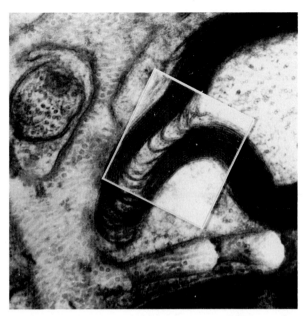

Figure 15-23 Use of an EEL spectrometer to diminish chromatic aberration in a thick section. Section on left is a conventional micrograph of a 70 nm ultrathin section of nerve tissue. Section on right is over seven times thicker and still usable since only a narrow wavelength of electrons was permitted to pass through the selecting slit. (Courtesy of Carl Zeiss, Inc.)

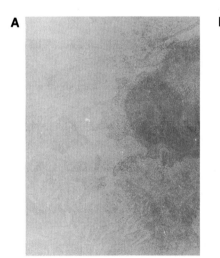

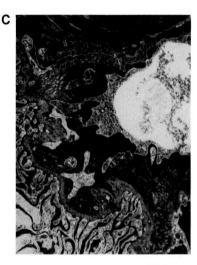

Figure 15-24 EEL spectrometers may be used to enhance contrast in unstained sections. (A) Unstained section viewed in conventional TEM mode. (B) Unstained section viewed in energy loss mode with spectrometer selecting only electrons that have not lost energy (zero energy loss). (C) In the spectroscopic mode, areas rich in phosphorous show up as bright areas indicating electrons that have lost 180 eV. (Courtesy of Carl Zeiss, Inc.)

Electron Diffraction

Even the most basic transmission electron microscope can generate a diffraction pattern from a specimen. This is because the diffraction pattern is always present in the back focal plane of the objective lens. Normally, most biologically oriented electron microscopists are interested in examining the *image* generated by the transmitted imaging electrons and therefore have no need for viewing the diffraction pattern. If one examines the ray diagram shown in Figure 15-26, it is apparent that the forward scattered diffracted electrons come to a focal point (this is the back focal plane) but are excluded by the objective aperture. As will be shown, one of the operational requirements to obtain diffraction patterns

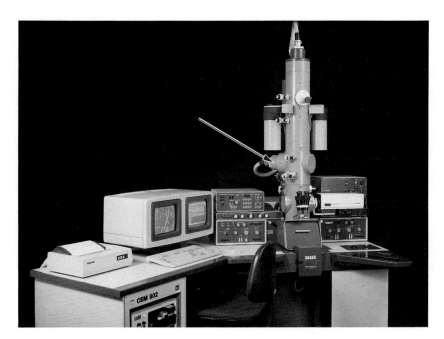

Figure 15-25 Commercial STEM equipped for electron energy loss spectroscopy, the Zeiss CEM 902. Arrow indicates location of spectrometer. (Courtesy of Carl Zeiss, Inc.)

may involve removal of the objective aperture or the use of a larger aperture.

Although diffraction patterns are generated by all specimens, some patterns have more information about the nature of the specimen than do others. For instance, specimens with randomly or nonperiodically oriented atoms (the majority of biological specimens) generate a diffuse electron diffraction pattern that simply confirms that the atoms of the specimen are not arranged in a repeating or periodic manner (Figure 15-27). By contrast, whenever the specimen or parts of the specimen consist of molecules or atoms with a *repeating periodicity* (as in a crystalline lattice), then a diffraction pattern is formed that may be useful in the identification of the crystal or molecule (Figure 15-28). Unlike the other analytical procedures of X-ray microanalysis and EELS, which identify and quantify amounts of individual elements present in an area, electron diffraction may give the spacing of the crystalline lattice and (since various crystals have unique lattice spacings and diffraction patterns) the chemical identity of the crystal. On the other hand, electron diffraction cannot be used to determine the quantity of a particular chemical that has been identified.

Formation of Diffraction Patterns

A crystalline object consists of identical atoms or molecules arranged as repeating units along certain planes. For example, imagine atoms or molecules arranged to form a single layer (like a sheet of paper). If one begins stacking additional layers upon the initial layer (a stack of paper sheets), one would generate an object with crystalline features. A model for such stacks of atoms or molecules is shown in Figure 15-29. Note that the crystal consists of the same basic repeating unit (atom or molecule) in various planes that are very precisely spaced relative to one another. Examples of crystalline specimens might include asbestos, sodium chloride, carotene, cholesterol stearate, ferritin, and myelin.

If a beam of electrons strikes a crystalline structure at an appropriate angle (so called Bragg angle) the electrons will be diffracted or "reflected" from the lattice planes. The reflection follows Bragg's law of diffraction that is summarized in:

Equation 15-1. Bragg's Equation

nλ = 2d sin θ

where n = an integer
λ = wavelength of the electron that is diffracted
d = crystalline lattice spacing
θ = angle of incidence and reflection of the electrons striking the crystal.

One will obtain a discrete diffraction pattern only when the incident electrons enter the crystalline lattice at the appropriate Bragg angle. With crystals, some of the electrons that enter the lattice at the proper angle will be reflected by the various

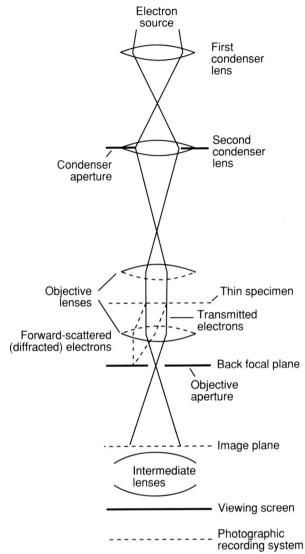

Electron
source

First
condenser
lens

Second
condenser
lens

Condenser
aperture

Objective
lenses

Thin specimen

Transmitted
electrons

Forward-scattered
(diffracted) electrons

Back focal plane

Objective
aperture

Image plane

Intermediate
lenses

Viewing screen

Photographic
recording system

Figure 15-26 Schematic of lenses in a transmission electron microscope. Note the dashed line indicating one group of diffracted electrons that converge in the back focal plane of the objective lens. (Courtesy of D. B. Williams and Philips Electronics Instruments.)

lattice planes in the same direction and at the same angle to come to focus in the back focal plane. This generates the diffraction pattern. In the case of an amorphous specimen, the electron beam that enters the specimen is diffracted in multiple directions and at various angles so that the electrons are unable to converge into a discrete spot and form a diffuse ring pattern instead. With a crystalline specimen, in order to obtain the proper Bragg angle, it is necessary to orient the specimen very precisely by tilting and rotating it relative to the electron beam until the diffraction pattern is obtained.

If one examines Figure 15-30, a number of phenomena may be observed. In this illustration, a crystalline object is struck by an electron beam so that some electrons pass through the specimen and are brought to a focus point (C or D) on an image plane (the first image plane). Note that the electrons that are brought together at the image points C and D are those electrons that were scattered from the same physical *location* in the specimen (A and B, respectively). Therefore in the standard imaging mode of the TEM, electrons that are scattered from the same *point* in the specimen come together at a common point in the image plane. These electrons impart information about the morphology of the specimen.

Again referring to Figure 15-30, one should notice another plane where electrons converge. It will be noted that in the back focal plane, some electrons converge at points X and Y. Careful examination of the ray diagram reveals that those electrons that were scattered in the same *direction* (rays that emerge from the specimen parallel to each other) converge along the back focal plane to form diffraction points. Therefore, in the diffraction mode those electrons that are scattered in the same *direction* (parallel to each other) converge at a common point in the back focal plane. These electron focal points when imaged on the viewing screen impart information about the atomic configuration and chemical identity of the specimen.

Single Crystal Versus Polycrystalline Specimens

A single crystal will generate a diffraction pattern consisting of spots (Figure 15-31, bottom), with the layout of the spots depending on the type of crystal lattice (14 different types exist) being illuminated and the orientation of the crystal to the beam. In practice, one photographs the diffraction pattern and, in a properly calibrated microscope, measures the distances and angles between the spots to determine the distance "d" between lattice planes. *Since the d spacings are unique for each crystal, one may look up the d spacings in a reference book (or computer system) and obtain an identification of the crystal.* If one has the capability of doing X-ray analysis, the identification of the unknown crystal will be greatly expedited since the possibilities will be limited to crystals composed only of those elements detected by the X-ray procedure.

In a polycrystalline specimen (Figure 15-31, top), many crystals are present all of which are gen-

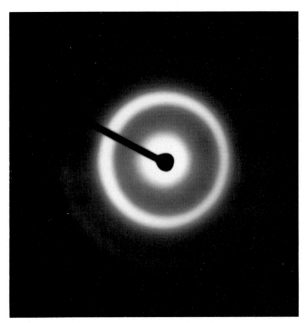

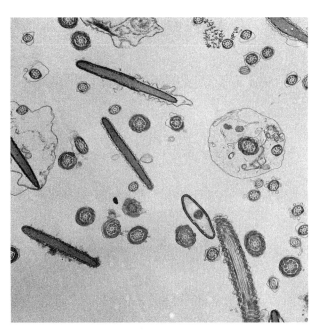

Figure 15-27   Electron diffraction pattern obtained from thin section shown to the right. The non-crystalline nature of the sectioned specimen generates a diffuse diffraction pattern.

erating spot patterns, so that the individual spots merge into rings surrounding a bright central spot (the undiffracted electrons). An example of a polycrystalline specimen would be an evaporated film of gold or aluminum where millions of tiny crystals of the metal have settled onto a plastic substrate such as Formvar. As in the previous example, the radius of the rings is related to lattice d spacings.

Determination of Spacings in a Crystalline Lattice

After one has recorded the spot or ring diffraction pattern, the negative is examined and distances are measured to be used in the following equation derived from the Bragg equation to calculate d spacing:

Equation 15-2. d Spacing in Crystalline Lattice

$$d = \frac{\lambda L}{R}$$

In this expression, R is the distance in millimeters from the central bright spot to one of the rings or spots, L is the *camera length* (distance in millimeters between specimen and photographic film), and lambda is the wavelength of the electron (based on the accelerating voltage: 50 kV = 0.00536 nm, 75 kV = 0.00433 nm, 100 kV = 0.00370 nm).

Equation 15-2 was derived from the Bragg equation 15-1 by substituting as follows. If one examines Figure 15-32, where R is the distance measured in millimeters (on the negative of the TEM diffraction pattern) from the central bright spot to the center of the diffracted spot and L is the camera length, then simple geometry tells us that:

$$R/L = \tan 2\theta$$

Furthermore, since the θ angles through which the electrons are diffracted are extremely small (about 1–2°), one may safely state the following:

$$\tan 2\theta = 2 \sin \theta$$

If one then recalls the Bragg equation,

$$\lambda = 2d \sin \theta,$$

upon following appropriate substitutions from the previous equations,

$$R/L = \lambda d,$$

one may obtain the final Equation 15-2,

$$d = \lambda L/R.$$

Since the critical three components in the equation are either known or may be measured from the electron micrograph of the diffraction pattern, it appears to be a simple process to determine the d spacings. As might be suspected, some background work must be done to confirm certain of these values.

The factor L, or camera length, is problematical since it may vary slightly every time a sample is examined. This may be due to variation in specimen positioning caused by bent grids, using a different specimen holder, a change in the vertical adjustment of the specimen holder in the stage (z axis control), and even cleaning of polepieces or readjustment of standard lens current settings. Conse-

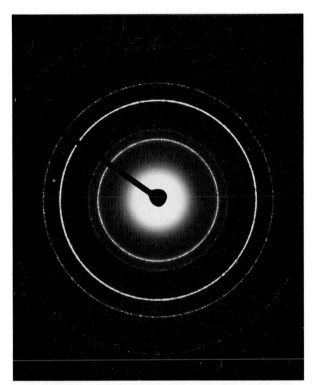

Figure 15-28 Electron diffraction pattern of a polycrystalline specimen. A thin film of gold was deposited onto a plastic film by evaporating the molten metal in a vacuum evaporator. The gold vapor settled onto the plastic and formed tiny crystals that are responsible for the pattern shown. (Courtesy of B. DeNeve.)

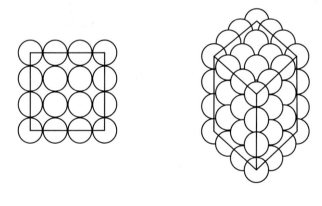

A　　　　　　　　　　　　**B**

Figure 15-29 (A) Array of atoms in a single plane. Each sphere equals one atom. (B) Stacks of planes generate the crystalline lattice. Here the crystal is viewed from one of the corners of the cube of atoms.

quently, for precise calculations, it is best to calibrate the camera length on a regular basis and to verify that the specimen is in the proper position (i.e., the eucentric position) following the specific TEM manufacturer's direc-

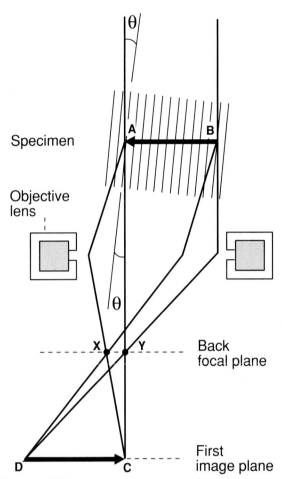

Figure 15-30 Diagram showing path that electrons take upon striking a crystalline object. Some rays (C, D) converge in the image plane while others (X, Y) converge in the back focal plane to generate the diffraction pattern.

tions. Once the camera length is satisfactorily determined, it is multiplied by the wavelength of electron used and the result is now termed the *camera constant* (λL).

Determination of Camera Constant (λL)

1. Insert a standard specimen with known d spacings into the TEM and focus the image. Some standards include evaporated thin films of gold, aluminum, or thallous chloride. As an example, we will use gold as the standard (see Figure 15-28 and Table 15-2).
2. Verify that the specimen is in the eucentric position (check manufacturer's directions). In some microscopes this may involve tilting the specimen +/– 10° and observing the centration of the focused image on the screen. If the image shifts off center during the tilting, then the proper stage adjustments (x,y,z movements) are made until the image shift is minimal.
3. Remove the objective aperture, insert the diffraction aperture, and center it using mechanical controls.

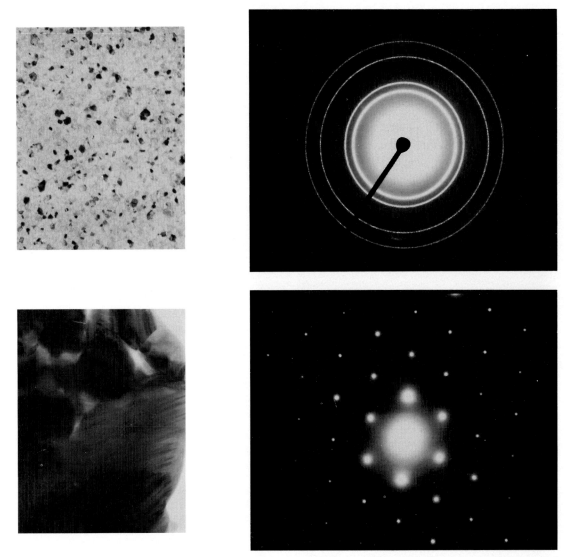

Figure 15-31 Comparison of diffraction pattern from a single crystal versus a polycrystalline specimen. The top left panel shows a standard TEM image of crystals magnified 4,500× and the top right panel shows the diffraction pattern obtained from this polycrystalline specimen. The lower left panel shows a TEM image of one of the crystals magnified at 90,000×. A diffraction pattern was obtained from the crystal and is shown in the bottom right panel. (Courtesy of P. Tlomak.)

4. Focus on the edge of the diffraction aperture and image while in the selected area diffraction mode (see Performing Selected Area Diffraction, in next section of this chapter).

5. Depending on the TEM model, obtain a focused diffraction pattern by varying the current to the proper lenses (intermediate or projector lenses).

6. Record the image of the diffraction pattern, as follows:
   a. adjust brightness with condenser lens 2 to barely see the faintest rings,
   b. insert the central beam stopper to block the bright central spot,
   c. expose the negative for 20 to 30 seconds,
   d. remove the beam stop during the final 3 to 5 seconds of exposure.

7. On a negative of the diffraction pattern:
   a. measure the diameter of each of the rings,
   b. divide the diameter by 2 to obtain the radius "R" in millimeters.

8. Look up the value for d_{hkl} for each of the rings (given in Å).
   NOTE 1: The designation d_{111} is a way of specifying a particular lattice plane (since many different planes are present in a crystal). It is important to be able to assign the d spacings to the proper set of lattice planes since the spacings of the various planes are different. The three subscripts in the expression d_{hkl}

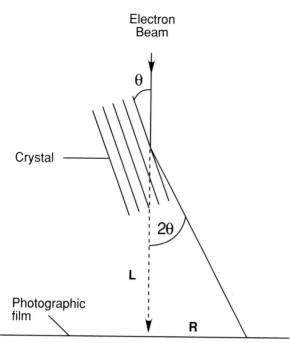

Electron Beam

θ

Crystal

2θ

L

Photographic film

R

Figure 15-32 Diagram showing the relationship between camera length (L) and distance (R) between central beam spot and a diffraction spot. Using simple geometry it may be determined that R/L = tan 2 θ.

Table 15-2   Gold Diffraction Standard

hkl	I	d	R	dR
111	100	2.355	_____	_____
200	52	2.039	_____	_____
220	32	1.442	_____	_____
311	36	1.230	_____	_____
222	12	1.1774	_____	_____
400	6	1.0196	_____	_____
331	23	0.9358	_____	_____
420	22	0.9120	_____	_____
422	23	0.8325	_____	_____
			Average	_____

are termed the *Miller indices* and the process of assigning d spacings to the proper lattice plane is termed *indexing*. The process of indexing single crystals is complicated and requires a knowledge of crystallography. Details of this procedure are outlined in the references cited at the end of this chapter.

NOTE 2: The rings represent various lattice planes in a polycrystalline specimen. In the gold specimen (Table 15-2) the rings are arranged in a particular order from the center outwards with the brightest ring, d_{111}, having a relative intensity "I" of 100%. Other rings are expressed in terms of relative brightness to d_{111} so that the d_{200} ring with I = 52 is nearly half as bright, etc.

9. Proceed to calculate camera constant for d_{111}:
   a. from Equation 15-2, we may derive the following expression for camera constant:

$$\lambda L = dR$$

   b. since we have already measured the radius R for the d_{111} ring from the negative, and since the actual d spacing for the d_{111} plane is known, simply multiply d by R to obtain the camera constant.
10. Repeat the calculation of camera constant for the remaining rings and fill in the values in Table 15-2 (make a photocopy copy of the page).
11. Average together all of the various camera constants (dR) in order to obtain a more accurate figure. This is the camera constant value to be used in subsequent determination of d spacings in unknown crystals. *Note: It is important to verify that the rings are symmetrical in order for the measurements to be accurate; therefore, several measurements must be made from various locations along the ring to verify the roundness of the rings.*

Once the camera constant has been accurately determined using the standard sample, one may then proceed to use one of the possible diffraction modes to determine the identity of an unknown crystal or group of crystals. After tilting and orienting the crystal to an appropriate zone axis to obtain the diffraction pattern, record the diffraction pattern of the unknown crystal(s). One must now determine the crystal structure by matching the diffraction pattern of the unidentified crystal to one of the patterns of the 14 lattice systems (consult diffraction references given at end of chapter). After determining the d spacings from the negative, the pattern is indexed and the values are looked up in a reference book (*Mineral Powder Diffraction File*, for example) or by computer program (the *Mineral Powder Diffraction File* is now available on CD-ROM) to identify the unknown crystal. Energy dispersive X-ray analysis will greatly expedite this procedure by identifying the various elements making up the crystal.

Types of Diffraction Modes

Depending on the design of the TEM or STEM being used, it is possible to operate the microscope in up to six different types of diffraction modes. Only those approaches that are used in biological studies will be discussed.

Selected Area Diffraction (SAD)

This is probably the most commonly used diffraction mode in TEM and STEM instruments. It can be used to generate diffraction patterns from crystalline

materials with lattice spacings smaller than 2.5 to 4.0 nm. In practice, one positions a diffraction aperture of the appropriate size over the area from which the diffraction is to be conducted (thereby "selecting an area") and then adjusts the microscope to obtain the pattern. To obtain diffraction from areas as small as 0.5 μm involves the use of very small diffraction apertures ($<$5–10 μm in diameter) that contaminate too rapidly. In addition, spherical aberration in the objective lens causes inaccuracies in the selection of an area in the specimen. If apertures smaller than 5 μm are used, one is never certain what area of the specimen is actually generating the diffraction pattern. Most of the time, therefore, SAD is conducted from specimen areas in the 10 to 100 μm range. The general steps for accomplishing SAD are outlined in the next section (specific procedures will vary from one microscope to another, so the manufacturer's manual should be consulted).

Performing Selected Area Diffraction

1. Insert specimen, locate area of interest, and focus in normal imaging mode using objective lens controls. A side entry stage with capabilities for tilting (and possibly rotating) the specimen is essential in order to orient the specimen for single crystal diffraction studies.
2. Insert and center the diffraction aperture. Focus the diffraction lens (intermediate or projector lens, depending on model of TEM) until the edge of the diffraction aperture is sharp. In some microscopes, this is done manually by focusing the diffraction lens, while in other microscopes, it may be achieved simply by pressing a button usually designated "SA" and then touching up the focus using an "SA Focus" knob.
3. Refocus the specimen image using the objective lens control. The objective aperture may help focusing by improving contrast and so may be left in place. A micrograph of the specimen may be taken at this point if desired. What one is doing at this point is focusing the image near the back focal plane (actually the plane of the diffraction aperture) so that one will be able to identify exactly the area in the specimen that is being diffracted.
4. Withdraw the objective aperture from the column.
5. Switch the microscope into the diffraction mode by either pressing the appropriate button or by manually adjusting the diffraction knob until the diffraction pattern comes into sharp focus.

   NOTE: Some microscopes have another adjustment called "Camera Length," which adjusts another projector lens to increase or decrease the camera length and thereby affect the final size of the diffraction pattern. There may be another smaller knob associated with the "Camera Length" knob that is used to care-

fully sharpen the focus of the diffraction pattern at each camera length.
6. Adjust condenser lens 2 ("Brightness") until the dimmest spot or ring is barely visible. This will also sharpen the diffraction pattern since a more coherent beam is being generated.
7. Insert the central beam blocker and expose the micrograph for 20 to 30 seconds. Withdraw the beam blocker during the last 3 to 4 seconds so that the bright central spot will be recorded at the proper density on the negative.

Microdiffraction

This term refers to selected area diffraction from very small areas (10 to 500 nm in diameter). Unlike conventional SAD, where apertures are used to outline the area to be diffracted, in microdiffraction one converges a beam of electrons on the area of interest to generate the diffraction pattern. In suitably equipped TEM and STEM instruments, it is possible to achieve even nanodiffraction patterns from areas as small as 1 nm in diameter.

The oldest method for achieving microdiffraction was described by Riecke in 1969 and is still useful for biologists—although materials scientists prefer to use newer *convergent beam diffraction* techniques since they yield more information about the crystalline lattice structure and are more readily performed in the latest generation of electron microscopes.

The Riecke method requires a third condenser lens (either a minilens inserted between condenser lens 2 and the objective lens or a combination condenser/objective lens). This lens effects a great demagnification of the C2 aperture. For example, if a C2 aperture of 5 μm was used and was demagnified 40$\times$ by the final condenser, then a spot of 125 nm could be formed on the specimen. Figure 15-33 shows the arrangement of lenses and positioning of specimen in an instrument capable of doing microdiffraction. An example of the utility of microdiffraction might include the identification of an unknown intracellular crystalline inclusion in a biopsy of human lung as an asbestos particle.

Performing Microdiffraction

1. Find the area of interest in the regular TEM mode and focus it carefully at 20,000 to 30,000$\times$. If a side entry adjustable stage is used, set the specimen height (z control) in the *eucentric* position so that adjustment of specimen tilt will not shift the specimen appreciably.
2. Same as step 2 in Performing Selected Area Diffraction.

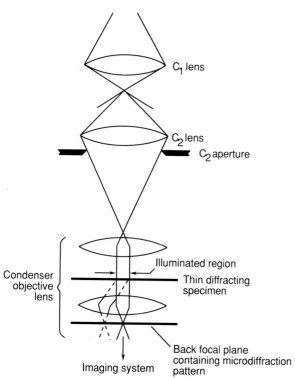

Figure 15-33 Arrangement of lenses and aperture needed to achieve microdiffraction by Riecke method. A parallel beam is generated by the upper polepiece of the condenser/objective lens. This narrow beam is used to illuminate the specimen area of interest to generate the microdiffraction pattern. (Courtesy of D. B. Williams and Philips Electronics Instruments.)

3. Remove diffraction aperture and return microscope to standard imaging mode.
4. Insert an appropriately sized C2 aperture (2 to 10 µm) and sharpen the image of the edge of the aperture using the objective lens control. Do not be concerned that the image will be thrown out of focus by this operation.
5. Bring the image back into focus by using the specimen height (z axis) control.

6. Put the microscope in the diffraction mode (press DIFF button) and observe the microdiffraction pattern.
7. Record the image as described in Performing Selected Area Diffraction, step 7.

Other Types of Diffraction

Although SAD and microdiffraction are the two most commonly used types of diffraction in biological electron microscopy, other procedures may be needed depending on the nature of the specimen being analyzed.

Low angle diffraction is used with biological and organic crystals with d spacings greater than 2.5 to 4.0 nm. It will be realized by examining Equation 15-2, that larger d spacings necessitate the use of a camera with a longer focal length. This is achieved by reducing the strength of the objective lens, thereby increasing its focal length. This necessitates moving the specimen further down the column, a condition that may not be possible on many microscopes that have limited z axis control (specimen height control). If this is a problem, special specimen stages placed lower in the electron microscope column (near the projector lenses) will be needed.

In the *high resolution diffraction* procedure, the specimen is inserted into special air-locked specimen holders that replace the SAD aperture or that are placed below the final projector lens. These specimen holders resemble a conventional aperture holder except that they have provisions for tilting and moving the specimen in x, y, and z axes. By moving the specimen so low in the microscope column, one is able to use all of the lenses to converge the electron beam to an extremely small spot. Since the aperture angle of the beam is quite small, a better separation of the diffraction pattern is achieved.

References

Beeston, B. E. P., R.W. Horne, and R. Markham. 1972. Electron diffraction and optical diffraction techniques. *Practical methods in electron microscopy*, Vol. 1, Pt. II, (A. M. Glauert, ed.), North-Holland/American Elsevier Publishing Co. 262 pp.

Chandler, J. A. 1977. X-ray microanalysis in the electron microscope. *Practical methods in electron microscopy*, Vol. 5, Pt. II (A. M. Glauert ed.), North-

Holland/American Elsevier Publishing Co.

Goldstein, J. I., D. E., Newbury, P., Echlin, D. C., Joy, C. Fiori, and E. Lifshin. 1981. *Scanning electron microscopy and X-ray microanalysis*. New York: Plenum Press. 673 pp.

Hren, J. J., J. I. Goldstein, and D. C. Joy, eds. 1979. *Introduction to analytical electron microscopy*. New York: Plenum Press. 601 pp.

Hutchinson, T. E., and A. P. Somlyo, eds. 1981. *Microprobe analysis of biological systems.* New York: Academic Press. 427 pp.

Mineral powder diffraction file. 1986. P. Bayliss, R. C. Erd, M. E. Mrose, A. P. Sabina, and D. K. Smith, eds. International Centre for Diffraction Data, 1601 Park Lane, Swarthmore, PA 19081.

Misell, D. L. and E. B. Brown. 1987. Electron diffraction: An introduction for biologists. *Practical methods in electron microscopy*, Vol. 12. A. M. Glauert, ed., North-Holland/American Elsevier Publishing Co.

Morgan, A. J. 1985. X-ray microanalysis in electron microscopy for biologists. *Royal microscopical society, handbook no. 5.* London: Oxford University Press.

Murr, L. E. 1982. *Electron and ion microscopy and microanalysis.* New York: Marcel Dekker, Inc. 793 pp.

Philibert, J., and R. Tixier. 1968. Some problems with quantitative electron probe microanalysis. In *Quantitative electron probe microanalysis.* K. F. J. Heinrich, ed. National Bureau of Standards special publication 298, Washington, D.C., p 13.

Riecke, W. D. 1969. Beugungsexperimente mit sehr feinen elektronenstrahlen (Diffraction experiments with very fine electron beams). Zeitschrift fur Angewandte Physik. 27:155–165.

Roomans, G. M., and J. D. Shelburne, eds. 1983. Basic methods in biological x-ray microanalysis. Scanning Electron Microscopy, Inc. (AMF O'Hare, IL). 284 pp.

Smith, D. K., M. C. Nichols, and M. E. Zolensky. 1983. A FORTRAN IV program for calculating X-ray powder diffraction patterns. College of Earth and Mineral Sciences, The Pennsylvania State University, University Park, PA.

Vaughn, D. 1983. Energy dispersive X-ray analysis: An introduction. Kevex Instruments, Inc. San Carlos, CA.

Williams, D. B. 1984. Practical analytical electron microscopy in materials science. Philips Electronics Instruments (New Jersey). Currently available through: Techbooks, 2600 Sesky Glen Court, Herndon, VA 22071.

Wong-Ng, W., C. R., Hubbart., J. K., Stalick, and E. H. Evans. 1988. Computerization of the ICDD powder diffraction database. Critical review of sets 1 to 32. *Pow Diff* 3:12–18.

Intermediate and High Voltage Microscopy

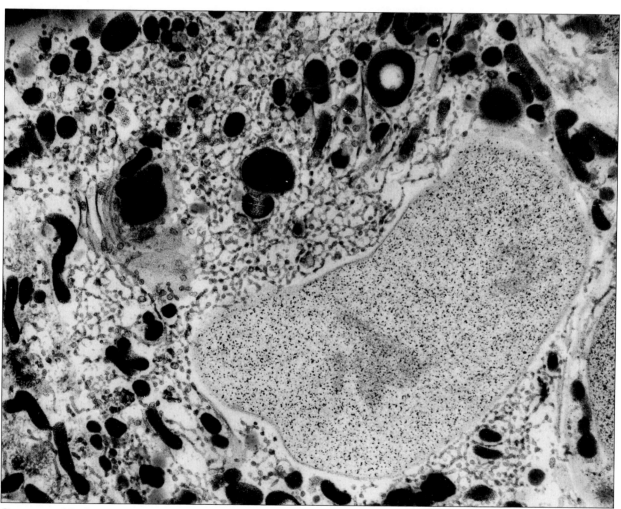

Courtesy of L. Hermo.

Conventional transmission electron microscopy employs accelerating voltages in the range of 25 to 125 kV (1 kV = 1,000 volts). Most commonly, 50 to 75 kV are used for routine work. Since most conventional microscopes have a limited voltage capability (up to 100 to 125 kV), they cannot be used for high voltage work. Instead, one would use a specially designed *high voltage electron microscope* (HVEM), which is essentially a standard transmission electron microscope with high voltage capabilities (Figure 16-1). Medium or *intermediate voltage electron microscopes* (IVEM) cover the voltage ranges from 200 up to 500 kV, while high voltage electron microscopes are used at accelerating voltages above 500 kV and well into the mega (million) voltage (MV) range.

There are four centralized facilities in the United States where biological investigators may share usage of a high voltage electron microscope: the University of Wisconsin (Madison), the University of Colorado (Boulder), New York State Department of Health (Albany), and the University of California (Berkeley; two instruments). The microscope with the highest voltage capability is in Toulouse, France. This microscope has a three megavolt (3 MV) capability. High voltage microscopes are extremely expensive and usually require housing in a specially designed building. Materials science has used high voltage microscopy extensively; however, it is now being used more frequently in biological investigations. Intermediate voltage microscopes that also can be used at the normal 50 to 75 kV are more commonly seen as replacements for outdated microscopes. Presently, many of these instruments are in service in university and industrial settings.

Historical Perspective

High voltage microscopy was a natural outgrowth of the desire to examine whole cells and obtain three-dimensional information. In the early days of electron microscopy, Porter and colleagues (1945), in a now classic paper, examined whole cells, especially the thin edges of fibroblasts, at relatively low voltages. Although valuable information was obtained, many of the cells features contained in the thicker portions of the preparation were impenetrable by the electron beam. For some time after the advent of good microtomes, investigators concentrated their efforts on obtaining information from the newly produced thin sections. The first high

voltage microscopes were constructed in the late 1940s and the 1950s. After the high voltage microscope was operational, Porter and colleagues returned to studying whole cells with this tool, this time with much more success (for examples, see Wolosewick and Porter, 1976; Porter and Tucker, 1981). Numerous other investigators have used the less than 50 high voltage electron microscopes scattered throughout the world. The Japanese, Americans, French, and English are especially active in biological applications of high voltage microscopy.

Advantages of High Voltage Microscopy

There are four main advantages to using voltages at or in excess of 100 kV. The first is the *increased resolution* obtainable at high voltages. High energy electrons have been shown to possess a shorter wavelength than lower energy electrons. As was demonstrated in Chapter 6, the shorter the wavelength of the electron, the higher (better) the resolution capability. Figure 16-2 illustrates the resolution difference obtained by changing the accelerating voltages. Unfortunately, the kind of thicker specimens usually used in the high voltage microscope partially offset this increase in resolving ability since there is an accompanying increase in *chromatic aberration*. Chromatic aberration is caused by the energy loss of electrons as they collide with the thicker tissue section (see Chapter 6). It is evident when the longer wavelength (lower energy) electrons do not focus at the same spot as electrons with higher energy and shorter wavelengths.

The second advantage of using higher voltage is the *increased specimen penetrating capability* of high energy electrons (Figure 16-3). Sections 3 μm in thickness may be penetrated by the beam of a high voltage microscope to produce quality micrographs. In a practical sense, penetration is determined by the fraction of the beam coming from the specimen that passes through the objective aperture. It is possible to collect more electrons if a larger objective aperture is used; however, contrast is sacrificed (see Chapter 6). The penetration of biological materials is greater when a nonembedded specimen is critical point dried and contains no embedding media (see Chapter 3) than when specimens have been conventionally processed and embedded. Moreover, each specimen has its unique features that affect penetrating capability of the

Figure 16-1   The National Institutes of Health-supported high voltage (1MV) microscope at the University of Colo- rado in Boulder. (Micrograph courtesy of R. McIntosh.)

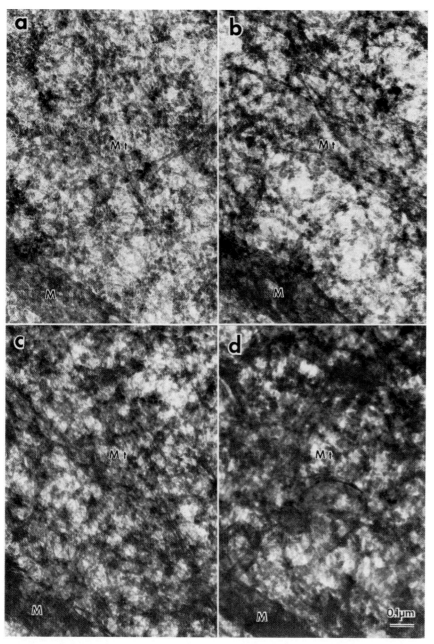

Figure 16-2 Resolution improvement in cytoplasmic structure at differing (500kV, 200kV[uHR], 200kV, and 100kV) accelerating voltages. M = mitochondria; Mt = microtubules. (Micrographs courtesy of K. Hama from *J Electron Microsc* 30:57–62. Used with permission from the publisher.)

beam. As accelerating voltage increases, the penetrating power of the beam does not increase proportionally; a 3MV microscope has only marginal advantage over a 1MV microscope (Figure 16-4).

The third advantage of high voltage microscopy is the *increased depth of information* as compared with the lower voltage microscope. Ironically, the high voltage microscope has less depth of field because the shorter wavelength electrons actually decrease the depth of field (see Chapter 6). In spite of this,

the information in the section is imaged at virtually all depths within the section, providing important information about three dimensionality (Figure 16-5). Because of the great depth of information exhibited by high voltage microscopes, micrographs may be taken in pairs in which one image is tilted slightly (about 15° to 20°). Such images are called *stereopairs* (Figure 16-6). Viewing structures from two different angles is the key to obtaining depth perception or three dimensionality. Stereopairs help

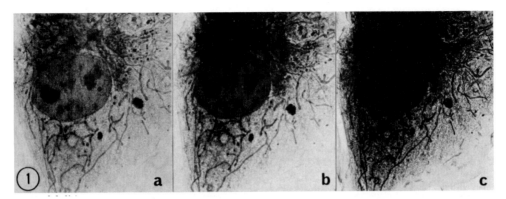

Figure 16-3 Series of micrographs taken at 1,000 (a), 200 (b), and 100 kV (c), demonstrating increased specimen resolution at the higher kVs. (Micrographs courtesy of K. Hama from *38th Ann Proc Elec Micros Soc Amer*, pp 802–5. Used with permission from the publisher.)

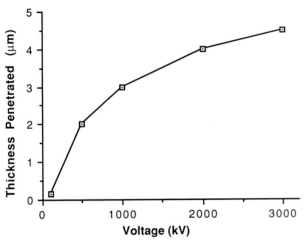

Figure 16-4 The relationship between accelerating voltage and the ability of the beam to penetrate specimens of various thickness.

ing advantage of both the increased penetrating capability and the overall reduction in specimen damage, investigators have made several attempts to view *living* or *wet specimens*. Only very limited information has been obtained from viewing live specimens. Many improvements are needed before the fine details of most living cells can be visualized.

One disadvantage of using higher accelerating voltages is that contrast is lessened. The decrease in contrast is partially offset by using stains or impregnation techniques that penetrate thicker sections and enhance contrast (Figure 16-7; Thiery and Rambourg, 1976).

Contributions of High Voltage Microscopy

Problems related to interpretations of three-dimensional images from two-dimensional electron micrographs are numerous (see Chapter 18). Short of extensive serial section reconstruction, high voltage microscopy has been the technique of choice to gain three-dimensional information. The complexities of the Golgi apparatus or the endoplasmic reticulum, the path taken by filaments and their relationships to other organelles have, among many other things, been visualized by high voltage microscopy. Complex relationships of one cell component to another have been realized. Serious attempts have been made at viewing living specimens, especially bacteria and viruses.

Most investigators agree that the relatively recent high voltage work by Porter and colleagues (Wolosewick and Porter, 1976; Porter, 1981) is a major step in our understanding of how cells are

resolve the confusion that results from superimposition of structures in extremely thick (0.3 to 3.0 μm) sections. If one is wearing glasses designed for stereo viewing, the effect is an image that has a three-dimensional appearance.

The fourth advantage is that the *damage to the specimen at very high accelerating voltages is proportionably less* than at lower voltages due to less electron interaction with the specimen. Since the electrons are traveling substantially faster at the higher accelerating voltages, fewer electrons are deflected by the specimen. (Ironically, the thicker specimens often viewed by the high voltage microscope are more susceptible to radiation damage than are thinner specimens because there is simply more tissue and embedment present to interact with electrons.) Tak-

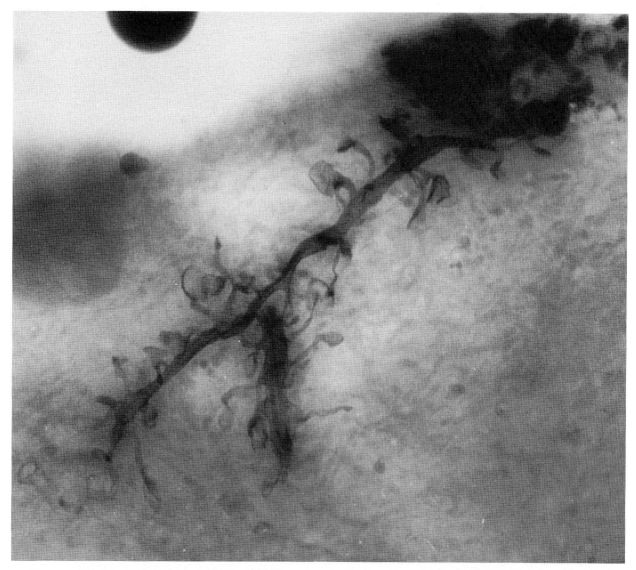

Figure 16-5 High voltage electron micrograph from a 3 μm thick plastic embedded specimen containing a dendrite stained by intracellular injection of horseradish peroxidase. The high resolution and penetration provided by the section thickness allow clear visualization of the fine dendritic appendages specialized for synaptic transmission. (Micrograph courtesy of C. Wilson.)

organized. These fascinating papers present evidence that there is a highly organized and intricate ground substance in the cell that mediates, regulates, and directs transport within the cell. The system of filamentous elements composing the ground substance has been termed the *microtrabecular system or lattice* (Figure 16-8). It is suggested that the microtrabecular lattice organizes most cell components into a unified structure, the *cytoblast*. This work is an important technical achievement advanced by high voltage microscopy, but most investigators feel that the hypotheses advanced by Porter and colleagues awaits further experimental testing.

The greatest advantages of high voltage microscopes have been achieved in the materials sciences. Microscopes currently available on the market are generally intermediate range voltage instruments, capable of imaging most biological specimen components that can be imaged by a high voltage microscope.

A

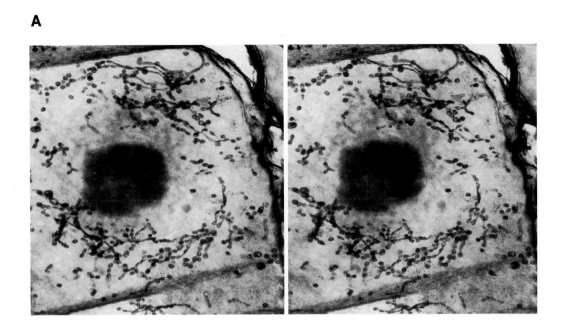

B

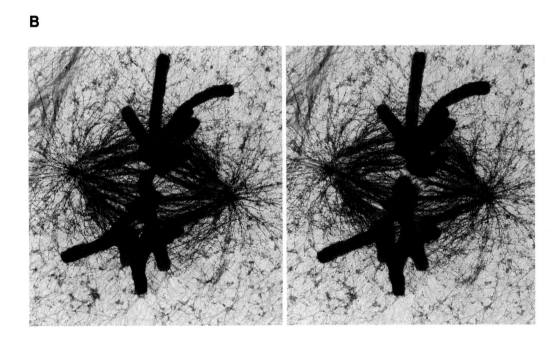

Figure 16-6   Stereopairs. (A) High voltage micrograph of a 2–3 μm section of a pea root tip; (B) Isolated mammalian cell in mitosis whose microtubules have been darkened by the binding of 20 nm particles of colloidal gold. Glasses designed for stereo viewing should be worn to visualize three dimensions. (Micrographs courtesy of P. Favard, N. Carosso and R. McIntosh.)

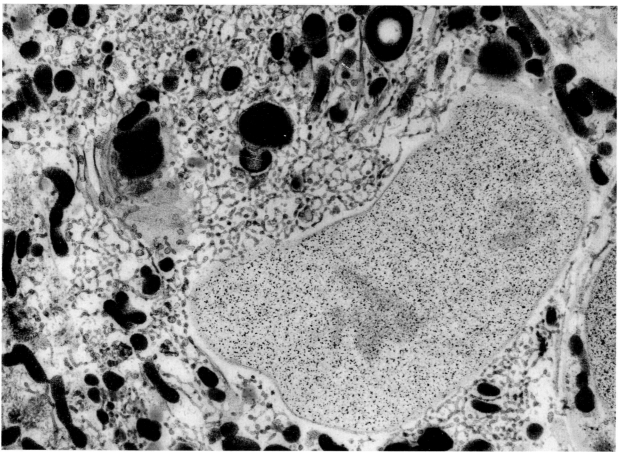

Figure 16-7 A 2 μm thick section impregnated with os- of the endoplasmic reticulum to be traced. (Micrograph cour-
mium and viewed using intermediate voltage conditions. tesy of L. Hermo.)
Note how the thickness of the section allows the continuity

Figure 16-8 The microtrabecular lattice (M). (Micrograph
courtesy of K. Hama from *J Electron Microsc* 30:57–62. Used with
permission from the publisher.)

References

Dupouy, G. 1985. Megavolt electron microscopy. In *Advances in electronics and electron physics*, Suppl. 16, pp 103–65.

Fujita, H. 1989. The research center for ultra-high voltage electron microscopy at Osaka University. *J Electron Microsc Tech* 12:201–18.

Hama, K. 1973. High voltage electron microscopy. In *Advanced techniques in biological electron microscopy*. J. K. Koehler, ed. New York: Springer-Verlag, pp 275–97.

Hama, K., and F. Nagata. 1970. A stereoscopic observation of tracheal epithelium of mouse by means of the high voltage electron microscope. *J Cell Biol* 45:654–9.

Mazzone, H. M., et al. 1968. The high voltage electron microscope in virology. *Adv Virus Res* 30:43–82.

Porter, K., and B. Tucker. 1981. The ground substance of living cells. *Sci Amer* 244:56–67.

Porter, K. R., et al. 1945. A study of tissue culture cells by electron microscopy. Methods and preliminary observations. *J Exptl Med* 97:727–50.

Ris, H. 1969. Use of the high voltage electron microscope for the study of thick biological specimens. *J Microscopie* 8:761–6.

Thiery, G., and A. Rambourg. 1976. A new staining technique for studying thick sections in the electron microscope. *J Microscopie Biol Cell* 26:103–6.

Wolosewick, J. J., and K. R. Porter. 1976. Stereo high-voltage electron microscopy of whole cells of the human diploid line, WI-38. *Am J Anat* 147:303–24.

Some Specific Tracers in Use

Cationic and Native Ferritin

Lanthanum

Horseradish Peroxidase

Lactoperoxidase

Ruthenium

References

Tracers

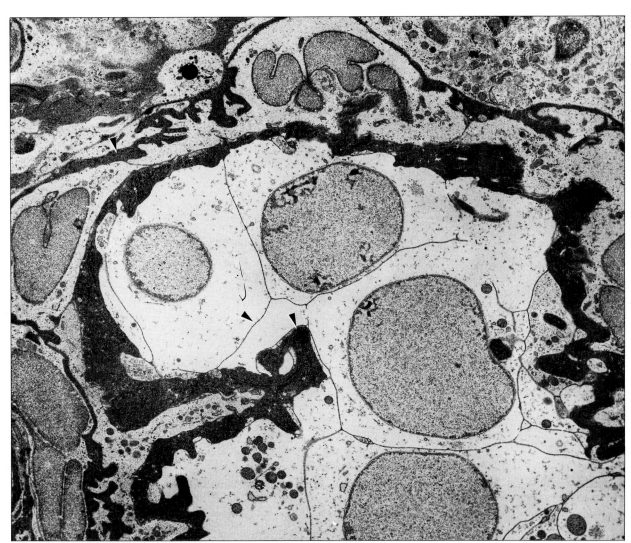

Courtesy of A. Mayerhofer.

Tracers are exogenously administered substances that, when visualized in electron microscope preparations, provide valuable information about cell compartments, junctional elements, and cell surfaces. Tracers are employed to delineate *extracellular spaces* and the limits to extracellular spaces. By highlighting extracellular spaces, tracers can be used to determine the site(s) of *physiologic barriers*. A tracer may be used to determine the *permeability* of the vascular system, and specifically, the permeability of the vascular endothelium. Tracers may be used to *follow the paths of molecules* in a physiological system (Figure 17-1). For example, an electron-dense marker such as ferritin is fre-

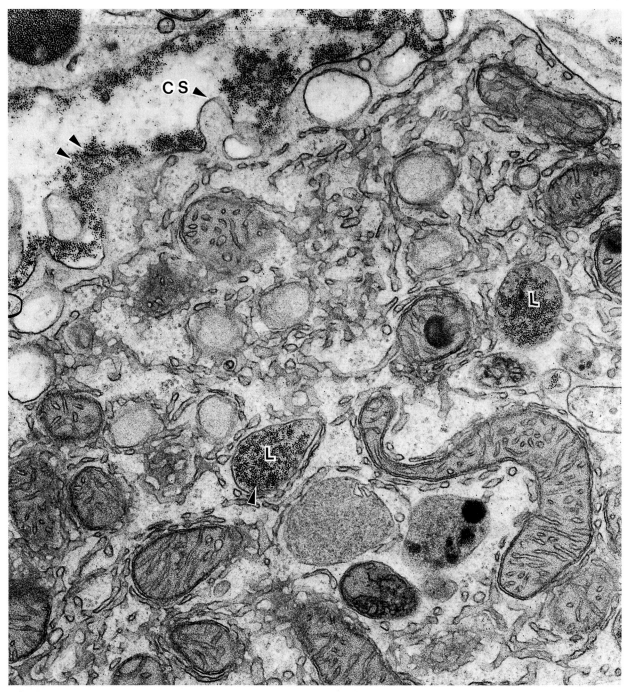

Figure 17-1 Cationic ferritin (arrowheads) used as a tracer. Ferritin is seen at the cell surface (CS) and has been internalized and incorporated into lysosomes (L). (Micrograph courtesy of L. Hermo.)

quently used to trace endocytic events and the fate of endocytosed materials. Since some tracers travel within the extracellular space, the extent of their excursion is often indicated by their binding to cell surfaces.

Tracers have been used to distinguish types of cell *junctions* (see Figures 19-4 through 19-21, Chapter 19) by virtue of their ability to delineate/highlight the extracellular space in the vicinity of the junction and to allow measurement of the intercellular space. For example, a tight or occluding junction excludes tracer, whereas a gap junction will permit a 2 to 4 nm wide deposition of tracer within the intercellular space (Figure 19-16). The width of the tracer between membranes forming the junction can thus be used to differentiate tight from gap junctions. A number of other junctional types are delineated and/or categorized by the use of tracers.

Some Specific Tracers in Use

Cationic and Native Ferritin

Cationic ferritin carries a positive charge and thus will bind to negatively charged moieties on cell surfaces. The internalization of bound substances, such as ferritin, is termed *adsorptive endocytosis*. It should be emphasized that this form of endocytosis carries specificity for the bound substance such as might be seen in the process of *receptor mediated endocytosis*. For many cells that show an endocytic process, vesicles form at the cell surface and pinch off within the cell. These contain the bound tracer. The cationic ferritin is successively shuttled to multivesicular bodies and then lysosomes. Figure 17-1 shows an example of cationic ferritin used as a tracer.

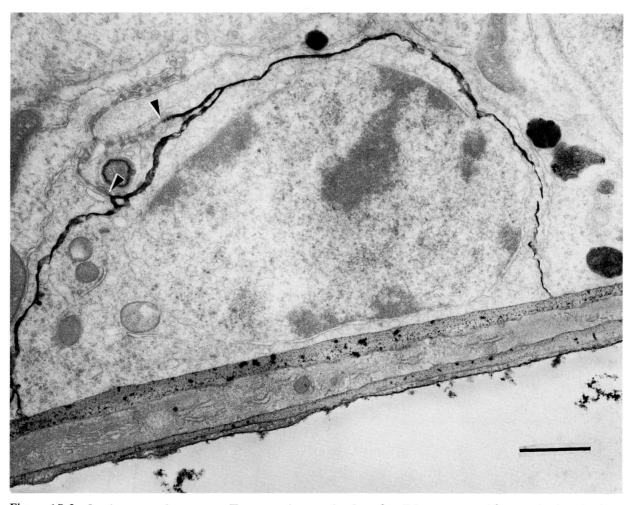

Figure 17-2 Lanthanum used as a tracer. Tracer was introduced into the vascular system. The tracer highlighted the borders of a cell, but was stopped from passing into the tissue by tight junctions at the site indicated by the arrowhead.

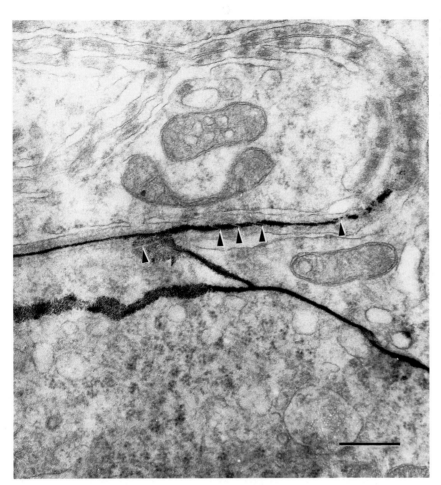

Figure 17-3 In a similar, but higher magnification micrograph to Figure 17-2, lanthanum was able to bypass some tight junctions, as evidenced by their presence in negative relief (arrowheads). Lanthanum was, however, prevented from penetrating deeply within the epithelium.

Not being a charged molecule, *native ferritin* is taken up, or endocytosed, by some cells in a similar manner as other molecules may be taken up in a nonspecific manner. This process has been termed *fluid-phase endocytosis*. In many cells, native ferritin is taken up during fluid-phase endocytosis by invaginations in the cell surface that later pinch off inside the cell. Tracer is next usually transferred to multivesicular bodies and then to lysosomes.

Lanthanum

Lanthanum, with an atomic number of 57, produces contrast under the electron microscope. Lanthanum is a trivalent cation and will bind to negatively charged ions on the cell surface, especially negatively charged glycoprotein moieties.

As the pH of a lanthanum hydroxide solution increases, it forms a colloidal suspension. A colloidal lanthanum suspension is most commonly used as a tracer rather than a stain. Lanthanum will follow an intercellular route until it is blocked by a specific structure such as a tight junction (Figures 17-2 and 17-3). Lanthanum may circumvent tight junctions at certain points if they are not continuous around the cell. In doing so, lanthanum will define the limits of the tight junction, which itself should appear free of lanthanum (Figure 17-3). By using lanthanum in this manner, it is possible to trace the pathway of molecules as they move through an epithelium and thus define the permeability of the epithelium. Junctions such as gap junctions and desmosomes will allow passage of lanthanum. Lanthanum will allow a clear definition of the width of the intercellular space and aid in the characterization of the junctional type.

Lanthanum cannot be used as a quantitative measure of permeability. It cannot be assumed that because lanthanum is blocked at a particular site that *all* substances are blocked under physiologic circumstances. Lanthanum grossly defines where a structural blockage exists within epithelia and whether or not an epithelium, after a treatment, has had a major change in its permeability to this substance.

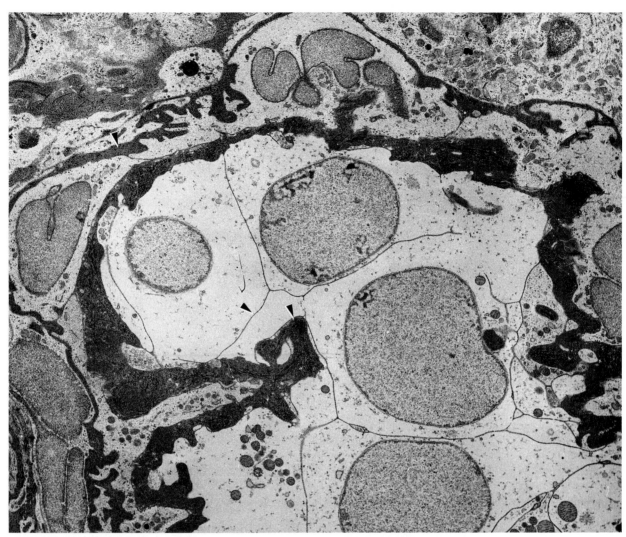

Figure 17-4 Horseradish peroxidase (HRP) used as a tracer. Horseradish peroxidase was injected intravascularly. After transport across the vascular system, it outlined spaces where it was free to flow. In this unstained tissue, HRP appears as the dense material in tissue spaces and between cells (arrowheads). (Micrograph courtesy of A. Mayerhofer.)

Horseradish Perioxidase

Horseradish peroxidase is an enzyme with a molecular weight of about 40,000. It is used in much the same way as lanthanum to outline the intercellular space and locate the site of permeability barriers. It differs, however, from lanthanum in that it is not inherently electron dense. In the presence of diaminobenzidine, H_2O_2, and later osmium, the site of deposition of horseradish peroxidase is revealed as an amorphous electron-dense deposit (see Chapter 9). This enzyme marker is amplified by the enzymatic reaction catalyzed by peroxidase, and thus the osmicated reaction product presents as a very electron-dense marker (Figure 17-4).

Horseradish peroxidase may be conjugated to colloidal gold to be used as a tracer for fluid-phase endocytosis in a manner similar to that described previously for native ferritin. In this instance, the gold and not the peroxidase becomes the marker visualized by the electron microscope (Figure 17-5).

Lactoperoxidase

Lactoperoxidase (MW 100,000) is a marker occasionally used for the demonstration of fluid-phase endocytosis (see native ferritin above). Demonstration of a dense reaction product is via diaminobenzidine, H_2O_2, and later osmium (see horseradish peroxidase, Chapter 9). The reaction product appears similar to that demonstrated for horseradish peroxidase shown in Figures 17-4 and 17-5.

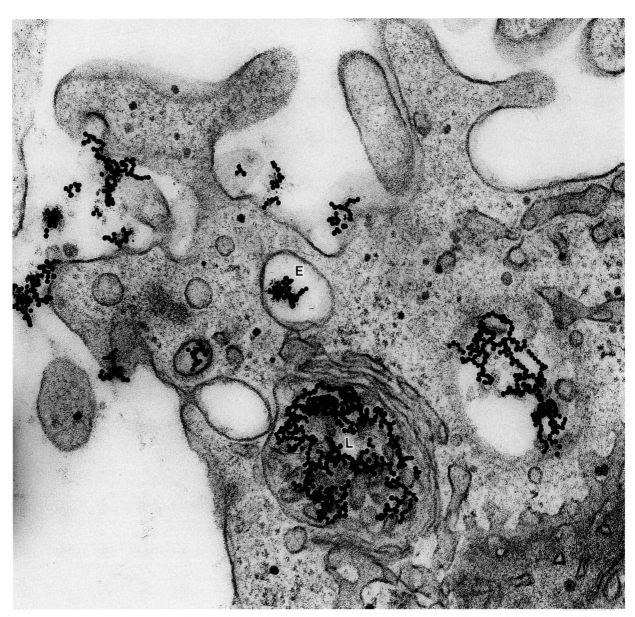

Figure 17-5   Horseradish peroxidase was coupled to colloidal gold and exposed to the cell. The tracer is seen at the cell surface and within endosomes (E) and lysosomes (L). (Micrograph courtesy of L. Hermo.)

Ruthenium

Like lanthanum, ruthenium carries a positive charge and consequently will bind to anionic sites on cell surfaces. Thus, it is capable of staining extracellular anionic sites. In doing so, it serves as a tracer since it only stains surfaces accessible to it. Like lanthanum, ruthenium will form a colloid and act as a tracer. When used in conjunction with osmium, it forms a ruthenium oxide (RuO_4), which is readily visualized under the electron microscope (Figure 17-6).

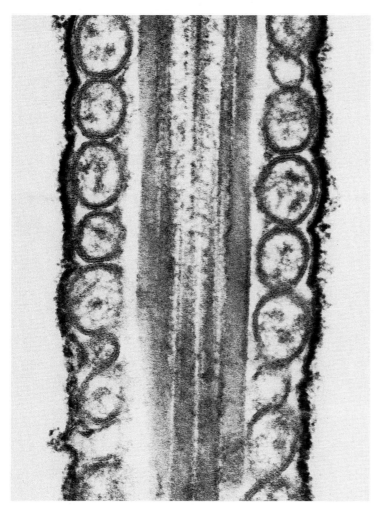

Figure 17-6 Ruthenium, often used as a tracer, stains the external surface of a sperm cell.

References

Blok, J., et al. 1981. Endocytosis in absorptive cells of cultured human small-intestinal tissue: Horseradish peroxidase, lactoperoxidase, and ferritin as markers. *Cell Tissue Res* 216:1–13.

Dannon, D., et al. 1972. Use of cationized ferritin as a label of negative charges. *J Ultrastruc Res* 38:500–10.

Dym, M., and D.W. Fawcett. 1970. The blood-testis barrier in the rat and the physiological compartmentation of the seminiferous epithelium. *Biol Reprod* 3:308–26.

Gould, B., et al. 1981. A comparative study of fluid-phase and adsorptive endocytosis of horseradish peroxidase in lymphoid cells. *Exptl Cell Res* 132:375–86.

Hayat, M. A. 1975. *Positive staining for electron microscopy.* New York: Van Nostrand-Reinhold.

Hermo, L., et al. 1985. Intercellular pathways of endocytosed tracers in Leydig cells of the rat. *J Androl* 6:213–24.

Karnovsky, M. J. 1967. The ultrastructural basis of capillary permeability studied with peroxidase as a tracer *J Cell Biol* 35:213–36.

Shaklai, M., and M. Tavassoli. 1982. Lanthanum as an electron microscopic stain. *J Histochem Cytochem* 12:1325–30.

Weber, J. E. 1988. Effects of cytochalasin D on the integrity of the Sertoli cell (blood-testis) barrier. *Am J Anat* 182:130–47.

Interpretation of Micrographs

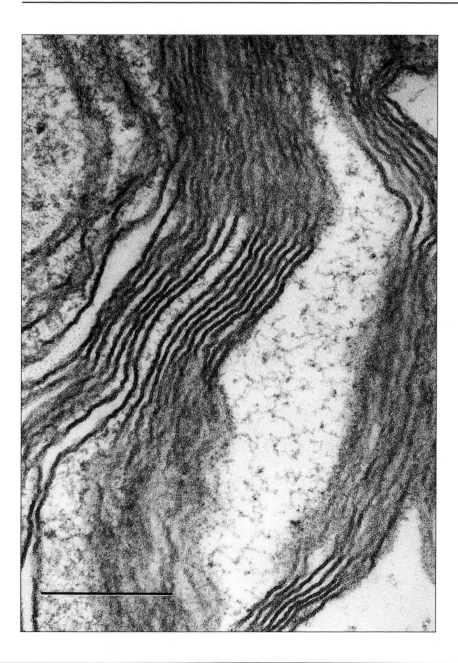

Micrographs are of little scientific value unless the investigator can interpret them accurately. Interpretation skills come from experience and a sound background in the fundamentals of cell structure and tissue architecture. Experience in interpreting micrographs is best gained in the laboratory setting with knowledgeable tutors. Interpretation skills take time to acquire.

The amount of information one is capable of obtaining from a micrograph also depends on the quality of the micrograph. Mistakes made in an electron microscope laboratory in tissue preparation and sectioning, and in the darkroom are inevitable and will be reflected in the quality of the final electron micrograph. It is important to be able to recognize which features of micrographs are caused by improper laboratory procedures. Undesirable micrographs are the best reminders that something needs to be changed to make the micrograph the best possible representation of tissue structure.

Introduction to Viewing Biological Electron Micrographs

A systematic approach to the interpretation of biological electron micrographs will make what, at first, appears to be a difficult experience, a pleasant and enjoyable one. It is not unusual for a novice to feel lost or bewildered at the first sight of an electron micrograph. There is a wealth of information in the average biological electron micrograph. With some experience, an individual will eagerly look forward to seeing something new and, it is hoped, aesthetically pleasing. It is important to obtain the greatest amount of information from the micrograph. For the novice, the process of micrograph interpretation should be tackled slowly in a step-by-step manner. A methodical approach follows.

1. **Determine the approximate magnification.** What one is capable of seeing and describing depends on the magnification of the micrograph. For example, it is highly unlikely that one could visualize the internal details of a cilium magnified to only 1,500 times. It is important to formulate an approximate idea of the magnification of a micrograph. Is the micrograph of *low, medium,* or *high magnification*? In this chapter, these three categories are assigned arbitrary magnification ranges (see Estimation of Micrograph Magnification); however, common sense is more important in estimating magnification. If numerous typical cells (e.g., 10 μm in diameter) can be seen in a single 8″ × 10″ micrograph of, for example, a vertebrate tissue, then the micrograph is printed at low magnification. If a quarter or more of one cell is visible, then the micrograph is printed at a medium range magnification. Finally, if the micrograph contains only a few organelles, or a portion of one organelle, the micrograph may be considered high magnification. At high magnification, the membrane is readily resolved into a bilayer (assuming proper fixation and thin sectioning). The magnifications between the three main categories may be given descriptions that imply they are between two of the ranges (e.g., medium high or medium low). Certainly, these broad magnification ranges are imprecisely defined and blend imperceptibly from one into another. Nevertheless, in a practical sense, one's mind must become accustomed to viewing micrographs with some knowledge of their magnification.

2. **Identify the basic features of the cells or tissues that are being examined.** The basic features are characteristic of most tissue types and of most cells, in general. Light microscope descriptions are available for most tissue types and should be used as reference to locate cell types and tissue compartments. Viewing thick sections, taken prior to thin sectioning, will facilitate identification of tissue elements and spaces. Since most cells are nucleated and the nucleus's appearance is characteristically different from the cytoplasm, low magnification micrographs should allow one to identify the nucleus, the nuclear membrane, the cytoplasm, and the plasma membrane. Identification of the nucleus is a practical initial step in electron micrograph interpretation. The viewer may next identify the cell border. Once this has been accomplished, one can systematically subdivide the cell into smaller and smaller compartments for analysis.

3. **Become familiar with the general features of cell structure.** If, for example, one knows the general architecture of the Golgi apparatus, the likelihood of identifying one in an unlabeled micrograph is greatly enhanced. Chapter 19 (Survey of Biological Ultrastructure) gets the novice on the road toward accomplishing this goal.

4. **Examine the technical aspects of the micrograph.** Important in this respect are fixation quality, sectioning and staining artifacts, and the photographic quality of the micrograph. This chapter describes many of the artifacts encountered in tissue preparation and staining. Chapter 8 describes the photographic procedure and resulting artifacts.

Rarely, if ever, is any micrograph considered perfect. The first micrographs made from recently taken negatives are usually printed for preliminary viewing. Such *work prints* often provide good examples of photographic technique inadequacies. Those that are published are selected for their high quality and are usually trimmed versions of much larger micrographs, which have been worked on for hours in the darkroom. With experience, a microscopist quickly scans a micrograph and perform the preceding four tasks automatically without any consideration of the extensive time and effort invested in learning how to do so.

It is important to be able to distinguish what is real from what is *artifact*. Some researchers openly state that the images obtained under the electron microscope are heavily artifactual. In one sense, this is true. When one considers the traumatic steps that a living cell has gone through in the process of fixation, dehydration, embedding, sectioning, and exposure to the electron beam, it is a wonder that the final image has the details of cell structure at all. Furthermore, electron micrographs are not pictures of the tissue itself, but are "shadowgrams" generated by the electron beam. It is also a wonder that shadows could portray tissue features adequately. Electron microscopists have devoted considerable effort to determining what constitutes the real properties of living tissue structure and what is artifact. For most tissue elements, there is a general consensus about what is real and what is artifact. A good feeling for what is representative of living tissue structure and what is artifact is gained with both time and experience.

This chapter serves as an introduction to the interpretation of micrographs. It provides guidance in interpreting normal elements in sectioned material that, for some reason, are not displayed as the eye is accustomed to seeing them. It will also illustrate and discuss some of the more common artifacts. It assumes a working knowledge of histology and basic cytology.

Interpretation of Normal Tissue Structure

Magnification and Resolution

Chapter 6 discusses the theoretical difference between magnification and resolution. While the theoretical resolving power of the transmission electron microscope is 0.2 to 0.3 nm, in a practical sense it is not possible to achieve this level of resolution on an electron micrograph displaying sectioned material. The section unavoidably contains overlapping material of the specimen and the embedment, which are in the path of the beam and compromise resolution. In addition, chromatic aberration occurs in sectioned material further limiting resolution. Although the microscope, as an instrument, may be capable of much better resolution, usually 1.5 to 2 nm is considered optimal resolving capability for most studies employing thin sections.

Resolution and magnification go hand in hand. At the lower magnification ranges, it is impractical to attempt to resolve 2 nm or thereabouts. Only in the higher magnification range does extremely fine resolution become a critical factor in micrograph quality. This is illustrated in Figure 18-1, where membranous structures (Golgi) are progressively magnified. Each lipid bilayer is about 8.5 nm, and the translucent space (hydrophobic region) in each bilayer is about 2.5 nm. At the lowest of magnifications, it is not possible to resolve even individual membranes of the Golgi stack due to the size and distribution of silver grains on the photographic paper and the resolving power of the unaided human eye (0.2 mm). The space in each bilayer is resolved only at the higher magnifications (see Membranes below).

To assist the human eye as it views micrographs, one must artificially increase its resolving power through enlargement of high quality negatives. A fine grain negative, as is used in electron microscopy, may be enlarged photographically as much as 10 times its original size to obtain the maximal information from it. To do this, the negative must be sharply in focus. For most electron microscope negatives, a 2 to 3 times enlargement will allow one to gain the great majority of information. Enlarging standard size negatives (3¼″ × 4″) about 2.6 times gives a final print magnification of approximately 8″ × 10″.

Membranes

Membranes are a commonly encountered feature in most electron micrographs. Staining of the lipid bilayer (by osmium and heavy metals) is observed primarily over the polar regions of phospholipid molecules causing them to appear dense, whereas the center of the membrane appears translucent (*trilaminar profile of a lipid bilayer*; Figure 18-2). The way

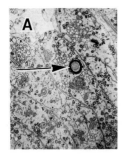

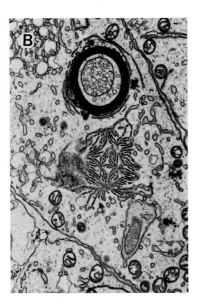

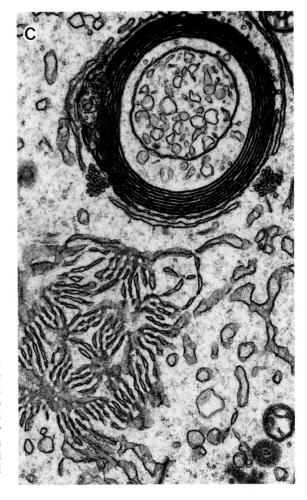

Figure 18-1   Series of five micrographs (A-E), taken from the same area of tissue showing the ability of the microscope to resolve structural detail as the magnification increases. The structure which appears in all micrographs and serves as an orientation guide to find structures in all of the micrographs is the Golgi apparatus (arrow). Only at the two highest magnifications is the trilaminar membrane of the Golgi resolved (arrowhead), although the section thickness is too great to obtain optimal membrane resolution. Had the section been thinner, the low magnification micrograph would have been less contrasty and, thus, barely visible.

membranes are sectioned leads to one of the most common causes of micrograph misinterpretation. Figure 18-2 shows a micrograph that depicts several unit membranes, each of which in some area shows the typical trilaminar profile of a lipid bilayer. The microtome may enter the membrane bilayer at an infinite number of angles. Our brain expects to see membranes sectioned perpendicularly, where they are resolved as distinct "railroad tracks." However, there are other possible ways that membranes may appear in sections. Their appearance depends on the angle at which the sectioning plane encounters the membrane and whether or not the membrane curves within the section (Figure 18-3). Keep in mind that the electron beam must pass through the entire thickness of a section and that any structural elements in the path of the beam will deflect electrons. When viewing micrographs, membranes may appear to be broken or missing in certain areas when, in reality, they are traveling parallel or nearly parallel to the plane of section. Look for a slight fuzziness

in the region of the missing membrane, which indicates the presence of a membrane (Figure 18-4). Usually, membranes that are really broken show sharp discontinuities (see Figure 18-13).

The less than ideal sectioning of membranes may be overcome to some extent by tilting the section on the specimen stage (see Chapter 2). In this way, the plane of the membrane bilayer may be placed parallel to the path of the electrons. This will overcome the fuzzy appearance of obliquely sectioned membranes. Figure 18-5 shows the effect tilting a specimen has on the appearance of membranes of a Golgi apparatus.

Shape, Kinds, and Number of Structures

It is important to remember that transmission micrographs are taken from three-dimensional material that has been finely sectioned in a single plane. It is almost impossible to ascertain from thin sections the (*Text continues on page 384*)

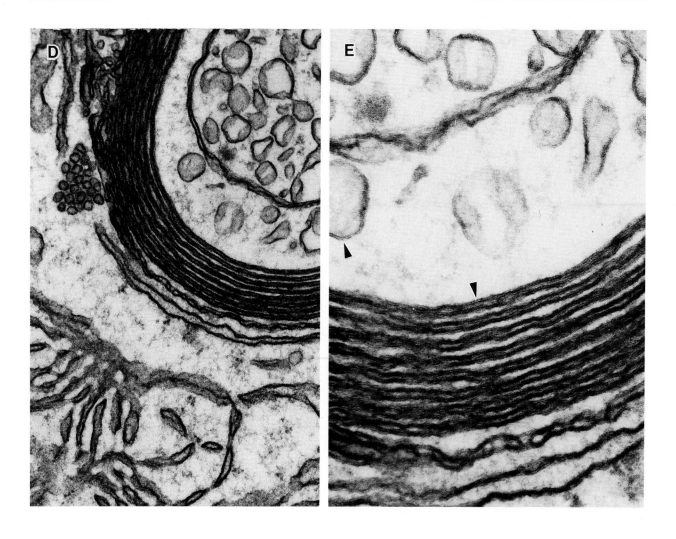

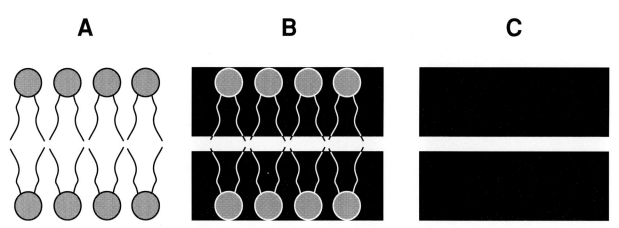

Figure 18-2   The phospholipid molecules of a membrane as shown in A are stained over their polar region and, to a large extent, over their non-polar tails as shown in B, giving rise to the trilaminar appearance of membranes in electron micrographs as shown in C.

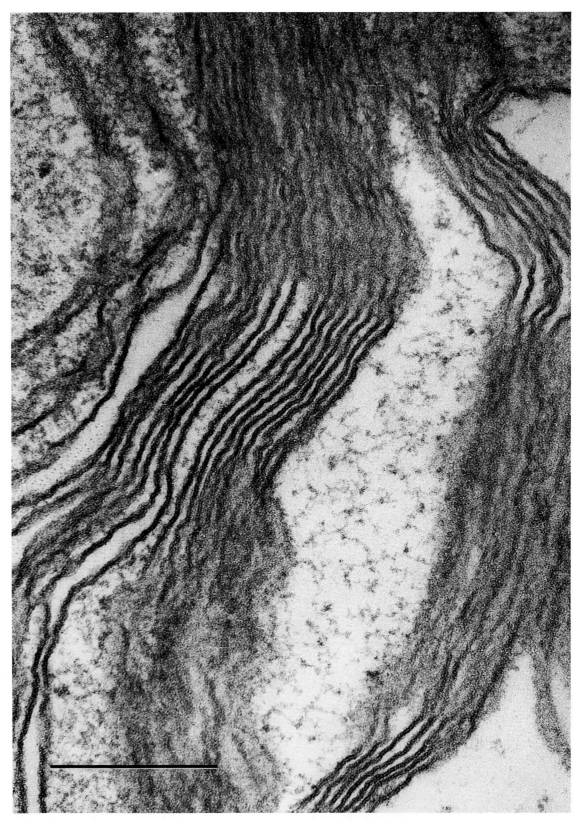

Figure 18-3 Unit membranes sectioned at various planes due to their change in orientation within thickness of the section. The membranes which are sectioned perpendicularly appear distinct and trilaminar, whereas, when those same membranes are sectioned obliquely and/or in the plane of the membrane (*en face*), they appear fuzzy. Membranes that change their orientation with respect to the plane of sectioning, may, at times, appear discontinuous. Bar = 0.25 μm.

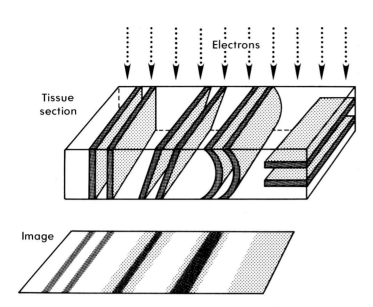

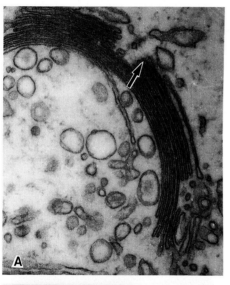

Figure 18-4 Drawing showing the various planes in which membranes may be sectioned and how their corresponding images would appear in electron micrographs.

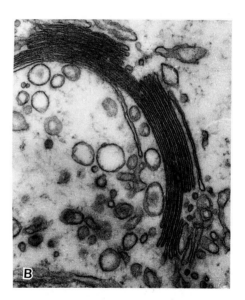

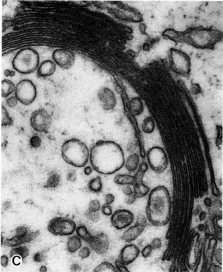

Figure 18-5 The effect of specimen tilting on the appearance of membranous elements of the Golgi apparatus. (A) The specimen that is in the normal position (0° angle) perpendicular to the electron beam shows regions where the membrane appears fuzzy (arrow). (B) After tilting the specimen 6° in one direction, the membranes in this region appear much sharper. (C) Tilting the specimen to a 12° angle makes them again appear fuzzy. However, in other regions they appear much sharper. The membrane bilayer is clearest when the membrane bilayer is in the plane of the electron beam. See Figure 18-4 above.

shape of a structure that is larger (and most structures are) than the thickness of the section (Figure 18-6). Because something appears rounded in a two-dimensional micrograph does not necessarily mean that the entire object is spherical in three dimensions. The shape and/or configuration of membranous organelles may be very deceptive. For example, rarely are sheets of smooth endoplasmic reticulum sectioned to reveal the configuration of the sheet as a whole. Fortuitous sections, or *en face* views, show the elaborate interconnecting pattern of endoplasmic reticulum, a pattern that could not easily be imagined from most thin sectioned areas (Figure 18-7).

Identification of the *kinds* of structures in sections is often made difficult by the plane of section. Figure 18-8 depicts what, at first sight, appears to be microvilli protruding into an internal cavity within a cell. Consider that the internal cavity may not actually be within the cell, but may instead be an indentation (invagination) into the cell. The site where the extracellular space joins the apparent internal cavity is not visualized in the plane of section; only the "internal cavity" is evident. The view that the cell is invaginated from the extracellular space is supported by the similarity of the extracellular material with the contents of the sectioned area. The nonuniform contour of the cell surface (microvilli) both outside the cell and also in the "internal cavity" indicates that the surface may have an invagination that is sectioned to make it appear intracellular when it is not.

Use of sectioned material may also lead to false interpretation of the *number* of structures encountered. If, for example, two profiles of a structure are seen within a section (similar to Figure 18-6F), can it be concluded that the cell has two structures that are separate entities? Not necessarily, because the structure depicted may be U-shaped. Only the limbs of the U may have been captured by the section. Long, thin objects may appear numerous whereas, in reality, they are few in number but configured tortuously to give this false impression.

When structures are frequently encountered in sections and, moreover, are sectioned in many planes, it is possible to gain confidence in one's interpretations about the kinds, numbers, and shapes of structures (see Figure 18-9). Otherwise it may be necessary to make these determinations with serial thin sections or thick 1 to 2 μm sections ex-

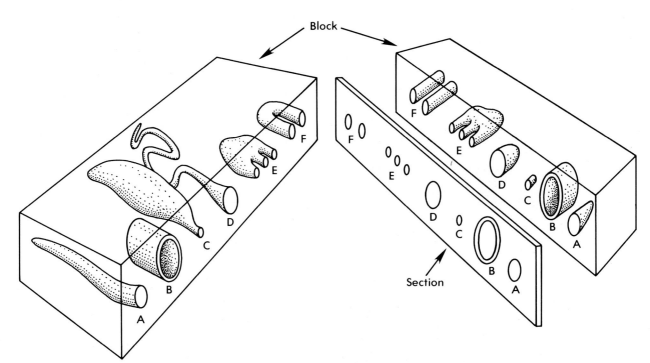

Figure 18-6 Sectioned profiles and their three-dimensional form in a tissue block: (A) A sphere seen in section is really an elongate structure; (B) a ring-shaped structure in section is really tubular in three dimension; (C) a structure which appears small in section is predominantly of large diameter; (D) a structure which appears large is predominantly of small diameter; (E) three structures in section are part of a larger single structure; (F) what appears as isolated structures are part of a single convoluted structure.

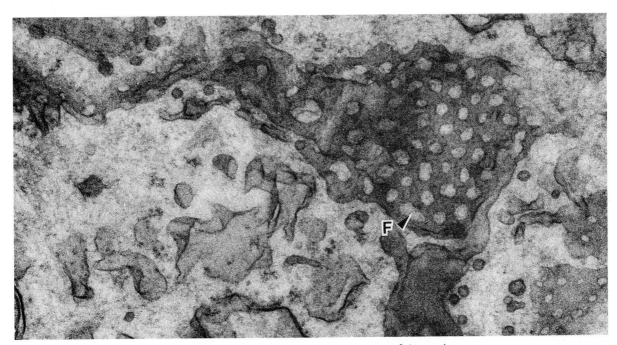

Figure 18-7   A cisternae of the Golgi apparatus captured within the plane of the section (*en face* view) shows the fenestrations (F) of the membranous saccule.

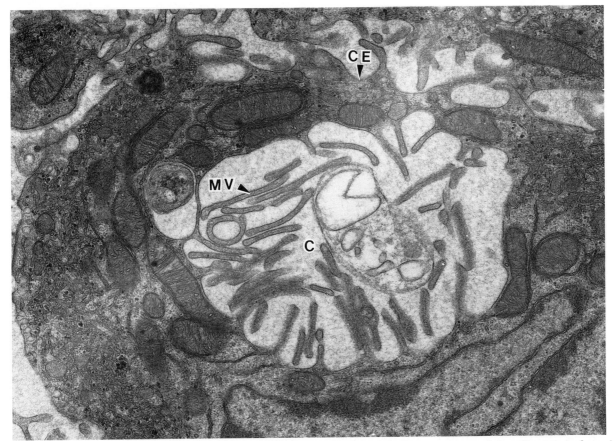

Figure 18-8   Structure appearing to be within a cell when it is actually outside the cell. This cell appears to have microvilli (MV) protruding from the cell interior into an internal cell cavity ("C") as well as from the cell exterior (CE). In reality, all of the microvilli of the cell protrude from the cell exterior; however, an inpocketing of the cell, gives the false impression that microvilli protrude from the cell interior.

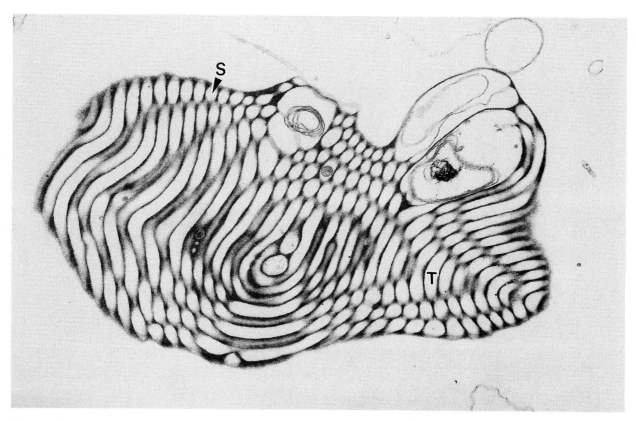

Figure 18-9 The shape of structures which appear frequently and that are sectioned from varying angles may be deduced by studying the various profiles seen. Although some profiles suggest that the intercellular structures shown are spheres (S), others suggest that this is an erroneous interpretation. It is likely that the plane of sectioning crossed structures that were tubular (T) in form and made them appear rounded. (Micrograph courtesy of R. Sprando.)

amined by high voltage electron microscopy, bulk examination of fractured specimens by scanning electron microscopy, or even more specialized techniques.

Fixation Artifacts

Fixation, in itself, is an artifactual process that distorts tissue structure from its living state. Cells are never completely free of fixation artifact. Even optimal fixation protocols for a particular tissue tend to emphasize certain features at the expense of others. For example, some fixation protocols emphasize membranes, but often the background matrix of the cell is virtually absent. Naturally, investigators will publish their best micrographs, thus emphasizing the areas of better fixation. The goal of the investigator is to minimize fixation artifact to the point that the tissue has as close a resemblance to its living state as possible. It is important for the investigator to recognize what constitutes acceptable fixation and what does not. Furthermore, to avoid the repetition of errors, different kinds of fixation artifacts should be recognized.

Fixation artifacts are of many different kinds. One type or a whole cadre of fixation artifacts may be evident in one micrograph. The artifacts described here, while not a complete cataloguing of artifacts that may be encountered, are some of the more common.

The method by which tissues are exposed to primary fixative may result in artifacts. *Immersion fixation*, for most tissues, is less desirable than *vascular perfusion fixation* (see Chapter 2). The animal's own vascular system is ideal to disseminate fixative to tissue as compared with the relatively slow penetration of fixative in tissue fragments that have been mechanically disturbed by excision. For example, by comparing Figure 18-10 A and B, which are micrographs of the same cell type from the same species fixed by immersion and perfusion, respectively, it is revealed that the smooth endoplasmic reticulum may lose its anastomosing tubular appearance and may become vesicular after immersion fixation. There

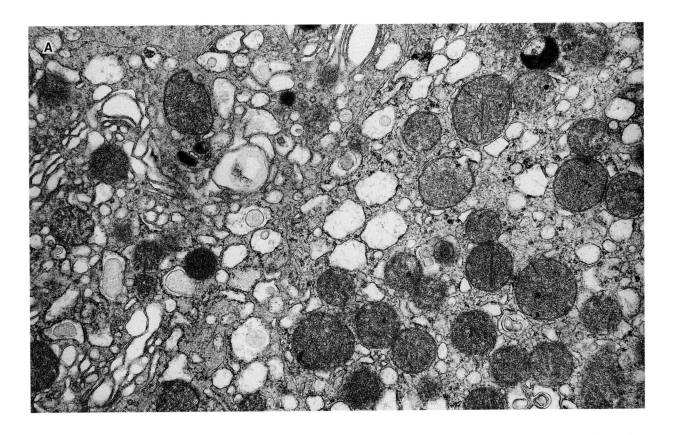

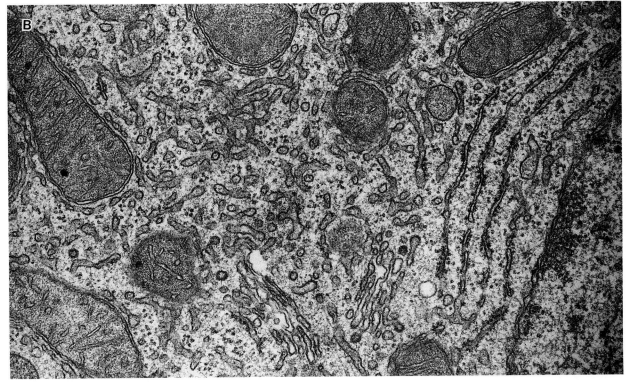

Figure 18-10 (A) Immersion-fixed Leydig cell and (B) perfusion-fixed Leydig cell showing the variable appearance of smooth endoplasmic reticulum. Note the apparent difference in density of the two cell types. Rat tissue.

are numerous other reasons to employ perfusion fixation whenever possible; however, the size of the animal being perfused often determines the accessibility to the vascular system. Thus, many animals cannot be perfused because they or their vessels are too small. Other large animals cannot be perfused because their vascular system would require an enormous volume of fixative and such an endeavor would not be practical in a laboratory setting.

Postfixation with osmium is important to obtain quality electron micrographs. The rate of penetration of osmium is very slow (less than 0.5 mm in 1 to 2 hours depending on the compactness of the tissue). Consequently, since osmium frequently does not reach the center of tissue blocks, the result is poor fixation of numerous elements, most notably membranes. Figure 18-11 shows an example of inadequate osmium penetration. Note that it is difficult to discern cell outlines due to poor membrane fix-

ation. Keeping tissue slices under 1 mm in any one axis will assure rapid penetration of osmium. If large blocks are used, there may be a temptation to over-fix tissues. Over-fixation will also result in poor ultrastructural visualization of membrane components or a leaching out of structural components of the cell. Ironically, the appearance of over-fixed micrographs is similar to those that have been under-fixed.

A common sign of poor fixation is swelling of mitochondria. Under conditions of poor fixation, the mitochondrial cristae often become peripherally positioned. The central region of the mitochondrion has a rarified matrix and/or swollen appearance (Figure 18-12), and for this reason is frequently referred to as "blown."

Membranes throughout the cell may also appear broken or discontinuous in certain areas (Figure 18-13). The membranes of the nuclear envelope may separate from each other causing a widening of the

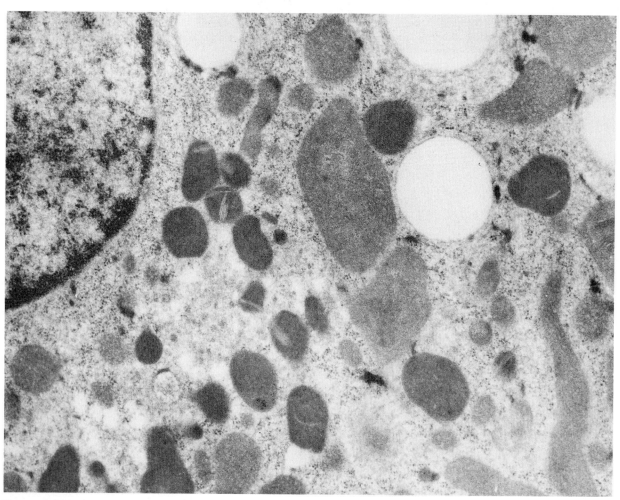

Figure 18-11   Appearance of cells that have not been adequately fixed with osmium. Note that membranes are not visualized.

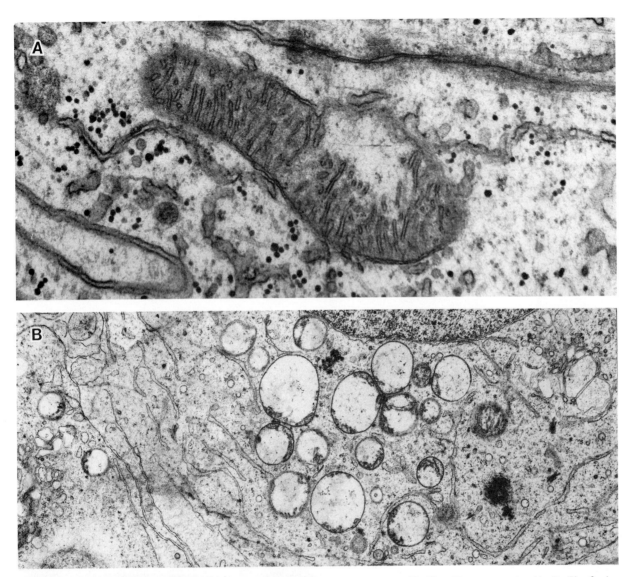

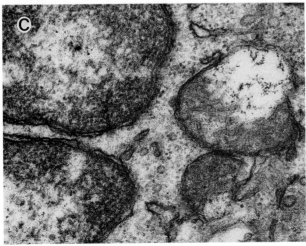

Figure 18-12(A-C)   Three examples (A, B, C) of mitochondrial swelling due to poor fixation. Portions of each of these mitochondria appear 'empty.'

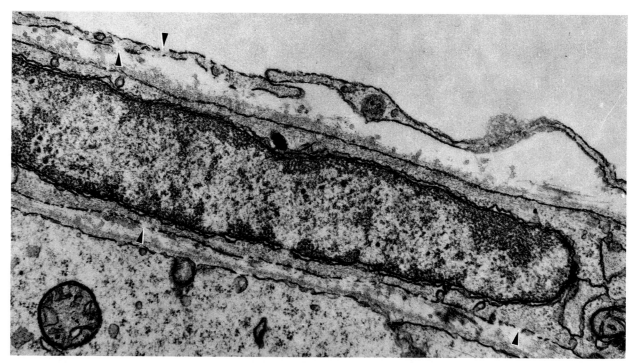

Figure 18-13   Membrane discontinuities (arrowheads) in this micrograph are due to poor fixation.

perinuclear cisternae (Figure 18-14). Some investigators add calcium chloride (1 to 2 mM) to the fixative to improve membrane preservation and to prevent mitochondrial swelling and swelling in the perinuclear cisternae of the nuclear envelope. However, in some cases, this may adversely affect the preservation of microtubules, especially in plant cells.

Fixation artifacts are frequently associated with the *osmotic strength* of the fixative. Exaggeration of the intercellular space is often the result of hyperosmolarity of the fixative. This can usually be remedied by decreasing the osmotic strength of the buffering system used with the fixative. Hypoosmotic solutions frequently cause swelling of cells and/or organelles.

Temperature, pH, duration of fixation, nature of the fixative, and *the nature of the buffer* are also important factors to consider when trying to prevent artifacts (see Glauert, 1978, for examples of the effects of buffers and fixatives). Since there is no standard way to fix all tissues, it is important to begin by following past scientific literature that is applicable to the specific tissue under consideration. Sometimes there is no general agreement on the fixative conditions, and one must modify the published methodology to obtain the desired results.

Dehydration, Infiltration, and Embedding Artifacts

Dehydration, infiltration, and embedding displace the water matrix of the specimen with an embedding material that is eventually polymerized into a hard plastic. Therefore, most artifacts are due to inadequate removal of either the water or the fluids used in the dehydration process. This usually results in *holes* in the tissue (Figure 18-15). A rare hole may be acceptable, but large numbers of holes like those illustrated should be cause for re-examination of the techniques used in embedding.

Holes may also be produced in tissues in zones of sharp transition from soft embedding material to a much harder tissue structure. This problem is usually solved by using a harder embedding resin with a hardness similar to the structure in question. Another solution is to use a low viscosity resin such as LR White, which penetrates hard tissues more readily than many other embedding media. Using a sharp diamond knife will usually reduce the severity of this problem.

Sectioning Artifacts

This category includes numerous artifacts associated with sectioning that are frequently bothersome to the investigator. The knife, whether it be glass or

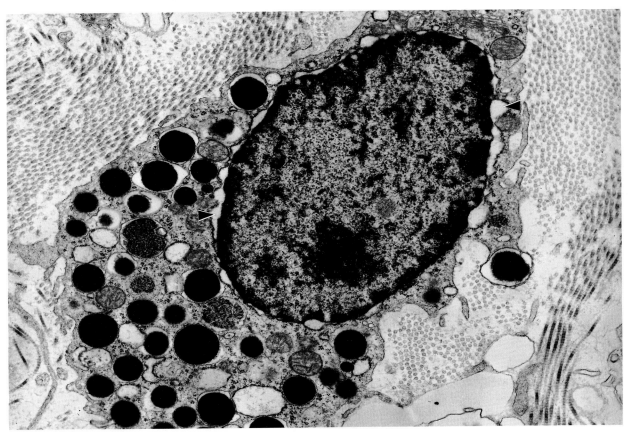

Figure 18-14   The nuclear envelope of this mast cell shows fixation artifact as evidenced by swelling of the perinuclear space (arrowheads).

diamond, may have lost its sharpness or be defective or dirty in certain areas. This leads to *knife marks or scrape marks* on the section, which appear as lines or tears perpendicular to the knife edge (Figure 18-16). The beam may enlarge small holes formed by knife marks (Figure 18-16).

An artifact called *chatter* is produced when vibrations in the block or knife edge occur as the result of improper sectioning speed or knife angle (Figure 18-17). Chatter may result from vibrations in the building in which the microtome is located. Chatter appears as parallel lines, or alternating thick and thin areas, parallel to the cutting edge of the knife. An easy way to distinguish chatter from knife marks is that chatter marks are quite regular and always spaced evenly from each other, whereas knife marks may or may not be repetitive, but are rarely evenly spaced. In situations where both knife marks and chatter are seen, knife marks always occur perpendicular to chatter (Figure 18-18). If chatter is caused by sectioning, changing the knife angle and/or sectioning speed will usually remedy chatter marks, but a different knife or clean cutting edge may be needed to remedy knife marks or scrapes.

Compression is produced when very soft or compressible materials are sectioned. The pressure exerted by the knife may compress the section such that its vertical dimension is less than that of the block face from which it was taken. Thus, tissue structure is slightly distorted in one dimension by as much as 30% (Figure 18-19). Some stretching of the section by warming or by exposure to vapors of organic solvents such as xylene or chloroform may restore the compressed surface to approximately its normal size and shape.

While not a sectioning artifact, per se, *section thickness* is important in determining the overall appearance of the tissue. Section thickness must be tailored to the specific needs of the investigator. As a general rule, studies that require low print magnifications (<6,000) ideally employ relatively thick sections, such as those displaying gold interference colors (90 to 120 nm thick). By using thicker sections at low magnification, it is possible to enhance contrast, and thus image quality, through the increased superimposition of dense tissue structures against nontissue elements. Studies at intermediate magnifications (print magnifications of 15,000 to

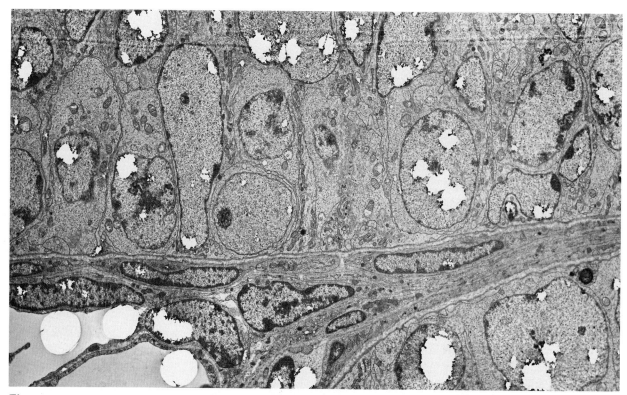

Figure 18-15 Holes in tissue due to failure to remove water or solvents prior to embedding.

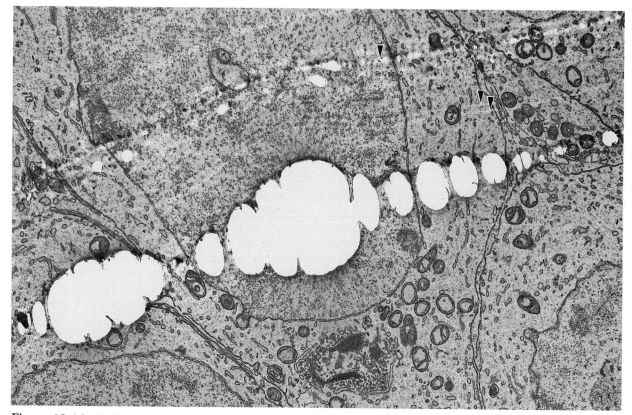

Figure 18-16 Knife marks (arrowheads) have scratched and caused irregularities in section thickness. The holes pro-duced by other knife marks are enlarged by the electron beam.

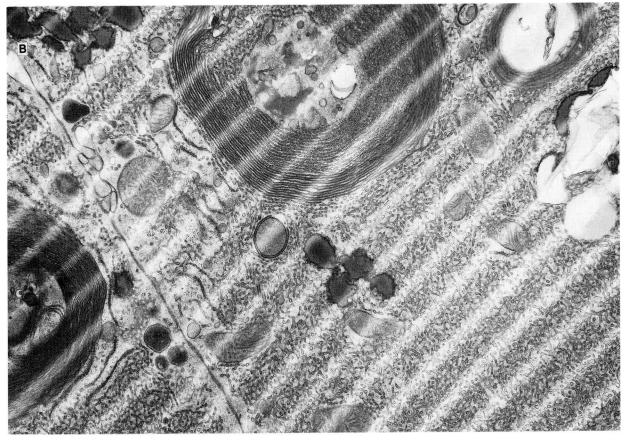

Figure 18-17(A-B) Two types of chatter, each showing alternating thick and thin areas. In A the chatter is severe. (Micrograph courtesy of students wishing to remain anonymous.)

Figure 18-18   Chatter with knife marks seen in a section of epoxy. Note that chatter and knife marks are perpendicular to each other.

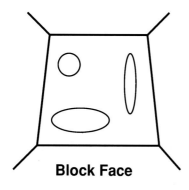

Block Face

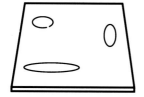

Section With Compression

Figure 18-19   Diagram of the shortening of one dimension of a section as compared with the block face due to compression. Note how tissue structures are compressed in the axis perpendicular to the knife face.

50,000) may require sections displaying silver interference colors (70 to 90 nm). For high resolution, high magnification (>80,000 print magnification) studies, it is desirable to have grey or silver-grey sections (40 to 70 nm). At the higher magnifications, resolution of biological materials is enhanced by decreased superimposition of structures. A practical rule for evaluating section thickness at moderately high and high magnifications is to look for the "railroad track" appearance of unit membranes. The better the railroad track is resolved, the thinner the sections (compare Figures 18-20 A and B). Very thin sections tend to break or distort easily under the electron beam. An example of "thick" and "thin" sections adjacent to each other on the same grid is provided in Figure 18-21.

If the boat used to pick up sections is not clean or the block is dirty prior to sectioning, the sections will undoubtedly show *boat contamination*. The appearances of this contamination are as varied as the nature of the contaminants themselves. The example provided in Figure 18-22 shows boat contamination produced by epoxy fragments remaining on the side of the block after it was ground with a rotating drill. Washing the block under a fast stream of water would have prevented this annoying artifact. Figure 18-23 shows *tissue folds*, produced either during sectioning or as the tissue was picked up on the grid. Tissue folds are easy to identify since the result is a shifting of the image caused by part of the tissue still being within the fold. Another type of fold may mimic a tissue fold, however, in this instance, the image is not shifted by the fold. This is a *fold in the support film* that occurred prior to sectioning when the film was placed on the grid (Figure 18-24).

If grids are not properly cleaned prior to retrieval of sections, the grids themselves may be a source of contamination often seen as an "oil slick" on the water surface. Cleaning the eyelash probe with the fingers or using oily forceps or troughs are common sources of oil problems.

For additional information on sectioning artifacts, see Chapter 4 on ultramicrotomy.

Staining Artifacts

Without a doubt, the most common and bothersome artifacts are those related to the staining of tissue. When other staining problems have been corrected, staining artifacts or *stain contamination* seem to occur again and again. Unfortunately, stain contamination seems to gravitate to tissue sections at the

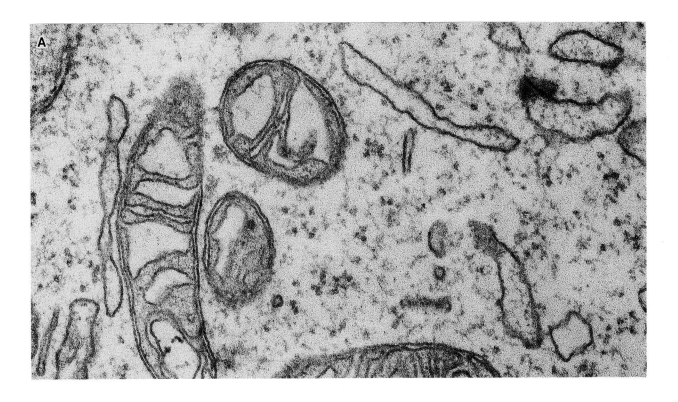

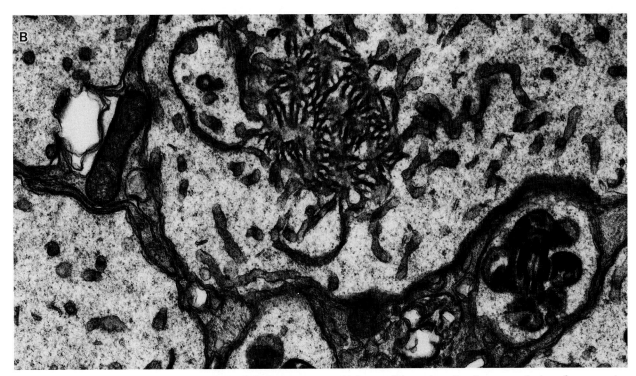

Figure 18-20 Sections at the extremes of thicknesses used in routine electron microscopy; (A) section showing a grey interference color indicating it is about 50–70 nm in thickness; (B) section showing a purple interference color indicating it is about 200 nm in thickness.

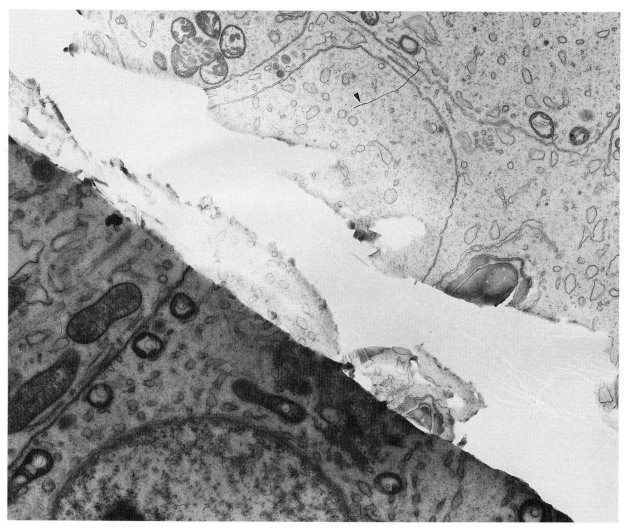

Figure 18-21   Two sections of greatly differing thickness are lying side-by-side on a Formvar-coated grid. One shows a silver-grey interference color and another is between gold and purple on the interference color chart. Note the difference in clarity of details. Dust particles (arrowhead) on the film during exposure appear in the photograph.

site considered most desirable for photography. Meticulous detail must be paid to the staining procedure; even then there is no guarantee of success.

Stain artifacts may take a variety of forms on electron micrographs. *Lead precipitate* may appear as fine grains ("pepper"), crystals, (Figure 18-25) or as dense spherical particles (Figure 18-26). *Uranyl acetate* contamination appears as dense amorphous aggregates or blotches of variable size (Figure 25 A-B; Figure 18-26A). It is always prudent to use double-distilled water in making stains, as properties of the water may cause heavy metal precipitation on tissue sections. The water used to make lead stains should be boiled before the stain is made to remove the CO_2 that will precipitate the lead as lead carbonate. The importance of washing tissues numer-

ous times after each stain application cannot be over emphasized.

When grids are used with support films, there frequently are defects in the support film that are caused by imperfectly cleaned slides. These imperfections cause holes in the support film that trap stain by the capillary action of the closely spaced film and underlying section (Figure 18-27).

Microscope Artifacts

Any artifact due to improper use of the microscope or misalignment of the microscope will result in a less than optimal image on a photographic plate. There are many causes for microscope artifacts and

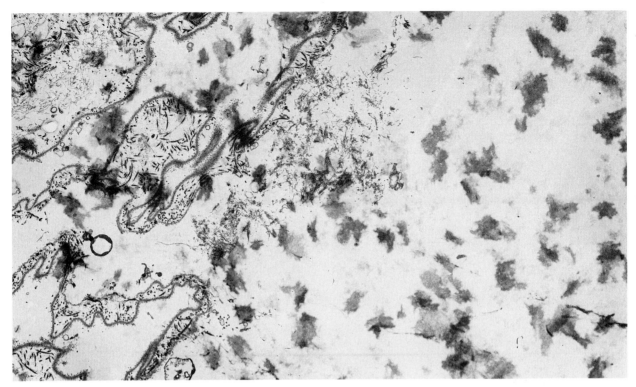

Figure 18-22   Boat contamination is evidenced by the numerous densities that overlay tissue components.

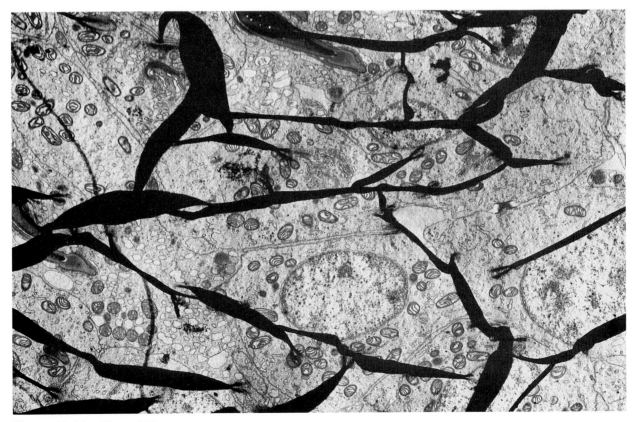

Figure 18-23   Tissue folds.

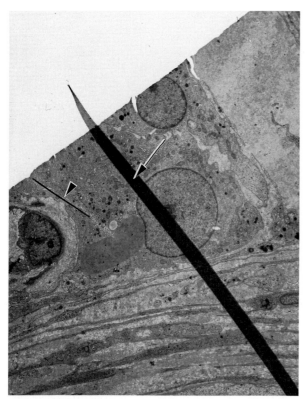

Figure 18-24 A fold in the support film (arrow) does not distort the image but only obscures it. A small tissue fold (arrowhead) is also seen.

most are discussed in Chapter 6; however, some of the more important causes follow.

Astigmatism produced by improper alignment of the microscope, and especially its apertures, is more noticeable as the magnification of the microscope is increased into the high magnification range (see Chapter 6 for causes of astigmatism). This is evidenced on the micrograph as a "streaking" of the information or streaking of the graininess of the embedding media in one direction. Numerous parallel streaks of the image or embedding media are evident, i.e., what was a point has now become a small line (Figure 18-28).

Focusing the *electron beam* on the section with the condensor lens adjustment knob (sometimes called intensity) is sufficient to alter the properties of that part of the section. If, for example, the electron beam is focused on the section at high magnification and the magnification is subsequently reduced to a very low scan magnification, then the former position of the beam will be evident (Figure 18-29). There is a certain amount of etching or removal of the plastic by the beam that changes its appearance in the microscope. A very low magni-

fication micrograph negative taken of such an area would be of uneven density and consequently difficult to print. It is usually desirable to take low magnification micrographs prior to going to high magnification. It this way the effect of the beam in any one area is minimized.

A high intensity beam often causes *section drift*, as well as *expansion* and *distortion* of tissue sections. It is best to allow a defocused beam to stabilize a section at low magnification prior to focusing on one particular area. Any movement of a section during photography will result in distorted negative images. If it is not possible to stabilize the tissue section, reducing the exposure to one-half second or less helps to minimize the artifact. Section drift with the resultant loss of clarity of the tissue detail is one of the most common causes of blurred micrographs at medium and high magnification.

Contamination within the microscope may interact with a focused beam to produce a rounded, dark spot on the specimen termed *beam contamination*. Usually hydrocarbons interacting with and being broken down by the electron beam are the source of the problem (Figure 18-30). Avoid concentrating the beam on any one area for a prolonged period. Modern microscopes are equipped with anticontamination devices or "cold fingers" (see Chapter 6), which aid in trapping contamination in the microscope column.

Photographic Artifacts

Ideally, the darkroom is more often used to eliminate, rather than to produce, tissue artifacts. Under normal circumstances, the contrast of tissue sections is usually less than desirable. In the process of making a negative and then a print from the same negative, one can usually remedy the low contrast of the tissue sections. Doing this requires using a grade of photographic paper designed to yield high contrast prints. Concentrated developing solutions, long low-light exposures, and a point-source enlarger can all be used to impart additional contrast to the final photographic print. Dodging of exposures can be used to counteract the uneven density of the negative. Dust on a negative (Figure 18-21) is readily remedied using a hand held gas dispenser or a fine brush. Artifacts due to improper photographic techniques are covered in Chapter 8.

Artifacts associated with photography usually involve procedural errors rather than actual artifacts inherent in the negative. They are overcome with

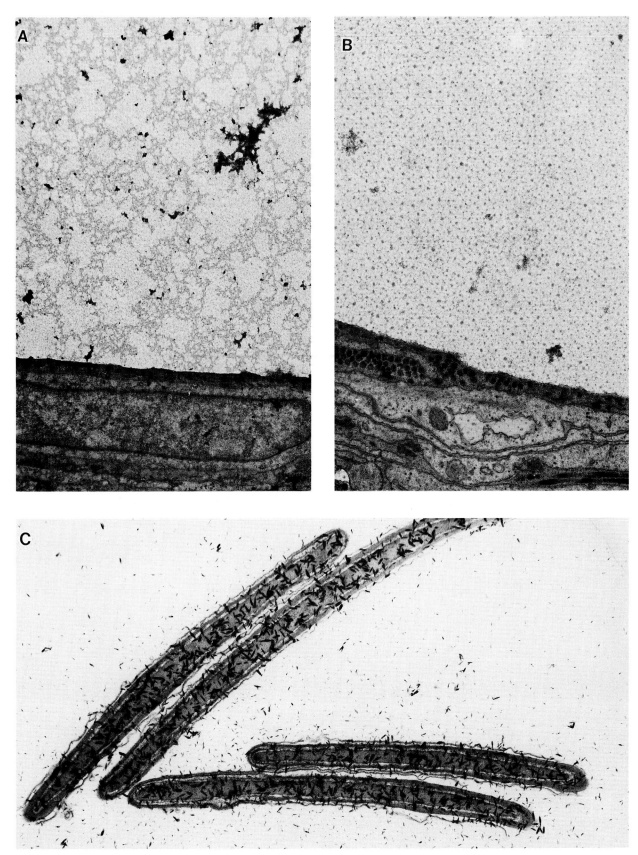

Figure 18-25(A-C) Various forms of lead contamination. Figure 25 (A and B) also depict uranyl acetate contamination (large irregular densities).

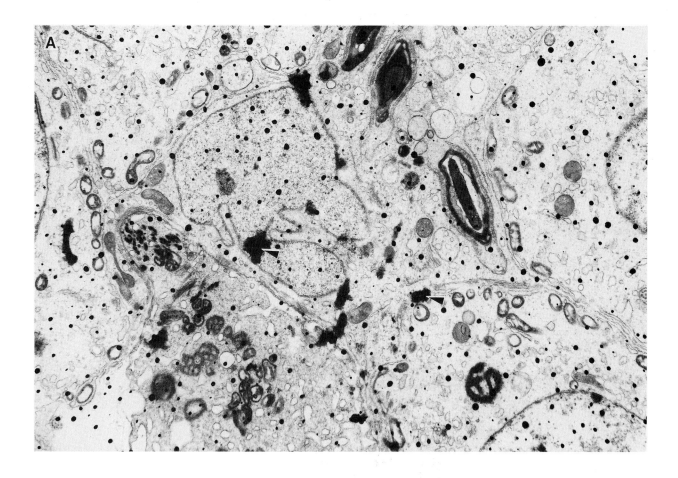

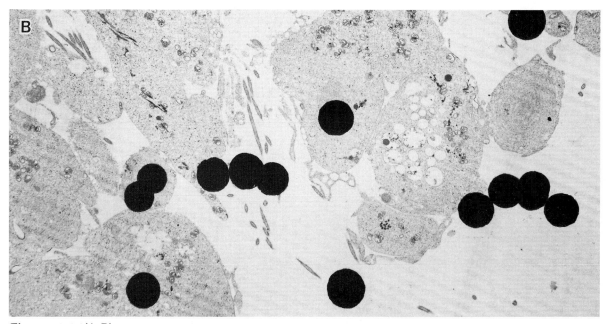

Figure 18-26(A-B)   Various forms of lead (spherical particles) and uranyl acetate (irregular densities) contamination.

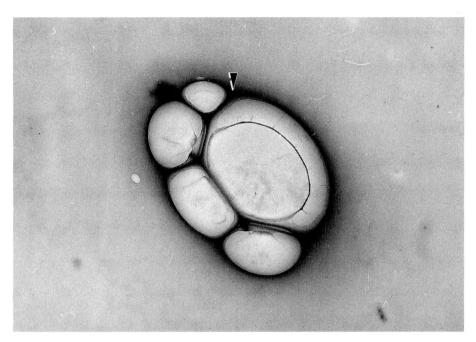

Figure 18-27   Stain contamination (arrowhead) is seen between holes in the support film and an epoxy section.

experience. Over- or underdevelopment, improper exposure, improper fixation during the photographic process, unsuitable contrast, scratching of the negative, etc., are common errors frequently made in developing and printing negatives and micrographs (see Chapter 8).

Interpreting Dynamic Processes from Static Images

When one wishes to follow events within biological structures, cautious interpretation of electron micrographs is recommended. Living tissue is in a dynamic state. Cell components are in constant movement and flux; growth and degradation processes are taking place. Ingestion and elimination of substances, cell division, and a variety of other processes characterize cellular activity. On the other hand, tissue prepared for electron microscopy is static because living processes have been purposefully halted in order to examine the tissue.

To conclude that some event is taking place when micrographs only *suggest* that it is happening is to extend the data beyond what is logically interpretable from the micrograph. Usually, several logical explanations are possible for how a given static image may participate in a dynamic process, which further limits one's ability to speculate. There are numerous examples in the literature of interpreta-

tions that have overextended the available data! The trend in analyzing dynamic processes is to use techniques that allow less room for error in interpretation, such as autoradiography (see Chapter 11) or specific labels. Tissues may be prepared at intervals to trace events from a particular starting point. Information can be quantitated at specific time points (see Chapter 13). In addition, other types of data (e.g., biochemical) may be used to support the interpretation. *It is highly questionable whether a conclusion about dynamic events in biological materials can be considered as fact if it is based strictly on descriptive information from micrographs.*

Estimation of Micrograph Magnification

The available range of electron microscope magnifications is great, extending from several hundred to over one million times. It is difficult, at first, to appreciate differences in this broad (thousand fold) range of magnifications. Modern electron microscopes print the negative magnification directly on the negative and/or display it on the scope panel. These features may not always be available to the individual. With experience, the investigator can provide an educated estimate of magnification. The terms *low, medium,* and *high magnification* are rel-

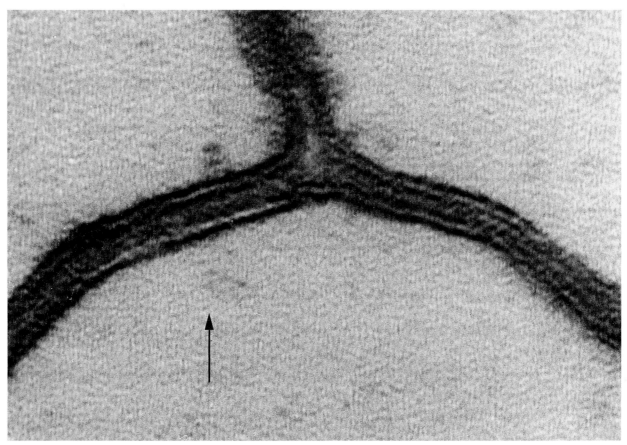

Figure 18-28   Astigmatism. Note there appear to be fine streaks on the micrograph in parallel with the direction indicated by the arrow.

ative terms and should be used in a context that is defined. Nevertheless, most experienced microscopists commonly describe micrographs as if there is a well understood definition. For our purposes, we have arbitrarily defined low magnification as a final image under ×6,000; medium magnification as being between ×15,000 and ×50,000; and high magnification as being above ×80,000. Gaps are intentionally left between the magnification ranges to indicate areas where they merge (e.g., medium low or medium high magnification).

If, for example, one is looking at an 8″ × 10″ micrograph of mammalian liver tissue, low magnification should reveal several cells within the micrograph. At medium magnifications, a single cell at most should be seen, or more likely a portion of a cell. At the lower range of high magnifications, one may see a few organelles. At the higher range, it is possible to visualize macromolecules.

Microscope magnification, as obtained from the settings on the microscope, is, of course, the magnification of the negative and not the print. To obtain the latter, measure the distance between two objects that are widely spaced on the negative, and measure the same distance between the two objects on the micrograph. The final magnification of the enlarged image is determined as follows:

$$\text{final magnification} = \frac{\text{measurement from print} \times \text{negative magnification}}{\text{measurement from negative}}$$

There are a number of possible errors (machine and human) that may lead to slight deviations of the calculated magnification obtained from the microscope digital readout as compared with the actual magnification of a printed micrograph. It is generally assumed that these errors exist, so the final expression of magnification is usually rounded off to the nearest 100 times. For example, ×35,700 is a more appropriate expression of magnification than is ×37,687. Sometimes when magnification is not critical to the interpretation, the final figure is rounded off to the nearest 500 or 1,000 times.

The ultimate source of the magnification of a negative is obtained from a gauge or digital display

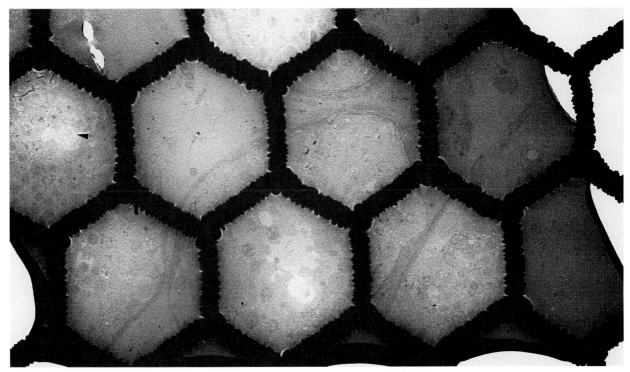

Figure 18-29 Artifacts on a tissue section (mounted on a hexagonal grid) which are viewed on a scan setting of the microscope by turning off the objective lens and removing the objective aperture. Areas in which the semi-condensed beam has focused on the section (prior to removal of the objective aperture) and changed the properties of the section appear lighter than areas not exposed to the electron beam.

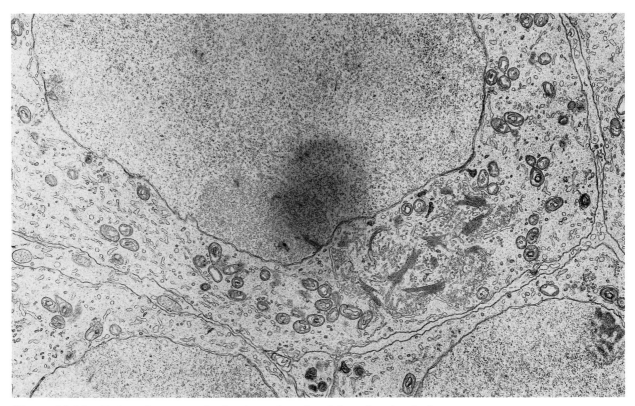

Figure 18-30 Hydrocarbon contamination (center) produced from focusing the beam too long on a single spot on the specimen.

on the microscope. If this information is not printed on the micrograph, a relatively accurate way to estimate micrograph magnifications is to use the knowledge of the size of cell structures to calculate magnification. Resolved unit membranes in their unmagnified state are 7 to 9 nm across, or 8 nm on the average. In micrographs, their size is directly proportional to the magnification. (Of course, the lipid bilayer is only resolved well at moderately high or high magnifications.) If, for example, a membrane (8 nm across) is measured at 1.2 mm using a fine ruler, it has been magnified about ×150,000 according to the following formula:

$$\text{magnification} = \frac{\text{measured size}}{\text{known organelle size}} \text{ or}$$

$$\frac{1.2 \times 10^{-3} \text{ m}}{8 \times 10^{-9} \text{ m}} = 1.5 \times 10^5 \text{ or } 150,000$$

Membranes are not the only structures of relatively constant size that may be measured in micrographs. Keep in mind that calculations of this type provide only a "ball park" estimation of micrograph magnification. Other organelles that may be measured to obtain a rough idea of magnification include:

cilia	0.2 μm across
centriole	0.15 μm across
glycogen particles	30 nm
intermediate filaments	10–12 nm
microfilaments (actin)	6–7 nm across
microtubules	22–25 nm across
ribosomes (mammalian)	20 nm

References

Crang, R. F. E., and K. L. Klomparens. 1988. *Artifacts in biological electron microscopy.* New York: Plenum Press.

Glauert, A. M. 1978. *Fixation, dehydration and embedding of biological specimens.* Amsterdam, The Netherlands: North Holland Pub. Co.

The Cell Surface
The Lipid Bilayer of the
 Plasmalemma
The Glycocalyx
Cell Junctions
Cell Surface Specializations

The Cytoskeleton
Microtubules
Microfilaments
Intermediate Filaments

The Nucleus
The Nuclear Envelope
Chromatin
The Nucleolus
Dividing Cells
The Synaptonemal Complex

Mitochondria

Protein Synthetic and Secretory Structures
Free Ribosomes
Membrane Bound Ribosomes
Rough Endoplasmic Reticulum
Smooth Endoplasmic Reticulum
The Golgi Apparatus
Secretory Products

Centrioles

Cilia and Flagella

The Lysosomal System
Lysosomes
Multivesicular Bodies

Microbodies

Annulate Lamellae

Cell Inclusions
Glycogen
Lipid
Crystalloids

Extracellular Material
Collagen
Basal Lamina
Matrix of Bone and Cartilage

Special Features of Plant Tissues
Chloroplasts
Vacuoles
The Cell Wall

Bacteria

Algae, Fungi, Yeast, and Protozoa

Viruses

References

Survey of Biological Ultrastructure

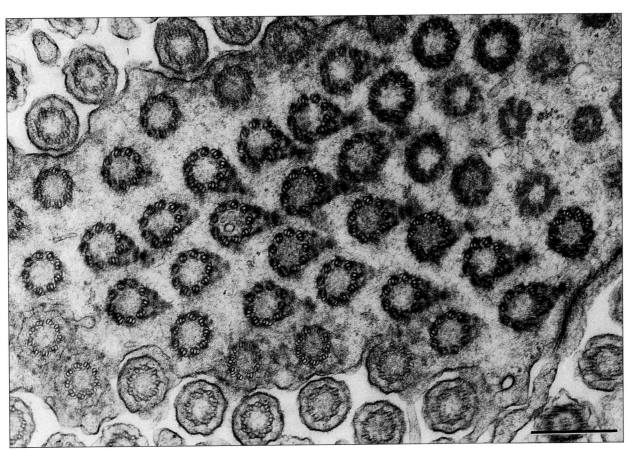

Courtesy of W. Dougherty.

It is important to understand the theoretical aspects of tissue preparation and image formation in the electron microscope. It is equally important, however, to be capable of *applying* this knowledge to obtain a better understanding of biological systems. This chapter introduces some of the more common features of biological ultrastructure. Other more extensive treatises and/or atlases of cell ultrastructure have been published and serve as excellent sources of illustrative and descriptive information. In addition, major textbooks of histology and cell biology are also good reference sources (see References at end of this chapter).

What follows is a brief but systematic introductory description of, and most importantly, a *guide* to the identification of basic biological ultrastructure in animal and plant cells. Some functional information is also included for each organelle or structure. It is important to associate the structure of each cell component with its function, since function is invariably a manifestation of structure. Finally, key references are included within each section to help the reader obtain more in-depth information about a particular structure or organelle.

The Cell Surface

In animal tissues, the cell surface is the plasmalemma or plasma membrane. In plant tissues, the surface is the cell wall, a structure covered later under Special Features of Plant Tissues. The surface of the cell, or plasmalemma, interacts with the extracellular environment. It regulates *transport* for both large and small molecules into and out of the cell. The surface participates in ingestion or elimination of large bodies of material as well as small molecules. The cell surface forms *junctions* with other cellular and noncellular elements that participate in attachment, cell-to-cell communication, and regulation of permeability between cells. The cell surface possesses *receptors* and recognition sites for binding various macromolecules. The binding event may then signal the cell interior to perform specific tasks or inform the cell about the identity of the other cell. The cell surface has configurational specializations that function to increase the overall surface area of the cell or to provide for motility of the entire cell or impart motility to one of the cell processes that protrude from the cell surface.

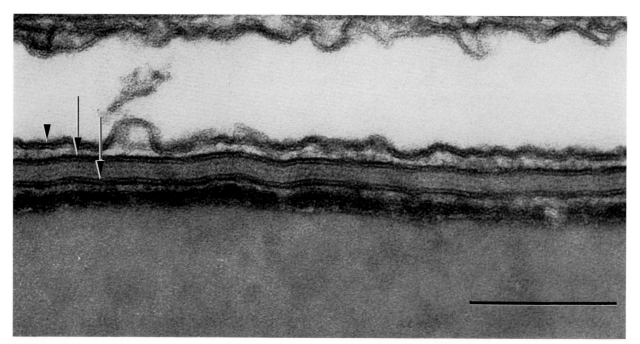

Figure 19-1   The surface membrane (arrowhead) or plasma membrane and its internal membranes (arrows) are shown at high magnification. The plasma membrane and the other membranes show a distinct tri-laminar or "railroad track" appearance. (Boar sperm) Bar = 0.05 μm.

The Lipid Bilayer of the Plasmalemma

At first glance, the cell surface may appear to be extremely simple, but a more detailed structural analysis reveals its extreme complexity. At low and medium magnifications, the cell membrane appears as an electron-dense line. At high magnification and with the appropriate fixation and staining protocols and sectioning perpendicular to the membrane, it, like all lipid membranes (see Chapter 18), appears as a *bilayer*. Membrane structure is emphasized by osmium and electron-dense stains, which are preferentially deposited over the polar heads of the phospholipid molecules (Figure 18-2). This 7 to 11 nm thick membrane is termed a *trilaminar membrane* because of the three visible lines one sees. There are two electron-dense lines separated by one electron-translucent line (Figure 19-1). Because the lines are parallel, it is often called a "railroad track." Chapter 18 describes how membranes appear when they are sectioned in planes other than perpendicular to the plane of the membrane.

The freeze fracture technique splits the lipid bilayer such that the two membrane halves, or faces, are visualized from their internal aspect (see Chapter 14). Molecules that span the width of the bilayer, or transmembrane proteins, appear as intramembranous particles (IMPs) in the freeze fracture micrograph (Figure 19-2). The term *intrinsic* proteins is used to designate proteins that span the lipid bilayer. It is in contrast to another term, *extrinsic*, which designates proteins that are bound only to the external surface.

The membrane bilayer provides a lipid interface between the aqueous environment of the cell exterior and that of the cell interior. It forms a framework to contain or bind molecules such as proteins. According to the *fluid mosaic model* of membrane structure (Singer and Nicolson, 1972), the membrane proteins are capable of movement in the semiliquid lipid bilayer. Plasma membrane molecules serve a variety of transport, receptor, and recognition functions.

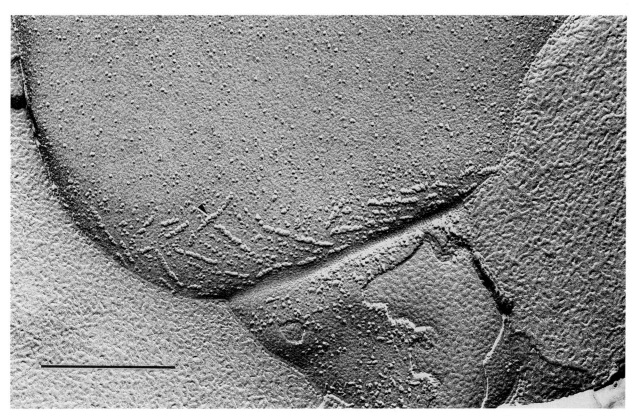

Figure 19-2 Intramembranous particles (IMPs) of various sizes are seen in a freeze fractured plasmalemma. The pattern and concentration of IMPs appear randomly distributed in some regions and in other regions discrete rows of particles (arrowhead) are seen. (Boar sperm.) Bar = 0.5 μm.

The Glycocalyx

Virtually every cell plasma membrane contains an exterior that appears fuzzy under the electron microscope. In some cells it is barely visualized and special stains are required to demonstrate its presence, but in other cells it forms a prominent fuzzy coat that extends many times the thickness of the membrane. Variously referred to as the *glycocalyx* or *cell coat*, it is composed of polysaccharides, especially those containing negatively charged sialic acid residues that are bound to the cell surface and are part of intrinsic proteins of the bilayer. Cell surface specializations that extend into a lumen frequently possess a well-developed glycocalyx (Figure 19-3).

The carbohydrates of the glycocalyx are important in cell-to-cell recognition and adhesion. Since the glycocalyx is the outermost aspect of the cell, it is assumed to function as a barrier to some substances. The net negative charge of the glycocalyx regulates, to some degree, the kinds of charged molecules that can approach the cell surface.

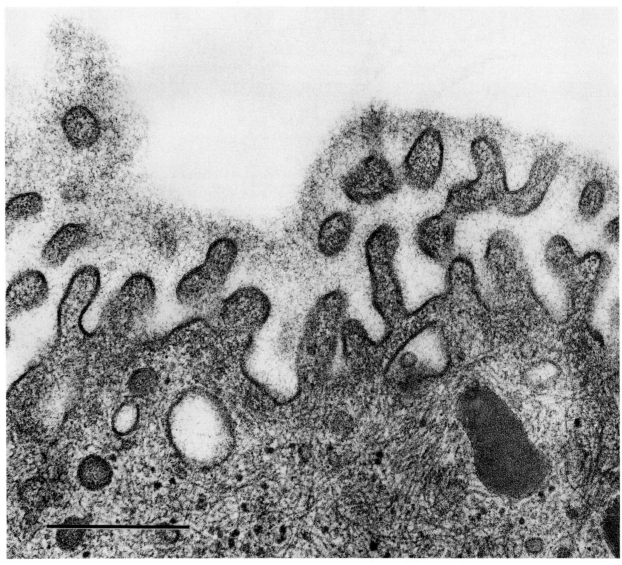

Figure 19-3 Microvillus projections from this epithelial cell show an extensive glycocalyx, which is prominent only on the lumenal surface of the cell. The glycocalyx appears as a fuzzy substance and on close inspection reveals a fine filamentous texture. (Sloughed epithelial cell in the human male reproductive tract.) Bar = 0.5 μm.

Cell Junctions

Cells may join to other cells or may be attached to connective tissue elements by specializations of their cell surface. A great number of variations of types of junctions have been described in biological materials. Only the most common of these will be illustrated. (The plasmodesmata of plant cells are considered under Special Features of Plant Tissues.)

Tight or occluding junctions, also classified as *zonula occludens*, are linear fusions of membrane between adjoining epithelial cells that usually extend around the entire circumference of an epithelial cell. Most commonly, tight junctions are sectioned such that they appear as a punctate fusion of the two plasma membranes, thus obliterating the intercellular space. At the junctional site, the resolved membranes give a five-layered or *pentalaminar* appearance (Figure 19-4). Some protocols (Figures 19-5

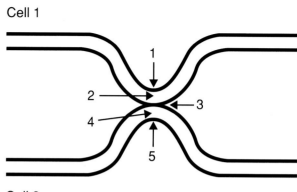

Tight Junction

Figure 19-4 Diagram showing the numbering of layers of a pentalaminar tight junction between two cells.

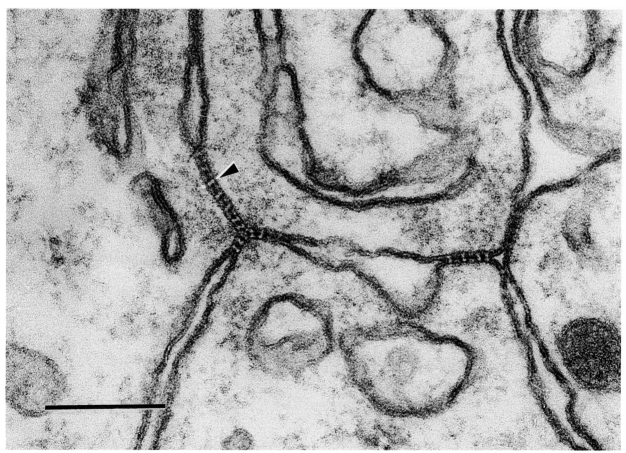

Figure 19-5 Tight junctions appear in this thin sectioned material as regions where the plasma membranes of two adjoining cells converge and join. At fusion sites, junctional particles are evident in negative relief as paired translucencies in the bilayer (arrowhead). (Sertoli cell junction from a rat.) Bar = 0.25 μm.

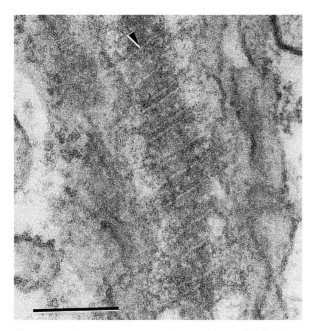

Figure 19-6 Grazing, or *en face*, sections of tight junctions shows junctional particles in negative relief as seen in Figure 19-5. Because of the grazing plane of section, the membranes forming the junction appear fuzzy and the junctional particles appear as linear arrays (arrowheads). (Sertoli cell junction.) Bar = 0.25 μm.

and 19-6) for fixing tissue show the junctional particles in negative relief forming the fusion site between two plasma membranes.

In freeze fracture replicas, occluding junctional particles are seen as linear rows of particles within the lipid bilayer. Particles generally predominate on one membrane face and complementary pits are apparent on the other face (Figure 19-7). Most epithelia contain multiple rows of junctions that either run parallel to each other or anastomose (join) with other rows to some degree.

Tight junctions have a major role in the regulation of paracellular transport of materials. Epithelia effectively regulate the environment of more deeply placed cells by acting in a relative way to exclude many substances from traveling along their exterior. Tracers (see Chapter 17) have been used to show the exclusion properties of tight junctions.

The *septate or continuous junction (zonula continua)* is the invertebrate counterpart of the tight junction. When viewed in thin section, the intercellular space (17 to 19 nm wide) does not narrow, but remains uniform and is traversed by numerous

Figure 19-7 Tight junctions appear in freeze-fracture replicas as extensive rows of intramembranous particles. Some tight junctional rows anastomose. (Sertoli cell junction.) Bar = 0.25 μm.

septa (Figure 19-8). The counterpart of septa in freeze fracture preparations are particle-rich ridges on one face and corresponding pits on complementary fracture face (Figure 19-9).

Some continuous junctions exclude tracers, others do not. Continuous junctions appear to have adhering properties and some appear to regulate paracellular transport.

Adherens junctions contain a variety of morphological types, two of which will be described. The *intermediate junction* or *zonula adherens* forms a continuous "belt" around the cell. At the junctional site, the intercellular space remains constant at about 25 nm (giving the appearance of rigidity) and may display a vague intercellular line running in parallel with the junctional plasma membranes. The cytoplasmic surface of the cell has a density that often receives actin filaments from areas far removed from the junctional site. It has an appearance similar to the desmosome described on page 414.

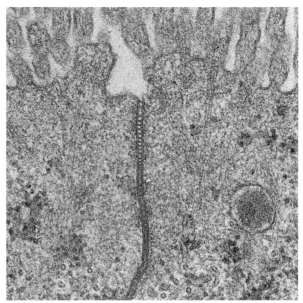

Figure 19-8 Continuous junction in a thin section of an invertebrate tissue. Numerous septa bridge the intercellular space between epithelial cells. (Micrograph courtesy of D. Friend.)

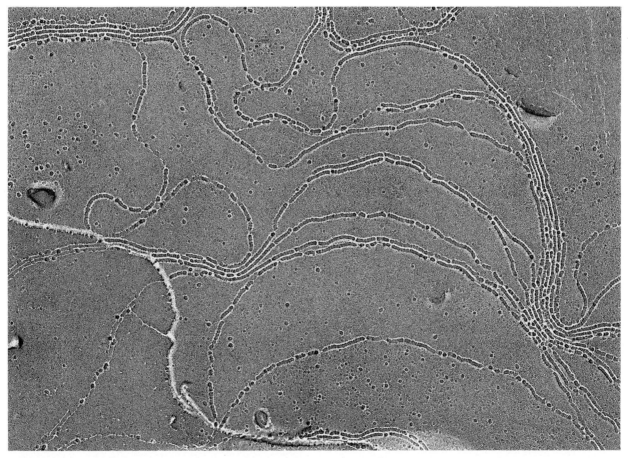

Figure 19-9 Continuous junctions in freeze fracture. This freeze-fracture micrograph of a larva moth intestine has been rotary shadowed (see Chapter 14). On one face particle rows predominate. On the other face, pits are in evidence. (Micrograph of a moth larva courtesy of D. Friend.)

The *desmosome* or *macula adherens* forms plaque-like structures that, in thin section, have an appearance similar to the zonula adherens, but the desmosome contains a definite intermediate dense line that lies within the intercellular space and parallels the cell surface. Intermediate filaments usually composed of keratin (*tonofilaments*) make hairpin turns at the junctional density (Figures 19-10 and 19-11). Freeze fracture of desmosomes reveals that they are characterized as aggregations of intramembranous particles of unequal size (Figure 19-12).

Hemidesmosomes or *half desmosomes* are cell surface specializations found at the interface of some epithelial cells and the underlying connective tissue acellular material such as a basal lamina (Figure 19-13). Their appearance at the cell surface is one whereby the junction resembles a desmosome joined to a basal lamina.

Adherens junctions, as the name implies, are important in cell-to-cell adhesion. They impart an overall cohesiveness to tissues, resisting forces that tend to pull cells apart. The function of hemidesmosomes is to anchor cells to the basal lamina.

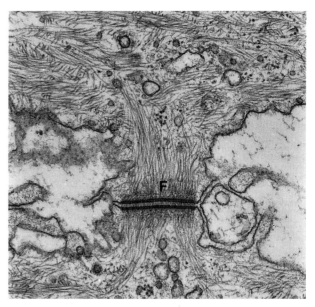

Figure 19-10   This desmosome junction shows the characteristic subsurface densities and the tonofilaments (f) which appear to insert into the densities (arrowhead). Tonofilaments actually make a sharp hairpin turn upon reaching the junctional density. (Micrograph of a newt desmosome courtesy of D. Kelly.)

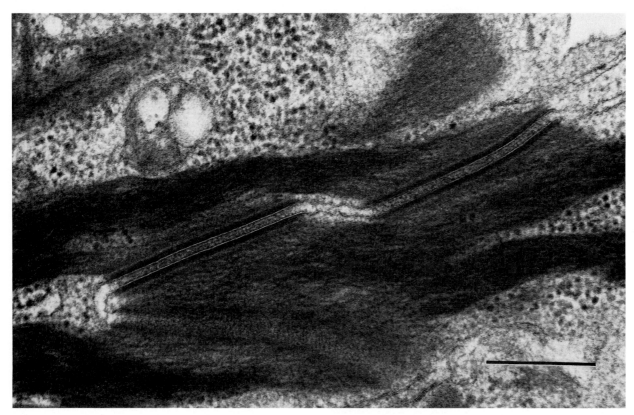

Figure 19-11   Features, similar to that depicted in the above micrograph, are depicted in this micrograph of a desmosome, however; the tonofilaments are tightly packed. (Micrograph courtesy of D. Kelly.) Bar = 1.0 μm.

Gap junctions or nexus and tight junctions were originally thought to be a single junctional type, but advances in both resolution capability, tissue preparation, and their visualization in freeze fracture have allowed them to be distinguished from one another. Gap junctions demonstrate a seven-layered (septalaminar) appearance, which suggests that the opposing membranes of the cell come close to each other, but do not fuse (Figure 19-14). The intercellular space appears to be reduced to 20 to 40 Å.

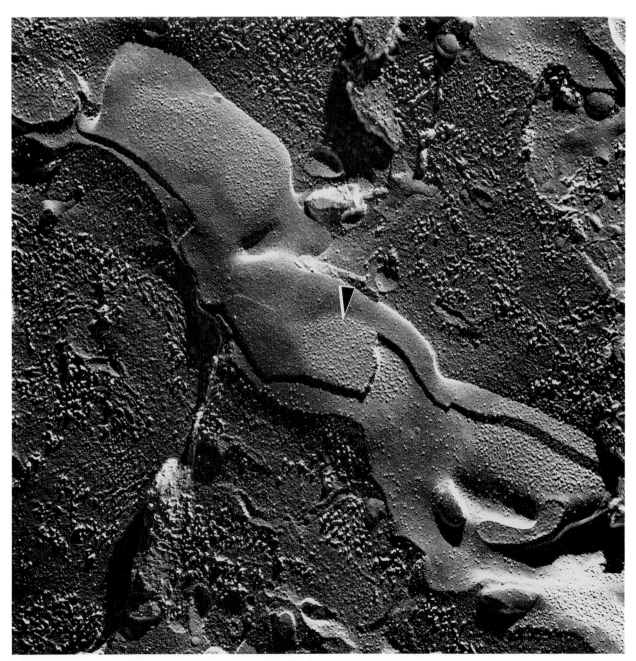

Figure 19-12 Freeze fracture image showing several desmosomes between two cells. The desmosome image (arrowhead) is recognized by the rounded aggregations of intra-membranous particles of non-uniform size. (Micrograph courtesy of D. Kelly.)

In many instances, the thinness of the section and the orientation are such that the intercellular space is resolved (Figure 19-15).

Gap junction particles (*connexons*) may be seen in appropriately stained *en face* thin sections (Figure 19-16). Gap junction particles are best seen in freeze fracture where they are most commonly found as plaque-like aggregations of hundreds of packed particles. Complementary pits are present on the opposite membrane face. Particles, measuring about 8 nm, are arranged hexagonally on one membrane face and are known to span the lipid bilayer of both membranes (Figure 19-17).

Gap junctions are generally regarded as sites of *intercellular communication*. Connexons contain pores that are thought to transport molecules of up to 1,200 molecular weight from one cell to another.

Junctional complexes join virtually all epithelial cells at their lateral surface near the apical aspect of the cell. Usually, these are present where the lateral surface of the epithelial cell meets a lumen such as seen in the gastrointestinal tract. The tight junction is the component closest to the lumen, followed by the intermediate junction, and then the desmosome, which is most basally positioned (Figure 19-18). Desmosomes may or may not be visualized in sectioned materials as part of the junctional complex since they are only periodic structures.

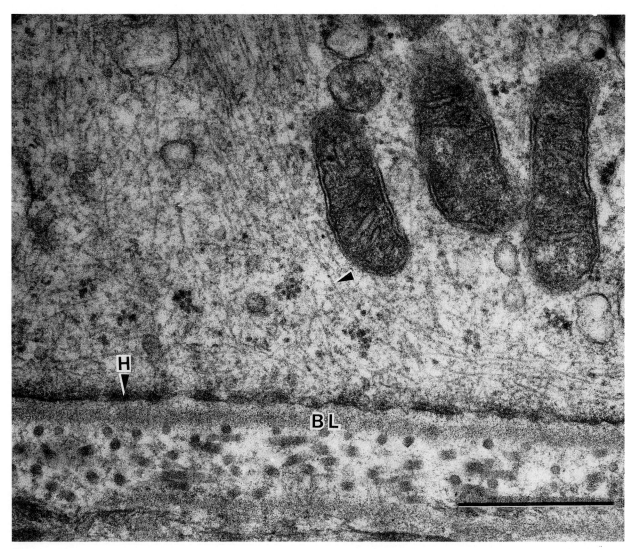

Figure 19-13 Transmission electron micrograph of a hemidesmosome (H). Numerous intermediate filaments (isolated arrowhead) converge to the cell surface in the region that it impacts the extracellular matrix of the basal lamina (BL). Bar = 0.5 μm.

Gap Junction

Cell 1

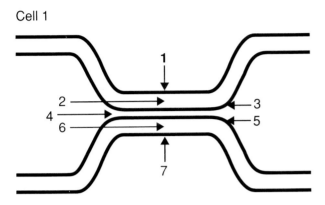

Cell 2

Figure 19-14 Drawing of a gap junction showing its seven-layered appearance.

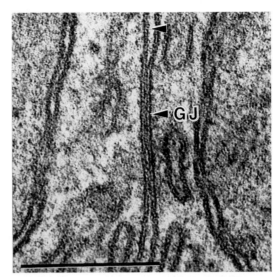

Figure 19-15 The intercellular space (arrowheads) at gap junction sites (GJ), narrows to the point that it is barely resolvable. Seven layers can be detected. (Micrograph of a liver gap junction.) Bar = 0.25 μm.

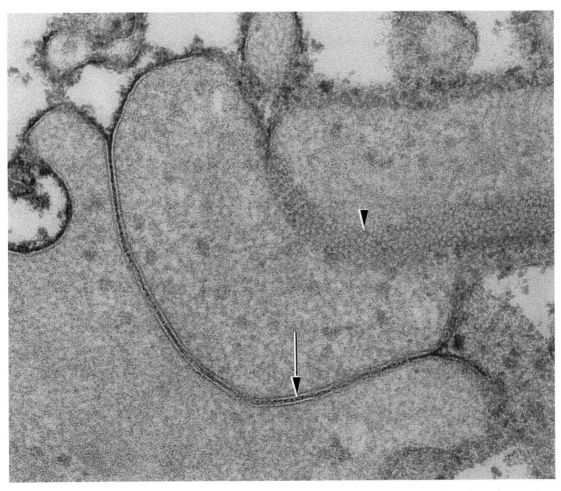

Figure 19-16 Gap junction particles in an *en face* thin section appear negatively stained (arrowhead) in this micro-graph in which a plasma membrane bound tracer (arrow) has been utilized. (Micrograph courtesy of D. Friend.)

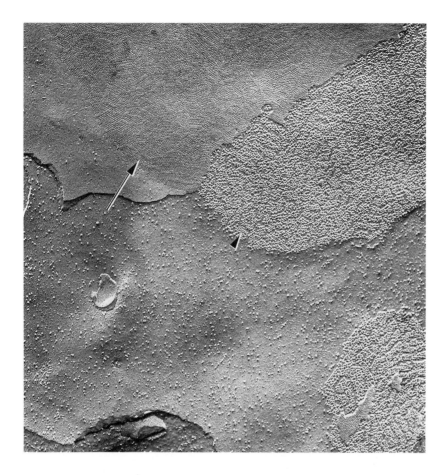

Figure 19-17 Gap junction in a freeze fracture preparation of chick skin. Closely packed gap junction particles (arrowhead) are seen on the P-face and corresponding pits (arrow) on the E-face. (See Chapter 14, Freeze Fracture Replication, for definition of fracture faces.)

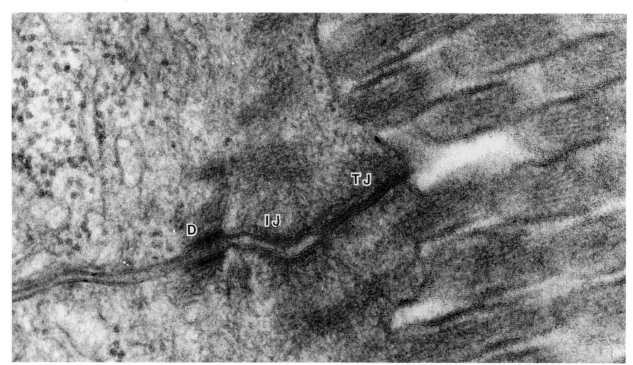

Figure 19-18 A junctional complex formed at the lateral surfaces of two epithelial cells near the luminal surface of the cells. The tight junction (TJ), intermediate junction (IJ), and the desmosome (D) are indicated. (Micrograph of intestinal cells courtesy of W. Dougherty.)

Cell Surface Specializations

Cells surfaces are variously configured to accomplish specific functions. Underlying cytoskeletal structures are responsible for the nonrounded appearance of many cells. These will be mentioned here, but described more fully in subsequent sections. Cell surface specializations are so numerous that it is impractical to describe all of them. A few more common forms will be illustrated. The scan-

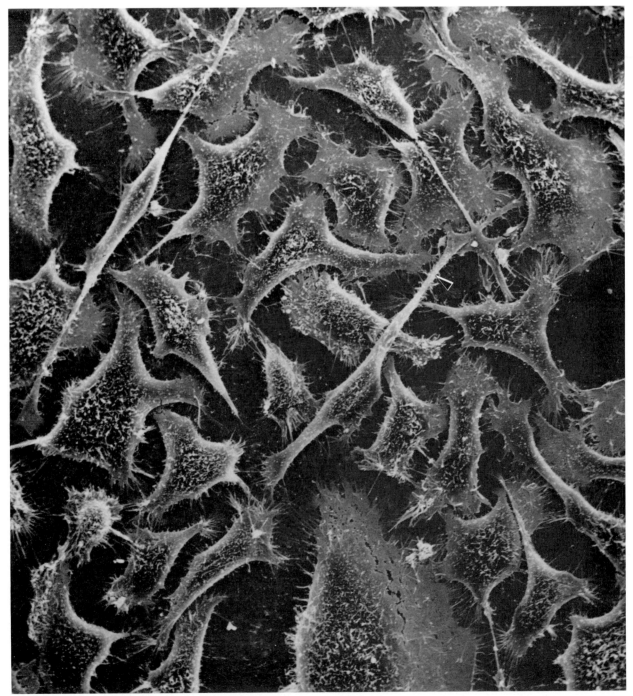

Figure 19-19 Cultured cells visualized by the scanning electron microscope. Processes of cells are numerous and are of various shapes, examples being sheet-like, needle-like and rod-like. Filipodia (arrowhead) are the long, bold processes seen in the figure. (ATCC CCL-13 Chang liver cells courtesy of W. Kournikakis.)

ning electron microscope and the freeze fracture technique are especially suited to visualize many of the three-dimensional aspects of complex cell surfaces. Cultured cells provide the most dramatic examples of cell surface irregularities (Figure 19-19).

Cilia are found protruding from cell surfaces at a lumen or into a fluid-filled space. Particular arrangements of microtubular elements at the cell surface grow from a centriole or basal body (see sections on Microtubules and Cilia) to provide numerous long evaginations of the cell surface. Several hundred such evaginations (up to 10 to 15 μm in length) may be present per cell (Figure 19-20). Cilia are actively motile and may function in propulsion of cells or movement of fluid at the cell surface.

Microvilli are long (2 to 3 μm), thin projections of the cell surface, which are seen primarily at the lumenal surface of most epithelial cells. Unlike cilia, their cores are filled almost entirely with actin filaments (Figures 19-21 and 19-22). Microvilli increase the surface area of many cells several fold and are thought to enhance adsorptive functions.

Stereocilia, cell surface projections, are structurally and functionally similar to microvilli (but not cilia) with the exception that they are much longer (5 to 10 μm) than microvilli. Good examples of stereocilia are seen in the male reproductive tract (Figure 19-23).

Lamellipodia are sheetlike extensions from the cell surface. They are prominent along the leading edge cells that are undergoing movement in a tissue culture system (Figures 19-24 and 19-25). Lamellipodia are prevented from forming by drugs that not only inhibit actin filaments, but also inhibit cell movements, suggesting a role for lamellipodia in cell locomotion.

Microspikes and filipodia are long needlelike specializations seen mainly in tissue culture cells that are adjusting to their new environment. They contain an actin core and may extend from 5 to 10 μm (microspikes; Figure 19-25) or from 10 to 50 μm (filipodia; Figure 19-19). These cell processes, like lamellipodia described above, are best viewed by scanning electron microscopy. Their actin core allows them to form and to retract rapidly. Both structures seem to be devices for exploring the environment around the cell.

Flagella have basic similarities to cilia, although they are a more highly modified cell surface specialization. Flagella are longer than cilia and there

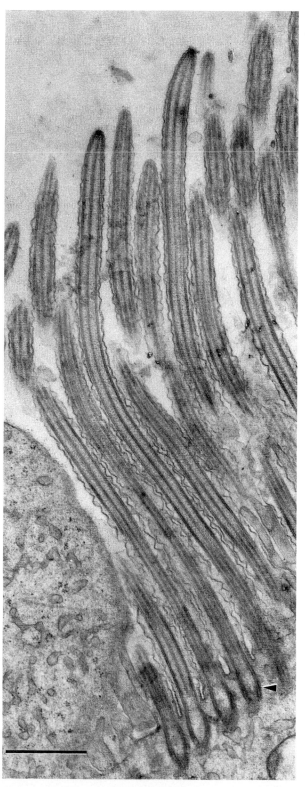

Figure 19-20 These cilia are shown in longitudinal and oblique section. Where the cilia join the body of the cell, basal bodies are seen (arrowhead). Bar = 1.0 μm. (Micrograph from the lining epithelium of the respiratory tract courtesy of W. Dougherty.)

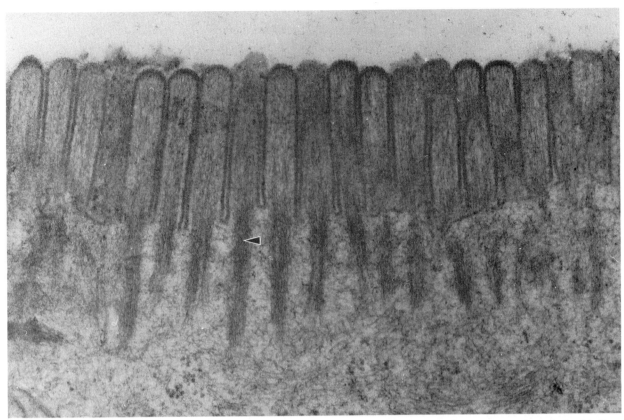

Figure 19-21 In these microvilli, actin filaments (arrowhead) are seen entering microvillus processes from within the cytoplasm. (Micrograph from the mouse intestinal epithelium courtesy of W. Dougherty.)

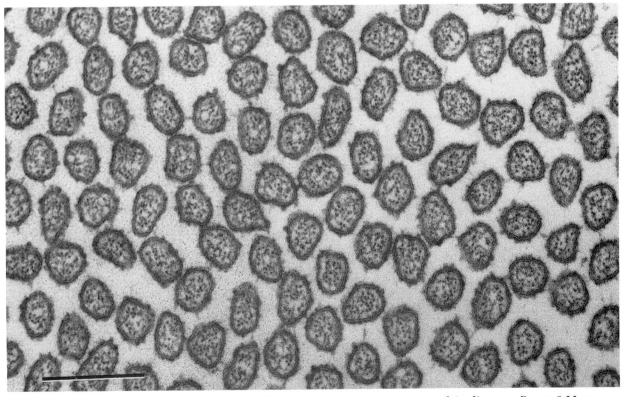

Figure 19-22 Microvilli in cross section. Actin filaments are seen within the cores of the filaments. Bar = 0.25 μm.

are generally fewer flagella per cell (Figure 19-26; see also Cilia and Flagella). Flagella are apparatuses for cell motility.

Micropinocytotic vesicles form as small, rounded pits that pinch off into the cell to form vesicles (Figure 19-27). These vesicles are nonselective transporters of materials, primarily transporting to sites within the cell or across the cell membrane to exit the cell (e.g., capillary endothelium).

Pinocytotic or endocytic vesicles are larger and often more irregularly shaped than micropinocytotic vesicles (Figure 19-28). Pinocytotic vesicles partic-ipate in *fluid phase endocytosis*, which is a continuous and nonspecific ingestion of extracellular materials.

Bristle-coated pits (or simply coated pits) and vesicles, appearing very similar to pinocytotic vesicles, show both an external and internal fuzzy substance that occurs on the surface of the pit or vesicles (Figure 19-29). Coated vesicles serve to ingest specific substances in the process of *receptor-mediated endocytosis*. They also shuttle various substances from one intracellular compartment to another or are involved in retrieval of membrane that has been added to the cell surface by phenomena such as secretion.

Figure 19-23   Stereocilia projecting from the apical surface of an epithelial cell. (Stereocilia from the efferent duct of the rat.)

Figure 19-24 Free lamellipodia are seen at one pole of this tissue culture cell. Much of the remainder of the cell shows smaller extensions of various types. (ATCC CCL-13 Chang liver cells courtesy of W. Kournikakis.)

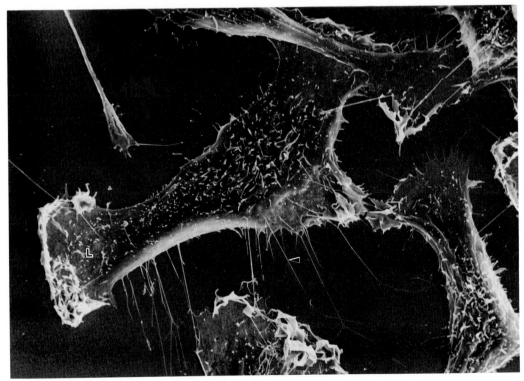

Figure 19-25 Flattened areas of two cultured cells form lamellipodia (L). Numerous microspikes (arrowhead) extend from the cell surface. (Micrograph courtesy of W. Kournikakis.)

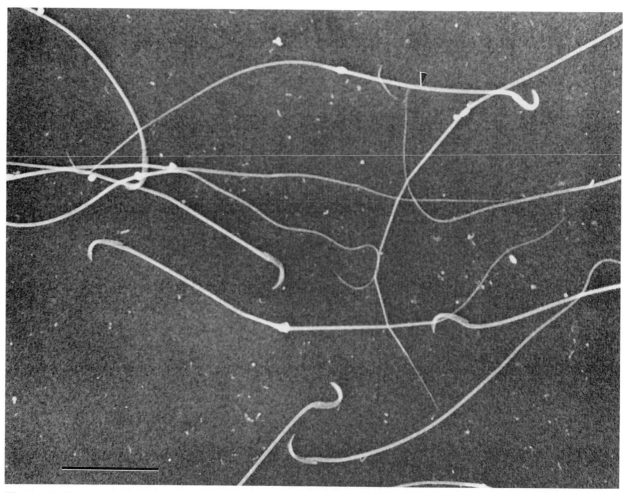

Figure 19-26   Long flagella of rat sperm attach to the sickle-shaped sperm head. Bar = 30 μm.

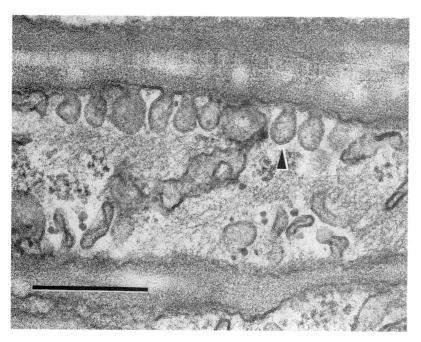

Figure 19-27   This flattened cell type shows numerous micropinocytotic vesicles (arrowheads) that invaginate the plasma membrane. Unlike bristle-coated pits, they show no fuzzy coat on either of their surfaces. (Myoid cell from the rat testis.) Bar = 0.25 μm.

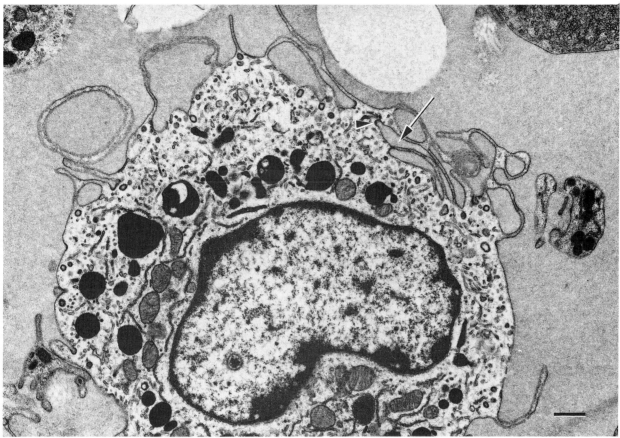

Figure 19-28 Pinocytotic vesicles (arrow) are seen near the surface of a macrophage. They have one region of their membrane that appears dense like the coated pits described below, indicating they were endocytozed by this specialized region of the plasma membrane. The density of their contents resembles that of the extracellular environment. Bar = 1.0 μm.

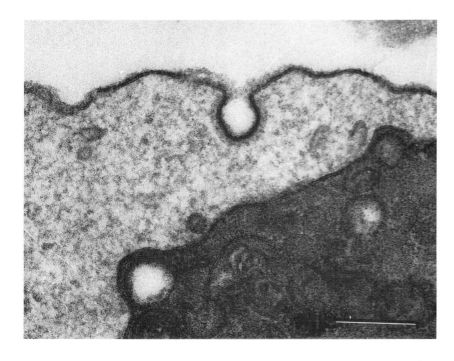

Figure 19-29 A single bristle coated pit is seen on the cell surface. The fuzzy coat is more prominent on the extracellular than the intracellular surface. Bar = 0.25 μm.

The Cytoskeleton

The cytoskeleton provides the architectural framework of the cell. It is also important in both maintenance and alteration of cell shape, in cell locomotion, and in cell organization and organelle movements. Composed primarily of filamentous elements, the cytoskeleton is a complex and dynamic system within the cell. The filaments have been classified according to size: *microtubules* are the largest; *microfilaments*, or actin filaments, are the smallest; and an intermediate-size class of filaments, not surprisingly, is named *intermediate filaments*. Other cytoskeletal elements have been suggested as present, but their validity as true cytoskeletal elements is not widely accepted at this time.

Microtubules

Microtubules are long, cylindrical elements having a diameter of about 25 nm and composed of protein subunits called *tubulin*. If sectioned longitudinally, they appear to take a straight or slightly curved course, giving the viewer the impression that they are relatively inflexible structures. The central region of microtubules appears less dense than the margins, a feature that, in addition to size, can be used to distinguish them in longitudinal section from other cytoskeletal elements (Figure 19-30). In cross section, the microtubules have a distinct translucent core (Figure 19-31).

Microtubules are best known for their role in chromosome movement during mitosis. Here they attach to structures known as *kinetochores* and, in some as yet undermined way, they are thought to act in concert with other cytoskeletal elements to affect chromosome movement (Figure 19-32).

Most cells contain at least a few scattered microtubules and many contain regions of organized bundles of microtubules. A certain subclass of microtubules is involved in laying down and orienting the cellulose of the plant cell wall. Microtubules function in the development and maintenance of an asymmetrical cell shape. They are thought to be responsible for organelle movements and function in secretory processes. Although they appear to be static structures in electron micrographs, they are usually in a state of flux and are capable of forming and dissociating rapidly.

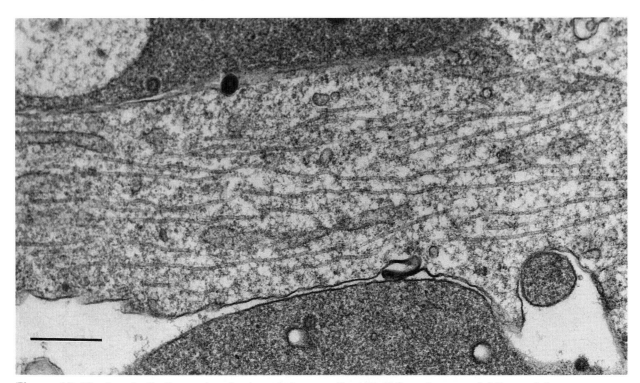

Figure 19-30 Longitudinally sectioned microtubules show a slightly lighter (electron-translucent) central region. (Sertoli cell from the rat testis.) Bar = 0.5 μm.

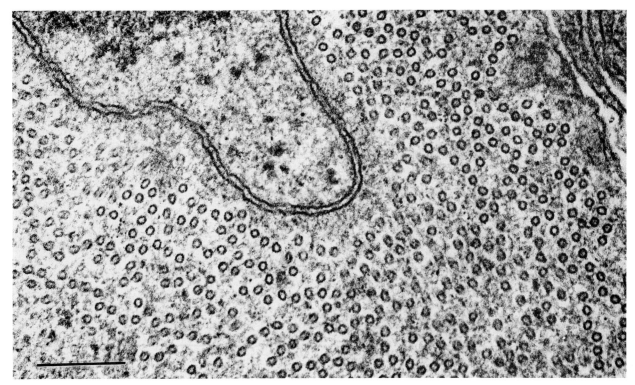

Figure 19-31 In cross sections, the microtubular profiles are rounded. The core of the microtubule is distinctly trans-lucent. (Manchette microtubules from the rat testis.) Bar = 0.25 µm.

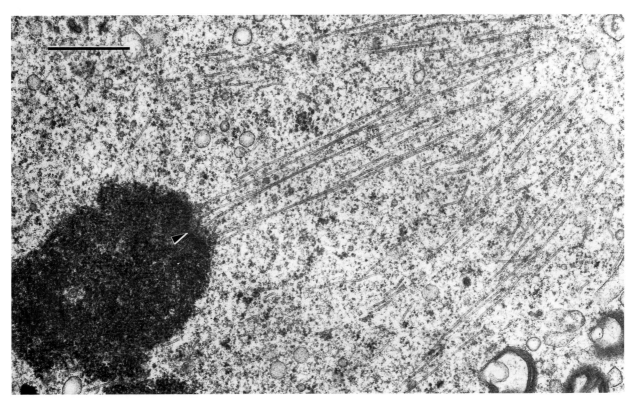

Figure 19-32 Microtubules of a metaphase cell are seen attaching to the kinetochore (arrowhead) of the chromo-some—a structure which appears slightly more dense than the chromatin.

Microfilaments

The smallest of the recognized cytoskeletal fila-mentous elements, microfilaments, are composed of polymerized *actin*. Microfilaments are notoriously difficult to preserve in standard fixation protocols and, as a consequence, may not appear as distinct filaments (Maupin-Szamier and Pollard,1978), but more as a fuzzy material (Figure 19-33). Tannic acid is a useful fixative additive for preservation of actin (Maupin and Pollard, 1983). If actin filaments are organized into geometric bundles, one's ability to preserve them is somewhat improved. Bundled actin filaments measure about 5 to 7 nm in diameter and may be seen scattered throughout the cell or in highly organized groupings (Figure 19-34). Actin filament groupings may take several forms within the cell. Actin may appear bundled in tissue culture cells, and as such are termed *stress fibers*. Commonly, actin is seen at the cortical region of cells. One of the most dramatic examples of cortical actin is the bundles of actin that form the cores of microvilli (Figures 19-21 and 19-22).

The ability of groups of actin filaments to as-sume so many forms in cells is due to their asso-ciation with proteins known as *actin-binding proteins* that may link them in several different ways. Actin-binding proteins also regulate the degree to which actin filaments polymerize and depolymerize. The myosin of skeletal or smooth muscle or other con-tractile systems may be regarded as an actin-binding protein.

Given the morphological diversity of groups of actin filaments and their plasticity within cells, it is not surprising that they perform a variety of func-tions within cells. Actin filaments may participate in contractile processes in concert with myosin, as in skeletal or smooth muscle contraction, or in the con-tractile ring of dividing cells that separates daughter cells during telophase. They are important in de-termining the shape of the cell surface, examples being the actin associated with microvilli, stereo-cilia, and a variety of other local surface specializ-ations. Actin filaments are involved in motile pro-cesses and cell locomotion.

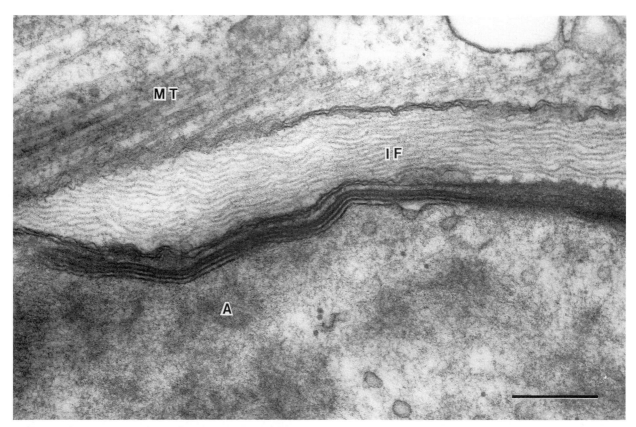

Figure 19-33   The indistinct appearance of actin filaments (A) in typical preparations is contrasted with that of inter-mediate filaments (IF) and longitudinally sectioned microtu-bules (MT). Intermediate filaments appear solid whereas mi-crotubules show a hollow core. (Opossum spermatid and adjoining Sertoli cell.) Bar = 0.25 μm.

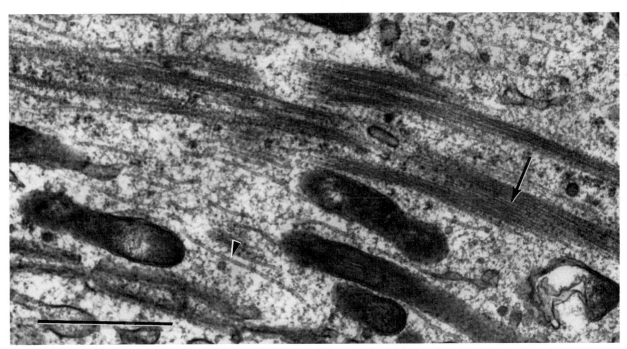

Figure 19-34 Bundles of well preserved actin filaments (arrow) are seen amongst a few microtubules (arrowhead). Bar = 1.0 μm.

Intermediate Filaments

These filaments, also known as *10 nm filaments*, actually represent molecular forms of filaments that are of slightly variable composition and size (about 10 to 12 nm in diameter). Tonofilaments of desmosomes (containing keratin), neurofilaments, vimentin-containing filaments, desmin-containing filaments, and glial filaments are the major intermediate filament types.

Intermediate filaments are not as difficult to preserve for ultrastructural examination as are actin filaments (Figure 19-33). They are from 30% to 50% larger than actin filaments and generally take a straight or slightly curved (sometimes wavy) course within the cytoplasm (Figure 19-35). Other than occasional bundles of intermediate filaments, there are fewer examples of organized arrays of intermediate filaments than there are of actin filaments. Unlike actin filaments, which prefer a cortical location, intermediate filaments often appear near the nucleus. In spite of these differences, the two filament types are easily confused by microscopists.

The precise function of intermediate filaments is not known. Their positioning at sites within the cell is subject to considerable stress, suggesting that they resist tension within cells and serve a structural role. They may anchor organelles or the nucleus at certain positions within the cell.

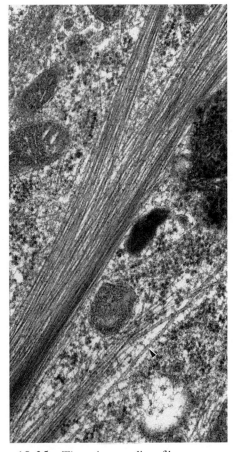

Figure 19-35 These intermediate filaments are present in a cultured cell. They appear to course throughout the cytoplasm in bundles. (Micrograph courtesy of A.W. Vogl.)

The Nucleus

The nucleus is the most easily recognizable feature in eukaryotic cells because of its large size and relative constancy of structural features. Genetic information is contained within the nucleus. Its main function is to duplicate the genetic information, as well as transcribe information necessary for synthetic processes. The size and shape of nuclei are highly variable. Many nuclei are spherical (Figure 19-40), others are ellipsoidal, flattened, lobed, or highly infolded (Figure 19-36).

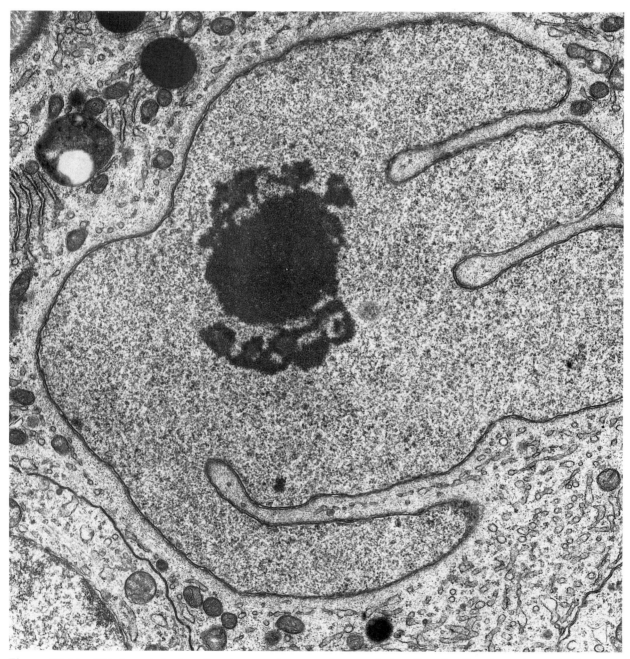

Figure 19-36   This irregularly shaped nucleus contains a prominent nucleolus. (Sertoli cell nucleus of a monkey.)

The Nuclear Envelope

The nuclear envelope, present in eukaryotic cells but not in prokaryotic cells (bacteria), is a double membrane consisting of *outer* and *inner membranes* that encompass the *nucleoplasm*. Rough endoplasmic reticulum is often said to be in direct continuity with the nuclear envelope, a feature seen more frequently in textbook drawings than in electron micrographs. At intervals, the two membranes of the nuclear envelope participate in forming *nuclear pores*. These structures are not open continuities between the nucleus and cytoplasm, but are each bridged by a *diaphragm* (Figure 19-37). The arrangement of nuclear pore complexes are seen to advantage in grazing sections of the nucleus and in freeze fracture preparations (Figures 19-38 and 19-39).

The nuclear envelope separates the DNA and RNA synthetic processes from that of protein synthesis, which occurs in the cytoplasm. Nuclear pores function as selective gates for entry and exit of molecules to and from the nucleus.

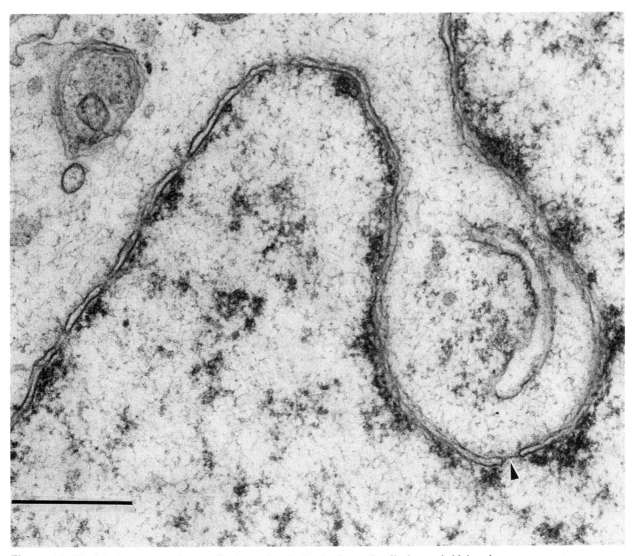

Figure 19-37   Nuclear pores. One profile (arrowhead) clearly shows the diaphragm bridging the pore. Bar = 0.25 μm.

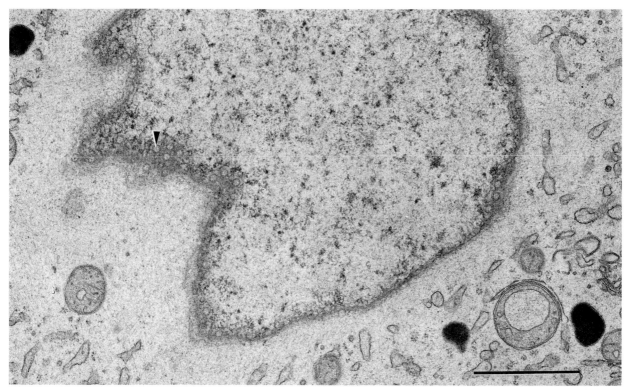

Figure 19-38 Nuclear pores. The indistinct appearance of the membranes of the nuclear envelope indicates that this nucleus was been grazed by the sectioning knife. Nuclear pore complexes are seen *en face* and appear spherical. Bar = 1.0 μm.

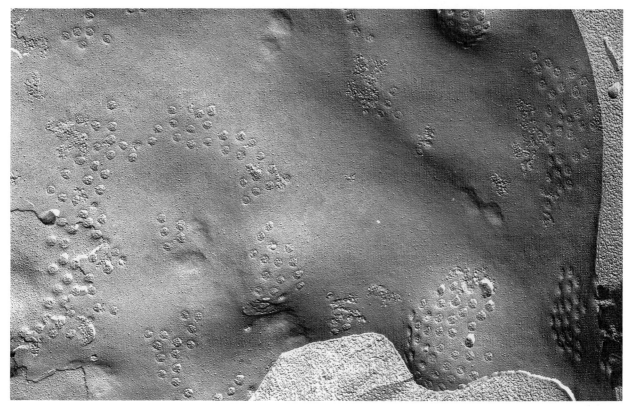

Figure 19-39 Freeze-fracture of a nuclear envelope shows that nuclear pores are not necessarily randomly distributed.

Chromatin

In structural terms, the condensed or visible form of chromatin is known as *heterochromatin*. The relatively clear areas of the *nuclear matrix* are assumed to be occupied by an uncondensed form of chromatin known as *euchromatin*. Upon close examination, heterochromatin is usually made up of fine 15 to 20 nm granules, but at low magnification the granules are not resolved (Figure 19-40). Heterochromatin is frequently seen lining the nuclear envelope, but absent from regions of the nuclear envelope containing pores.

In eukaryotes, heterochromatin is assumed to be complexed with histones, whereas euchromatin is presumed to be actively synthetic. Thus, cells with large proportions of euchromatin are thought to be active in synthesizing substances.

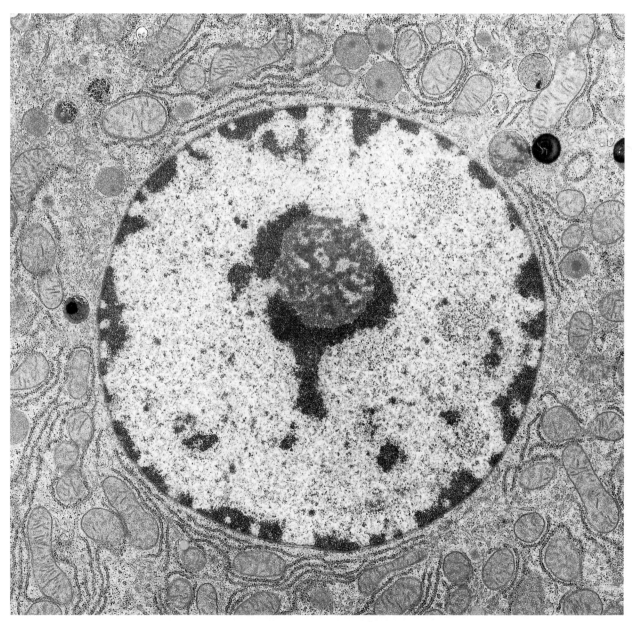

Figure 19-40 This nucleus displays areas that are relatively clear and occupied by euchromatin. Dense heterochromatin is associated with the nuclear envelope and the nucleolus. (Micrograph of a hepatocyte courtesy of W. Dougherty.)

The Nucleolus

Of variable size and number, the nucleoli occupy only a portion of the nucleus and are thus seen in only some thin sections of the nucleus. Although the organization of the components of the nucleolus differ widely from one cell type to another, there are three major recognizable structural components (Figures 19-41, 19-42, and 19-43). The *pars fibrosa*

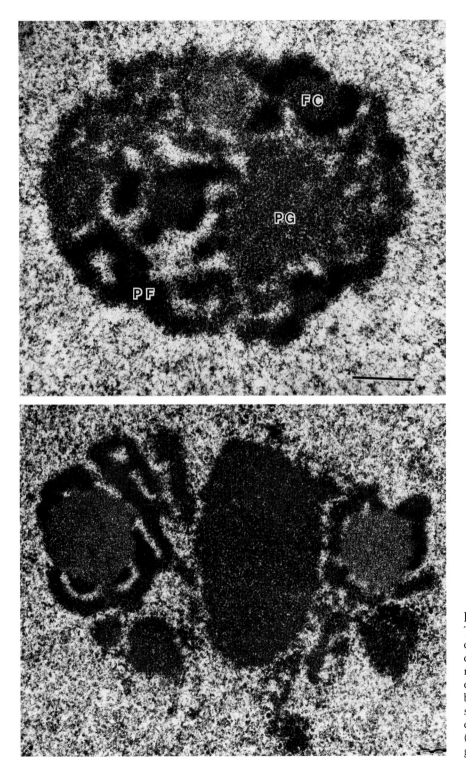

Figure 19-41 and 19-42
Three components of the nucleolus can be distinguished in these micrographs. The pars fibrosa (PF) is represented by the electron dense cords of the nucleolus. The pars fibrosa partially encloses areas of low stain intensity called the fibrillar centers (FC). The pars granulosa (PG) appears as an aggregation of granules. Bar = 0.5 μm.

or *nucleonema* is represented as strands of dense material that form a spongy network. Associated with the pars fibrosa are the *fibrillar centers*, which are usually finely granular, rounded mass. The third component, the *pars granulosa*, is usually near the pars fibrosa and appears as an aggregation of finely granular material. RNA synthesis takes place in the nucleolus.

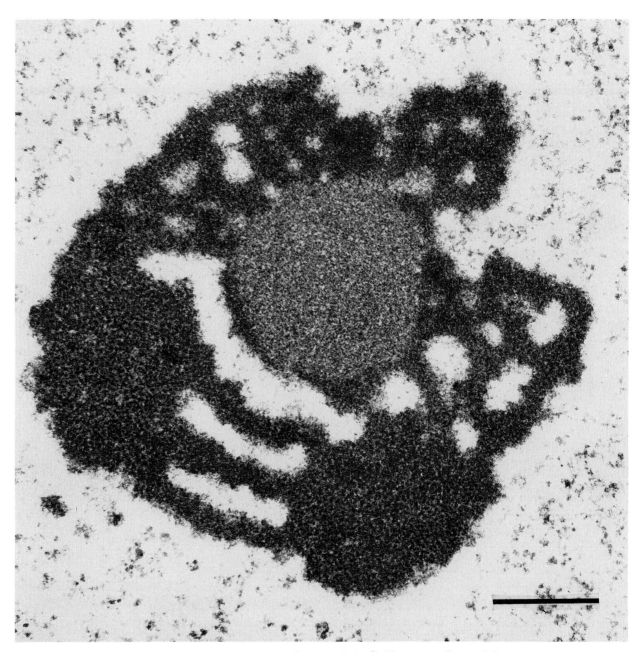

Figure 19-43 This nucleolus clearly shows the pars fibrosa and the fibrillar center. Bar = 0.5 μm.

Dividing Cells

During cell division, the nuclear envelope is resorbed, the nucleolus dissipates, and the cell goes through the classic phases of cell division: *prophase* (Figure 19-44), *metaphase* (Figure 19-45), *anaphase* (Figure 19-46) and *telophase* (Figures 19-47 and 19-48).

In germ cells of the testis, the cleavage furrow never pinches off in telophase, but remains as an *intercellular bridge* (Figure 19-49). Such bridges may connect hundreds of germ cells and are most likely involved in synchronous development of these cells.

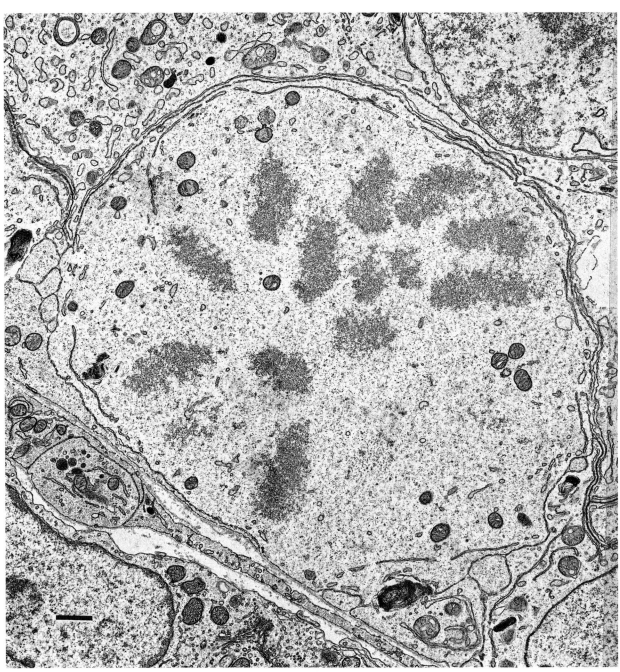

Figure 19-44 Prophase cell. The nuclear envelope is in the process of fragmenting. Individual chromosomes are visible in the section. Bar = 1.0 μm.

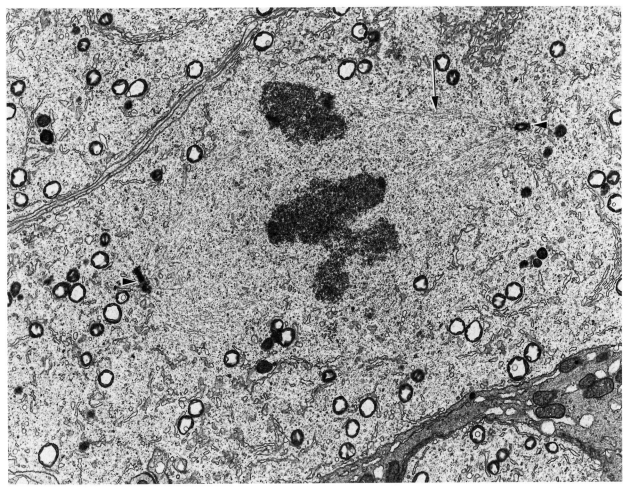

Figure 19-45   In metaphase, the chromosomes align on an equatorial plate. The centrioles (arrowhead) are apparent and the spindle apparatus (arrow) is visible. Bar = 1.0 μm.

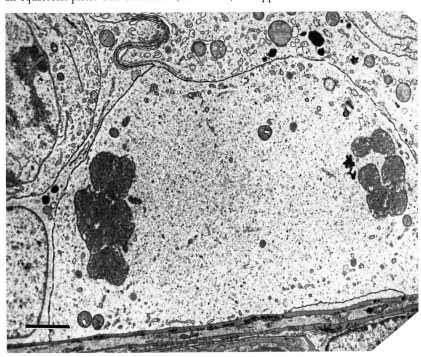

Figure 19-46   In late anaphase, the chromosomes have just reached the poles of the cell. Bar = 2.0 μm.

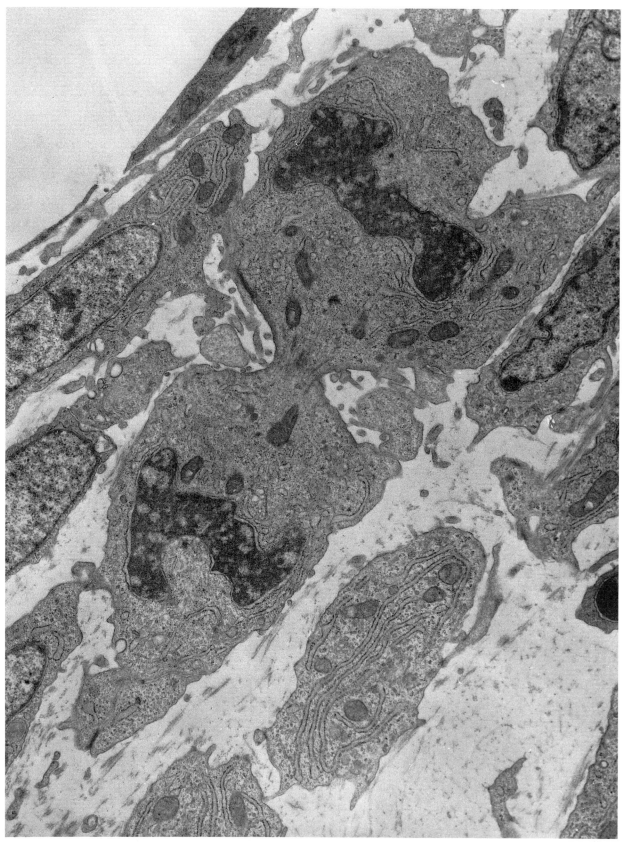

Figure 19-47 Telophase cell. A cleavage furrow has begun to divide the two daughter cells and the nuclear envelope has reformed. (Micrograph courtesy of W. Dougherty.)

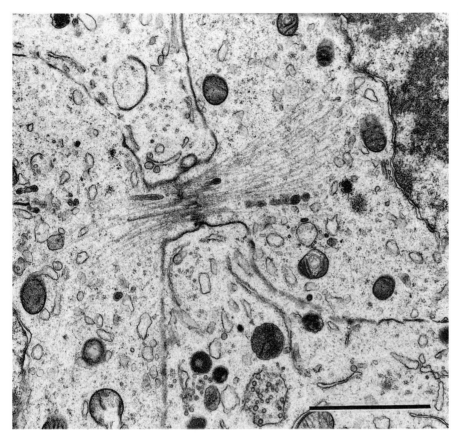

Figure 19-48 In this late telophase division, the nuclear envelope has reformed but the two cells remain connected by a small cytoplasmic channel which was formed by the cleavage furrow and which contains microtubules of the spindle apparatus. Bar = 2.0 μm.

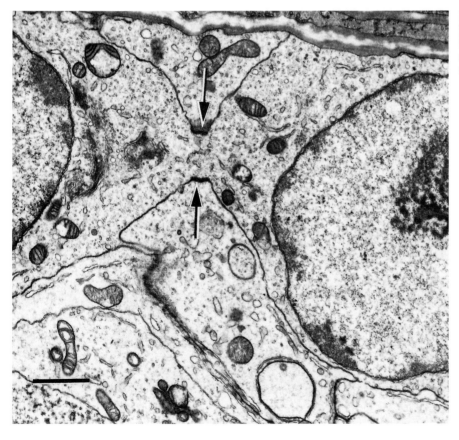

Figure 19-49 The intercellular bridge between two spermatogonia is about 1 μm in width. Two densities (arrows) surround the bridge channel. Bar = 1.0 μm.

The Synaptonemal Complex

During the prophase of meiosis in both male and female, numerous threadlike structures appear in the nucleus of spermatocytes. Upon close examination, these structures are composed of two dense parallel lines and a much less dense central line. Known as *synaptonemal complexes*, there are as many synaptonemal complexes within the nucleus as there are chromosome pairs. They insert into the nuclear membrane through a structure called the *basal knob*. Since longitudinal profiles of synaptonemal complexes are rarely seen for any distance, it is assumed that they take an irregular course within the nucleoplasm (Figure 19-50). In special preparations, chromosomes of a pair may be seen external to the dense lines of the synaptonemal complex (Figure 19-51).

Synaptonemal complexes are the structural unit controlling the process of synapsis. During synapsis, there often is exchange of genetic material between homologous pairs of chromosomes, a process called *crossing over*.

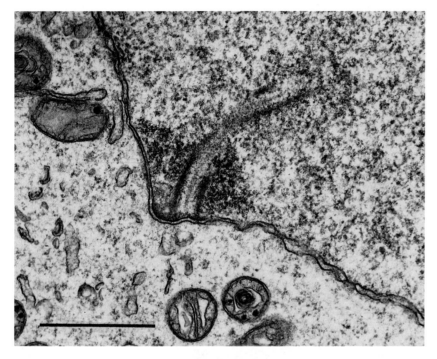

Figure 19-50 This synaptonemal complex appears as a tripartite structure which inserts into the nuclear envelope at the basal knob. (Spermatocyte from the rat testis.) Bar = 1.0 μm.

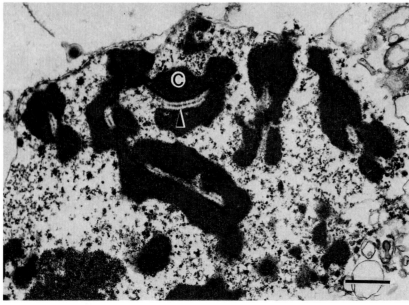

Figure 19-51 In specially prepared tissue, several synaptonemal complexes (C) are seen flanked by the dense heterochromatin of chromosome pairs. (Spermatocyte from the rat testis.) Bar = 1.0 μm.

Mitochondria

Nearly all eukaryotic cells contain mitochondria. Mitochondria have a simple structural plan that may take many forms in different cell types. Some mitochondria are spherical, while others are rod shaped (Figure 19-52) or highly irregularly shaped. In addition, there may be great size variations among mitochondria. The mitochondrion is bounded by an *outer mitochondrial membrane*. A second membrane, the *inner mitochondrial membrane*, generally paral-

lels the contour of the outer membrane. However, at intervals, the inner mitochondrial membrane involutes into the central region of the mitochondrion to form the *cristae*. These membrane folds are numerous and are sectioned in various planes within the mitochondrion. However, in sections, their connection to the inner membrane as it begins to fold inward is not always observed. The cristae may be lamellar (sheetlike) or tubular or may take other geometrical forms. Branching and anastomosing cristae have also been observed. The space between the

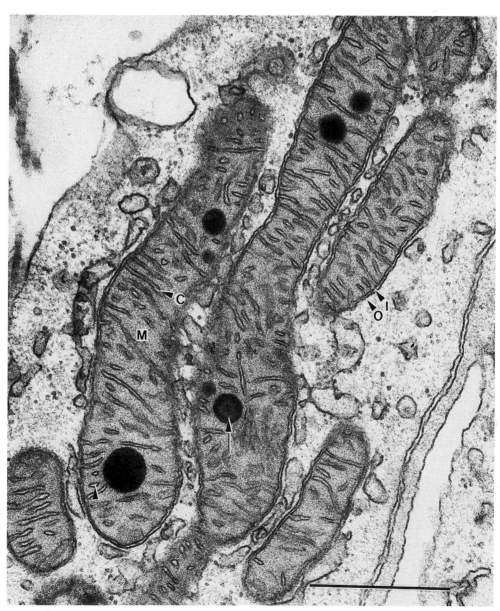

Figure 19-52 These animal cell mitochondria (M) are elongate or rod shaped. Structural features include: outer mitochondrial membrane (O), inner mitochondrial membrane (I), intracristal space, mitochondrial matrix and matrix granules (arrows). The cristae (C) are a mixture of lamellar and tubular, as evidenced in certain areas of the micrograph where they are cut transversely and thus appear as flattened circular profiles (arrowheads). (Leydig cell from the rat testis.) Bar = 1.0 μm.

outer and inner mitochondrial membranes is the *intracristal space* and extends inward with the cristae. The *mitochondrial matrix* lies internal to the inner mitochondrial membrane and appears homogeneous and finely granular, except for the occasional presence of *matrix granules*. These very dense bodies are calcium salts.

The structural features of mitochondria as well as variations in size, shape, and internal structure are illustrated in Figures 19-52 through 19-61.

The mitochondrion contains enzymes used in oxidative phosphorylation to provide energy to the cell. The inner mitochondrial membrane contains the *respiratory chain of enzymes*, which oxidizes substrates to form ATP. Mitochondria also contain DNA and RNA, which are biochemically different from the chromosomal DNA and cytoplasmic RNA. Mitochondria divide autonomously (independent of chromosomal DNA) and are also involved in lipid synthesis.

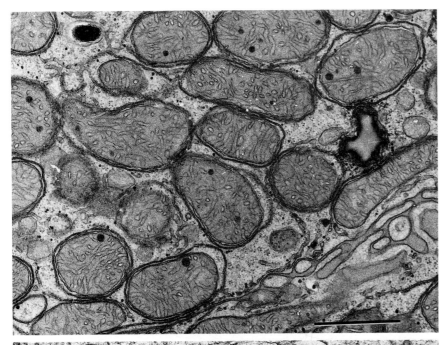

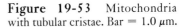

Figure 19-53 Mitochondria with tubular cristae. Bar = 1.0 µm.

Figure 19-54 Mitochondria with exclusively lamellar cristae.

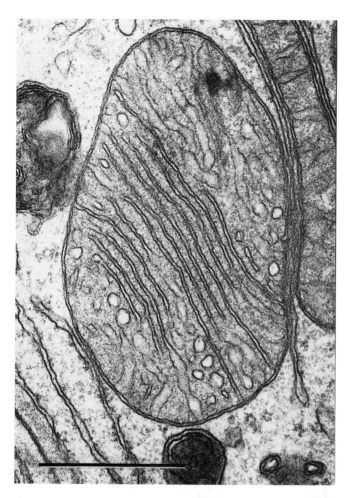

Figure 19-55 Mitochondria showing both lamellar cristae and tubular cristae. Bar = 0.25 µm.

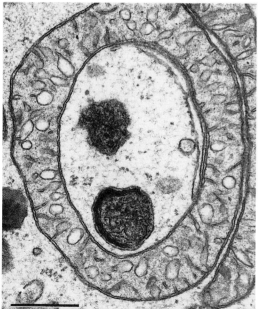

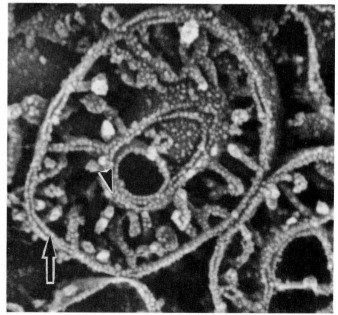

Figure 19-56 and **19-57** The mitochondrion in 19-56 (left) is cup-shaped, but here appears ring-shaped since only the lip of the cup has been sectioned. Its cristae are distinctly tubular and expanded at their ends. In Figure 19-57 (right) the same mitochondrial type is visualized by scanning microscopy after it has been broken open. The outer and inner mitochondria membrane (arrow and arrowhead, respectively) and cristae are visualized. (Scanning micrograph courtesy of I. Fritz.)

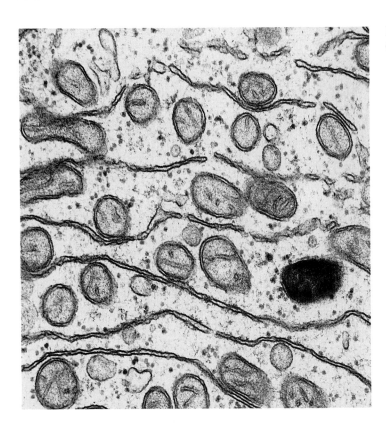

Figure 19-58   These small, spherical mitochondria contain only a few cristae.

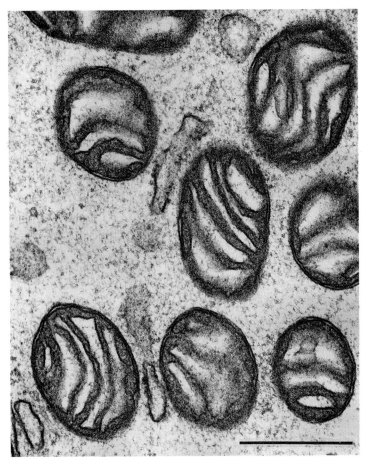

Figure 19-59   The intracristal space is greatly dilated in cristae of these mitochondria. Although they superficially resemble poorly fixed mitochondria described in Chapter 18, the modifications of these mitochondria are seen in well fixed tissues. Bar = 0.5 μm.

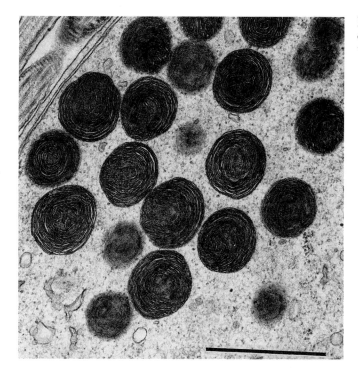

Figure 19-60   The cristae of these mitochondria are difficult to visualize since they are extensively spiraled within the mitochondrial matrix. Bar = 1.0 μm.

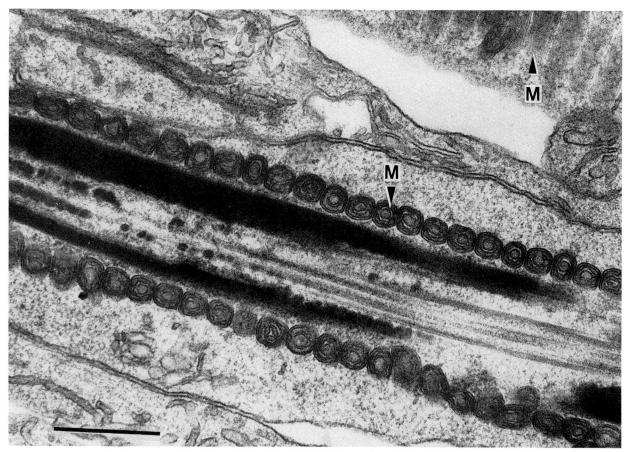

Figure 19-61   Mitochondria (M) are characteristically spiraled around the middle piece of the flagellum of mammalian spermatids and sperm. A transverse section (extending from upper left to below) shows their internal structure and a grazing section (upper right) indicates their spiral arrangement around the flagellum. The intracristal space is the less dense area of these mitochondria. Bar = 0.5 μm.

Protein Synthetic and Secretory Structures

Free Ribosomes

In the cytoplasm, ribosomes may be seen singly or in groups (*polysomes*) in virtually all cell types. Individual mammalian ribosomes measure about 20 nm in diameter. There are some difficulties in obtaining precise measurements of ribosomes in sectioned material, since ribosomes are not entirely spherical entities and their surfaces are not sharply demarcated from the surrounding cytoplasm (Figure 19-62).

Polysomes function in the synthesis of protein that remains as an internal constituent of the cell, i.e., not exported as a packaged secretory product. Polysomes are linked by messenger RNA to produce several identical copies of a protein from one copy of messenger RNA.

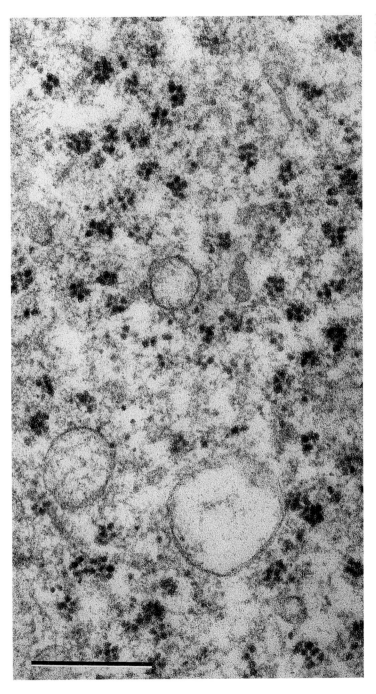

Figure 19-62 Ribosomes. These ribosomes are free within the cytoplasm and seen in groups known as *polysomes*. Bar = 0.5 μm.

Membrane Bound Ribosomes

Ribosomes bound to membranes are frequently found in spirals that are bound to the external aspect of the endoplasmic reticulum (Figures 19-63 and 19-64). Their protein secretory product is threaded through the membrane of the endoplasmic reticulum and eventually cleaved to leave the protein secretory product inside the rough endoplasmic reticulum.

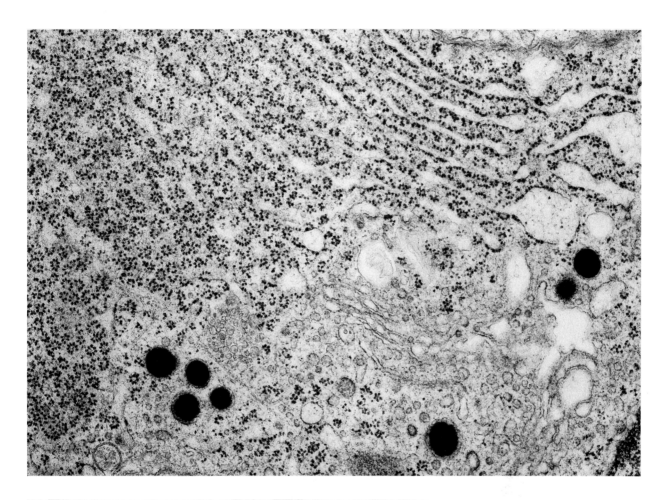

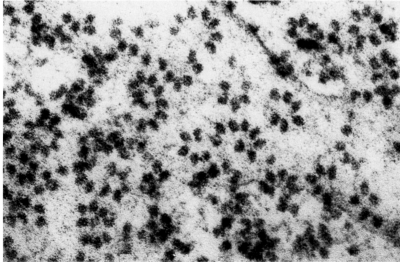

Figure 19-63 and **19-64** Membrane bound ribosomes. Although the membranes of the rough endoplasmic reticulum are sectioned *en face* and do not appear obvious, the numerous bound ribosomes are seen. The high magnification micrograph (left) shows that many of the bound ribosomes are organized in spirals. (Micrographs courtesy of A. K. Christensen.)

Rough Endoplasmic Reticulum

Of the various flattened enclosed membranous elements (*cisternae*) within the cytoplasm, those studded with ribosomes are termed *rough endoplasmic reticulum*. Ribosomes are seen on the external surface of the membrane of the cisternae and may be numerous or sparsely cover its surface. Individual cisternae, as seen in thin section, are considered to be interconnected and part of one totally enclosed membrane system, although the plane of section does not often reveal the continuities between sectioned cisternae. The *lumen* of the rough endoplasmic reticulum often appears dense due the presence of secreted protein. Occasionally, the reticulum is seen to be in continuity with the two membranes forming the nuclear envelope, although, as indicated above, this is rarely seen in micrographs. The rough

and smooth endoplasmic reticula (Figure 19-65) are more commonly seen to be in direct continuity. In epithelial cells, the rough endoplasmic reticulum is usually concentrated in parallel cisternae at the base of the cell in a juxtanuclear position. Rough endoplasmic reticulum may be dilated or may contain dense material to reflect the storage of proteinaceous material in its lumen (Figures 19-66 and 67).

The rough endoplasmic reticulum segregates secreted material from the cytosol and is the site of synthesis of secretory proteins destined for export from the cell. The bound ribosomes congregate with both messenger RNA and specific transfer RNAs to produce a polypeptide chain that is extruded into the lumen of the rough endoplasmic reticulum and that eventually is cleaved off into its lumen.

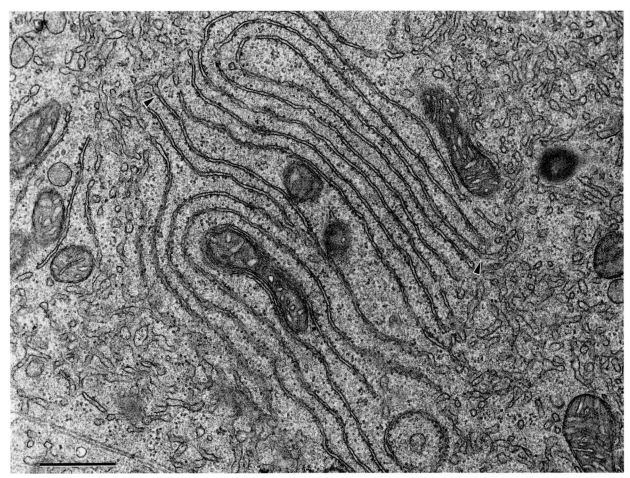

Figure 19-65 Ribosomes sparsely populate the external surface of the stacked cisternae of rough endoplasmic reticulum. The cisternae show a relatively collapsed lumen indicating that they contain relatively little secretory material.

Communication of the rough endoplasmic reticulum with the smooth endoplasmic reticulum is indicated by the arrowheads. Bar = 0.5 μm.

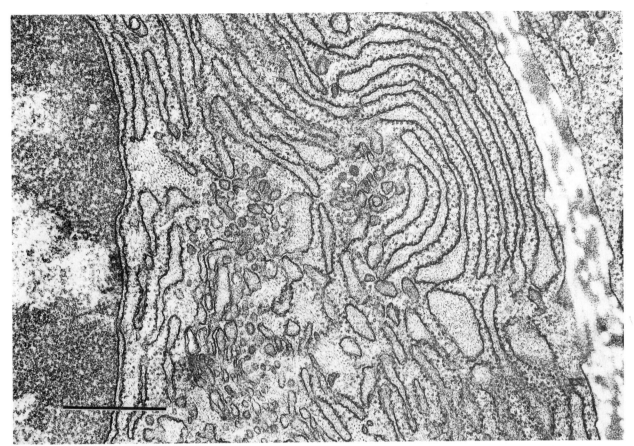

Figure 19-66   In contrast to that shown in Figure 19-65, this rough endoplasmic reticulum is heavily populated with ribosomes and its lumen is swollen. Bar = 0.5 μm.

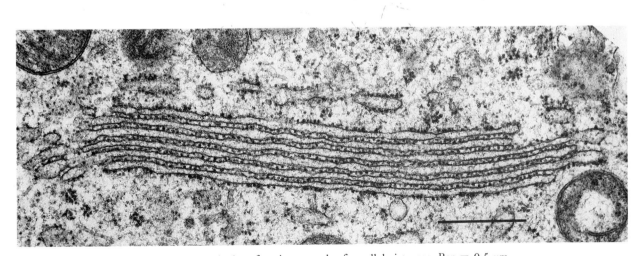

Figure 19-67   Rough endoplasmic reticulum forming a stack of parallel cisternae. Bar = 0.5 μm.

Smooth Endoplasmic Reticulum

There is a dangerous tendency to refer to all of the nonribosome bound membranous vesicles within the cytoplasm as smooth endoplasmic reticulum. In a strict sense, the smooth endoplasmic reticulum is defined not only by the absence of ribosomes from its membranes, but also by the enzymes that are associated with its membranes. There are other vesicular elements within the cytoplasm, such as transport vesicles, which do not contain ribosomes and cannot be classified as smooth endoplasmic reticulum.

In most instances, the smooth reticulum displays characteristic morphological features and, for practical purposes, can be distinguished as such. Because it is usually composed of anastomosing tubules of irregular shape and diameter, it appears in micrographs cut in various planes as a "contorted" network of smooth membranous elements (Figures 19-68 and 19-100; also see lipid section). Tubules of about 100 nm in diameter form a continuous membrane system that weaves in and out of the plane of section. Occasionally, in some cells, the smooth reticulum forms flattened cisternae in concentric swirls. These may surround lipid droplets (Figure 19-69).

Smooth endoplasmic reticulum performs a variety of functions. It contains enzymes necessary for cholesterol and steroid synthesis. This structure is involved in lipid metabolism and metabolism of toxic substances as well as the breakdown of glycogen.

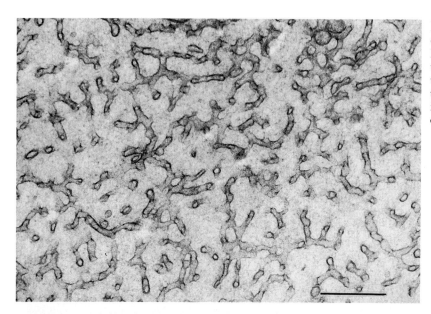

Figure 19-68 High magnification micrograph of smooth endoplasmic reticulum in a steroid secreting cell depicting the characteristic anastomotic nature of the reticulum. The continuity of the reticulum can be traced for some distance in this planar micrograph. (Rat Leydig cell.) Bar = 0.5 μm.

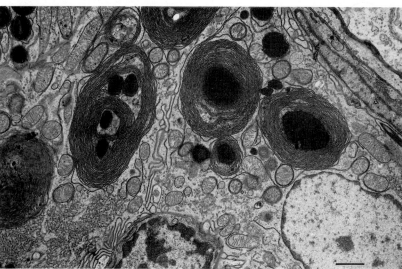

Figure 19-69 Swirls of smooth endoplasmic reticulum organized around lipid droplets predominate in this tissue, although major aggregations of the tubular reticulum are also found. (Mouse Leydig cell.) Bar = 1.0 μm.

The Golgi Apparatus

The Golgi apparatus appears as a group of stacked membrane cisternae (*Golgi stack* or *dictyosome*) that together form a slightly curved or U-shaped structure. Associated with the Golgi stacks are numerous small *Golgi vesicles*, some of which are smooth surfaced and others of which display a bristle or fuzzy coat.

The Golgi complex is generally located near the nucleus and, in epithelial cells, is positioned between the nucleus and the apex of the cell, i. e., *supranuclear location*. All synthetic cells have one or many Golgi stacks that are generally interconnected. Individual stacks contain a variable number of cisternae, but generally less than ten. The Golgi stack is polarized. The convex face is known as the *forming* or *cis face* and the concave face is known as the *maturing* or *trans face*. Near the trans face of many secretory cells are seen vacuoles containing dense material. They are slightly larger than the majority of those associated with the Golgi apparatus. These *condensing vacuoles* are considered as immature secretory granules. A variety of Golgi types are illustrated (Figures 19-70 through 19-74).

The Golgi is involved in modifying, packaging, and sorting proteins newly synthesized in the rough endoplasmic reticulum. Sugar residues are added and also modified in the Golgi. Golgi cisternae within a particular stack are not similar biochemically and, thus their functions are compartmentalized. There is considerable experimental evidence that the workings of the Golgi apparatus are very complex.

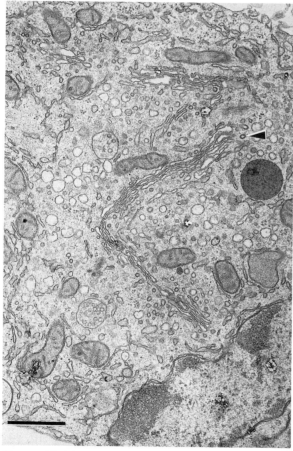

Figure 19-70 The typical Golgi complex is U- or V-shaped with a convex (cis) and concave (trans) face. The latter is frequently directed towards the nucleus (lower right). Numerous Golgi associated vesicles are seen. Some of these contain a bristle coat. Bar = 1.0 μm.

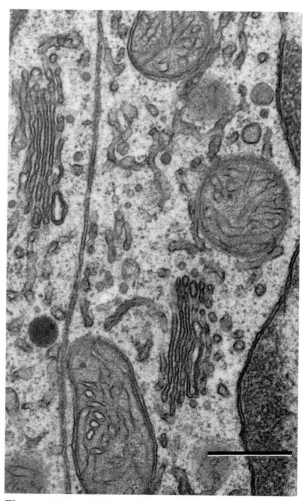

Figure 19-71 Simple stacks of the Golgi apparatus as seen in two adjoining cells. Bar = 0.5 μm.

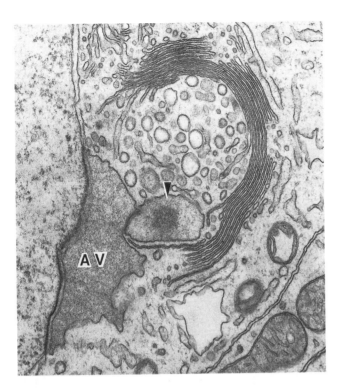

Figure 19-72 This Golgi apparatus from a spermatid shows a condensing vacuole with secretory product. The condensing vacuole is in the process of joining the acrosomal vesicle (AV), a structure which was produced previously by the Golgi. The secreted material will eventually produce the acrosome of the sperm.

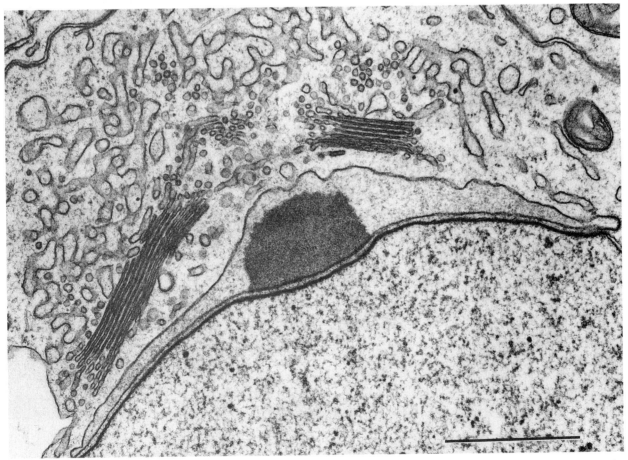

Figure 19-73   Golgi apparatus from a spermatid showing secretory material which has been deposited from the Golgi. Note the anastomotic smooth endoplasmic reticulum at the cis face of the Golgi. Bar = 1.0 μm.

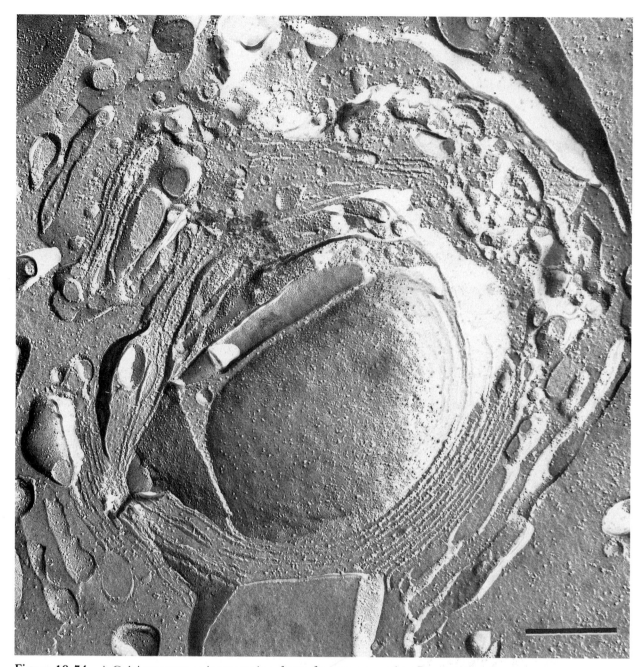

Figure 19-74   A Golgi apparatus as it appears in a freeze fracture preparation. Bar = 1.0 μm.

Secretory Products

Secretory granules are membrane-bound structures, almost always with dense appearing interiors. Unequivocal proof that a membrane-bound structure is a secretory granule, and not, for example, a lysosome destined for internal use, rests with the biochemical identification of its contents as a secretory product. The kinds of material being secreted are varied. In a practical sense, secretory products are abundant within cells and generally clustered, whereas lysosomes are rarely clustered. In epithelial cells, secretory granules are usually positioned between the Golgi apparatus and the apical surface of epithelial secretory cells. The content of secretory granules is most often homogeneous (Figures 19-75 and 19-76).

Secretory granules are the storage site for secretory material produced in the Golgi apparatus. Under the appropriate stimulus, they exit the cell by a process known as *exocytosis*. Fusion of the limiting membrane of the granule with the plasma membrane allows the contents of the vesicle to escape, whereas the bounding membrane of the secretory granule becomes part of the plasmalemma and is recycled.

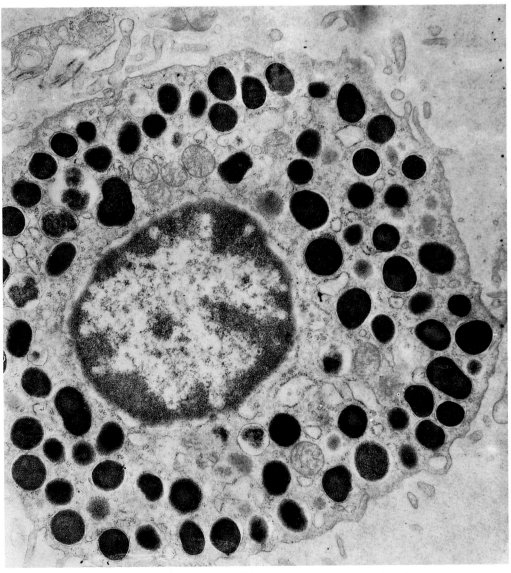

Figure 19-75   A mast cell containing numerous packets of dense secretory material (primarily histamine), each packet individually bound by a membrane.

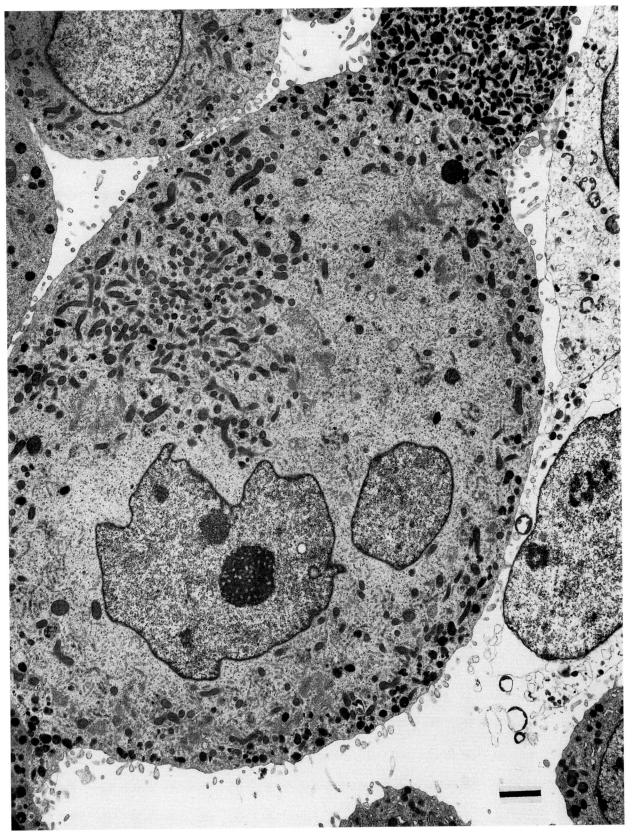

Figure 19-76 Melanin is the dense 'packaged' material within a melanin-secreting tumor cell grown in culture. Bar = 1.0 μm.

Centrioles

Centrioles are short (up to 0.5μm long), cylindrical (0.15μm) structures that are frequently seen in pairs called *diplosomes*. The centrioles of the diplosome frequently lie at right angles to one another. Centrioles are usually positioned near the nucleus close by the Golgi apparatus. In animal cells, this area is termed the *centrosome* (Figure 19-77). The wall of the centriole cylinder is best seen in a cross section of the centriole. It is composed of microtubular elements, or *sub-fibrils*, in fused triplets. Nine sets of triplets are arranged in a pinwheel fashion (Figure 19-78). The designation of individual sub-fibrils of a triplet is by letters; the innermost circular sub-fibril *A*, the middle *B*, and the outermost *C*. Dense material is often seen radiating outward from the triplets (*pericentriolar satellites*). In longitudinal section, the short, parallel, microtubular walls readily identify the structure as a centriole.

Centrioles function in organizing microtubules as, for example, in the development of the mitotic spindle or in the development of cilia or the flagellum. Prokaryotes and plant cells lack centrioles, but do have comparable centrosomal areas at their poles for regulation of cell division.

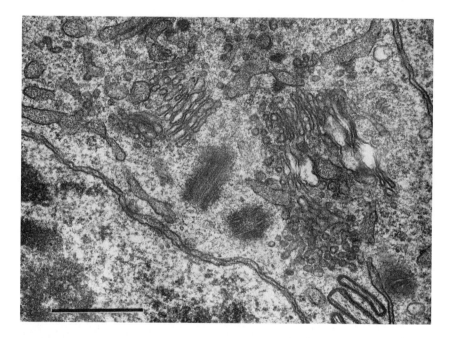

Figure 19-77 This pair of centrioles, or diplosome, is situated near the nucleus, sandwiched between it and the Golgi apparatus. This general area is known as the centrosome. The long axis of the centrioles lie at approximate right angles to one another and in neither centriole is its internal structure distinct, a result of how these centrioles were sectioned. Bar = 1.0 μm.

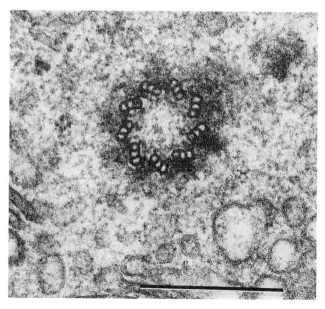

Figure 19-78 A centriole is sectioned transversely and displays the nine triplets, each containing three sub-fibrils. The dense material, or pericentriolar satellites, lying external to the centriole is closely associated with each triplet. Bar = 0.5 μm.

Cilia and Flagella

Cilia and flagella are structures that arise from centrioles and are specialized for the purpose of motility. Generally, there are fewer flagella (usually one) on a per cell basis than there are cilia. Flagella are almost always longer than cilia (Figure 19-26), but because they contain more accessory structures, they may be considered a more elaborate form of a cilium.

Cilia

About 0.2 µm across and 10 to 15 µm long (also see the description in the Cell Surface) cilia are projections from the cell surface usually into a fluid-filled cavity. The detailed structure of cilia is best examined in cross sectioned profiles of its shaft. Cilia are often referred to as 9+2 structures, meaning that they are composed of 9 fused peripheral doublets (*microtubule sub-fibrils*) and a central unfused pair of microtubules. The 9+2 arrangement is called an *axoneme* (Figure 19-79). Sub-fibril *A* of the cilium appears to be a complete microtubule, whereas sub-fibril *B* joins sub-fibril A and may be described as C-shaped. Cells bearing cilia are common in the respiratory epithelium; however, cilia may develop from many cell types in an organism not normally known to possess cilia. There is considerable sub-structure to the cilium that is not apparent upon routine examination.

At the base of the cilium is a *basal body*, which resembles a centriole that has been modified in some regions (Figure 19-80). There is considerable species variability in the structure of the basal body. In some species, the centriole undergoes little modification, whereas in others there are appendages (e.g., striated rootlets) that extend more deeply into the cytoplasm and act as anchors for the cilium.

Cilia beat in a wavelike motion. They move fluids or other structures, such as an ovum, over the surface of an epithelium.

Flagella

Flagella are seen in many phyla of the animal kingdom. They are commonly found in bacteria and also are seen in mammals. A prime example of a flagellum in mammals forms the tail of the spermatozoon. The core of the sperm flagellum resembles the 9+2 microtubule organization (*axoneme*) of the cilium (see Figure 19-79). Accessory structures are seen that characterize the flagellum regionally and divide it into various parts, the *middle, principal,* and *end pieces* (Figures 19-81 and 19-82).

The wave motion of the flagellum is distinctly different from that of the cilium. Flagella usually propel the entire cell through a fluid medium.

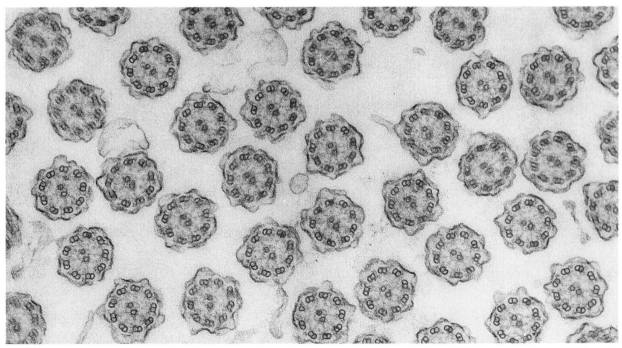

Figure 19-79 These ciliary axonemes from respiratory epithelium are sectioned transversely showing their 9+2 pattern of microtubules. (Micrograph courtesy of W. Dougherty.)

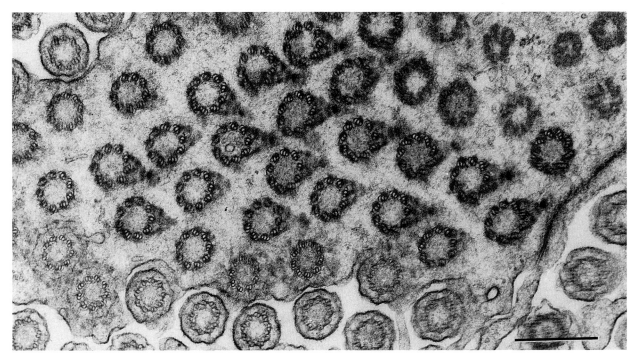

Figure 19-80 Basal bodies within the cytoplasm are cross sectioned. Each basal body resembles a centriole, but gives rise to a cilium. (Micrograph of respiratory epithelium courtesy of W. Dougherty.) Bar = 0.5 μm.

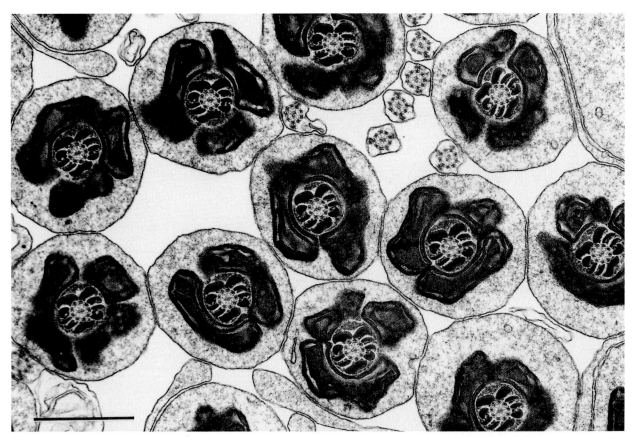

Figure 19-81 Several flagella of rat spermatids are illustrated. Sections are taken through both the middle piece of almost fully formed flagella and from the axonemal core of newly-forming flagella. (From Russell et al., 1990; Histological and Histopathological Evaluation of the Testis. Cache River Press, Clearwater, FL.) Bar = 1.0 μm.

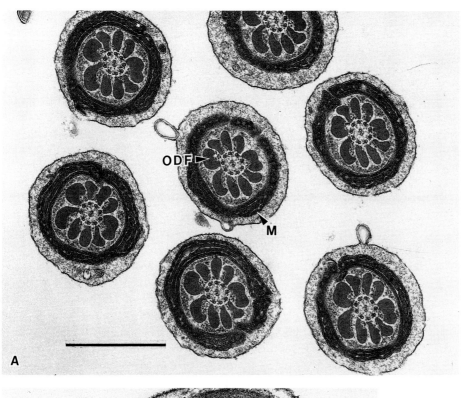

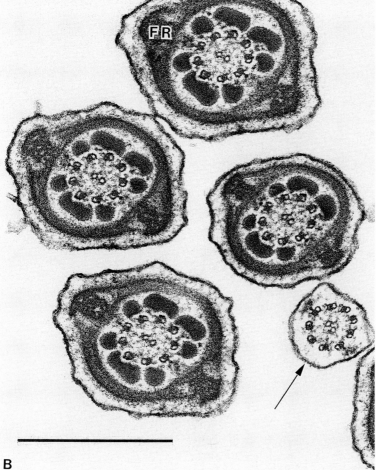

Figure 19-82 Flagella from the different regions of the rat sperm tail are shown. In the middle piece (19-83A), mitochondria (M) spiral around the axoneme. Dense bodies termed outer dense fibers (ODF) are also seen. Mitochondria are no longer seen in the principal piece (19-83B), but, instead, a structure known as the fibrous ring (FR) is seen. In the end piece (arrow), the axoneme is about the only internal structural feature noted. Bars = 1.0 μm.

The Lysosomal System

This assorted grouping of structures is related to the internalization and degradation of internalized materials.

Lysosomes

Bounded by a single membrane, lysosomes are strictly defined by the presence of hydrolytic enzymes within a bounding membrane rather than by their structural characteristics. Technically, a lysosome and a secretion granule cannot be distinguished by structure alone. Nevertheless, lysosomes can usually be identified within the cytoplasm as distinct entities. *Primary lysosomes* are generally newly formed, small and, having digested no material, they demonstrate a more or less homogeneous, but granular interior. Usually, there are some similar-appearing lysosomes, only containing digested material, that provide a clue that may be used to distinguish lysosomes from secretory granules (the latter generally being more homogeneous as a class).

In addition, lysosomes are frequently sparsely scattered throughout the cell, whereas secretory granules lie near the Golgi or in an epithelial cell clustered in a supranuclear position (Figure 19-83).

Secondary lysosomes are generally larger than primary lysosomes and their contents are not homogeneous, as evidenced by the presence of remnants of ingested material. This morphological criteria is distinguishing in thin sections (Figure 19-84), although the functional differentiation of primary from secondary lysosomes must take into account not only the contents of sectioned part of the lysosome, but also the entire lysosome (Figure 19-85). Frequently, the two types of lysosomes are intermingled allowing one to see the spectrum of lysosomal types (Figure 19-86).

Lysosomes function to break down substances within a cell (autophagy) or substances that have been taken in from outside the cell (heterophagy). They are important in maintenance of body defenses against outside organisms. In addition, lysosomes may cause death of the cell (autolysis) by emptying their contents into the cytoplasm.

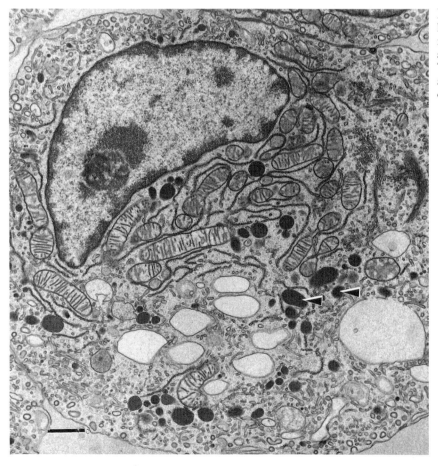

Figure 19-83 Lysosomes (arrowhead) within a macrophage are homogeneous appearing structures and are scattered throughout this cell. Their bounding membrane is not resolved in this low magnification micrograph. Bar = 1.0 μm.

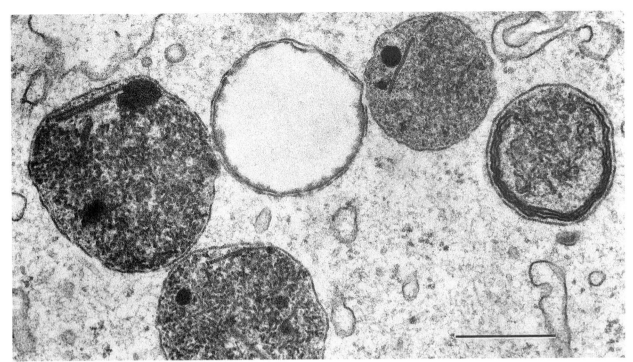

Figure 19-84 At high magnification, these secondary lysosomes contain a coarse granular and/or filamentous matrix in addition to densely stained material and membranous elements. A membrane bounds each lysosome. Bar = 0.5 μm.

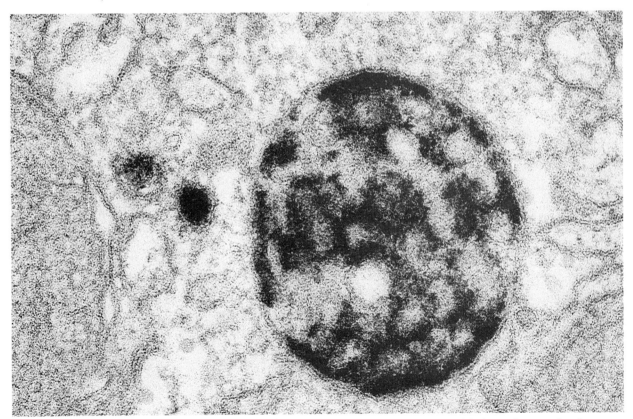

Figure 19-85 Two distinct lysosomal types are shown in this acid phosphatase preparation. The smaller is homogeneous and can be termed a primary lysosome. The larger is heterogeneous and designated a secondary lysosome. (Micrograph from Alberts et al. 1989, courtesy of D. Friend and used with permission of Garland Pub. Co.)

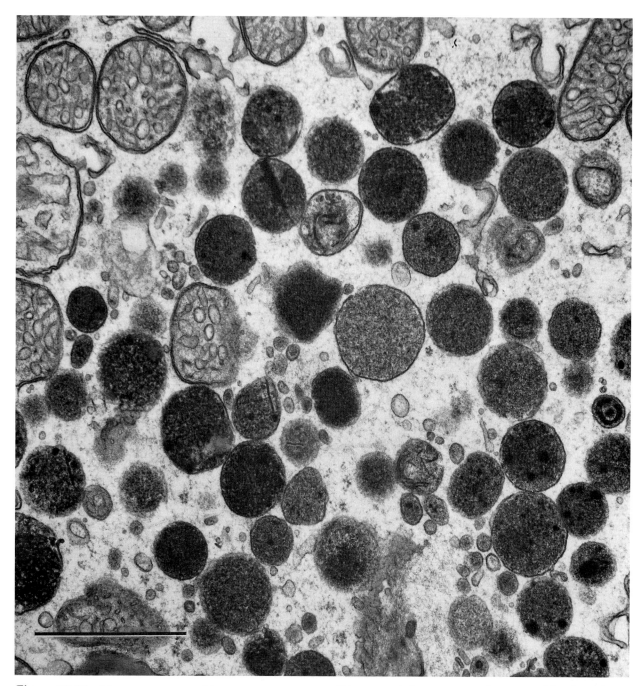

Figure 19-86 Numerous aggregated lysosomes are seen, some of which are relatively small and homogeneous and, in thin sections, could be classified as primary lysosomes. Others contain heterogeneous dense material and membrane and as such should be termed secondary lysosomes. Bar = 0.5 μm.

Multivesicular Bodies

The *multivesicular body* is bounded by a single membrane and contains numerous smaller membrane-bounded vesicles that are generally rounded and of relatively uniform size (Figure 19-87 through 19-89).

Digestive vacuoles are large structures that contain a large volume of ingested material. Generally, the material within the vacuole is partially digested and extremely heterogeneous (Figure 19-90).

Autophagic vacuoles form as the result of self-ingestion of constituents of the cell (Figure 19-91).

Heterophagic vacuoles contain other cells or material from other cells (Figure 19-92).

Residual bodies contain the remnants of undigested material. A form of residual body is the *lipofuscin granule*, which contains pigmented material (*lipochrome pigment*). Usually, a portion of the lipofuscin granule has material that appears like lipid (Figure 19-93).

Lipofuscin is regarded as undigested material left over from lysosomal activity and is seen in tissues as the result of the aging process.

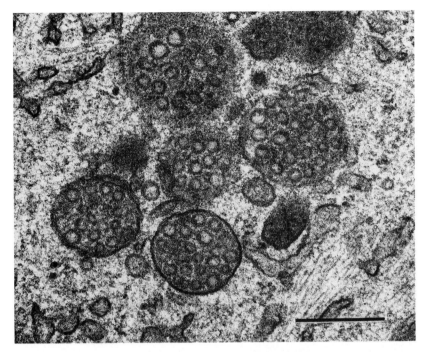

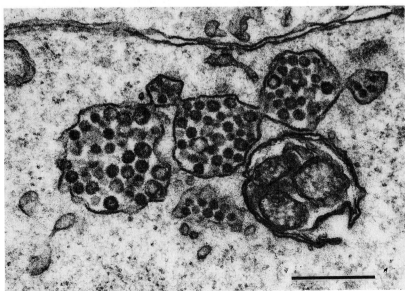

Figure 19-87 and **19-88** Multivesicular bodies are identified by the multiple small vesicles enclosed by a single membrane. Bars = 0.5 μm.

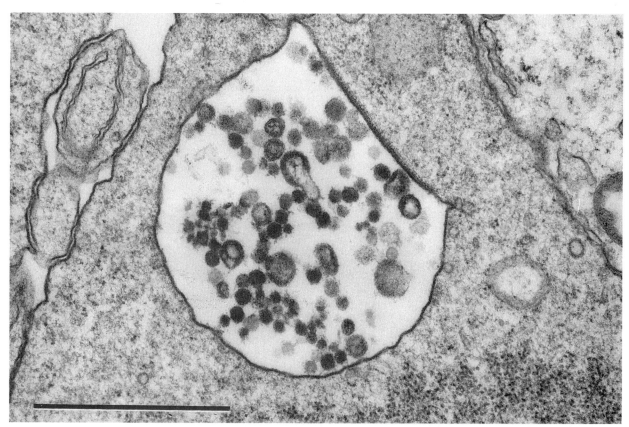

Figure 19-89   This large multivesicular body is bounded by a single membrane, part of which shows a fuzzy coat. Numerous small vesicles are seen within its interior. Bar = 0.5 μm.

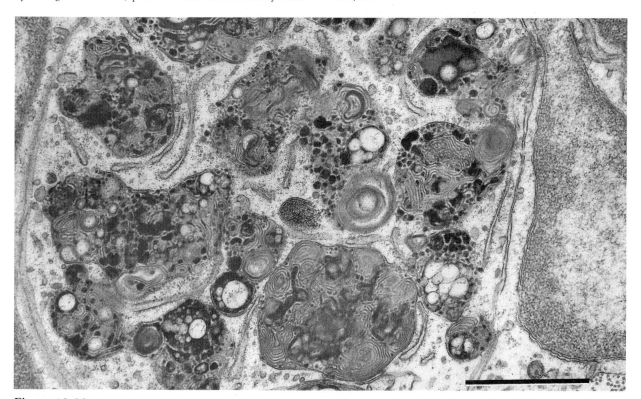

Figure 19-90   Several membrane bounded digestive vacuoles are seen within the cell. These comprise a large volume of this cell. (Micrograph courtesy of R. Sprando.) Bar = 1.0 μm.

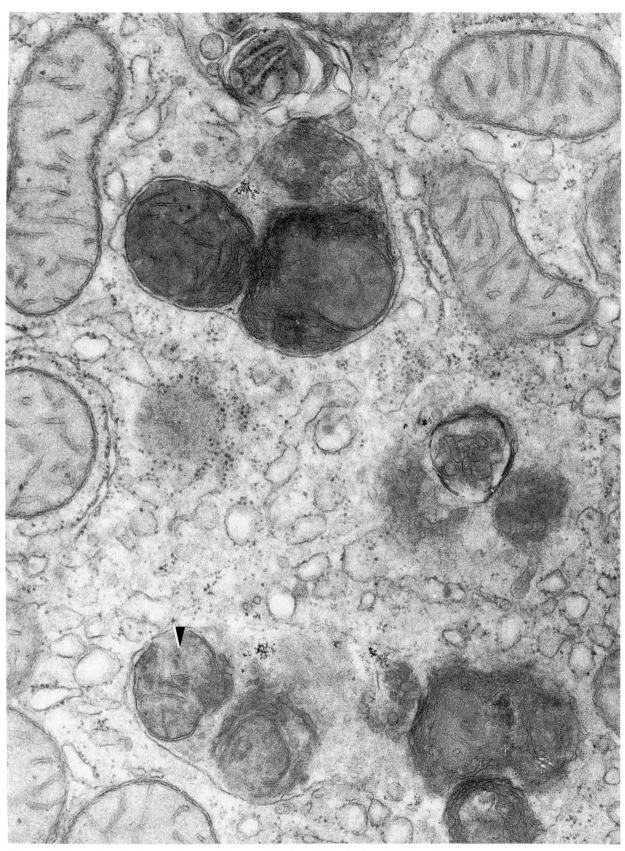

Figure 19-91   Autophagic vacuole containing partially degraded mitochondria from the same cell. (Micrograph courtesy of W. Dougherty.)

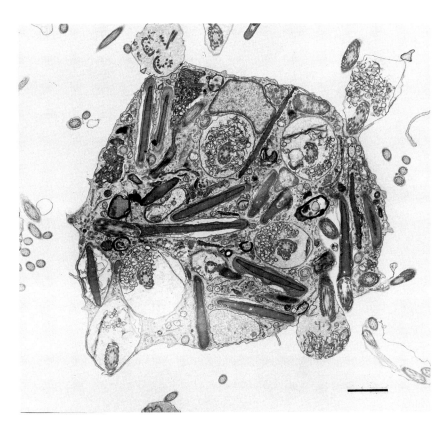

Figure 19-92 This macrophage in the reproductive tract of a rabbit has engulfed and walled off several sperm within heterophagic vacuoles. Bar = 2.0 μm.

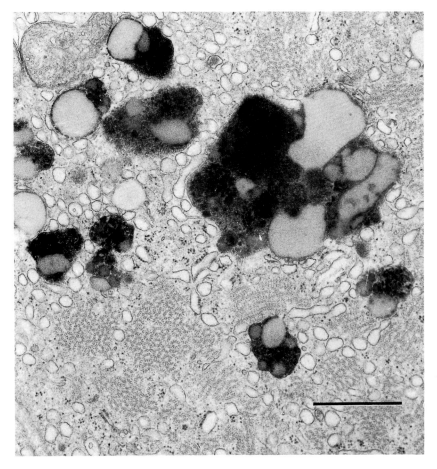

Figure 19-93 Lipofuscin granules, like these shown, are irregularly shaped and bounded by a membrane. Their contents are extremely heterogeneous, usually containing dense lipochrome pigment and sometimes less dense lipid droplets. (Human testis.)

Microbodies

Peroxisomes are one form of microbody. Like the lysosome, the peroxisome is defined by its constituent enzymes (catalase and oxidases) and is positively identified only by the localization of these enzymes. These membrane-bounded structures resemble lysosomes, but usually possess a less dense matrix than do lysosomes (Figure 19-94).

In some cells, peroxisomes contain crystalloids, called *nucleoids*, which identify them as peroxisomes (Figure 19-95).

The function of peroxisomes in animal cells is not clear. They use oxygen and may participate as detoxifying agents. Their role is clearer in plant tissues where they are involved in photorespiration. They serve as enzymes that are involved in oxidation of the side products of cellular respiration. *Glyoxysomes* appear similar to peroxisomes and are another form of microbody in plants. They are seen almost exclusively in germinating seeds and have a structure resembling the peroxisome. Glyoxysomes function in the conversion of fatty acids into sugars.

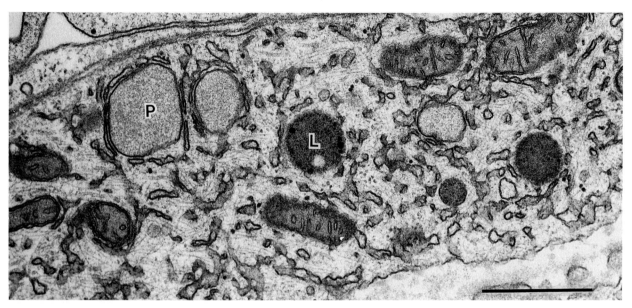

Figure 19-94   Peroxisomes and lysosomes are illustrated. The peroxisome (P) displays a finely granular matrix whereas the lysosome (L) is more densely granular and contains ingested material. (Rat Leydig cell.) Bar = 1.0 μm.

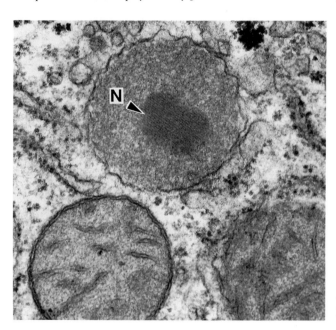

Figure 19-95   A nucleoid (N) is located centrally within this peroxisome. (Micrograph courtesy of W. Dougherty.)

Annulate Lamellae

This infrequently found membranous complex appears as stacked cisternae with pore complexes that appear much like pores of the nuclear envelope. The cisternae of annulate lamellae may be continuous with the smooth and/or rough endoplasmic reticulum (Figure 19-96).

The function of annulate lamellae is not known. Annulate lamellae are frequently seen in rapidly dividing cells and germ cells. Whether or not this membranous complex arises from the nuclear envelope is disputed.

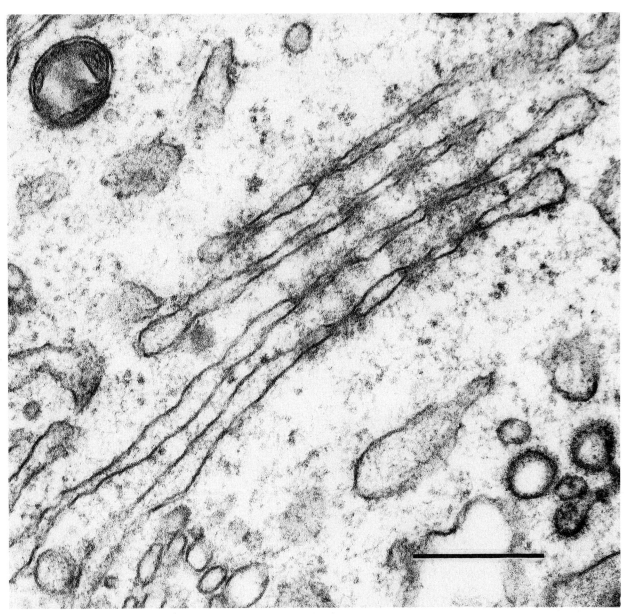

Figure 19-96 Annulate lamellae. Four cisternae of annulate lamellae are seen with their characteristic pore complexes. At one end they are continuous with the smooth endoplasmic reticulum and at the other end the rough reticulum. (From a germ cell in the rat testis.)

Cell Inclusions

Glycogen

Glycogen may be present as isolated granules or in aggregations. The isolated glycogen particle (*beta particle*) is usually about 30 nm in diameter and is often not well preserved, if at all, in most electron micrographs. When well preserved, individual par- ticles are rounded and have discrete borders in con- trast to the slightly irregular shape of the ribosome and its indistinct borders (Figure 19-97). Beta par- ticles may form aggregations within the cell. Rosette aggregations of glycogen are called *alpha particles* (Figures 19-98 and 19-99).

Glycogen is a storage form of carbohydrate pri- marily made of glucose subunits. Glycogen is com- monly used as an energy source.

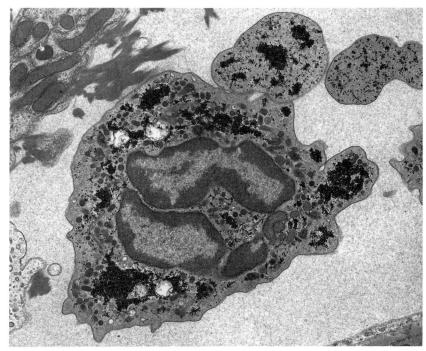

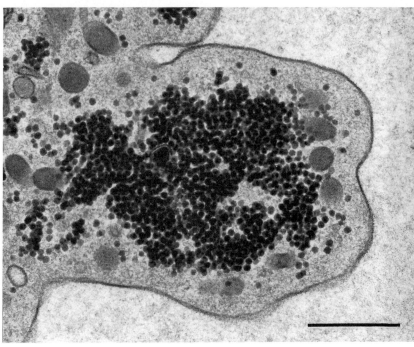

Figure 19-97 This leucocyte shows glycogen deposits throughout the cell; however, dense regional ac- cumulations of particles are seen. The figure below is a higher magnification of one of the cell processes and shows the accumulations of glycogen. Bar in lower figure = 0.5 μm.

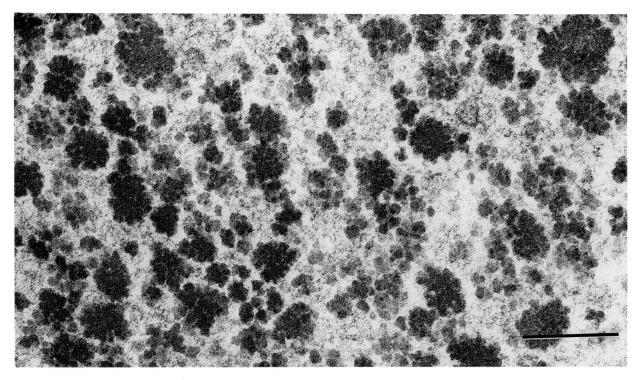

Figure 19-98   Aggregations of glycogen. Bar = 0.25 µm.

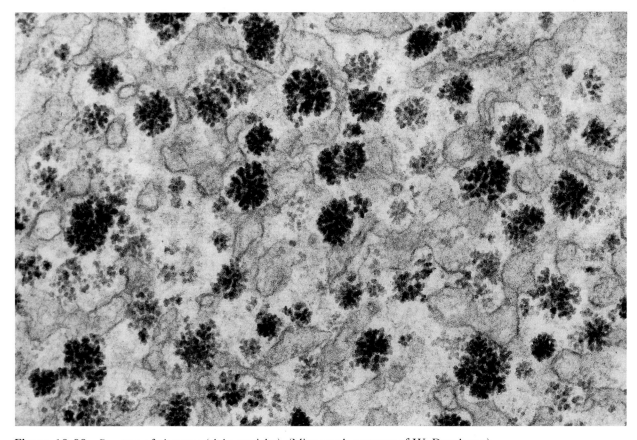

Figure 19-99   Rosettes of glycogen (alpha particles). (Micrograph courtesy of W. Dougherty.)

Lipid

Lipids show great morphological heterogeneity within cells. The size of individual droplets is highly variable, generally less than 1 μm in diameter, but in some instances (fat) exceeding 100 μm. Lipid droplets have no bounding membrane and appear to be held together by their hydrophobic interaction with the aqueous environment. Their lack of a membrane is a distinguishing feature, but it is sometimes difficult to discern whether or not a membrane is present at low magnification, especially if the lipid droplet has a constituent that stains more densely at its periphery and, thus, mimics a membrane (Figure 19-100).

Lipid droplets are usually rounded, but may appear in a variety of configurations. The contents of lipid droplets may be leached out by the dehydration process, especially if osmium penetration is poor. Moreover, the kind of lipid determines its staining properties with osmium. Thus, lipid may appear very electron dense or may stain hardly at all. Variations in lipid droplet form and configuration are illustrated in Figures 19-101 and 19-102.

The content of lipid appears less granular or has a more homogeneous texture than other dense staining structures within the cytoplasm, such as lysosomes or secretion granules. If sectioning artifacts such as chatter or knife marks are present, they are very prominently displayed within lipid droplets, whereas they may not be noticed elsewhere in the section. These features may sometimes be used to identify lipid from other organelles (Figure 19-103).

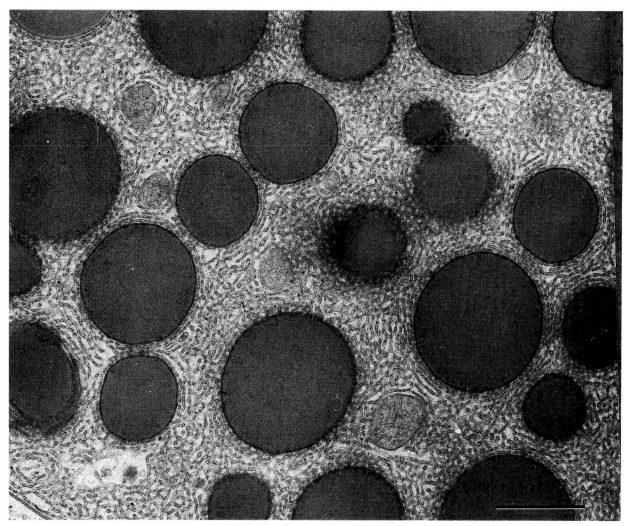

Figure 19-100 These lipid droplets are found among smooth endoplasmic reticulum. The interiors of the lipid droplets are finely granular, giving them a very homogeneous appearance. At its periphery, the lipid droplet has a denser staining band that might be easily mistaken for a membrane, except no tri-laminar structure is apparent. Bar = 1.0 μm.

Extracellular Material

Collagen

Collagen is the most abundant protein in the body. The Type IV collagen *fibril* has a characteristic appearance, being of variable length (10 to 250 nm in diameter) and demonstrating major cross banding at intervals of about 67 nm (Figure 19-106). Many col-

lagen fibrils form a bundle, visible in the light microscope as a collagen *fiber*. Collagen is also organized into sheets of fibrils that lie over other sheets, approximately perpendicular to the first layer (Figure 19-107).

The collagens are a family of proteins secreted by fibroblasts that add tensile strength to tissues and organs and hold them together.

Figure 19-106 Collagen fibrils at high magnification. Note the cross banding of individual fibrils. (Micrograph courtesy of W. Dougherty.)

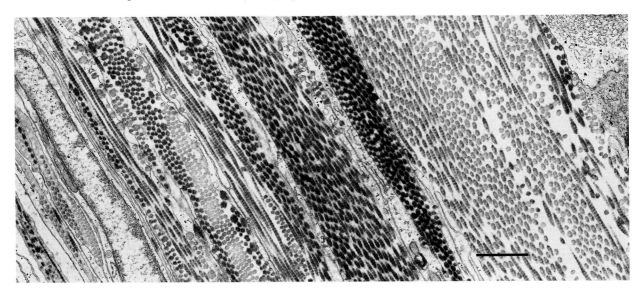

Figure 19-107 Sheets of collagen are shown in which the orientation of the fibrils is at approximate right angles to the adjacent sheet. Flattened cellular elements are occasionally interposed. (Capsule of the rat testis.) Bar = 1.0 μm.

Basal Lamina

At the base of epithelial cells, the basal lamina is a continuous sheet (about 150 nm thick) of extracellular material that is formed primarily by the epithelial cells and separates these cells from the underlying connective tissue. The basal lamina generally follows the undulating contours of the epithelial cell. At low and medium magnifications, it has a homogeneous appearance, but at high magnification, a fine fibrillar appearance is evident (Figure 19-108).

The function of the basal lamina is primarily structural, forming a framework on which epithelial cells rest. The basal lamina may act as a partial filter or participate in tissue restructuring. Its composition and associations with other connective tissue elements is complex.

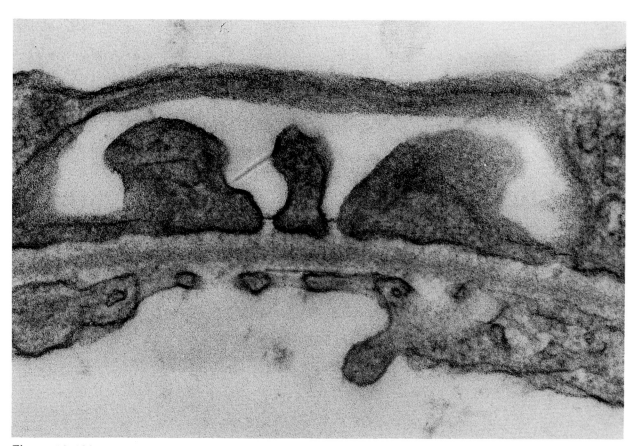

Figure 19-108 Basal lamina of the kidney glomerulus. This acellular structure separates the endothelium of the capillary from the podocytes of the glomerulus. (Micrograph courtesy of W. Dougherty.)

Matrix of Bone and Cartilage

Osteocytes (bone cells) and chondrocytes (cartilage cells) are present in a great volume of extracellular matrix. Bone is very hard and its matrix must be decalcified before it can be thin sectioned. The matrix of both cartilage and decalcified bone is complex, but may be recognized by the numerous collagenous (nonstriated in cartilage) fibrils running in various planes. In bone the fibrils are collagenous, being organized in layers or *lamellae* that run at right angles to each other. In cartilage the fibrils are less well organized than in bone (Figures 19-109 and 19-110) and of a different type (Figure 19-111).

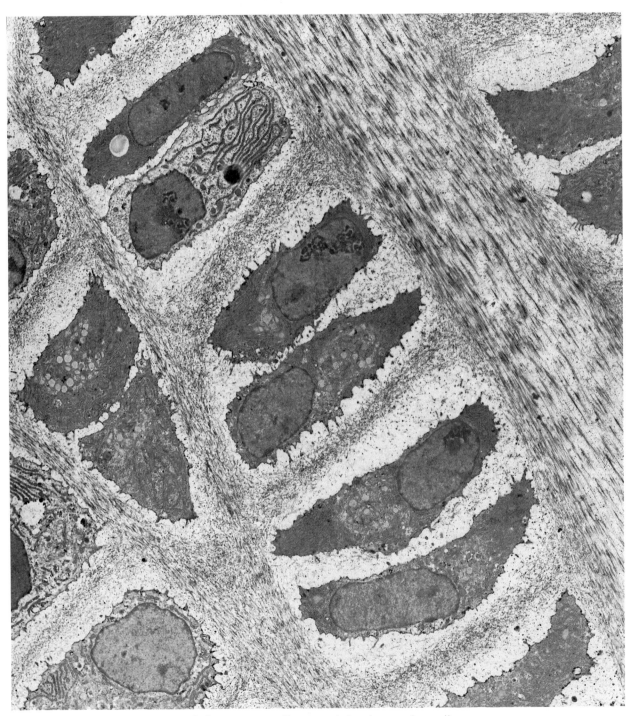

Figure 19-109 Extracellular matrix is seen surrounding several chondrocytes in cartilage.

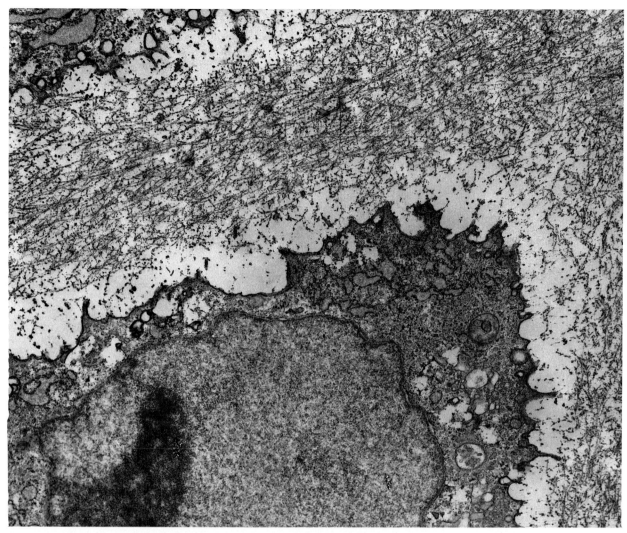

Figure 19-110   The relationship of the chondrocyte to the matrix is shown. Collagenous fibers of cartilage. (Pig cartilage.)

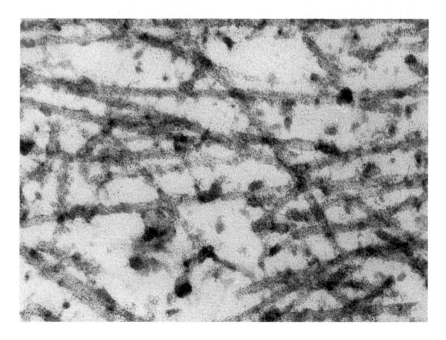

Figure 19-111   The collagenous fibrils of cartilage are of a type which are not typically striated.

Special Features of Plant Tissues

Plant cells are similar, in many ways, to animal cells, but several structural variations are worthy of mention.

Chloroplasts

The chloroplast is bounded by a double membrane. The majority of the internal aspect of the chloroplast is termed the *stroma*. Within the stroma are closed sacs called *thylakoids*, which do not connect to the bounding membrane of the chloroplast, but which do interconnect to form a space called the *thylakoid space*. Thylakoids are usually arranged in parallel stacks called *grana* (Figures 19-112 and 19-113).

The chloroplast is an essential energy-producing component of plants that functions in the process of *photosynthesis*. Besides generation of ATP, choloroplasts use this energy source to convert six CO_2 molecules into one glucose molecule. Often, the glucose is polymerized and stored as starch grains in the chloroplast. Mitochondria are also present in plant cells and may sometimes be mistaken for chloroplasts. However, mitochondria are smaller and their cristae are connected to the inner mitochondrial membrane.

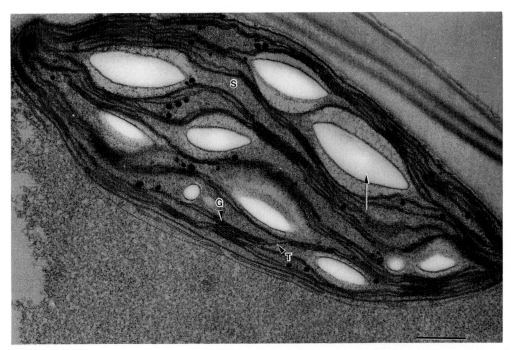

Figure 19-112 A chloroplast from the plant, *Takakia*. Indicated are the thylakoid (T), grana (G), and stroma (S). Accumulations of starch (arrow) are evident. (Micrograph courtesy of B. Stotler.) Bar = 0.5 μm.

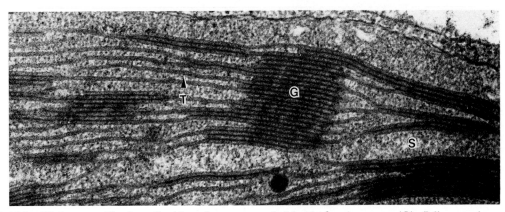

Figure 19-113 Higher magnification of a chloroplast revealing the detail of the stroma (S), thylakoids (T), and the thylakoids forming grana (G). (Micrograph courtesy of M. Gillott.)

Vacuoles

The plant vacuole is bounded by a single membrane called a *tonoplast*. Vacuoles are generally large structures, sometimes occupying over half of the cell volume, and in many plant cells they are so large as to cause the cytoplasm to be confined to a narrow area at the periphery of the cell just along the cell wall (Figure 19-114).

Vacuoles are storage elements largely for water, but also for waste. As the plant cell grows, the vacuole is the primary structure that is enlarged.

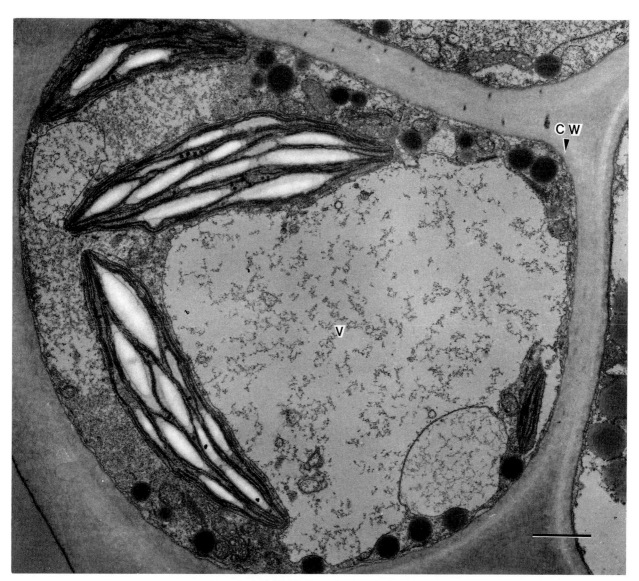

Figure 19-114 A plant cell containing a large vacuole (V) and a thick cell wall (CW). Chloroplasts are also evident. (Micrograph of *Takakia* courtesy of B. Stotler.) Bar = 1.0 µm.

The Cell Wall

Each plant cell has an external surrounding wall that appears amorphous in structure (Figure 19-115). The thickness of the wall may vary from a fraction of a micron to many microns. Small cell-to-cell channels called *plasmodesmata* pierce the cell wall between adjoining cells to provide areas of cell-to-cell communication between many adjoining plant cells (Figure 19-115).

Cell walls are rigid structures that do not allow cell shape changes, but do act as restraining structures as cells osmotically swell up against them (*tur-gor*). Being semipermeable structures, cell walls passively regulate many of the substances that reach plant cells and also may form functional fluid channels within plants. They are composed primarily of cellulose synthesized by the cell, which is exported to the outside of the cytoplasmic membrane. During the process of cell division, the separation of the nucleus and the cell proper (by means of a cross wall) may occur so closely in time as to appear to be one event. However, mitosis ends just before the final dividing cross wall is laid down by a specialized spindle structure, the *phragmoplast* (Figure 19-116).

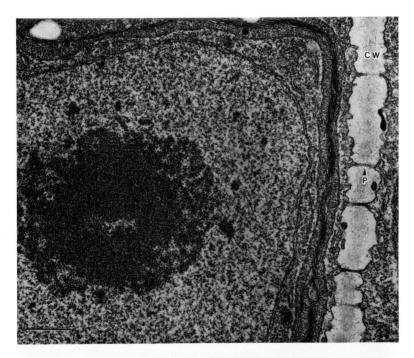

Figure 19-115 Two plant cells are depicted. In one cell, the nucleus, containing a nucleolus, occupies most of the figure. At the right of the figure the cell wall (CW) is seen. Plasmodesmata (P) are seen penetrating the cell wall to join the two cells. (Micrograph of *Takakia* courtesy of B. Stotler.) Bar = 0.5 μm.

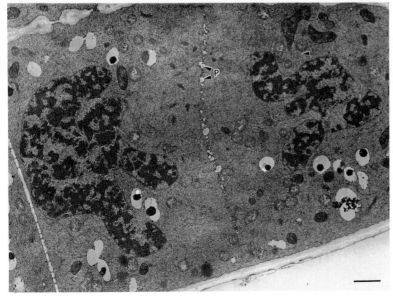

Figure 19-116 *Allium* cell completing mitosis. The phragmoplast (P) appears as a series of vesicles that will eventually coalesce to separate the two daughter cells. (Micrograph courtesy of D. Molsen and L. Hanzely.) Bar = 2.0 μm.

Bacteria

Bacterial cells are composed of *genetic material (DNA), ribosomes,* a *cell membrane* and a *cell wall* (except *mycoplasmas,* which are similar to bacteria but do not have a cell wall). A diagrammatic representation of a bacterium and its various components is shown in Figure 19-117. A typical thin section view is also shown (Figure 19-118).

Bacteria lack a nuclear envelope and for this reason are considered prokaryotes in contrast to eukaryotes, which have their nuclear material packaged by a nuclear envelope. Some bacteria have a flagellum and thus are motile. The typical bacterium ranges from a fraction of a micron to a few microns long (range 0.1 to 5 μm). The shape (spherical, rod shaped, or spiral) and organization (sheets, chains, or clusters) of groups of bacteria is diverse.

Bacteria may be divided into two major types: gram positive or gram negative (indicated in Figure 19-118). This classification is based on a staining reaction wherein gram positive cells retain the dye, crystal violet, while gram negative cells lose the dye upon decolorization with ethanol. Retention of the dye is based on a thicker gram positive cell wall that becomes impervious upon treatment with an iodine reagent used just before the decolorization step. A gram positive bacterial cell is shown in Figure 19-119. Some bacteria secrete extracellular material (Figure 19-120).

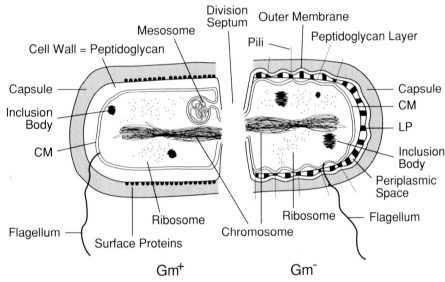

Figure 19-117 A diagrammatic representation of a bacterium and its various components. (Modified from Joklik et al, from Zinsser Microbiology, used with permission of Appleton-Century Crofts.)

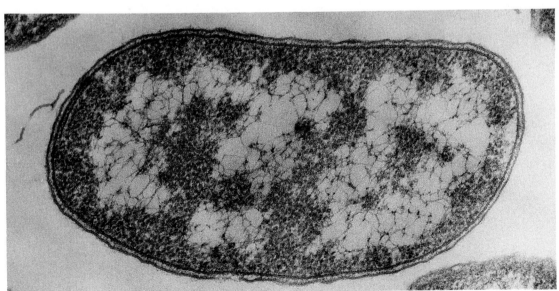

Figure 19-118 A thin section through a gram negative bacterium. (Micrograph courtesy of G. Brewer.)

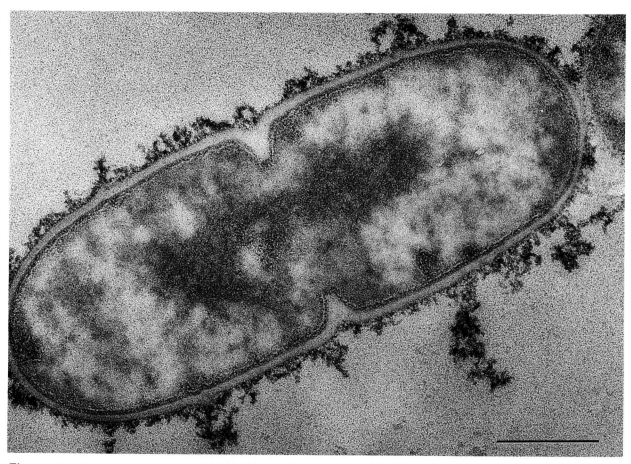

Figure 19-119   Ultra thin section of the gram positive bacterial cell, *Streptococcus mutans*, responsible for causing tooth decay. Bar = 0.25 μm.

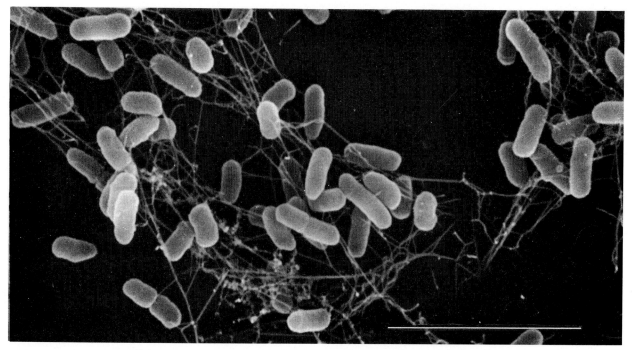

Figure 19-120   Scanning electron micrograph of pathogenic bacteria of the genus *Xanthomonas*. These bacteria secrete a stringy appearing extracellular substance or slime. Bar = 5.0 μm.

Algae, Fungi, Yeast, and Protozoa

Algae contain chloroplasts (see plants above, Figure 19-121). Algae may be unicellular or arranged in aggregates or in long branched or unbranched chains. Some divisions of algae (kelps and red algae) can become quite large with differentiated tissues approaching the complexity of higher plants.

Fungi are sometimes thought of as nongreen plants that lack chloroplasts and may consist of a plant body composed of threadlike structures called *hyphae*. In some classification schemes, fungi are viewed as a separate Kingdom. In fungi, food is absorbed from the external environment. The cell walls usually contain the polysaccharide chitin (also present in insect and crustacean exoskeletons), but usually not cellulose. Fungi are ubiquitous and may be found in extreme environments. Some are pathogenic for animals or plants but most are saprophytic organisms involved in recycling of biomass in nature (Figures 19-122, 19-123, 19-124).

Yeasts are one of the families of fungi that are widely distributed in substrates that are moist or contain sugar (Figures 19-125, 19-126, 19-127).

Protozoa are motile, unicellular organisms such as amoeba or paramecia that may be found in stagnant waters. They lack cell walls and chlorophyll. Some are quite simple cells, while other cells have rudimentary digestive apparati (mouth pores, gullets) that ingest food particles. Digestion takes place in cytoplasmic vacuoles and digested residue is excreted through an anal pore in the cell membrane. Most are actively motile by means of cilia, flagella, or ameboid movement using pseudopodia.

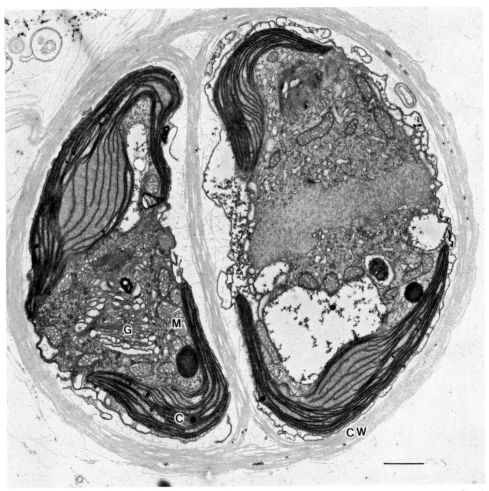

Figure 19-121 Algae. These algal cells show a cell wall (CW) of cellulosic scales as well as chloroplasts (C), mitochondria (M), and a Golgi apparatus (G). It is interesting to note that the scales originate from the cisternae of the Golgi apparatus and are deposited on the cell surface by exocytosis. Micrograph of *Pleurochrysis scherfelii*. (Courtesy of M. Brown, Jr.) Bar = 1.0 μm.

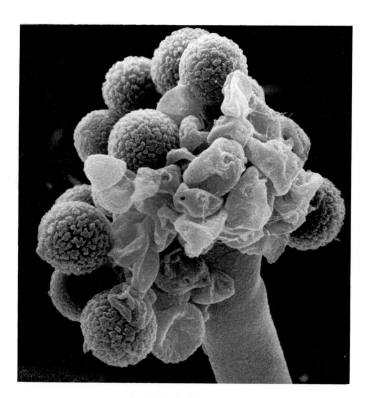

Figure 19-122   Scanning electron micrograph of an unidentified saprophytic fungus.

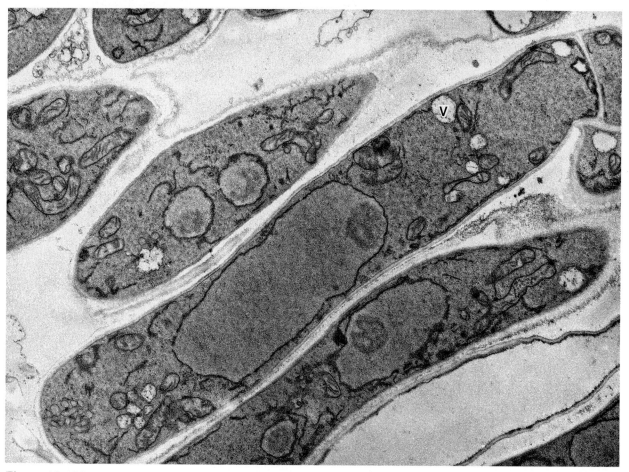

Figure 19-123   A longitudinal section through a diploid fungal basidium (*Schizophyllum commune*) shows the nucleus, nucleolus, vacuoles (V), mitochondria, and other organelles. (Micrograph courtesy of W. J. Sundberg.)

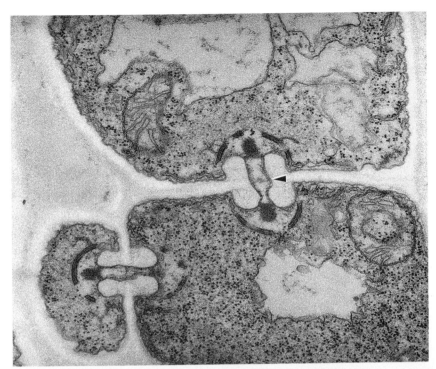

Figure 19-124 Transmission electron micrograph showing three fungal cells (*Schizophyllum commune*) connected by septal pores (arrowhead). Septal pores are larger and more structurally complex than the plasmodesmata described above and are generally seen in higher fungi. (Micrograph courtesy of W. J. Sundberg.)

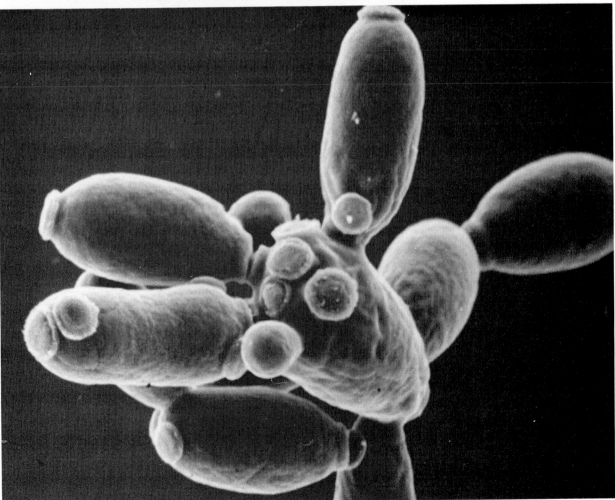

Figure 19-125   Scanning electron micrograph of a colony of *Candida albicans* yeast cells.

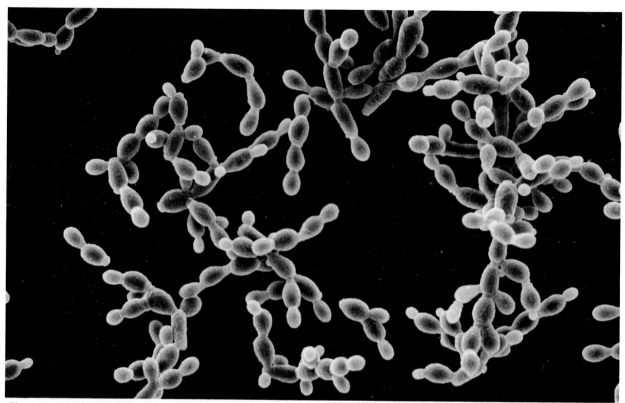

Figure 19-126   Scanning electron micrograph of common pathogenic yeast, *Candida albicans*.

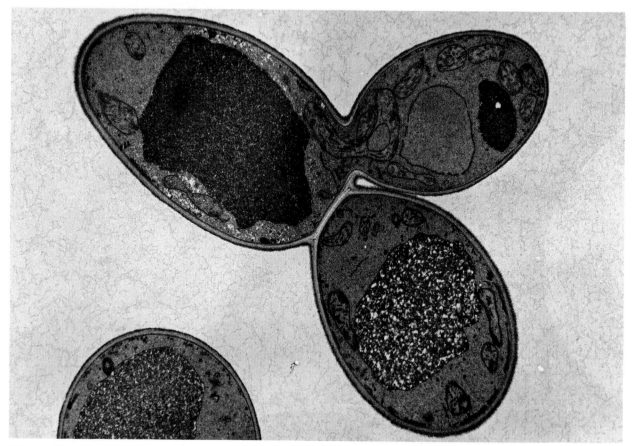

Figure 19-127   This thin section shows *Candida albicans* in the process of budding. (Fixed with potassium permanganate.)

Viruses

Electron microscopy is a requirement to visualize most viruses, since their diameter ranges from about 0.02 μm to 0.3 μm. The basic structure of viruses is relatively simple. They are composed of genetic material (DNA or RNA) covered by a protective coat. The variability of virus shape within this general plan is great. Often viruses take rounded or geometrical shapes (Figure 19-128).

The coat material of the virus, the *capsid*, is largely protein with some lipid. The form the coat may take is diverse. Some viruses may have a fuzzy appearing coat, whereas in others a lipid bilayer is visualized.

Electron microscopy is a powerful aid in the classification of viruses since it is possible to correctly identify the family of the virus based on a number of morphological features (size, symmetry of capsid, presence or absence of an envelope or unit membrane, surface projections, etc.). When combined with specific immune sera, the technique can give rapid and specific identification of the viruses to the species level. Very often, electron microscopy is the only technique available for the identification of newly discovered viruses such as the retroviruses causing AIDS and the virus responsible for hepatitis B.

Viruses gain entry into other cells by binding to highly specific cell *surface receptors*. After entrance and uncoating of the capsid, they multiply within the cell by using the cell's own anabolic metabolism. The *host cell* may be a bacterium, animal, or plant cell. Viruses may cause the death of the host cell when they are released from the cell (Figure 19-129).

The usual method to see detail within isolated viruses is by employing a negative staining technique (see Chapter 5; Figure 19-130).

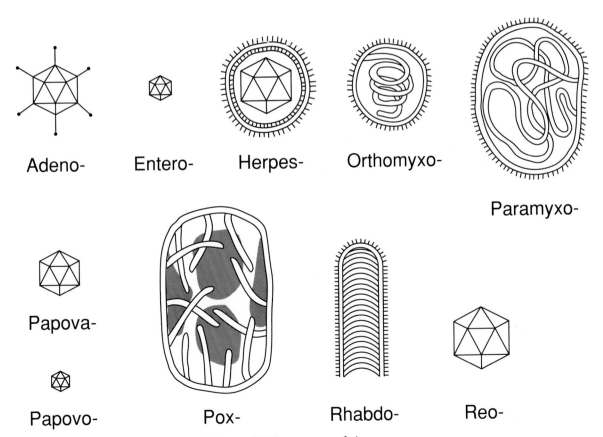

Adeno-   Entero-   Herpes-   Orthomyxo-

Paramyxo-

Papova-

Papovo-   Pox-   Rhabdo-   Reo-

Figure 19-128   Variability in shapes and forms of different types of viruses.

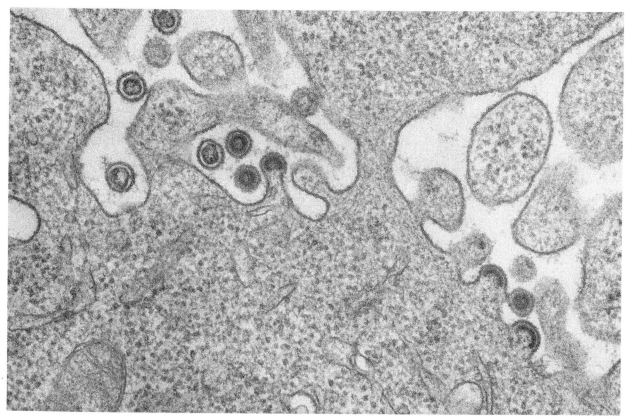

Figure 19-129 Budding retroviruses from a leukemia cell. (Micrograph courtesy of D. Friend.)

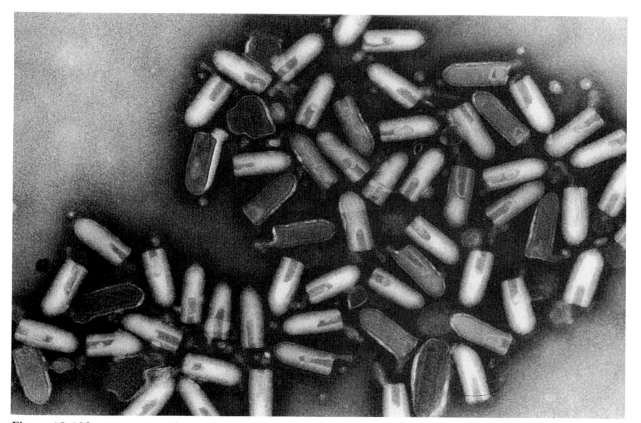

Figure 19-130 Aggregation of vesicular stomatitus viruses as seen in a negative stain preparation.

References

Atlases of Ultrastructure, Histology Texts, Cell Biology Texts

Alberts, B., D. Bray, J. Lewis, M. Raff, K. Roberts, and J. D. Watson. 1983. *Molecular biology of the cell*, 2nd ed. New York: Garland Pub. Co.

Cormack, D. H. 1987. *Ham's histology*, 9th ed. Philadelphia: Lippincott.

Fawcett, D. W. 1981. *The cell*, 2nd ed. Philadelphia: W. B. Saunders Co.

Fawcett, D. W. 1986. *A textbook of histology*, 11th ed. Philadelphia: W. B. Saunders Co.

Geneser, F. K. 1986. *Textbook of histology*, 1st ed. Philadelphia: Munksgard/Lea Febiger.

Kelly, D. E., R. L. Wood, and A. C. Enders. 1984. *Bailey's textbook of histology*. Baltimore: Williams and Wilkins.

Kessel, R. G., and R. H. Kardon. 1979. *Tissues and organs*. New York: W. H. Freeman.

Krstic, R. V. 1979. *Ultrastructure of the mammalian cell: an atlas*. Berlin: Springer-Verlag.

Lentz, T. L. 1971. *Cell fine structure. an atlas of drawings of whole-cell structure*. Philadelphia: W. B. Saunders Co.

Plattner, H. 1989. *Electron microscopy of subcellular dynamics*. Boca Raton, FL: CRC Press.

Porter, K. R., and M. A. Bonneville. 1973. *Fine structure of cells and tissues*. Philadelphia: Lea and Febiger.

Rhodin, J. A. G. 1974. *Histology: a text and atlas*. New York: Oxford University Press.

Ross, M. H., and L. J. Romrell. 1989. *Histology: a text and atlas*. Baltimore: Williams and Wilkins.

Shih, G., and R. Kessel. 1982. *Living images*. Boston: Jones Bartlett.

Weiss, L. 1988. *Histology: cell and tissue biology*, 6th ed. New York: Elsevier Biomedical.

Wheater, P. R., H. G. Burkeitt, and V. G. Daniels. 1987. *Functional histology*, 2nd ed. Edinburgh: Churchill Livingstone.

The Cell Surface

The Lipid Bilayer of the Plasmalemma

Branton, D. 1969. Membrane structure. *Annu Rev Plant Physiol* 20:209–38.

Bretcher, M. S. 1985. The molecules of the cell membrane. *Sci Am* 253:100–9.

Eisenberg, D. 1984. Three-dimensional structure of membranes and surface proteins. *Annu Rev Biochem* 53:595–623.

Hendler, R. W. 1971. Biological membrane ultrastructure. *Physiol Rev* 51:66–97.

Kleinfeld, A. 1987. Current views of membrane structure. *Curr Top Memb Transp* 29:1–27.

Singer, S. J., and G. L. Nicolson. 1972. The fluid mosaic model of the structure of cell membranes. *Science* 175:720–31.

Yeagle, P. 1987. *The membranes of cells*. Orlando, FL: Academic Press.

The Glycocalyx

Alberts, B., D. Bray, J. Lewis, M. Raff, K. Roberts, and J. D. Watson. 1989. *Molecular biology of the cell*, 2nd ed. New York: Garland Pub. Co. pp 298–300.

Rambourg, A., M. Neutra, and C. P. Leblond. 1966. Presence of a cell coat rich in carbohydrate at the surface of cells in rat. *Anat Rec* 154:41–52.

Cell Junctions

Dewey, M. M., and L. Barr. 1964. A study of the distribution of the nexus. *J Cell Biol* 23:553–60.

Farquhar, M. G., and G. E. Palade. 1963. Junctional complexes in various epithelia. *J Cell Biol* 17:375–412.

Garrod, D. R. 1986. Desmosomes, cell adhesion molecules and the adhesive properties of cell in tissues. *J Cell Sci* Suppl 4:221–37.

Kelly, D. E. 1966. The fine structure of desmosomes, hemidesmosomes and an adepidermal globular layer in developing newt epidermis. *J Cell Biol* 28:51–72.

Lane, N. J., H. Le B. Skaer, and L. S. Swales. 1977. Intercellular junctions in the nervous system of insects. *J Cell Sci* 26:175–99.

Pitts, J. D., and M. E. Finbow. 1986. The gap junction. *J Cell Sci* Suppl 4:239–66.

Staehelin, L. A. 1974. Structure and functions of intercellular junctions. *Int Rev Cytol* 39:191–283.

Staehelin, L. A., and B. E. Hull. 1978. Junctions between living cells. *Sci Am* 238:141–52.

Cell Surface Specializations

Abercrombie, M. 1980. The crawling movement of metazoan cells. *Proc R Soc Lond* (Biol) 207:129–47.

Fawcett, D. W. 1975. The mammalian spermatozoon. *Dev Biol* 44:394–436.

Goldstein, J. L., G. W. Anderson, and M. S. Brown. 1979. Coated pits, coated vesicles and receptor mediated endocytosis. A review. *Nature* 279:679–85.

Roth, Y. F., and K. R. Porter. 1964. Yolk protein uptake in the oocyte of the mosquito, *Aedes aegypti*. *J Cell Biol* 20:313–32.

Small, J. B., G. Rinnerthaler, and H. Hinssen. 1982. Organization of actin meshworks in cultured cells: the leading edge. *Cold Spring Harbor Symp Quant Biol* 46:599–611.

Steinman, R. M., I. S. Mellman, W. A. Muller, and Z. Cohn 1983. Endocytosis and recycling of plasma membrane. *J Cell Biol* 96:1–11.

Sullivan, A. L., A. Grasso, and L. R. Weintraub. 1976. Micropinocytosis of transferrin by developing red cells: an electron microscope study utilizing ferritin-conjugated transferrin and ferritin-conjugated antibody to transferrin. *Blood* 47:133–43.

The Cytoskeleton

Brinkley, B. R. 1990. Toward a structural and molecular definition of the kinetochore. *Cell Motility and the Cytoskeleton* 16:104–9.

Microtubules

Dustin, P. 1984. *Microtubules*, 2nd ed. Berlin: Springer-Verlag.

Inou'e, S. 1981. Cell division and the mitotic spindle. *J Cell Biol* 91:131s–47s.

Olmstead, J. B., and G. G. Borisy. 1973. Microtubules. *Annu Rev Biochem* 42:507–40.

Rieder, C. L. 1982. The formation, structure and composition of the mammalian kinetochore fiber. *Int Rev Cytol* 79:1–58.

Roberts, K., and J. S. Hyams. 1979. *Microtubules*. New York: Academic Press.

Stephens, R. E., and K. Y. Edds. 1976. Microtubules: Structure, chemistry and function. *Physiol Rev* 56:709–77.

Microfilaments

Boyles, J., L. Anderson, and P. Hutcherson. 1985. A new fixative for the preservation of actin filaments. *J Histochem Cytochem* 33:1116–28.

Byers, H. R., and K. Fujiwara. 1982. Stress fibers in cells *in situ*: immunofluorescence visualization with anti-actin and anti-myosin and anti-alpha actinin. *J Cell Biol* 93:804–11.

Maupin, P., and T. D. Pollard. 1983. Improved preservation and staining of HeLa cell actin filaments, clathrin-coated membranes, and other cytoplasmic structures by tannic acid-glutaraldehyde-saponin fixation. *J Cell Biol* 96:51–62.

Maupin-Szamier, P., and T. D. Pollard. 1978. Actin filament destruction by osmium tetroxide. *J Cell Biol* 77:837–52.

Mooseker, M. S., and L. G. Tilney. 1975. Organization of an actin-filament membrane complex. Filament polarity and membrane attachment in the microvilli of intestinal epithelial cells. *J Cell Biol* 67:725–43.

Weeds, A. 1982. Actin-binding proteins-regulators of cell architecture and motility. *Nature* 296:811–16.

Intermediate Filaments

Anderton, B. H. 1981. Intermediate filaments: a family of homologous structures. *J Muscle Res Cell Motil* 2:141–66.

Geiger, B. 1987. Intermediate filaments: looking for a function. *Nature* 329:392–3.

Hynes, R. O., and A. T. Destree. 1978. Ten nanometer filaments in normal and transformed cells. *Cell* 13:151–63.

Lazarides, E. 1980. Intermediate filaments as mechanical integrators of cellular space. *Nature* 283:249–56.

Traub, P. 1985. *Intermediate filaments: a review*. Berlin: Springer-Verlag.

Wang, E., D. Fischman, R. K. H. Liem, and T.-T. Sun. 1985. Intermediate filaments. *Ann N Y Acad Sci*, Vol. 455.

The Nucleus

The Nuclear Envelope

Franke, W. W., U. Scheer, G. Krohne, and E. D. Jarasch. 1981. The nuclear envelope and the architecture of the nuclear periphery. *J Cell Biol* 91:39s–50s.

Kessel, R. G. 1973. Structure, biochemistry and functions of the nuclear envelope and related cytomembranes. *Prog Surface Membrane Sci* 6:243–329.

Newport, J. W., and D. J. Forbes. 1987. The nucleus: structure, function, and dynamics. *Annu Rev Biochem* 56:535–65.

Pappas, G. D. 1956. The fine structure of the nuclear envelope of *Amoeba proteus*. *J Biophys Biochem Cytol* 2 (Suppl):431–4.

Weiner, J., D. Spiro, and W. R. Lowenstein. 1965. Ultrastructure and permeability of nuclear membranes. *J Cell Biol* 27:107–17.

Chromatin

Eissenberg, J. C., I. L. Cartwright, G. H. Themas, and S. D. Elgin. 1985. Selected topics in chromatin structure. *Annu Rev Genet* 19:485–536.

Ris, H. 1975. Chromosome structure as seen by electron microscopy. In *Structure and function of chromatin, Ciba Symposium* (No. 28), pp 7–28, Amsterdam, The Netherlands: Assoc. Sci. Pub.

Ris, H., and D. Kubai. 1970. Chromosome structure. *Annu Rev Genet* 4:263–94.

Stubblefield, E. 1973. The structure of mammalian chromosomes. *Int Rev Cytol* 1–60.

The Nucleolus

Bernhard, W., and N. Granboulan. 1968. Electron microscopy of the nucleolus in vertebrate cells. In *Ultrastructure in biological systems*, Vol. 3. A. J. Dalton and F. Haguenau, eds. New York: Academic Press, pp 81–149.

Hay, E. D. 1968. The structure and function of the nucleolus in developing cells. In *Ultrastructure in biological systems*, Vol. 3. A. J. Dalton and F. Haguenau, eds. New York: Academic Press, pp 2–79.

Jordan, E. G., and C. A. Cullis. 1982. *The nucleolus.* Cambridge: Cambridge University Press.

Sommerville, J. 1986. Nucleolar structure and ribosome biogenesis. *Trends Biochem Sci* 11:438–42.

Dividing Cells

Lloyd, D., R. K. Poole, and S. W. Edwards. 1982. *The cell division cycle.* New York: Academic Press.

Weber, J. E., and L. D. Russell. 1987. A study of intercellular bridges during spermatogenesis in the rat. *Am J Anat* 180:1–24.

The Synaptonemal Complex

Moses, M. J. 1968. Synaptonemal complex. *Annu Rev Genet* 2:363–412.

von Wettstein, D., S. W. Rasmussen, and P. B. Holm. 1984. The synaptonemal complex in genetic segregation. *Annu Rev Genet* 18:331–43.

Mitochondria

Attardi, G., and G. Schatz. 1988. Biogenesis of mitochondria. *Annu Rev Cell Biol* 4:289–333.

Ernster, L., and Z. Drahota. 1969. *Mitochondria, structure and function.* New York: Academic Press.

Tzagoloff, A. 1982. *Mitochondria.* New York: Plenum Press.

Protein Synthetic and Secretory Structures

Free Ribosomes

Lake, J. A. 1981. The ribosome. *Sci Am* 245:84–97.

Rich, A. 1963. Polyribosomes. *Sci Am* 209:44–53.

Spirin, A. S. 1986. *Ribosome structure and protein synthesis.* Menlo Park, CA: Benjamin-Cummings.

Wool, I. G. 1979. The structure and function of eukaryotic ribosomes. *Annu Rev Biochem* 48:719–54.

Rough Endoplasmic Reticulum

Blöbel, G., and B. Doberstein. 1975. Transfer of proteins across membranes. I. Presence of proteolytically processed and unprocessed nascent immunoglobulin light chains on membrane-bound ribosomes of murine myeloma. *J Cell Biol* 67:835–51.

De Pierre, J. W., and G. Dallner. 1975. Structural aspects of the membrane of the endoplasmic reticulum. *Biochem Biophys Acta* 415:411–72.

Hortsch, M., D. Avossa, and D. I. Meyer. 1986. Characterization of secretory protein translocation: ribosome-membrane interaction in endoplasmic reticulum. *J Cell Biol* 103:241–53.

Smooth Endoplasmic Reticulum

Emans, J. B., and A. L. Jones. 1968. Hypertrophy of liver cell smooth endoplasmic reticulum following progesterone administration. *J Histochem Cytochem* 16:561–71.

Higgins, J. A., and R. J. Barrnett. 1972. Studies on the biogenesis of smooth endoplasmic reticulum membranes in livers of phenobarbital treated rats. *J Cell Biol* 55:282–98.

Mori, H., and A. K. Christensen. 1980. Morphometric analysis of Leydig cells in the normal rat testis. *J Cell Biol* 84:340–54. (Smooth endoplasmic reticulum is related to steroid synthesis in this article.)

The Golgi Apparatus

Farquhar, M., and G. Palade. 1981. The Golgi apparatus (complex)-(1954–1981) from artifact to center stage. *J Cell Biol* 91:77s–103s.

Mollenhauer, H. H., and D. J. Morré. 1991. Perspectives on Golgi apparatus form and function. *J Elect Micros Techn* 17:2–14.

Pavelka, M. 1987. Functional morphology of the Golgi apparatus. *Adv Anat Embryol Cell Biol* 106:1–94.

Rambourg, A., and Y. Clermont. 1990. Three-dimensional electron microscopy: structure of the Golgi apparatus. European *J Cell Biol* 51:189–200.

Rothman, J. E. 1985. The compartmental organization of the Golgi apparatus. *Sci Am* 253:74–89.

Whaley, W. G. 1975. *The Golgi apparatus*. Heidelberg: Springer-Verlag.

Secretory Products

Burgess, T. L., and R. B. Kelly. 1987. Constitutive and regulated secretion of proteins. *Annu Rev Cell Biol* 3:343–93.

Jamieson, J., and G. E. Palade. 1971. Synthesis, intracellular transport and discharge of secretory proteins in stimulated pancreatic exocrine cells. *J Cell Biol* 50:135–58.

Centrioles

Brinkley, B. R. 1990. Toward a structural and molecular definition of the kinetochore. *Cell Motil and Cytoskel* 16:104.

Fulton, C. 1971. In *Centrioles. Origin and continuity of cell organelles*, Vol. 2. J. Reinert and H. Ursprung, eds. Springer-Verlag, pp 170–221.

Karsenti, E., and B. Maro. 1986. Centrosomes and the spatial distribution of microtubules in animal cells. *Trends Biochem Sci* 11:460–63.

Stubblefield, E., and B. R. Brinkley. 1967. Architecture and function of the mammalian centriole. *Symp Int Soc Cell Biol* 6:175–218.

Vorobjev, I. A., and Y. S. Chentsov. 1982. Centrioles in the cell cycle. I. Epithelial cells. *J Cell Biol* 93:938–49.

Cilia and Flagella

Cilia

Gibbons, I. R. 1981. Cilia and flagella of eukaryotes. *J Cell Biol* 91:107s–24s.

Steinman, R. M. 1968. An electron microscopic study of ciliogenesis in developing epidermis and trachea in embryos of Xenopus. *Am J Anat* 122:19–55.

Wolfe, J. 1972. Basal body fine structure and chemistry. *Adv Cell Molec Biol* 2:151–92.

Flagella

Fawcett, D. W. 1975. The mammalian spermatozoon. A review. *Dev Biol* 44:394–436.

The Lysosomal System

Multivesicular Bodies

Bainton, D. 1981. The discovery of lysosomes. *J Cell Biol* 91:66s–76s.

De Duve, C. 1963. The lysosome. *Sci Am* 208:64–72.

Essner, E. and A. B. Novikoff. 1961. Localization of acid phosphatase activity in hepatic lysosomes by means of electron microscopy. *J Biophys Biochem Cytol* 9:773–84.

Helenius, A., I. Mellman, D. Wall, and A. Hubbard. 1983. Endosomes. *Trends Biochem Sci* 8:245–50.

Hottzman, E. 1976. *Lysosomes, a survey.* Springer-Verlag, New York.

Microbodies

de Duve, C. 1983. Microbodies in the living cell. *Sci Am* 248:74–84.

de Duve, C., and P. Baudhuin. 1966. Peroxisomes (microbodies and related particles). *Physiol Rev* 46:323–57.

Fahimi, H. D., and H. Sies. 1987. *Peroxisomes in biology and medicine.* Heidelberg: Springer-Verlag.

Frederick, S. E., P. J. Gruber, and E. H. Newcomb. 1975. Plant microbodies. *Protoplasma* 84:1–29.

Lazarow, P. B., and Y. Fujiki. 1985. Biogenesis of peroxisomes. *Annu Rev Cell Biol* 1:498–530.

Tolbert, N. E., and E. Essner. 1981. Microbodies: peroxisomes and glyoxysomes. *J Cell Biol* 91:271s–83s.

Annulate Lamellae

Chen, T-Y, and E. M. Merisko. 1988. Annulate lamellae: Comparison of antigenic epitopes of annulate lamellae membranes with the nuclear envelope. *J Cell Biol* 107:1299–1306.

Kessel, R. G. 1985. The structure and function of annulate lamellae: porus cytoplasmic and intercellular membranes. *Int Rev Cytol* 82:181–303.

Merisko, E. M. 1989. Annulate lamellae: An organelle in search of a function. *Tissue Cell* 21:343–54.

Cell Inclusions

Glycogen

Geddes, R. 1986. Glycogen: a metabolic viewpoint. *Biosci Rep* 6:415–28.

Revel, J. P. 1964. Electron microscopy of glycogen. *J Histochem Cytochem* 12:104–14.

Revel, J. P., L. Napolitano, and D. W. Fawcett. 1960. Identification of glycogen in electron micrographs of thin tissue sections. *J Biophys Biochem Cytol* 8:575–89.

Lipid

Ashworth, C. T., J. S. Leonard, E. H. Eigenbrodt, and F. J. Wrightsman. 1966. Hepatic intracellular osmiophilic droplets. Effect of lipid solvents during tissue preparation. *J Cell Biol* 31:301–18.

Suter, E. 1969. The fine structure of brown adipose tissue. *Lab Invest*: 21:246–58.

Crystalloids

Glusker, J. P., and K. N. Trueblood. 1985. *Crystal structure analysis: a primer.* Oxford: Oxford University Press.

Nagano, T., and I. Ohtsuki. 1971. Reinvestigation of the fine structure of Reinke's crystal in the human testicular interstitial cell. *J Cell Biol* 51:148–61.

Extracellular Material

Collagen

Kefalides, N. A. 1975. Basement membranous: Structural and biosynthetic considerations. *J Invest Dermatol* 65:85–92.

Martin, G. R., and R. Timpl. 1987. Laminin and other basement membrane components. *Annu Rev Cell Biol* 3:57–85.

Watt, F. M. 1986. The extracellular matrix and cell shape. *Trends Biochem Sci* 11:482–5.

Matrix of Bone and Cartilage

Cormick, D. 1987. *Ham's histology*, 9th ed. Philadelphia: Lippincott, pp 264–323.

Jamde, S. S. 1971. Fine structural study of osteocytes and their surrounding bone matrix with respect to their age in young chicks. *J Ultrastruct Res* 37:279–300.

Special Features of Plant Tissues

Cutter, E. G. 1978. *Plant anatomy*, 2nd ed. Part 1, Cells and Tissues. London: Arnolds.

Chloroplasts

Alberts et al., 1989. Special features of plant cells. In *Molecular biology of the cell*, 2nd ed. New York: Garland, pp 1137–86.

Anderson, J. M. 1986. Photoregulation of the composition, function, and structure of thylakoid membranes. *Annu Rev Plant Physiol* 37:93–136.

Basic, A., P. J. Harris, and B. A. Stone. 1988. Structure and function of plant cell walls. In *Biochemistry of plants: a comprehensive treatise*, J. Preiss, ed. Vol. 14. Carbohydrates, San Diego, CA: Academic Press, pp 298–371.

Bogorad, L. 1981. Chloroplasts. *J Cell Biol* 91:256s–70s

Gunning, B. E. S., and M. W. Steer. 1975. *Ultrastructure and the biology of plant cells*. London: Arnolds.

Hoober, J. K. 1984. *Chloroplasts*. New York: Plenum Press.

Miller, K. R. 1979. The photosynthetic membrane. *Sci Am* 241:102–13.

The Cell Wall

Gunning, B. E. S., and R. L. Overall. 1983. *Plasmodesmata*. New York: Springer-Verlag.

McNeil, M., A. G. Darvill, S. C. Fry, and P. Alfersheim. 1984. Structure and function of the primary cell walls of plants. *Annu Rev Biochem* 53:625–63.

Vacuoles

Matile, P. 1978. Biochemistry and function of plant vacuoles. *Annu Rev Plant Physiol* 29:193–213.

Bacteria

Fuller, R., and D. W. Lovelock. 1976. *Microbiol ultrastructure*. The *use of the electron microscope*. New York: Academic Press, 331 pp.

Algae, Fungi, Yeast, and Protozoa

Bold, H. C., and M. J. Wynne.1978. *Introduction to the algea: structure and reproduction*. Prentice-Hall, Englewood Cliffs, NJ.

Levandowsky, M., and S. H. Hutner. 1980. *Biochemistry and Physiology Protozoa*. Academic Press, New York.

Prescott. D. M. 1988. *Cells*. Jones and Bartlett Publishers, Boston.

Viruses

Dalton, A. J., and F. Haguenae. 1973. *Ultrastructure of animal viruses and bacteriophages, an atlas*. New York: Academic Press, 413 pp.

Field, A. M. 1982. Diagnostic virology using electron microscopic techniques. *Advanced Virus Res*, 27:1–69.

Fraenkel-Conrat, H. 1985. *The viruses. Catalogue, characterization, and classification*. New York: Plenum Pub. Co.

Hsiung, G. D., and C. K. V. Fong. 1982. Diagnostic virology illustrated by light and electron microscopy. *J Electron Microsc Tech* 4:265–301.

Hsiung, G. D., C. K. V. Fong, and M. J. August. 1979. The use of electron microscopy for diagnosis of virus infections: an overview. *Prog Med Virol* 25:133–59.

Palmer, E. L., and M. L. Martin. 1988. *Electron microscopy in viral diagnosis*. Boca Raton, FL: CRC Press, 208 pp.

Simmons, K., H. Garoff, and A. Helenius. 1982. How an animal virus gets into and out of its host cell. *Sci Am* 246:58–66.

Personal Safety in the Labortory
Safety Apparatus and Safe Practices
Pathogens and Radioisotopes

Chemical Safety
Handling Chemicals in a
    Safe Manner
Storage of Chemicals
Some Chemicals Commonly Used
    in Electron Microscopy
Disposal of Spent Chemicals
Safety Monitoring
Cleaning Up Hazardous Spills
Exposure to Chemicals
New OSHA Standard

Fire Safety
Preventing Fires
Stopping Fires

Electrical Safety
Darkrooms
Vacuum Evaporators and Sputter
    Coaters
Proper Grounding of Equipment
Servicing of Electron Microscopes
    and Small Equipment
First Aid for Shock Victims

Physical and Mechanical Hazards
Sharp Objects
Critical Point Dryers
Vacuum Pumps
Vacuum Evaporators
Sputter Coaters
Ovens
Cryogenic Gases and Vacuum
    Dewars
Compressed Gas Safety
Microwave Ovens
Radiation
Centrifuges
Pipetting
Falls

Training and Orientation Programs

Hotlines and Other Resources

References

Safety in the Electron Microscope Laboratory

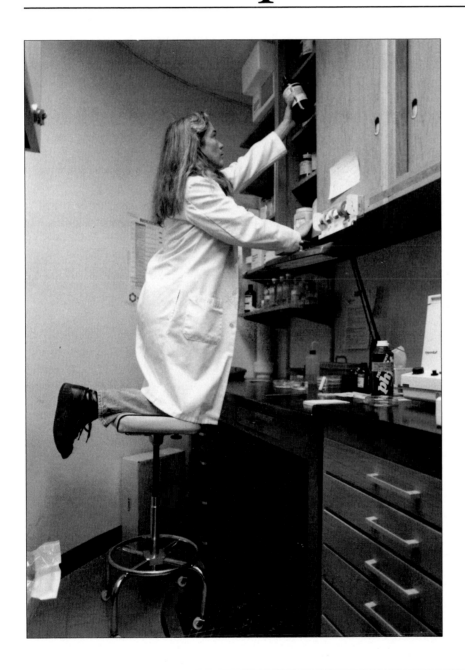

It is important for researchers not only to be trained in the proper use of all equipment and reagents in the electron microscope laboratory, but also to be aware of potential hazards (fire, chemical, electrical, physical) associated with these items. Microscopists must know the proper procedures to follow in all hazardous situations and the proper method of disposing of dangerous wastes. The environmental impact of using various chemicals (and even equipment) must be discussed so that microscopists can make informed decisions about pursueing particular lines of research. Although electron microscopists and students often become engrossed in learning new techniques and making scientific discoveries, they must not lose track of the importance of safety in conducting their discipline. They must be aware of the consequences of their actions and be prepared to correct their mistakes in a manner that is consistent with the preservation of the quality of life and the environment.

Personal Safety in the Laboratory

All electron microscopists should be aware of basic operating conditions or guidelines to protect the safety of coworkers, equipment, and themselves. Even though each electron microscope laboratory will undoubtedly develop its own specific sets of rules for safe operation, certain basic safety principles and practices are covered here for the benefit of those who may be considering establishing or refining a set of safe operational procedures in their laboratory.

Safety Apparatus and Safe Practices

1. **Fume hoods** (Figure 20-1) are needed in all electron microscope laboratories since most of the chemicals used in electron microscopy are toxic and some are known carcinogens. A quick way to check if a fume hood is functioning properly is to place several drops of an odorous chemical such as beta mercaptoethanol on a filter paper in a petri dish inside of the operating hood for about 30 minutes. If one detects the odor outside of the confines of the cabinet, then the hood should be checked for proper operation. In one such check, an electron microscopist determined that the main duct of a fume hood was leaking into the overhead ceiling space of an adjacent office. Fume hoods should

be evaluated regularly by trained personnel to ascertain that they are preventing the escape of fumes into the work environment.

Chemicals and apparatus should not be stored permanently in the working fume hood since they might interfere with the efficiency of the ventilation system. If one chooses to store dangerous chemicals in a fume hood, a special hood or vented cabinet should be designated for storage only and checked regularly for proper operation. Compatibilities of the chemicals stored in the hood must be checked to avoid placing unsafe combinations in the same location.

A surgeon's mask is not a substitute for a fume hood. Masks will not protect the wearer from noxious fumes or finely powdered chemicals (such as lead citrate). Surgical masks are designed only to prevent the liberation of aerosols of infected droplets from the wearer.

2. **Gloves** or finger cots should be worn whenever handling potentially toxic chemicals. Perhaps the most useful gloves are tightly fitting latex and polyvinyl chloride. Such gloves are resistant to most chemicals used in electron microscopy and do not interfere with the dexterity needed for most operations. For safety, however, one must verify that the glove is appropriate for the situation prior to using the chemical.

Care should be taken not to contaminate objects with soiled gloves. A common mistake is to touch books, papers, reagent bottles, telephones, or doorknobs with soiled gloves, thereby contaminating them. One researcher working late at night ran out of one of the components of an epoxy embedding medium. In an attempt to locate more of the reagent, the researcher entered several rooms while wearing gloves contaminated with resin monomer. The following day, another researcher entering one of the rooms noticed a sticky substance on the doorknob. Fortunately, they were able to deduce the nature of the substance and wash their hands and the doorknob using hot water and a strong detergent.

Gloves should be removed carefully so as not to contaminate the hands. One way of doing this is to remove the glove from the less dominant hand so that the glove is turned inside out. The inverted glove may then be used to take hold of and similarly invert the other glove on the dominant hand. One glove may then be

Figure 20-1 Fume hoods are essential pieces of safety equipment for working with chemicals used in electron microscopy. Hoods should be evaluated on a regular basis to verify that they are operating properly. Gloves are still necessary to protect the hands against contact with the chemicals.

placed inside of the other and tied into a knot to seal the contamination inside of the gloves. Be aware that most latex gloves are powdered with talc, a substance to which some individuals may be sensitive. Unpowdered (or minimally powdered) gloves may be substituted. If a dangerous chemical is being used, most savvy researchers use doubly gloved hands and change the outer glove as it becomes contaminated.

3. **Eye protection** may be necessary when working with cryogenic agents such as liquid nitrogen and when handling most chemicals. Goggles or shields should be used to cover glasses or contact lenses since they do not provide adequate eye protection. In fact, regular contact lenses may actually trap material between the lens and eye and interfere with the expeditious removal of materials that have splashed into the eye. Soft contact lenses may absorb toxic fumes and transfer them onto the eye surface. If possible, substitute readily removable eyeglasses for contact lenses when handling chemicals such as osmium tetroxide, formaldehyde, acrolein, and other volatiles.

One should know how to operate the safety eyewash stations. In an emergency, it may be necessary to locate and operate the station with one or both eyes closed, so practice with both eyes closed. If the eyes come into contact with chemicals, they must be held open and exposed to the tempered stream of water for 15 minutes!

4. **Suitable clothing** should be worn. Laboratory coats, aprons, and arm protectors may be needed when handling certain chemicals or cryogens. Snap, rather than button, closures are desirable in laboratory clothing to permit its rapid removal if contaminated or on fire. Always check with the laboratory supervisor if there are any doubts about proper safety attire. If clothing becomes contaminated, remove it immediately and dispose of or launder it in the proper manner. Open-toed shoes or sandals are unsafe and should not be allowed in the laboratory. Do not wear expensive or hard to clean clothing when handling laboratory chemicals. It is a good idea to have a change of clothing nearby. Remove jewelry such as watches, rings, bracelets, and necklaces to prevent their contact with chemicals and electrical sources.

5. **Verify new or untried operational procedures or protocols** with laboratory supervisors before using them. This may save time and reagents as well as ensure the researcher's safety. Researchers should never work alone on untried protocols unless they are totally aware of all the dangers and potential problems likely to be encountered.

6. **Eating, drinking, and smoking are forbidden** in the electron microscopy laboratory. This is not only for the operator's protection, but also to protect equipment from contamination. Do

not store food or drinks in a laboratory refrigerator. Do not use laboratory glassware for food or beverages. Special rooms and refrigerators should be set aside for the consumption and storage of food and drinks.

7. **Mouth pipetting is not permitted.** Use bulbs or other suitable equipment for such procedures.

8. **Work areas must be kept clean, uncluttered, and free of physical obstruction.** Put away unneeded reagents or apparatus when the procedure is finished and clean the work area thoroughly to remove any traces of chemicals. Dispose of contaminated materials in a safe and ecologically sound manner. Do not toss dangerous chemicals or broken objects into conventional trash bins since cleaning and maintenance people may be harmed.

9. **Be familiar with the location and operation of all safety equipment** or reagents such as safety showers, eyewash stations, spill control units, fire extinguishers, and first aid kits.

10. **Post important telephone numbers near all phones.** Such numbers should include fire department, ambulance, physician or nurse, poison control center, pollution control, power plant, security office, and the home and office numbers of the laboratory supervisor.

11. **Label all reagents** with complete chemical names of contents, date prepared, and with the initials or name of the preparer. Indicate any special precautions, such as need for refrigeration or freezing, avoidance of light and mechanical shocks, electrical hazards, etc.

12. **Post material safety data sheets** (Figure 20-2), normally supplied by the manufacturer close to where the chemicals will be used, and require all users to read the sheets. These sheets detail the characteristics of the chemicals and the precautions to be followed when handling the reagents.

13. **Arrange for regular safety inspections and drills.** This not only helps identify and eliminate potential hazards, but also gives practice in the proper actions to take if an accident occurs. *A plan of action* is essential for all electron microscope laboratories. Everyone must be familiar with proper safety procedures regarding exits, chemicals and spills, fires, electricity, and equipment.

14. **Wash hands and arms before leaving the laboratory**, especially after handling chemicals.

MATERIAL SAFETY DATA SHEET
(Adapted from USDL Form LSD-005-4)

SECTION I. IDENTIFICATION OF PRODUCT	
CHEMICAL NAME Lead Acetate	FORMULA $Pb(C_2H_3O_2)$ 3 H_2O

SYNONYM OR CROSS REFERENCE
none available

SECTION II. HAZARDOUS INGREDIENTS	
MATERIAL Lead Acetate	NATURE OF HAZARD Irritant, poisonous

SECTION III. PHYSICAL DATA	
BOILING POINT decomposes at 100 C	MELTING POINT 75 C
VAPOR PRESSURE not available	SPECIFIC GRAVITY 2.55
VAPOR DENSITY not available	% VOLATILE BY VOLUME not available
WATER SOLUBILITY 60 gm in 100 gm water	EVAPORATION RATE not available
APPEARANCE white crystalline granules	ODOR slightly acetic

SECTION IV. FIRE AND EXPLOSION HAZARD DATA	
FLASH POINT not flammable	FLAMMABLE LIMITS not applicable

FIRE EXTINGUISHING MEDIA
as appropriate for surrounding fire
SPECIAL FIRE-FIGHTING PROCEDURES
protective clothing, self contained breathing apparatus with full faceplate; may decompose to acetic acid, carbon monoxide and toxic fumes of lead oxide
UNUSUAL FIRE AND EXPLOSION HAZARD
not considered to be explosive; may combine with azides to form explosive compounds (e.g., lead azide)

Figure 20-2 Material safety data sheet (first page of two) for lead acetate stain. It is advisable to post these sheets on a bulletin board in the work area where the chemicals are to be used.

15. **Verify the security of the laboratory before leaving:** (a) extinguish all flames or sources of ignition; (b) turn off all unnecessary gases, water, vacuum, and electricity; (c) lower the sash to the fume hood; (d) turn off all nonessential lights; (e) lock all doors.

Pathogens and Radioisotopes

Most electron microscope laboratories are not prepared to handle materials contaminated by microbes or radioisotopes. Normally, such specimens are best handled outside of the electron microscope facility with the fixed specimens then taken to the electron microscope facility where they may be further processed. If pathogens or radioisotopes are to be used in an electron microscope facility, prior clearance with the laboratory supervisor must be obtained. Clear guidelines for the safe handling of pathogens and radioisotopes have been established by professional organizations and have been mandated by federal law. For example, one must be certified by a radiological control officer prior to handling radioisotopes, and an infection control officer may become involved in the handling of human pathogens. Federal approval is required before any of these materials can be transported either privately or commercially.

Staff must be aware of the subtle precautions to take with contaminated materials that enter the electron microscope laboratory. For example, negatively stained clinical specimens are still potentially infectious, and forceps as well as specimen cartridges from the electron microscope may become contaminated by contact with them. Even radioisotopes contained in specimens that have been embedded in plastic are dangerous since it is possible to inhale fragments of the specimen blocks generated during the trimming steps for ultramicrotomy.

Chemical Safety

Nearly all chemicals used in electron microscopy are toxic to humans and dangerous to the environment unless handled and disposed of properly. All electron microscopists must be acquainted with the potential dangers associated with the chemicals they are about to use, dispose of, or store.

Handling Chemicals in a Safe Manner

The first step in the safe handling of chemicals is to read the *material safety data sheet* (MSDS) for the chemical provided by the supplier. Most EM supply houses have these sheets, which give information on the dangerous properties of the chemical. Typical information in the MSDS may include name (chemical, synonyms, formula), physical data (boiling and freezing points, appearance and odor, etc.), fire and explosion hazard data (as well as fire fighting information), health hazard data (effects of exposure, emergency and first aid procedures), reactivity data (chemical compatibilities), ingredients (if other than 100%), spill or leak procedures, special protection information (respirators, gloves, eye protection, etc.), special precautions (handling and storage), and regulatory information. It is advisable to post such sheets on a bulletin board in the area where the chemicals are likely to be used and to insist that all first-time users read and understand the information

on the sheet. Some discussion of the sheet is probably necessary to insure compliance.

Many chemical manufacturers also label reagent bottles using the "704 Diamond" system recommended by the National Fire Protection Association. This diamond is subdivided into quadrants that classify the chemical in terms of its flammability (top quadrant, usually red background), reactivity (right quadrant, usually yellow background), health hazard (left quadrant, usually blue background), and specific hazards (bottom quadrant, usually white background). Higher numbers (range = 0 to 4) within the quadrants indicate greater risks. In Figure 20-3, for example, we see the safety diamond for acetone. In some institutions, safety or fire officers may place such a diamond on the door of each room to indicate the presence of various types of chemicals in each location.

With few exceptions, all operations involving chemicals should be carried out in a properly operating fume hood and while wearing gloves. This recommendation is frequently ignored during the weighing out of chemicals, since the fume hood may set up a draft that will affect the weighing and may actually spread the chemicals being measured. Nonetheless, it is usually possible to adjust the sash of the fume hood to minimize the drafts or to place a wind screen around the scale during the weighing process. Another possible approach is to place the scale inside of a glove bag and to carry out the operation inside of the plastic shroud. After weighing, the plastic or glassine disposable weighing dish should be rinsed into an appropriate waste container and the dish disposed of in the trash. Do not discard contaminated weighing dishes in the regular trash but place

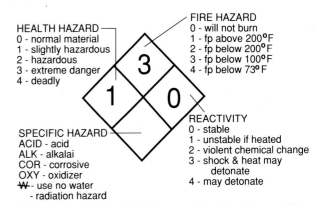

Figure 20-3 The "704 Diamond" of the National Fire Protection Association. This system is used to identify the various physical and health hazards of different chemicals. This is the type of symbol one would expect to encounter for acetone.

them in the containers designated for contaminated solid wastes.

Be aware that a number of chemicals may be readily absorbed through the skin (acrylamide, benzene, benzidine, carbon tetrachloride, dimethylsulfoxide, dinitro-benzene and -toluene, dioxane, mercury, nicotine, nitrobenzene, phenol, picric acid, sodium cacodylate, o-toluidine, trichloroethane—to name a few) as well as by breathing the vapors they give off.

A newly arrived chemical should be marked with the date received and initialed by the person who received it. Before using any chemical, read the labels twice to verify that it is the proper chemical. Some chemicals have similar names (sodium chlorite versus sodium chloride), but their effects are quite different. Do not use chemicals that are ambiguously labeled or whose labels have deteriorated. Check that the reagent is not too old to use. If no expiration dates are posted on the label, check with the laboratory supervisor to determine if it is still usable. If a stopper or lid is stuck, attempt to open it very carefully and seek assistance after several attempts. Stuck lids (especially on plastic embedding monomers) are normally caused by not wiping off the threads or cap of the reagent bottle after use.

When removing chemicals from their container, use a clean spatula for dry chemicals or a clean pipette for light liquids. Pour more viscous liquids, such as plastic monomers, slowly into the working container and take care not to generate airlocks in the bottle that could lead to splashes during pouring. If possible, wipe off the outside of the stock container using either a dry paper towel (for liquids) or a towel moistened in water (for dry reagents). But take care not to contaminate the contents or to damage the label during the wiping. Remove only the minimal amount needed and do not return any unused materials back to the stock bottle; dispose of any remaining chemical after determining that no one else needs it. Materials that have been stored in the refrigerator or freezer usually should be allowed to come to room temperature before opening to prevent moisture condensation on the contents.

Always pour contents slowly into water or into less concentrated solutions while stirring the solution. Remember that concentrated acids are added very slowly to the larger volume of water as the water is being stirred. Never look down into an open vessel unless it is empty; otherwise one risks splashing material into the face or coming in contact with dangerous fumes. Neither taste the contents of an unknown solution nor sniff it by placing the nose directly over the opening. If one may safely sniff the contents, do so by gently wafting a tiny amount towards the nose using the cupped hand. During the sniffing, the container should be on a table and not held in the hand in case there is an involuntary response to a noxious substance.

Return the stock bottle to its proper location in the laboratory and inform the supervisor if any irregularities were noted (cracked or chipped container, wrong lid, discoloration, absorbed water, decomposition, nearly exhausted, etc.). If one has prepared a reagent, it should be completely labeled to indicate contents, date prepared, date expired, and name of individual who prepared it. After using the chemicals, wash hands (and face, if necessary) and leave the work area clean and in proper order.

Storage of Chemicals

For convenience, most laboratories store chemicals in alphabetical order on shelves. Although this may be adequate in some instances, it is not the safest method of storage since it is possible to store incompatible chemicals (sodium nitrite and sodium thiosulfate, for example) in close proximity to each other. As a guide to the placement of chemicals, refer to Table 20-1.

Rayburn (1990) developed a set of recommendations for storing chemicals. A modified version of his recommendations is included in the panel that follows.

Storage Recommendations (Rayburn, 1990)

- Do not store incompatible chemicals in close proximity to each other.
- Store acids in a special cabinet designed for corrosives with nitric acid isolated from the others. The containers should rest on resistant trays capable of retaining the acids in case of leakage.
- Store flammable chemicals in a specially designed cabinet that is vented.
- Do not store chemicals in a fume hood that is used for routine work.
- Use shelving that is anchored to the wall rather than free-standing shelves.
- Do not store chemicals on top shelves or above eye level.
- Provide restraints at the edges of shelving to keep chemicals from falling off.
- Do not store chemicals on the floor. An exception is large storage drums that are kept in storage rooms.

Table 20-1 Some Incompatible Chemicals

Chemical	Incompatible with
Acetic acid	Chromic acid, nitric acid, hydroxyl compounds, ethylene glycol, perchloric acid, peroxides, permanganates
Acetone	Concentrated nitric and sulfuric acid mixtures
Anhydrous ammonia	Mercury, chlorine, calcium, hypochlorite, iodine, bromine, hydrofluoric acid
Ammonium nitrate	Acids, powdered metals, flammable liquids, chlorates, nitrates, sulfur, finely ground organics or combustible compounds
Aniline	Nitric acid, hydrogen peroxide
Arsenic compounds	Any reducing agent, certain acids
Azides	Acids, lead compounds such as lead nitrate or citrate
Calcium oxide	Water
Carbon (activated)	Hypochlorites, all oxidizers
Chlorates	Ammonium salts, all acids, powdered metals, sulfur, finely ground organics or combustibles
Cyanides	Acids
Flammable liquids	Ammonium nitrate, chromic acid, all peroxides, nitric acid, halogens (chlorine, bromine, etc.)
Hydrocarbons (butane, propane, benzene, toluene)	Chromic acid, peroxides, halogens
Hydrogen peroxide	Alcohols, acetone, organics, aniline, copper, chromium, iron, most metals or their salts, all combustibles
Hypochlorites	Acids, activated carbon
Iodine	Ammonia, hydrogen, acetylene
Mercury	Ammonia, acetylene, fulminic acid
Nitrates	Sulfuric acid
Nitrites	Thiosulfates
Nitric acid (conc)	Acetic acid, aniline, chromic acid, hydrogen sulfide, flammable liquids and gases, copper, brass, heavy metals
Oxalic acid	Silver, mercury
Oxygen	Oils, grease, hydrogen, flammable liquids, solids or gases
Perchloric acid	Acetic anhydride, bismuth and alloys, alcohol, paper, wood, grease, oils
Peroxides	Acids, all combustibles (avoid friction, store cold)
Phosphorous (white)	Air, oxygen, alkalis, reducing agents
Potassium permanganate	Glycerol, ethylene glycol, benzaldehyde, sulfuric acid
Sodium nitrate	Ammonium compounds
Sulfides	All acids
Sulfuric Acid	Chlorates, permanganates

Source: National Academy of Sciences, National Research Council. 1980. *Prudent practices for handling hazardous chemicals in laboratories.* Committee on Hazardous Substances in the Laboratory. Washington: National Academy Press.

- Store very toxic chemicals or controlled substances in a locked area.
- Keep chemicals away from sunlight, heat, moisture and flames.
- Minimize the amounts of chemicals stored in the laboratory. Use a specially designated, restricted storage area for large volumes.

Some Chemicals Commonly Used in Electron Microscopy

Due to ever changing guidelines and information regarding the toxicity and handling of chemicals, it is not possible to discuss all of the chemicals used in electron microscopy. Instead, users of chemicals should familiarize themselves with the material safety data sheets included with the chemicals and assume that *all* chemicals are dangerous.

In electron microscopy, there are some chemicals, however, that are particularly dangerous for one reason or another. A common protocol for specimen preparation might include fixation in a formaldehyde/glutaraldehyde mixture buffered in cacodylate, postfixation in osmium tetroxide, dehydration in acetone or ethanol, clearing in propylene oxide, embedding in an epoxy resin (possibly Spurr's), followed by staining of the sections in uranyl and lead salts. Since these are commonly used chemicals, they will be discussed in some detail.

Cacodylate salts are approximately 50% arsenic by weight and should be used only under very special circumstances, since they are very toxic, probably carcinogens, and will produce an allergic sensitization. They are readily absorbed through the skin (entry into the bloodstream is evidenced by a metallic or garlicky taste in the mouth). When possible, *substitute other salts* (phosphates, for example) for cacodylate in the production of buffers.

Formaldehyde is known to cause nasal cavity squamous cell carcinomas upon inhalation, as well as other types of skin cancers upon continued contact. In addition, formaldehyde (and probably glutaraldehyde) may sensitize one to the aldehyde itself or to molecules that have reacted with the aldehyde, leading to various allergies. Sensitization to formaldehyde may lead to an allergic reaction to an immunochemically related substance such as Urotropin, Mandelamine, Urised, or Methanamine (Fisher, 1976). One should minimize contact with the gas liberated from formaldehyde solutions by working in a fume hood and by wearing gloves.

Osmium tetroxide solutions are toxic, volatile, and highly irritating to the mucous membranes. Contact with skin and membrane surfaces must be avoided by using gloves and working in a fume hood. Wash contacted areas immediately with soap and water for 10 to 15 minutes.

Acetone and ethanol are both used as solvents, dehydrants, and general cleaning agents in all electron microscope laboratories. Both are flammable, toxic upon inhalation, and capable of facilitating the penetration through the skin of chemicals that have been dissolved in them.

Propylene oxide is highly flammable and has been shown to cause cancers of the nasal cavity as well as other types of cancers. If possible, less toxic solvents (ethanol, acetone) should be substituted in its place.

Most of the *embedding resins* are allergenic and some are probably carcinogens (epoxy components). The amines present in the epoxy resins are immunochemically related to aminophylline and ethylenediamine antihistamine so that allergic cross reactions may result when these medications are used.

The *heavy metal salts* (lead, uranium) used commonly to enhance the contrast of sectioned materials are highly toxic, and uranium salts are alpha and beta emitters of radiation. Lead acetate and lead phosphate are suspected carcinogens (and uranium is probably also carcinogenic due to the radioactivity likely to be present). These salts must be weighed in an enclosed environment, used in a manner that minimizes contact with the dissolved salts (especially if dimethyl sulfoxide has been added), and disposed of in accordance with local and federal guidelines.

Most of the chemicals used in *cytochemical reactions* are toxic or carcinogenic (diaminobenzidine, toluene diisocyanate, lead salts, etc.).

Disposal of Spent Chemicals

After use, most chemicals must be disposed of in accordance with specified local and federal guidelines. It is important to discriminate between spent or useless chemicals and those for which one has no further need. Most institutions have clearinghouses for donating unwanted chemicals to other researchers in the same or even in other institutions, if permitted. On the other hand, spent or otherwise unusable chemicals must be disposed of following certain procedures. Use the *minimum amount* of reagent feasible. This will not only conserve expensive chemicals, but also generate less waste to dispose of later. Use *less toxic substitutes* whenever possible, e.g., phosphate instead of cacodylate buffer, LR White instead of Spurr's embedding resin.

Whenever possible keep the wastes in separate containers rather than mixing them together. This will not only prevent the combination of incompatible chemicals, but also greatly facilitates possible recycling of the chemicals.

Solid chemical wastes must be placed in individual, sealed containers, labeled as to content and amount, and sent to appropriate disposal agencies. Small amounts of nontoxic substances, such as sodium chloride, may be disposed of in the trash providing that this does not violate any local ordinances. Large volumes of even nontoxic substances should not be discarded in such a manner since they may be put into landfills and eventually contaminate the water supply.

Liquid wastes should be kept in separate, chemically resistant containers rather than mixed together in a single container. Leakproof caps must be provided, and the containers must be completely labeled. It may be possible to pour small amounts (less than 100 ml) of water soluble nontoxic liquids (certain buffers, ethanol, etc.) down the drain if followed by flushing with running water for several minutes. If the liquids have been contaminated with toxins (osmium, for example), then this is not permitted. Some communities do not permit pouring any flammable materials (including ethanol) down the drain. It may be possible for certain recyclers to purify or reuse certain liquids such as acetone, ethanol, methanol, etc.; however, this is only possible if they are kept separate and notes are kept as to the presence of other components in the liquid.

Solutions containing heavy metals should be kept in separate containers and precipitated whenever practical. Uranyl and lead salts may be precipitated with phosphate buffers. Osmium tetroxide solutions may be mixed with corn oil until the oil blackens and the osmium is reduced. Carcinogens and very toxic materials should be placed in tightly sealed bottles or containers and then placed in a large, heavy duty outer container of absorbent materials adequate to absorb the contents of the inner container in case of leakage. Mark the container with the name of the contents, date and any precautionary notes needed ("Danger: Carcinogen" or "Danger: Highly Toxic").

Used rotary pump oils should be stored in their original containers, labeled as used, and given to an oil recycler. Be careful, however, that the oils have not been contaminated with toxic materials such as epoxy resin components or heavy metals. Diffusion pump oils should be kept separate from rotary pump oils in their original containers, which are subsequently sealed, labeled, and taken to a disposal agency.

Embedding resins should be mixed and polymerized in an oven rather than disposed of as liquid wastes. A good way of dealing with epoxy resins (especially the toxic components in Spurr's resin) is to prepare the resin mixture in a large container (500 ml glass bottle, for example) without adding the catalyst. Dispense small amounts of the incomplete resin mixture into preweighed glass scintillation vials, cap the vials, and place them in a freezer. As needed, thaw a vial, add the appropriate amount of

catalyst, and mix together in the scintillation vial. Any unused resin should be polymerized by placing the loosely capped vial into an oven. Collect used resin in a glass container and polymerize the resin after each experiment. When the container is filled with the polymerized layers, seal and label it, and dispose of the polymer in accordance with institutional policy.

Mercury should be picked up using special agents that chemically combine with elemental mercury and complex it. The complexed mercury is then sealed inside of a glass container, labeled, and taken to an approved disposal agency.

Safety Monitoring

A number of devices are available for monitoring exposure to formaldehyde, organic vapors, and mercury vapors. Although they are expensive and "after the fact" devices, they may be useful to determine if unsuspected exposures are taking place. Detector/alarms are also available for smoke, heat, radioactivity, natural gas, and even water (floods in the lab). These devices may be obtained from various safety supply houses and occasionally from general laboratory supply catalogs.

Cleaning Up Hazardous Spills

If materials that are only slightly hazardous have been spilled (weak acids or bases, oils, developers, etc.), one should absorb the spill using paper towels, newspapers, or spill-control pillows. The absorbed wastes can then be transferred into a plastic bag and placed into a container for solid wastes. The floor should be thoroughly cleaned with soap and water and the area dried to prevent slips and falls by other laboratory workers.

If elemental mercury has been spilled, try to contain the spill in a small area and warn coworkers not to walk over the spill. Use gloves and remove any sources of heat that might volatilize the mercury. Do not use ordinary vacuum cleaners; they will only aerosolize and spread the mercury all over the laboratory. If the mercury has gotten into porous surfaces (floors, lab benches, etc.), then special chemical inactivators must be used. One decontaminant, called "Hg-x" (available from most laboratory supply houses), is mixed with water and the contaminated area wiped with the mixture. Alternatively, zinc powder moistened with 5% to 10% sulfuric acid to form a paste is spread over the contaminated area and allowed to dry. Both chemicals complex the mercury to facilitate its removal. After the materials have been removed, the area should be checked for the presence of mercury using an appropriate monitor or "mercury sniffer." Several cleanings may be necessary to remove all of the mercury.

If acids or alkalies have been spilled, it is best to contain the spill using spill-control pillows designed for acids and bases. Small volumes of dilute acids/bases may be neutralized using commercially available spill-control products. It is not a good idea to try to neutralize large volumes of acids/bases (especially concentrated ones) since this may only liberate heat and lead to the splattering of concentrated acids/bases. One should wear heavy gloves and take care not to step on any of the spilled materials since they will be very slippery. After the spill has been absorbed with the appropriate agents, the residues should be placed into a plastic bag, labeled, and transferred into the chemical disposal bin. The area should be cleaned with soap and water and dried thoroughly.

Flammable or volatile chemicals pose a special problem. Immediately extinguish all open flames and shut down all sources of ignition (sparking equipment, incubators, etc.). Alert other workers present in the lab of the situation and seek assistance if necessary. Open any windows and the fume hood sash to help dissipate the fumes. Absorb the spill using spill-control pillows or other absorbent materials designed for volatile agents. Place the saturated pillow into a plastic bag and move the bag into a fume hood or outside of the main laboratory if safely possible. Return to the site and, after the remaining material has evaporated, clean up the contaminated area using soap and water if possible. If cleanup procedures are in doubt, mark the contaminated area with a warning sign (also place a warning on the door to the room) and check with the laboratory supervisor.

A *spill control cart* should be placed in a convenient location so that necessary supplies can be moved quickly to the contaminated area. Such a cart can be made using a standard laboratory cart outfitted with a number of provisions such as heavy gauge neoprene gloves, disposable plastic shoe covers or booties, rubber or vinyl aprons, disposable coveralls, safety eye goggles and face shield, disposable respirators (acid/gas, dust/mist), commercial acid and base neutralizer powders, mercury absorbers, spill-control pillows for acids/bases and volatiles, several plastic dustpans, whisk broom, squeegee with rubber blade, sponge mops, detergent, several plastic buckets, plastic bags, flashlight, paper towels, and warning signs.

Exposure to Chemicals

Exposure to a chemical may occur by direct contact with the hands, inhalation of fumes or powders, splashing into eyes or mucous membranes, and ingestion. If one contacts the chemical with the unprotected hand, then the immediate response should be to wash the hands using warm water and a strong soap such as Lava or a dishwashing detergent. Do not use alcohol, acetone, or other solvents, since they may only facilitate absorption of the chemical through the skin.

Splashes into the eye, nose, or lips must be removed immediately using an eyewash station or by placing the face under a stream of warm running water. A mild soap may be used for the lips and nose, but only running water is to be used for the eye. It is important that the eye be held open in the stream of water and that contact lenses be removed to facilitate cleansing of the eye. Roll the eye during the rinsing, which should continue for 15 minutes. Seek medical attention immediately.

If a chemical has been spilled onto the clothing or other parts of the body, remove any contaminated clothing (modesty aside) or jewelry immediately and wash the body part with warm water and soap for at least 10 minutes. Contact with phenol, DMSO-containing solutions, or cacodylate acid is especially worrisome since these compounds pass through the skin and into the bloodstream rapidly. Dispose of the contaminated clothing in the same manner that chemical wastes are treated. Seek medical attention.

Inhalation of a toxic substance is serious since the substance is absorbed quite rapidly into the bloodstream. Those personally involved should leave the contaminated area and seek medical attention. Attempt to cough up as much bodily secretions as possible and rinse the mouth with water. Remove stunned or unconscious victims from the contaminated area taking care not to inhale any materials in the process yourself. It may be necessary to administer artificial respiration, but take care not to inhale any residual toxins that may be expelled from the victim. Leave warning signs on the door to the laboratory and indicate the contaminated area within the laboratory.

If materials have been ingested, rinse the mouth with warm water repeatedly. If corrosives were ingested, do not attempt to neutralize the chemical since this will only generate heat that may further damage sensitive tissues. Most physicians recommend diluting the corrosive materials using large quantities of water or milk. Do not attempt to induce vomiting if corrosives have been swallowed since this may rupture the esophagus.

If other toxic substances have been swallowed, drink large amounts of water or milk to dilute the toxin. Vomiting may be induced only if corrosives or hydrocarbons (petroleum distillates) have *not* been swallowed. Seek medical attention and be prepared to inform the medical team about the nature and approximate amount of chemical that was ingested.

New OSHA Standard

On January 31, 1990, the Occupational Safety and Health Administration (OSHA) published a set of rules in the Federal Register for dealing with dangerous chemicals in the workplace. Entitled *Occupational Exposures to Hazardous Chemicals in Laboratories*, this standard must be complied with by all academic, clinical and industrial laboratories by January 31, 1991. The purpose of the standard is not to issue new exposure limits, but to define a so-called *chemical hygiene plan*, or CHP, that establishes practices and procedures for the protection of the workers in a laboratory. An important part of the CHP is the requirement that laboratories develop a set of *standard operating procedures* (SOPs) that cover general safety precautions while using a chemical, exposure control measures (fume hoods, respirators, gloves, etc.), spill control measures, accident responses, and disposal methods to follow upon completion of the procedure. A chemical hygiene officer must be designated, and the employer is required to verify that fume hoods and other protective equipment are in proper operating condition. In essence, the employer is required to provide training and information on

- the chemical hygiene plan,
- protective measures the employees should take,
- specific hazards (health, physical, fire, etc.) associated with the chemicals,
- procedures or equipment used to detect the presence of the chemicals in the workplace.

OSHA has estimated employers' costs for compliance to the standard to be approximately $15 million, but that such compliance will result in at least a 10% reduction in illness and injuries caused by the misuse of chemicals in the workplace.

Fire Safety

Three elements are necessary to start a fire: *fuel*, *oxygen*, and an *ignition source*. Once a fire has started, heat generated by the burning process further decomposes the fuel into highly reactive free radicals that rapidly combine with available oxygen to generate even more heat. This results in an unrestrained *chain reaction* that will continue to increase in intensity as long as the three elements are available. A fire may be prevented or stopped by removing any of the three necessary elements and interfering with the chain reaction.

Preventing Fires

One may *restrict the fuel* by using only minimal amounts of flammable materials. For instance, instead of using a quart or pint container of propylene oxide or acetone during the embedding process, use 100 ml working containers and keep the larger stock containers in an approved safety cabinet. Limit access to the stock solvents to authorized personnel trained in the proper dispensing of the solvents.

Limit access to oxygen by keeping all containers capped when not in use. Especially with volatile solvents, make sure that the containers are closed immediately after dispensing the necessary amounts. When embedding, after transferring the solvent from the working container into the specimen vials, cap both systems and remove the working container from the work area after it is no longer needed. Limit storage of combustible materials in the immediate work vicinity to prevent the development of a fuel stockpile.

Keep ignition sources away from available fuels. Although it should be obvious that an ignition source never should be near combustible materials, researchers may become so engrossed in the procedure being followed that the ignition sources may be overlooked. For example, it may not be uncommon to use solvents in close proximity to electrically powered equipment such as stirrers, ultrasonic baths, or even ovens. Since solvent vapors may travel some distance, a spark or heating coil may ignite the vapor phase, which may then serve as a conduit back to the fluid phase. One should never store solvents in a standard refrigerator since a spark from the thermostat, compressor, or other relays may cause an explosion. Instead, a specially designed *labora-tory-safe refrigerator* may be used to store volatile chemicals. Self-defrosting refrigerators cannot be safely modified to store volatile materials.

Even experienced individuals may occasionally make blunders that result in damage to the laboratory and trauma to the person. The author recalls an incident that occurred in his own laboratory during the routine maintenance of an electron microscope. A coworker was using a bunsen burner to flame platinum apertures some six feet away from a 300 ml open beaker of acetone. The vapors from the acetone drifted over to the flame, igniting the vapor trail back to the fluid phase. Literally in a flash, the acetone in the beaker burst into flames. In an attempt to cover the beaker to limit access to oxygen, it was tipped over and burning acetone was spilled along the countertop and down onto the floor. Not only was the countertop, floor, and a valuable camera damaged by the burning acetone, but the coworker sustained burns along his forearms and shin as he attempted to cover and stamp out the fire. In this instance, the fire was limited by the lack of additional fuel.

Stopping Fires

Once a fire has been started, proper measures must be taken rapidly to put it out by interfering with the chain reaction. Since it may not always be possible to remove the fuel from an established fire, fire fighting is done most often by limiting the source of oxygen and lowering the heat that sustains the reaction. This may be accomplished by using the appropriate fire extinguisher (Figure 20-4), depending on the class of fire in progress.

Classes of Fires and Proper Extinguishers

Class A Fires: Wood, cloth, paper, or other common combustibles are involved. This type of fire is best extinguished with a pressurized tank usually containing either plain water or water in combination with wetting agents and aqueous film-forming foam (AFFF). This type of extinguisher is often found in corridors and in offices. A special fire blanket may also be used.

Class B Fires: Flammable liquids are involved. These fires are extinguished by using pressurized carbon dioxide, halogenated compounds such as Halon (bromo-trifluoromethane, bromochlorodifluoromethane), dry chemicals (ammonium phosphate, sodium/potassium bicarbonates, potassium chloride), or possibly by means of AFFF.

Figure 20-4 Fire extinguishers used for various types of fires. Class A: aqueous solution (left); Class B/C: carbon dioxide (middle); Class B/C: dry powder (right). Not shown is the Class D type of extinguisher used for certain types of metals such as magnesium or sodium.

Class C Fires: Live electrical circuits are involved either as the cause of the fire or as simply being near the fire. Use carbon dioxide, Halons, or dry chemical extinguishers. Aqueous-based or electrically conductive reagents must not be used to fight such fires.

Class D Fires: Combustible metals such as magnesium, sodium or potassium are involved. Extinguishers contain dry chemicals such as sodium chloride containing a thermoplastic binder that forms a solid suffocating crust over the fire.

When selecting a fire extinguisher, consider not only the class of fire but also whether or not valuable equipment may be damaged by the extinguisher. For instance, if a fire were to occur in the electron microscope room (or darkroom) where live electrical circuits are involved, select a Class C fire extinguisher. Unfortunately, dry chemicals may damage delicate electronic components by corroding and short circuiting them, while carbon dioxide may damage the components with the extreme cold generated. In this case, Halon would be the best choice.

To use the fire extinguisher, direct the spray at the *base of the fire*, not at the flames. Remember, the object is to suffocate and cool the chain reaction that is occurring at the level of the fuel. Slowly sweep the extinguishing agent over the base of the fire until all flames are extinguished. Even after the flames have subsided, continue to apply the agent to prevent a flare up of the fire.

Should a fire break out in a laboratory, prompt action must be taken to prevent the spread. Therefore, one must *have a plan* of procedures to follow.

Plan of Action for Fire Fighting

1. Evacuate all persons to a safe location. Leave the building, if necessary. Everyone should gather at a prearranged assembly point. This is important to determine if someone is still behind in the building. To avoid further injury, do not move injured persons unless the fire appears to be out of control.
2. If it is possible to extinguish the fire yourself, do so immediately. After determining the type of fire, select the proper fire extinguisher and position yourself between the fire and the room exit. Do not permit the fire to block your exit. After the fire is safely out, notify the fire department so that they may evaluate the need for further measures.
3. If it is not possible to quickly extinguish the fire, close off the room and activate the fire alarm to alert others in the building. Telephone the fire department, giving the exact location of the fire (street address, building name, floor, and room number). Indicate the nature of the fire (electrical, flammable liquids, etc.) and if dangerous chemicals are involved. Proceed to the assembly point and be prepared to meet the fire department with further directions as needed.

Electrical Safety

The electron microscope laboratory has numerous pieces of equipment, besides the microscope itself, that use electrical power. Any of these smaller pieces of equipment may produce a potentially damaging or lethal shock or serve to ignite flammable materials. Although such shocks and fires may result from defective equipment, most often they result from the unsafe practices of the user.

Darkrooms

The darkroom is a particularly dangerous location since electrical equipment is used in close proximity to water. It is important to keep such equipment well away from the processing sinks so they do not tumble into the sinks. Cords should be checked regularly for fraying, and they must not be allowed to dangle over the sinks. Safelights and timers must be secured so that they will not fall into the sink. Bat-

tery or spring-driven timers are much safer to use in the darkroom.

Darkroom users must be cautioned not to handle electrical equipment with wet hands and to keep the area around the sink dry. It is dangerous to contact the enlarger and the sink at the same time since the body may be providing an electrical bridge between a defective enlarger/timer and the grounded sink. If an energized item should drop into the sink, avoid the instinct to retrieve it. Instead, unplug it first or, better yet, cut off the circuit breaker and then unplug it. The retrieved equipment should be labeled as potentially dangerous, and it should not be used until it has been evaluated by a trained individual.

Vacuum Evaporators and Sputter Coaters

These systems use large amounts of current to thermally evaporate noble metals or high voltages to sputter metal targets. Although most contain safety interlocks, caution must be exercised during their use and maintenance. If electrodes are not properly attached and safe grounding procedures are not followed, a short circuit to the equipment case or operator may be established. Always follow the manufacturer's directions when working with such equipment. Always unplug the equipment when cleaning or servicing it.

Proper Grounding of Equipment

All electrical circuits in laboratories that use electrical equipment near water should have *ground fault interrupt devices* that will discontinue the electrical connection when an individual completes the circuit from the equipment ground to the earth ground (e.g., contacts a defective piece of electrical equipment and a sink at the same time). In addition, all electrical equipment should use a three-pronged or other properly grounded connector. One should not use two-to-three pronged adapters without ascertaining that a proper ground has been established (i.e., check with an electrician). Users of equipment must be shown the location and proper use of circuit breakers or fuse boards to cut off the power to equipment in an emergency.

Servicing of Electron Microscopes and Small Equipment

Except for individuals with the proper training, users should not attempt to service electrical components. If a problem is suspected, disconnect the equipment, mark it as potentially defective, and notify the laboratory supervisor. Although most microscopists do not service the electrical boards inside of the electron microscopes, be aware that electrical dangers exist even while doing routine maintenance on the microscope.

There is a story about a service engineer who was performing routine maintenance on a TEM. In a hurry to replace the filament, he attempted to remove the cathode without first discharging the high voltage built up in the gun. Two other observers who were being instructed by the first engineer witnessed a bluish white discharge arc from the electron gun over to the instructor who had grounded himself on the console of the microscope. Although visibly shaken, the instructor was able to continue routine maintenance of the microscope after a short discussion on electrical safety. In another similar incident, a service engineer claims to have had the soles of his shoes welded to the floor as he worked on an undischarged high voltage tank. In this instance, the engineer was so shaken by the incident that he transferred to an administrative position where the potential hazards were of a different type.

First Aid for Shock Victims

If contact with a live circuit has taken place, the first step should be to cut off the current at the breaker box. If this is not possible, separate the victim from the electrical source using a nonconductive material such as a lab coat, wooden chair, or rubber hose. Make sure that you do not become involved in the circuit (i.e., avoid water spills or making contact with grounded objects such as water faucets or other electrical equipment). If victims are unconscious, attempt to arouse them by gentle shaking. If breathing has stopped or a pulse cannot be felt, apply cardiopulmonary resuscitation (CPR) following approved methods. At least one person in the laboratory, in addition to the instructor, should be versed in CPR. Send someone to call for medical help.

Physical and Mechanical Hazards

The safe and proper handling of sharp objects and other potentially hazardous equipment or laboratory facilities are covered in some detail in the following section.

Sharp Objects

A number of objects used in electron microscopy may inflict cuts or lacerations if improperly handled. Used glass knives and pipettes should be kept with other shards of glass in a thick-walled box that is lined with a plastic bag to contain the smaller glass fragments. When the box is filled, it should be taped shut and the appropriate disposal officer contacted.

Used razor blades and syringe tips also should be collected in a box and disposed of in accordance with local regulations. Contaminated syringes should not be placed in the box until they are sterilized. Housekeeping staff should be warned about the dangers associated with these disposal areas in the laboratory.

Critical Point Dryers (CPDs)

Since dangerous pressures build up inside of these units, it is important that they be shielded. One method is to place a Lexan sheet between the pressure chamber and the operator so that it will deflect material away from the face if the vessel explodes. Never look directly into the viewing window of such chambers, but use a metal mirror located between the Lexan sheet and the chamber. The minimum precaution that should be taken with CPDs is to wear a full face shield made of Lexan or polycarbonate.

The CPD should be checked for damage or bends to the high pressure lines from the tank to the chamber, and the seals should be checked and replaced as necessary. No lubricants should be used near the door or on any gaskets associated with the door seal.

Avoid inhalation of the fumes exhausted from such units (CO_2, amyl acetate, Freons) since they may asphyxiate or sicken the operator. The hoses used to vent gas from such units should be constructed of copper, stainless steel, or other cryogen-safe materials. The hose must be firmly anchored and vented into a fume hood.

Vacuum Pumps

Rotary and diffusion pumps may present several hazards. The rapidly spinning pulley and belts of older types of rotary pumps should be shielded to prevent entanglement of operators' clothing or pinching of the fingers. Both types of pumps become quite hot during operation, and it is very easy to burn oneself when working with them. The used oils from such pumps must be considered hazardous waste (potential carcinogens) and disposed of in a manner consistent with local pollution control guidelines.

Rotary pumps should not be permitted to exhaust into the room. Even mist traps placed on the outlet of rotary pumps are not 100% efficient all of the time. A better method is to exhaust the pumps into a fume hood or outside the building (Figure 20-5). A few diffusion pumps use mercury that must be very carefully disposed of after consultation with experts.

Vacuum Evaporators

The major risks associated with such units are implosion of the vacuum chamber bell jar, electrical shocks, burns to the hand or damage to the eye caused by heated filaments, and trauma caused by contact with the belt or pulley of the spinning mechanical pump. Since most evaporators use a glass bell jar to enclose the chamber, it is essential to

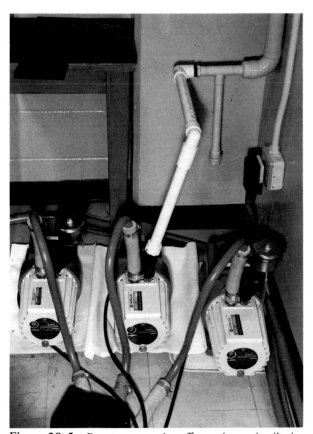

Figure 20-5 Rotary pumps give off a carcinogenic oil mist that must be efficiently trapped by special filters placed over the pump outlet as shown on the two pumps on the left and right. A safer approach is to vent the fumes to the outside by means of pipes (middle pump).

check the glass before each use for nicks, hairline fractures, or scratches. If any flaws are detected, do not use the jar and check with the laboratory supervisor. The bell jar should be surrounded by a plastic shroud or wire shield to contain glass fragments in case of implosion. Eye protection should be worn as an additional safeguard against flying glass. To avoid eye damage, never stare directly into the extremely bright, heated filament. Instead, use welder's goggles or a fogged photographic film to view the evaporation process. Ordinary sunglasses usually are not dark enough to protect the eyes.

Sputter Coaters

The risks associated with sputter coaters are less from implosion of the vacuum chamber than from electrical shocks from the high voltage sources used to generate the plasma. Consequently, safety interlocks should be checked periodically to insure that one cannot activate the high voltage when the chamber is open. Inspect the glass chamber enclosure (bell jar) to determine if it has developed scratches or nicks along the sealing edge or if it is cracked and in need of replacement.

Ovens

Several problem areas exist when laboratory ovens are used. For example, a fire may start if volatile reagents such as acetone, propylene oxide, or ether are placed in most ovens. Most ovens vent directly into the working area rather than into a fume hood, releasing fumes and chemical vapors in the work area. Remember that some resin components (Spurr's epoxy resin, for example) are potential carcinogens and that most are allergenic. It is dangerous to use a vacuum oven to outgas resins since the more volatile components of the resin will be distilled over into the vacuum pump oil and possibly liberated into the room if the pump is not vented to the outside.

Cryogenic Gases and Vacuum Dewars

One of the obvious dangers of working with liquefied cryogenic agents is the potential for frostbite upon contact with cold liquid, gases, or chilled surfaces. Liquid nitrogen, helium, and carbon dioxide may all cause serious burns on contact with living tissues. Whenever dispensing such cryogens, properly insulated gloves should be used and a face shield (rather than goggles) should be worn. Sandals or

If any problems are encountered during the use of the compressed gas, turn off the main tank valve first before attempting to resolve the problem. Wear appropriate eye, hand, and body protection when working with the tanks and use no lubricants on the tank or the pressure regulators. It is best to discontinue using the tank before it is completely empty to prevent contaminants from entering the tank during the refilling step at the suppliers. Remove the original necklace label and replace it with another label indicating that the tank is empty. Have the empty tank removed as soon as possible to prevent others from using it and to remove clutter from the laboratory. The tank must be transported back to storage on the tank cart with the cylinder cap in place.

Microwave Ovens

Electron microscopists are making increasing use of microwave ovens for fixation, embedding, and staining purposes as well as to warm up various reagents. Just as in the home, such ovens should be regularly checked to make sure the door seal is intact using inexpensive monitoring devices. The oven should not be used to heat foods or drinks, since the inside may be contaminated with toxic chemicals. The ovens should be kept clean to prevent the carryover of reagents from one experiment to another. Solutions containing azides, picric acid, or other potentially explosive chemicals should not be used inside of the ovens. For example, if the oven had been contaminated during lead citrate staining and another researcher were to use the oven to warm up a solution containing the preservative sodium azide, it may be possible to generate lead azide, the explosive chemical used in blasting caps. In order to contain dangerous chemicals, it is advisable to place them inside of a microwave-safe receptacle inside of a sealable plastic bag.

CAUTION: Verify that the bag will not expand and explode due to the evolution of vapors. The bag and container should be opened in a fume hood.

Radiation

Most modern electron microscopes are extremely well shielded against x-ray leakage. Nonetheless, an instrument should be inspected for leakage with a beta-gamma counter if the instrument has been re-cently installed or moved, after additional detectors have been installed, or after major modifications to the instrument. Although older instruments are more likely to show leakage, newer instruments should be suspected until proven otherwise. Annual monitoring of an instrument is recommended even if no major changes to the instrument have taken place during the year. This may be achieved by operating the instrument at the highest accelerating voltage with the apertures slightly out of alignment and the beam expanded. Check the gun, column, specimen chamber, and viewing screen especially in the areas of gaskets. Essentially, zero counts above background should be expected. Leakage greater than 0.5 mR/hr measured at 5 cm from the microscope is considered significantly hazardous by the EMSA Radiation Committee of 1973. A vacuum evaporator equipped with an electron beam gun type of electrode may generate significant levels of X rays and should be evaluated before extensive use.

All uranium salts are radioactive to various degrees since several radioisotopes may be present (^{238}U, ^{235}U, ^{234}U). Even though most uranyl salts are said to be radioactivity "depleted" since the last two isotopes have been removed, one should still consider the salt to be radioactive until checked with an appropriate counter (alpha and beta particles are emitted). Most workers probably will be shocked by the results of the counting process.

The toxic effects of the heavy metal uranium are as much of a concern as the radioactive emanations. Soluble uranium salts (uranyl acetate, nitrate, etc.) as used in electron microscopy are less worrisome than the insoluble compounds (UO_2 and U_3O_8) since soluble salts are more rapidly cleared from the body. Most ingested uranium is excreted in the urine within 24 hours (Hursh and Spoor, 1973) with only 2% to 10% becoming incorporated into bone. Therefore, use caution when preparing stains so that the powders are not inhaled; prepare the stains in a fume hood and wear rubber gloves. All spills must be cleaned up immediately using a wet paper towel. One should check with regional authorities on the proper ways of disposing of uranyl salts since some localities consider it to be radioactive while others classify it only as a toxic waste. To reduce volume, one may precipitate the aqueous uranyl solutions using a phosphate buffer and collect the sediment for disposal.

Centrifuges

The major problems associated with using centrifuges have to do with the improper balancing of centrifuge tubes, which results in broken tubes, contamination of the work environment with the tube contents, and potential damage to the equipment. If proper operating procedures are followed, such accidents may be avoided.

Before using the centrifuge, check if it is in proper operating condition by asking other users and by noting the log book. Make certain that the proper rotor is being used. Different brands of rotors may resemble each other, and it may even be possible to install the wrong rotor on the centrifuge. In addition, make sure that the rotor is rated for the speeds that you desire. Verify that the centrifuge tubes are proper for the rotor and that they can withstand the g forces that will be generated and the chemicals that will be placed in them. If a swinging bucket rotor is used, make certain that the buckets are properly attached, balanced, and free to swing out during the centrifugation. The tubes should not be so long as to interfere with the free swing of the buckets.

Stand clear of the rotor as it is accelerating or slowing down. Take care to keep long hair or jewelry away from the spinning rotor. It is very poor practice to slow down the rotor by using one's hand or other object pressed against the rotor. Clean up any spills inside of the centrifuge and especially in the rotor. If contaminated, follow proper procedures to clean the rotor and store it in the proper manner. Note any unusual noises or operational problems in a log book (or attach a sign to the equipment to alert other users).

Pipetting

The use of mouth pipetting is strictly forbidden in all laboratories. The danger of ingesting toxic liquids or fumes from the liquids is simply too great to risk such a procedure. There are numerous bulbs and other devices for pipetting that are always available in the laboratory (Figure 20-7). If such devices cannot be found, ask for assistance and wait until one is found before proceeding. The author recalls an incident in which a senior researcher was mouth pipetting some 2% glutaraldehyde when the phone rang, startling the researcher. Even though the small volume of fixative was quickly expelled from the mouth, followed by rinsing with water, the researcher still sustained temporary tissue damage to the soft palate.

Volatile liquids have a tendency to expand when taken into a pipette for the first time. This may result in the expulsion of liquid from the tip of the pipette.

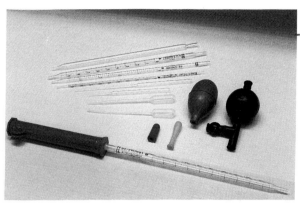

Figure 20-7 Various types of pipetting aids that may be used in the electron microscope laboratory.

One should be prepared for this phenomenon and take care not to point the pipette in the direction of another researcher.

Glass pipettes should be disposed of in a proper manner when broken or after use. Unless contaminated with toxic chemicals, they are best placed in a box containing other broken glass and sent to a recycler.

NOTE: Recyclers may not accept Pyrex types of glass, only softer varieties. Check with the recycler first.

Falls

Although less dramatic than other accidents, falls are probably second only to cuts and burns as major types of accidents in the laboratory. Cluttered laboratory aisles and liquids spilled on floors are common reasons for falls in laboratories. Standing on chairs or other unsafe surfaces also results in many accidents each year (Figure 20-8). Surprisingly, nearly all laboratories in which these incidents occurred had ladders or stools that should have been used for such procedures.

Training and Orientation Programs

Most electron microscopists recognize the importance of proper training in the safe manner to work in the laboratory environment. Most have an established set of rules that define the procedures associated with working in a particular laboratory. Unfortunately, simply posting the rules in the laboratory will not always ensure that everyone is

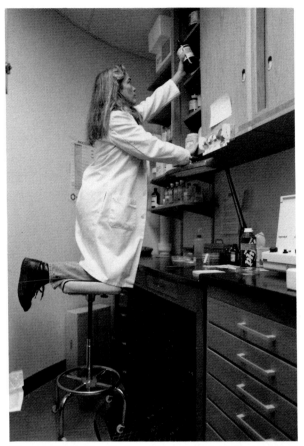

Figure 20-8 Reach for objects placed high on shelves using the proper safety equipment. Do not use stools or chairs for such purposes.

aware of, let alone willing to comply with them. Some formal training is needed.

In the electron microscopy courses given at the author's institution, the first period is spent instructing the students in such basic topics as:

- location of exits and routes to follow in various situations
- location and proper use of fire extinguishers and first aid kit
- proper methods of handling and disposing of chemicals
- safe operation of all equipment
- electrical safety in the darkroom
- the use of fire alarms and telephones to alert others
- importance of proper laboratory attire
- the operational rules for the laboratory

It is important to incorporate an awareness of safety consistently throughout training in electron microscopy. In addition to having a presentation on overall safety, it is important to reinforce safety concepts each time a new area is covered. As modified

from the safety book by Rayburn (1990), several questions to ask each time *before* embarking on a procedure are:

1. What are the potential hazards associated with this procedure and the materials being used in it?
2. What can one do to minimize the risks in this procedure?
3. What are the *worst* events that could occur during the procedure?
4. If an incident does occur, what is the plan of action?
5. What safety equipment is needed? Is it nearby, operational, and does everyone know how to operate it?

A regular safety inspection is necessary in order to assure compliance with established rules and to spot any neglected dangerous situations before an accident occurs. A checklist, such as the one shown in Figure 20-9, is useful for such inspections since it prevents one from overlooking some areas.

Hotlines and Other Resources

Telephones Numbers

American Chemical Society	202-872-4515
(Health and Safety Referral Service)	202-872-4511
Chemical Manufacturers Association	800-262-8200
Chemical Transportation Emergency Center	800-424-9300
	202-483-7616
Environmental Protection Agency	
(Wastes)	800-424-8802
(Conservation)	800-424-9346
(Toxins)	800-424-9065
National Institute of Occupational Health and Safety (NIOSH)	800-356-4674
Consumer Product Safety Commission	
	800-638-2772

Addresses

Center for Disease Control
1600 Clifton Road, N.E.
Atlanta, GA 30333
404-329-3535

terminology and equipment used for measurement of vacuum levels.

9. Discuss the two main types of pumps used in the electron microscope in terms of: basic principles of operation, vacuum ranges achieved, precautions or problems in their use, and how one would utilize the two in an integrated system to evacuate various parts of an electron microscope.

10. Discuss the design of the illuminating system of the transmission electron microscope. Be able to explain the functions of each part.

11. Discuss the functions of each of the imaging lenses in the transmission electron microscope.

12. What are apertures and aperture angles? Discuss the functions of the condenser and objective apertures in the electron microscope.

13. Discuss how two condenser lenses are used to vary the amount of illumination striking a specimen. Specify the functions of each lens.

14. What is depth of field and depth of focus? Why are they important in electron microscopy?

15. What are anticontaminators? Describe the function of anticontaminators in two very different locations in the electron microscope.

16. What is alignment? Why must one align an electron microscope?

17. Outline the general steps that must be done in the alignment process.

18. Discuss the following terms: translational lens movement, lens tilt, optical axis, current and voltage centration.

19. How may one correct astigmatism in an electromagnetic lens? Discuss the design of stigmators.

20. Discuss the four major operational modes of the transmission electron microscope. List the conditions necessary to achieve each of the modes.

21. Why is it necessary to calibrate the magnification and determine the resolving power of an electron microscope? Discuss at least one way to determine each of these capabilities.

22. Describe one method to simultaneously determine the magnification and resolving power of an electron microscope.

Chapter 7

1. Compare the SEM to the TEM and light microscopes in terms of: types of images obtained, resolution capabilities, types of specimens examined, specimen preparation procedures, use of vacuums inside the instrument, lenses, mode of viewing and recording.

2. Discuss the concept that the lenses of the SEM are not image-forming lenses.

3. How is magnification varied in the SEM? What effect does this have on spot size?

4. Discuss considerations for operating the SEM in the high resolution mode.

5. Is it possible to obtain both high resolution and great depth of field? Discuss your answer.

6. Compare the types of images obtained in the secondary versus the backscattered imaging mode of the SEM.

7. Consult chapters 2 and 3 and list the similarities and differences in the processing of specimens for TEM and SEM.

8. What factors will influence resolution in the SEM?

9. Of what value are stereomicrographs?

10. Discuss factors that contribute to the three dimensionality of the SEM image.

11. Discuss why astigmatism, chromatic and spherical aberration degrade resolution of the SEM.

12. What is "noise" and how can it be minimized in the SEM?

Chapter 8

1. In the darkroom, produce a contrast series and exposure series similar to those shown in this chapter. How much variability in contrast and exposure is acceptable in your eyes.? Obtain other opinions.

2. Review the previous chapters dealing with the transmission and scanning electron microscopes and generate a list of all possible ways that one may increase contrast in an electron micrograph (i.e., instrument adjustments and darkroom procedures). How can one decrease contrast?

3. How does the exposure of a photographic emulsion to photons differ from the exposure of the

emulsion to electrons? Consider the terms speed, resolution, contrast, and efficiency in this discussion.

4. What factors will affect the speed and resolution of an emulsion? Are high speeds and high resolution mutually exclusive when referring to electron micrographic emulsions?

5. Why is an underexposed electron micrograph often mistakenly said to be "grainy"?

6. Why is it better to err on the side of overexposure versus underexposure of the negative?

7. Suppose you examine one of your TEM negatives and determine that the margins of the film (which are normally clear) have a gray cast to them. What might have caused this grayness?

8. How can one determine how much a negative was enlarged on a print? How does one determine magnification of the final print?

Chapter 9

1. When should immunocytochemistry be used in place of other localization methods?

2. Divide into two groups. One group will give a technical problem with one aspect of immunocytochemistry, and the other group will respond on how the problem may be overcome or how another approach will help. Reverse roles.

3. From the list of current localization reports given above, select two to describe the purpose, methods, results, and impact of the experiment.

4. How do preembedding labeling and postembedding labeling differ?

5. Obtain the current issue of *Journal of Histochemistry and Cytochemistry* and review the approaches to current immunocytochemical localization experiments.

6. Describe how antibodies are obtained for immunocytochemistry.

7. Why is protein-A sometimes called a pseudo-immunocytochemical tag?

8. What is the major difference between the direct and indirect methods of immunocytochemical labeling?

9. How do the various immunocytochemical tags appear under the electron microscope?

Chapter 10

1. How is a sound knowledge of enzyme histochemistry an important preparation for understanding enzyme cytochemistry?

2. Review two recent reports using enzyme cytochemistry. Examine the methods section to determine the basis for the localization, and outline the procedure.

3. In the protocol described above (under A Typical Protocol) for demonstration of acid phosphatase activity, why was phosphate buffer not utilized in the incubation medium? How would one determine which buffers would be suitable?

4. Are methods, other than enzyme cytochemistry, available to localize enzymes? Discuss the reason for your answer.

5. Could you envision the use of an enzyme to localize a substrate at the electron microscope level? If so, outline a method by which this could be accomplished.

Chapter 11

1. What is the difference between an autoradiograph and an ordinary photograph taken with a hand-held camera?

2. What kinds of information does autoradiography provide that make it a suitable technique to answer biological problems?

3. What are the basic steps in performing autoradiography?

4. Autoradiography, at the electron microscope level, is not in widespread use. What factors limit the technique's usefulness and popularity?

5. Obtain one or two of the recent reports (see Chapter 11 for a partial list), and describe the details of the purpose of the experiment, the methods, the results, and the significance. Discuss whether some other technique would have answered the question.

Chapter 12

1. Consider how the inability to use colored dyes at the electron microscope level has limited the number of techniques available for electron microscopy.

2. Obtain a recent volume of the *Journal of Histochemistry and Cytochemistry* and categorize the basis

for each localization technique used (e.g., immunologic, enzymatic, autoradiographic, etc.).

Chapter 13

1. What is the difference between obtaining relative and absolute stereological data? If one wishes to obtain absolute data, then how does one perform stereology as compared with obtaining relative data?

2. What are the different parameters measured in stereology, and what test system is used for each? What formula is used for each parameter?

3. Choose a body tissue other than the one illustrated, and compose a chart for the various compartments through the organelle level.

4. Two key, but difficult, procedures in stereology are the determination of nuclear diameter, especially in nonspherical cells, and the determination of section thickness. Summarize how to perform these procedures.

5. Design a *detailed* protocol to determine mitochondrial volume (V) in a cell type of your choice. Consider what must be done at each level of the experiment.

6. How would unbiased sampling be obtained in question 5?

7. How does one determine if sufficient data has been gathered from a test system?

8. Take an 8½″ × 11″ sheet of paper and draw about thirty irregularly shaped enclosed structures that will occupy about 1/3 to 1/2 of the surface area. Photocopy the paper twice and use one sheet for point counting, another for digitizing (if a digitizer is available), and another for cutting and weighing to determine volume density. Compare the results from the three methods.

9. Obtain one or more of the recent references cited previously and be able to show how the final results were obtained. Pay particular attention to the methodology section of the reference.

10. Using Figure 13-8, determine the volume percentage of lipid and mitochondria in the micrograph.

Chapter 14

1. Rotate freeze fracture micrographs in Figures 14-15 and 14-16 and describe the three-dimensional features. Do they appear different when rotated

from their original orientation? Why are arrows sometimes placed on micrographs to indicate the direction of shadowing?

2. What is the difference between freeze fracturing and freeze etching? Would it not be better to say that etching is an additional step in the freeze-fracturing process?

3. List the steps necessary to produce a replica.

4. What rules must be followed to determine membrane faces?

5. What are complementary replicas?

6. Select two of the recent publications provided in this chapter and summarize them.

7. Have someone else cover over the labeling and legends to Figures 14-15 and 14-16. Use the instructions provided in the chapter to determine areas of the cytoplasm and the membrane faces.

8. In Figure A1 (a) determine the orientation of the micrograph, (b) identify the membranes that have been numbered, and (c) provide membrane face designations for each membrane.

Chapter 15

1. Could a standard transmission electron microscope be considered an analytical instrument? Explain your answer.

2. Review the designs of the various (secondary, backscattered, and transmitted electron, cathodoluminescent) detectors used in SEM and TEM instruments. What types of information may be obtained from each detector type?

3. Compare the designs of the standard SEM and TEM to the STEM instrument. What are the capabilities of each instrument in terms of: resolving power, accelerating voltage, magnification, ability to be fitted with analytical detectors, and types of specimens that may be examined?

4. Outline the events that take place (on the atomic level) to yield an X-ray for analytical purposes.

5. Compare characteristic X rays to continuum X rays in terms of: mode of generation, energy levels, information obtainable, and mode of detection.

6. Diagram and explain the basic features of an EDX unit versus a WDX unit.

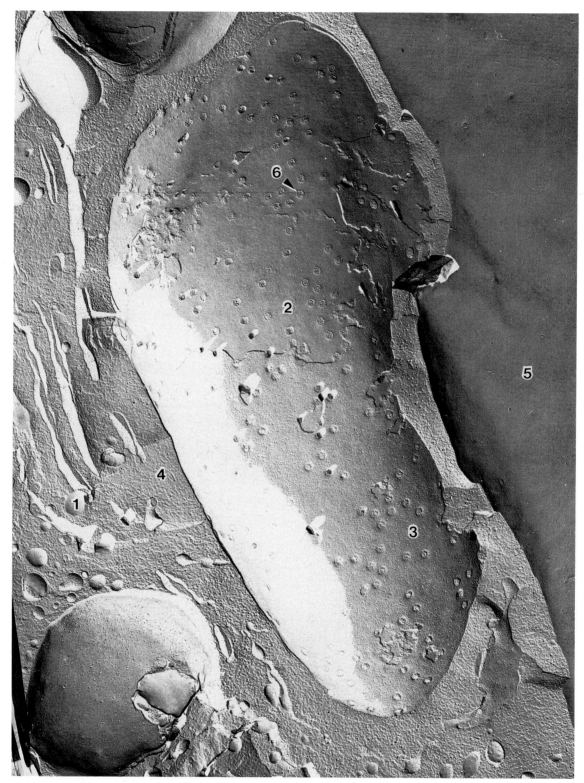

Figure A1   Micrographs for self interpretation. Identify each numbered surface by giving
the name of the membrane and its faces from the following list.
   E-face of the inner leaflet of the nuclear membrane
   P-face of the outer leaflet of the nuclear membrane
   P-face of the plasma membrane
   organelle membranes
   nuclear pores
   cytoplasm

7. What are the strong points and weak points of each of the two types of X-ray detection systems (EDX, WDX)?

8. Why is it not possible to obtain both excellent structural preservation and accurate retention of intracellular ions?

9. Describe a procedure that will allow adequate ultrastructural preservation and permit accurate localization of intracellular ions using conventional procedures.

10. Discuss why quantitative microanalysis is so difficult in bulk versus thin sectioned specimens.

11. Discuss the basic principles involved in electron energy loss spectroscopy.

12. Compare the use of X-ray microanalysis and EELS for the detection of light elements (atomic number under 11).

13. Explain the basic phenomenon involved in electron diffraction.

14. Tell how you would prepare a suspension of cells that contained water insoluble crystals for examination by electron diffraction. It is important that the cell ultrastructure be as well preserved as possible to show where in the cell the crystals are located.

15. Why must one check the camera length of a diffraction camera on a regular basis?

16. Calculate the camera constant for the electron microscope system that obtained the data in Table 15-2 (the gold diffraction standard). Make your measurements from the gold diffraction pattern shown in Figure 15-28.

17. Suppose you are given unknown biological crystals taken from the spleens and brains of individuals suffering from Alzheimer's disease. Tell how you would use analytical techniques to: (a) determine if the two types of crystals were the same, (b) determine the chemical identity of the crystals, (c) determine if aluminum is present in the crystal.

Chapter 16

1. What are the advantages of using a high voltage microscope? How has the high voltage microscope contributed to our knowledge of biological structure?

2. Read the papers published by Porter in 1945 and by Wolosewick and Porter published 30 years later (1976). Compare them and discuss how the high voltage microscope contributed to the latter paper.

3. Read the journal article by Porter and Tucker (1981) and discuss how the investigators tested their hypothesis of the existence of a microtrabecular lattice.

Chapter 17

1. Define a tracer by giving three of its applications.

2. Describe the different types of procedures, in this and other chapters, that have employed horseradish peroxidase.

3. What junctional types (tight, gap, and desmosome) would allow a tracer to pass through the junction and which would not? What junctional types would have their features highlighted by the tracer?

Chapter 18

1. Go to the literature of electron microscopy, especially the earlier literature, and find examples of four of the artifacts described in this chapter.

2. In the section entitled Survey of Biological Ultrastructure (Chapter 19), attempt to approximate the magnifications of the figures based on the general description of cell size and membrane appearance provided in this chapter. In the low magnification range, your estimate should be equal or better than 50% to 70% of the actual magnification. At medium and higher magnification ranges, try to guess within 50% of the actual magnification.

3. In published micrographs, measure structures as described above, and determine how close your calculated magnification comes to the published magnification. You are doing well if you are within 50% to 100% of the actual magnification given for the published figure (assuming the published magnification is accurate).

4. Without referring to the text, list (1) as many specific artifacts that one might find on a micrograph as possible, (2) the problem each causes, and (3) the solution(s) to the problem.

5. The micrograph shown (Figure A2) was produced by a student who had celebrated the evening before sectioning and staining the tissue. How many artifacts can be found, and what is the specific cause of each?

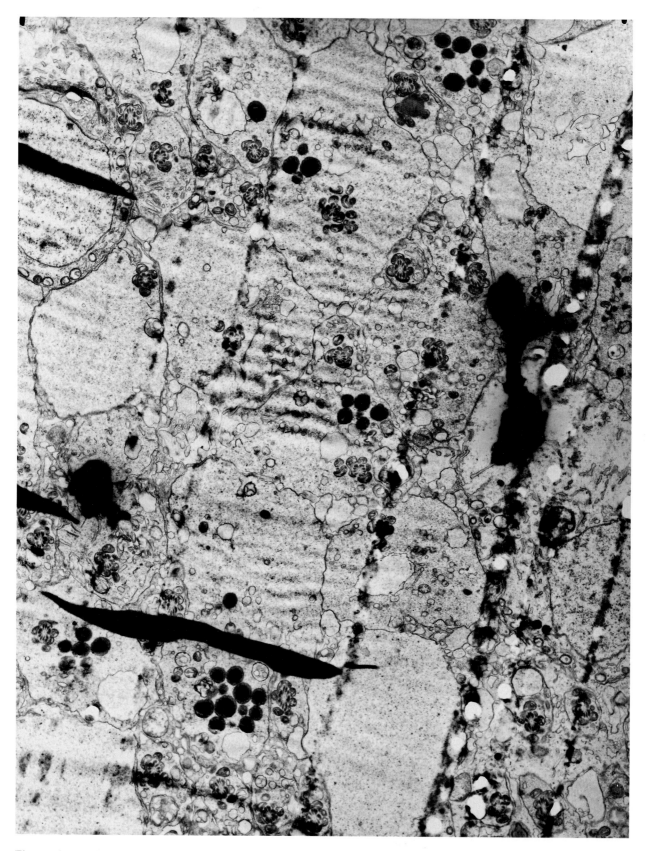

Figure A2 Student micrograph.

Chapter 20

1. Perform a safety inspection on your research environment and indicate any potential problems that were discovered. Discuss how the problems will be remedied.

2. Outline which types of fire extinguishers are to be used for the various types of fires. Check the rooms in the electron microscope laboratory for the appropriateness of the fire extinguishers.

3. What are the advantages and disadvantages of the various types of fire extinguishers?

5. On an established fire, what steps should be taken to stop the fire? Arrange for a demonstration (or film to be shown) by the local fire department or institutional safety officers.

6. Study the material safety data sheets for the chemicals in use in the electron microscope laboratory. Are there any chemicals for which data sheets are not available? If so, obtain and post them in the laboratory.

7. Develop an evacuation plan in the event of a fire or accident involving chemicals used in the EM laboratory.

8. How does one deal with spilled osmium (salt and liquids), cacodylate buffer, propylene oxide, epoxy resins? Develop a plan of action with the laboratory supervisor.

9. Make a list of chemicals that would be classified as fire hazards, toxins, or carcinogens.

10. Discuss the proper way of disposing of specific chemicals used in the EM laboratory.

11. Evaluate the fume hoods for proper functioning and the presence of clutter.

12. Evaluate several types of gloves for resistance to the commonly used solvents in the EM laboratory.

13. Arrange to meet with recyclers to discuss the re-utilization of various laboratory reagents or apparatuses.

Appendix B
Standard Units of Measurement

An international plan of measurement, designated "SI" (Systeme International d'Unites), based upon the metric system is currently in use by most scientists. Basic multipliers, or prefixes, are placed before the appropriate base unit of length, mass, time, etc to arrive at the final unit. Some final examples are shown in the table entitled Physical Measurements.

Prefixes

Name	Symbol	Multiplication Factor
kilo	k	10^3
hecto	h	10^2
deca	da	10^1
deci	d	10^{-1}
centi	c	10^{-2}
milli	m	10^{-3}
micro	μ	10^{-6}
nano	n	10^{-9}
pico	p	10^{-12}
femto	f	10^{-15}
atto	a	10^{-18}

Base Units

Name	Symbol	Quantity Measured
meter	m	length
liter	l	volume
second	s	time
ampere	A	electric current
kelvin	K	temperature
mole	mol	mole
pascal	Pa	pressure
watt	W	power
volt	V	electric potential
ohm	Ω	electric resistance

Physical Measurements

1 nanometer (nm)	= 10 Angstroms (Å)		1 millimeter^3 (mm^3)	= 10^9 μm^3
1 micrometer (μm)	= 1000 nanometers		1 centimeter^3 (cm^3)	= 1000 mm^3
1 millimeter (mm)	= 1000 micrometers		1 microliter (μl)	= 1 mm^3
1 centimeter (cm)	= 10 millimeters		1 milliliter (ml)	= 1000 μl
1 meter (m)	= 100 centimeters		1 liter	= 1000 ml

1 nanogram (ng) = 1000 picograms (pg)
1 microgram (μm) = 1000 nanograms
1 milligram (mg) = 1000 micrograms
1 gram (gm) = 1000 milligrams
1 kilogram (kg) = 1000 grams

Index

Appendix A
Review Questions and Problems

Chapter 1

1. Obtain both early and more recent issues of *Journal of Biophysical and Biochemical Cytology* (now entitled *Journal of Cell Biology*). What are the differences in quality of electron micrographs shown? What are different approaches to biological problems in the two issues? Are there descriptive studies in both issues?

2. Why did advances in biological electron microscopy not proceed rapidly until the early 1950s?

Chapter 2

1. Which are the toxic chemicals used in the preparation of tissues for electron microscopy? How should these be handled safely? Which chemicals are volatile? What precautions should be taken with them? What procedures are in place in your institution to insure safety for those working with these chemicals? What procedures are in place to dispose of electron microscope-related toxicants?

2. What is the purpose of each step in the tissue preparation protocol? List the possible negative consequences if there is a problem with any one of the steps in tissue preparation.

3. What is the generally held chemical basis for fixation in (a) glutaraldehyde and for fixation in (b) osmium tetroxide?

4. Without referring to the chapter, outline a standard tissue processing protocol. If you have difficulty, refer to Table 2-1.

5. What are the factors that must be considered in performing immersion fixation versus perfusion fixation?

6. Determine the best fixation protocol for one selected tissue (e.g., liver) using literature sources. Justify your choice by demonstrating fixation quality in published micrographs from these literature sources.

7. Design a protocol for preparing a specific tissue of your own choosing. After consulting current literature, give all necessary details starting with specimen acquisition and ending with a polymerized specimen block.

Chapter 3

1. How are the techniques used to prepare specimens for SEM similar to those used for TEM? Discuss how and why they differ.

2. Discuss the advantages and disadvantages of critical point and freeze-drying. List some criteria one would use to determine which method to employ.

3. Why can water not be used as a transitional fluid in critical point drying?

4. Why must cryoprotectants be employed when freezing specimens for SEM or TEM? What is being protected? Name some cryoprotectants.

5. What is the advantage of using a fluorocarbon drying procedure versus the critical point or freeze-drying methods?

6. Discuss the operational principles of the three types of drying methods (critical point drying, freeze-drying, and fluorocarbon drying).

7. When should one consider the use of a fracturing technique for specimen preparation?

8. Discuss the advantages and disadvantages of using the conventional methods of specimen coating (sputter coating and thermal evaporation) compared to the noncoating techniques.

9. Suppose that you had a precious biological sample that must not be damaged. Discuss some methods that can be employed to observe the specimen at reasonably high resolutions and magnifications in a SEM.

Chapter 4

1. Describe the two methods of specimen advancement to affect the section thickness in ultramicrotomes. Under what circumstances would one method be advantageous over another?

2. What are the advantages and disadvantages of glass, sapphire, and diamond knives? How is each used?

3. When would it be appropriate to use grids containing a plastic or carbon film?

4. Discuss the two general methods of preparing plastic films.

5. Why are grids made of different metals and available in different mesh openings?

6. Go to the library and scan older publications for published micrographs of sectioned materials. Compare these images to micrographs published within the last two years. Can you account for any differences in quality between the two, based on how they were sectioned?

7. Summarize the major systems in a modern ultramicrotome.

8. Without reading the captions, scan the electron micrographs in Chapter 18. Determine which artifacts were the result of sectioning problems. Verify this by reading the captions.

Chapter 5

1. Discuss the advantages and disadvantages of pre- and postembedding staining of tissues for transmission electron microscopy. Which method would you use on a unique specimen that could not be replaced? Why?

2. Why are electron micrographic stains devoid of color?

3. Compare and contrast positive and negative stains.

4. What are some precautions to observe when working with uranyl acetate solutions? Consider the stain itself as well as the researcher using the stain.

5. Consult local pollution control and radiological control officers for guidance in the proper disposal of all wastes generated in the staining procedures. Post the guidelines over areas where stains are likely to be used.

6. Survey the recent literature for articles dealing with staining of sectioned materials as well as those articles dealing with negative staining procedures. Outline any differences in procedure or new staining combinations used in the articles. Which stains appear to be the most commonly used?

7. When would one use negative staining versus metal shadowing procedures for contrasting biological macromolecules?

8. Suppose one wished to evaporate 5.0 nm of platinum onto a specimen. If the evaporating source is located 12 cm above the level of the specimen and the specimen is located 14 cm away from a point directly under the filament source, what weight of platinum should be used. (Density of platinum = 21.45 gm/cm²). Hint: Barring the use of trigonometry, the simplest way is to draw this relationship on paper in the proper scale and simply make measurements of distances and angles to use in the formula.

9. If one measured the shadow cast by the setup described in question 8 and found it to equal 350 nm in length, what would be the height of the object?

10. Examine the transmission electron micrographs of positively and negatively stained specimens in Chapter 19 to compare the contrasts obtained using various stains. What factors besides the stain itself could affect the final contrast of the micrographs? Consider all factors from specimen preparation to final printing of the negative.

Chapter 6

1. Why does the diffraction phenomenon degrade resolving power?

2. Why is the equation for the radius of an Airy disc (Equation 6-1) identical to the equation for resolving power?

3. Discuss how one may take advantage of the various terms in Equation 6-2 to achieve the very best resolving power.

4. What is the difference between magnification and resolving power? Can one have one without the other? Discuss.

5. Discuss how the various lens aberrations (chromatic, spherical, astigmatism) degrade resolving power.

6. If one had a microscope with an optimum resolving power of 2.5 nm, what would be the maximum expected or useful magnification at which one could expect to operate?

7. Compare a light microscope to a transmission electron microscope in terms of: lens design, arrangement of order of lenses and placement of specimen, illumination, resolution and magnification ranges, and specimen restrictions.

8. Why are vacuums necessary in the electron microscope? How are vacuums measured? Discuss the

Technical Resources, Inc. 1989. Fifth annual report on carcinogens. Summary 1989. U.S. Dept. of Health and Human Services Public Health Service. NTP 89-239. Technical Resources, Inc. Rockville, MD 20852.

Weakley, B. S. 1977. How dangerous is sodium cacodylate? *J Microsc* 109:249–51.

Young, J. A. 1987. *Improving safety in the chemical laboratory: a practical guide*. New York: John Wiley and Sons.